Lecture Notes in Physics

Springer-Verlag
Berlin Heidelberg GmbH

The Editorial Policy for Proceedings

The series Lecture Notes in Physics reports new developments in physical research and teaching – quickly, informally, and at a high level. The proceedings to be considered for publication in this series should be limited to only a few areas of research, and these should be closely related to each other. The contributions should be of a high standard and should avoid lengthy redraftings of papers already published or about to be published elsewhere. As a whole, the proceedings should aim for a balanced presentation of the theme of the conference including a description of the techniques used and enough motivation for a broad readership. It should not be assumed that the published proceedings must reflect the conference in its entirety. (A listing or abstracts of papers presented at the meeting but not included in the proceedings could be added as an appendix.)
When applying for publication in the series Lecture Notes in Physics the volume's editor(s) should submit sufficient material to enable the series editors and their referees to make a fairly accurate evaluation (e.g. a complete list of speakers and titles of papers to be presented and abstracts). If, based on this information, the proceedings are (tentatively) accepted, the volume's editor(s), whose name(s) will appear on the title pages, should select the papers suitable for publication and have them refereed (as for a journal) when appropriate. As a rule discussions will not be accepted. The series editors and Springer-Verlag will normally not interfere with the detailed editing except in fairly obvious cases or on technical matters.
Final acceptance is expressed by the series editor in charge, in consultation with Springer-Verlag only after receiving the complete manuscript. It might help to send a copy of the authors' manuscripts in advance to the editor in charge to discuss possible revisions with him. As a general rule, the series editor will confirm his tentative acceptance if the final manuscript corresponds to the original concept discussed, if the quality of the contribution meets the requirements of the series, and if the final size of the manuscript does not greatly exceed the number of pages originally agreed upon. The manuscript should be forwarded to Springer-Verlag shortly after the meeting. In cases of extreme delay (more than six months after the conference) the series editors will check once more the timeliness of the papers. Therefore, the volume's editor(s) should establish strict deadlines, or collect the articles during the conference and have them revised on the spot. If a delay is unavoidable, one should encourage the authors to update their contributions if appropriate. The editors of proceedings are strongly advised to inform contributors about these points at an early stage.
The final manuscript should contain a table of contents and an informative introduction accessible also to readers not particularly familiar with the topic of the conference. The contributions should be in English. The volume's editor(s) should check the contributions for the correct use of language. At Springer-Verlag only the prefaces will be checked by a copy-editor for language and style. Grave linguistic or technical shortcomings may lead to the rejection of contributions by the series editors. A conference report should not exceed a total of 500 pages. Keeping the size within this bound should be achieved by a stricter selection of articles and not by imposing an upper limit to the length of the individual papers. Editors receive jointly 30 complimentary copies of their book. They are entitled to purchase further copies of their book at a reduced rate. As a rule no reprints of individual contributions can be supplied. No royalty is paid on Lecture Notes in Physics volumes. Commitment to publish is made by letter of interest rather than by signing a formal contract. Springer-Verlag secures the copyright for each volume.

The Production Process

The books are hardbound, and the publisher will select quality paper appropriate to the needs of the author(s). Publication time is about ten weeks. More than twenty years of experience guarantee authors the best possible service. To reach the goal of rapid publication at a low price the technique of photographic reproduction from a camera-ready manuscript was chosen. This process shifts the main responsibility for the technical quality considerably from the publisher to the authors. We therefore urge all authors and editors of proceedings to observe very carefully the essentials for the preparation of camera-ready manuscripts, which we will supply on request. This applies especially to the quality of figures and halftones submitted for publication. In addition, it might be useful to look at some of the volumes already published. As a special service, we offer free of charge LATEX and TEX macro packages to format the text according to Springer-Verlag's quality require-ments. We strongly recommend that you make use of this offer, since the result will be a book of considerably improved technical quality. To avoid mistakes and time-consuming correspondence during the production period the conference editors should request special instructions from the publisher well before the beginning of the conference. Manuscripts not meeting the technical standard of the series will have to be returned for improvement.

For further information please contact Springer-Verlag, Physics Editorial Department II, Tiergartenstrasse 17, D-69121 Heidelberg, Germany

Arno Bohm Heinz-Dietrich Doebner
Piotr Kielanowski (Eds.)

Irreversibility and Causality

Semigroups and Rigged Hilbert Spaces

A Selection of Articles
Presented at the 21st International Colloquium
on Group Theoretical Methods in Physics (ICGTMP)
at Goslar, Germany, July 16–21, 1996

Springer

Editors

Arno Bohm
Department of Physics
University of Texas
Austin, TX 78712, USA

Heinz-Dietrich Doebner
Arnold Sommerfeld Institut
Leibnizstrasse 10
D-38678 Clausthal-Zellerfeld, Germany

Piotr Kielanowski
Departamento de Física
CINVESTAV
Ap. Postal 14-740
07000 Mexico D.F., Mexico

Cataloging-in-Publication Data applied for.

Die Deutsche Bibliothek - CIP-Einheitsaufnahme

Irreversibility and causality : semigroups and rigged Hilbert space ;
a selection of articles presented at the 21st International Colloquium
on Group Theoretical Methods in Physics (ICGTMP) at Goslar,
Germany, July 16 - 21, 1996 / Arno Bohm ... (ed.).
  (Lecture notes in physics ; 504)
  ISBN 978-3-662-14196-0   ISBN 978-3-540-69725-1 (eBook)
  DOI 10.1007/978-3-540-69725-1

ISSN 0075-8450
ISBN 978-3-662-14196-0

© Springer-Verlag Berlin Heidelberg 1998
Originally published by Springer-Verlag Berlin Heidelberg New York in 1998
Softcover reprint of the hardcover 1st edition 1998

The use of general descriptive names, registered names, trademarks, etc. in this publication does not imply, even in the absence of a specific statement, that such names are exempt
from the relevant protective laws and regulations and therefore free for general use.

Typesetting: Camera-ready by the authors/editors
Cover design: *design & production* GmbH, Heidelberg
SPIN: 10644076   55/3144-543210 - Printed on acid-free paper

We dedicate this volume to

Günther Ludwig

on the occasion of his 80th birthday

The Editors

Preface

This volume has its origin in the Semigroup Symposium which was organized in connection with the 21st International Colloquium on Group Theoretical Methods in Physics (ICGTMP) at Goslar, Germany, July 16–21, 1996.

Just as groups are important tools for the description of reversible physical processes, including stationary states, semigroups are indispensable tools in the description of irreversible physical processes in which a direction of time is distinguished. There is ample evidence of time asymmetry in the microphysical world, such as the decay of excited states of molecules, atoms, nuclei, hadrons, etc. The desire to go beyond the few, overemphasized, stationary systems and to improve upon approximate descriptions in which excited states do not decay has generated much recent effort and discussion regarding the application of semigroups to time asymmetric processes. Thus semigroups were given a prominent place at the 21st ICGTMP.

This volume is not the traditional proceedings volume of a symposium. Since the subject is relatively new, and several aspects of time asymmetry discussed at the Semigroup Symposium were not widely known, we agreed with the wishes of the publishers and expanded some of the talks into review papers, in order to make the volume accessible to a wider audience of non-specialists. This dual purpose explains the varying length of the articles.

The Colloquium and the Semigroup Symposium were sponsored and supported by the following institutions and agencies:

International Union for Pure and Applied Physics IUPAP
Niedersächsisches Ministerium für Wissenschaft und Kultur
Deutsche Forschungsgemeinschaft DFG
Alexander von Humboldt Stiftung AvH
Deutscher Akademischer Austauschdienst DAAD
Technische Universität Clausthal
Arnold Sommerfeld Institut ASI

We also gratefully acknowledge support from NATO and the Welch Foundation.

January 1998

A. Bohm
H.-D. Doebner
P. Kielanowski

Contents

Chapter I. Quantum Theory Beyond Hilbert Space

Chapter III. Resonances

**Chapter V. Exact Quantum Theory of the Neutral K-System –
Hilbert Space Versus Rigged Hilbert Space Formulation**

Chapter VI. Lax-Phillips Semigroup

The Lax-Phillips Semigroup
of the Unstable Quantum System
E. Eisenberg, L.P. Horwitz, and Y. Strauss 323

A Geometrical Approach to Calculating Determinants
of Wiener-Hopf Operators
J.P. MacCormick and B.S. Pavlov 333

Chapter VII. Irreversibility and Measurement
in Quantum Mechanics

Time Scale, Objectivity and Irreversibility
in Quantum Mechanics
L. Lanz and O. Melsheimer 345

Chapter VIII. Semigroups Operator Theory

List of Contributors

J.-P. Antoine
Institut de Physique Théorique
Université Catholique de Louvain
B-1348 Louvain-la-Neuve, Belgium
Email: antoine@fyma.ucl.ac.be

I. Antoniou
International Solvay Institutes
for Physics and Chemistry
Campus Plaine ULB C.P.231
Boulevard du Triomphe
Brussels 1050, Belgium
Email: isi@solvayins.ulb.ac.be

V.P. Belavkin
Mathematics Department
University of Nottingham
Nottingham NG7 2RD, UK
Email: vpb@maths.nott.ac.uk

F. Benatti
Dipartimento di Fisica Teorica
Università di Trieste
Strada Costiera 11
I-34100 Trieste,Italy
Email: benatti@trieste.infn.it

A. Bohm
Department of Physics
University of Texas at Austin
Austin, TX 78712, USA.
E-mail: bohm@physics.utexas.edu

C.G. Bollini
Department of Physics
University of La Plata,
C.C.67, La Plata (1900), Argentina

E. Brändas
Department of Quantum Chemistry
Uppsala University
Box 518
S-751 20 Uppsala, Sweden
Email: erkki@kvac.uu.se

M.A. Castagnino
Instituto de Astronomía y Física
del Espacio
Casilla de Correos 67, Sucursal 28,
1428 Buenos Aires, Argentina
Email: castagni@iafe.uba.ar

O. Civitarese
Department of Physics
University of La Plata,
C.C.67, La Plata (1900), Argentina
Email: civitare@venus.fisica.unlp.edu.ar

A.L. De Paoli
Department of Physics
University of La Plata,
C.C.67, La Plata (1900), Argentina
Email: depaoli@venus.fisica.unlp.edu.ar

E. Eisenberg
Department of Physics
Bar-Ilan University
Ramat-Gan 52900, Israel
Email: eisenber@egoz.cc.biu.ac.il

M. Gadella
Departamento de Física Teórica
Facultad de Ciencias
E-46011 Valladolid, Spain
Email: gadella@cpd.uva.es

K. Gustafson
Department of Mathematics
CB 395
University of Colorado,
Boulder, CO 80309-0395, USA
Email: gustafs@euclid.colorado.edu

N.L. Harshman
Department of Physics
University of Texas at Austin
Austin, TX 78712, USA.
Email: harshman@physics.utexas.edu

G.C. Hegerfeldt
Institut für Theoretische Physik
Universität Göttingen
Bunsenstr. 9
D-37073 Göttingen, Germany
Email: hegerf@theorie.physik.
uni-goettingen.de

E. Hernández
Instituto de Física, UNAM
Apartado Postal 20-364
01000 México DF, Mexico
Email: queta@ft.ifisicacu.unam.mx

B. Hessmo
Department of Quantum Chemistry
Uppsala University
Box 518
S-751 20 Uppsala, Sweden
Email: hessmo@kvac.uu.se

A.S. Holevo
Steklov Mathematical Institute
Vavilova 42
117966 Moscow, Russia
E-mail: holevo@class.mi.ras.ru

L.P. Horwitz
School of Physics and Astronomy,
Raymond and Beverly Sackler Faculty of Exact Sciences
Tel-Aviv University
Ramat-Aviv 69978, Israel
Email: horwitz@taunivm.tau.ac.il

L.A. Khalfin
St.Petersburg Department Steklov
Mathematical Institute
Russian Academy of Sciences
Fontanka 27
St. Petersburg 191011, Russia
Email: khalfin@pdmi.ras.ru

P. Kielanowski
Departamento de Física
CINVESTAV
Ap. Postal 14-740
07000 México DF, Mexico
E-mail: kiel@physics.utexas.edu

A. Kossakowski
Institute of Physics
Nicholas Copernicus University
87-100 Toruń, Poland

L. Lanz
Istituto Nazionale di Fisica Nucleare
Sezione di Milano
Via Celoria 16, I-20133, Milan, Italy
Email: Ludovico.Lanz@mi.infn.it

B.S. Pavlov
Department of Mathematics
University of Auckland
Private Bag 92019
Auckland, New Zealand
Email: pavlov@math.auckland.ac.nz

G. López Castro
CINVESTAV, Departamento de Física
Ap. Postal 14-740
07000 México DF, Mexico
E-mail: Gabriel.Lopez@fis.cinvestav.mx

T. Riemann
DESY – Institut für Hochenergiephysik
Platanenallee 6, D-15738 Zeuthen,
Germany
Email: riemann@ifh.de

J.P. MacCormick
Department of Mathematics
University of Auckland
Private Bag 92019
Auckland, New Zealand
Email: jmac@robots.oxford.ac.uk

M.C. Rocca
Department of Physics
University of La Plata,
C.C.67, La Plata (1900), Argentina
Email: rocca@venus.fisica.unlp.edu.ar

Yu. Melnikov
Laboratory of Complex
Systems Theory
St.Petersburg State University
St.Petersburg 198904, Russia
Email: melnikov@solvayins.ulb.ac.be

Y. Strauss
School of Physics and Astronomy,
Raymond and Beverly Sackler
Faculty of Exact Sciences
Tel-Aviv University
Ramat-Aviv 69978, Israel
Email: straussy@ccsg.tau.ac.il

O. Melsheimer
Fachbereich Physik Philipps-Universität
Renthof 7
D-35032 Marburg, Germany
Email:
melsheim@mailer.uni-marburg.de

Z. Suchanecki
Hugo Steinhaus Center
and Institute of Mathematics
Wrocław Technical University
ul. Wybrzeże Wyspiańskiego 27
50-370 Wrocław, Poland
Email: zsuchane@ulb.ac.be

A. Mondragón
Instituto de Física, UNAM
Apartado Postal 20-364
01000 México DF, Mexico
Email: mondra@sysul1.ifisicacu.unam.mx

Chapter I

Quantum Theory Beyond Hilbert Space

Quantum Mechanics Beyond Hilbert Space

J.-P. Antoine

Institut de Physique Théorique, Université Catholique de Louvain, B-1348 Louvain-la-Neuve, Belgium

Summary. When describing a quantum mechanical system, it is convenient to consider state vectors that do not belong to the Hilbert space. In the first part of this paper, we survey the various formalisms have been introduced for giving a rigorous mathematical justification to this procedure: rigged Hilbert spaces (RHS), scales or lattices of Hilbert spaces (LHS), nested Hilbert spaces, partial inner product spaces. Then we present three types of applications in quantum mechanics, all of them involving spaces of analytic functions. First we present a LHS built around the Bargmann space, thus giving a natural frame for the Fock-Bargmann (or phase space) representation. Then we review the RHS approach to scattering theory (resonances, Gamow vectors, etc.). Finally, we reformulate the Weinberg-van Winter integral equation approach to scattering in the LHS language, and this allows us to prove that it is in fact a particular case of the familiar complex scaling method.

1. Going Beyond Hilbert Space

1.1 Why?

In the conventional von Neumann-Mackey formulation, a quantum system is described in the language of Hilbert space, states being represented by unit rays and observables by self-adjoint operators.

Yet it is a long time now that some physicists have felt this framework as too narrow [5, 17, 55, 56]. On one hand, Hilbert space is 'too small', in the sense that objects which are singular, but of great practical importance, often cannot be made to belong to it. For instance, neither a plane wave, nor a delta function belong to $L^2(\mathbb{R})$. More generally, the 'eigenvectors' associated to the points of the continuous spectrum of a self-adjoint operator do not belong to the space. Also the Hilbert space cannot support very singular operators, they end up with a domain reduced to $\{0\}$. An example is an unsmeared field operator. On the other hand, all separable Hilbert spaces are isomorphic, but all quantum systems are not: something is missing in the description. Also it does not make physical sense to take every self-adjoint operator as an observable. This concept requires a certain operational content, which is of a nonmathematical nature.

Therefore, one would like to go beyond Hilbert space in order to be able to incorporate very singular objects (functions, operators). But at the same time, one wants to keep the good geometrical structure of Hilbert space, and the spectral theory as well, that fits so neatly with the interpretation of quantum mechanics. The answer is to consider a structure built *around* a Hilbert space, in the spirit of distribution theory [40].

1.2 The Different Formalisms

Several formalisms are available to that effect, all built around a central Hilbert space.

- *Rigged Hilbert space*
  This framework, originating from distribution theory, consists in a triplet

$$(1) \qquad \Phi \subset \mathcal{H} \subset \Phi^{\times},$$

  where $\mathcal{H}$ is a Hilbert space, Φ a dense subspace of $\mathcal{H}$, endowed with a topology finer than the norm topology, and $\Phi^{\times}$ is the space of all continuous antilinear functionals on Φ.
- *Scales of Hilbert or Banach spaces*
  In the simplest case, this is an increasing scale of Hilbert spaces

$$(2) \qquad \ldots \mathcal{H}_2 \subset \mathcal{H}_1 \subset \mathcal{H}_0 \subset \mathcal{H}_{-1} \subset \mathcal{H}_{-2} \subset \ldots,$$

  each one dense into the next one. The scale carries an involution $\mathcal{H}_n \leftrightarrow \mathcal{H}_{-n}$, such that $\mathcal{H}_{-n} = \mathcal{H}_n^{\times}$, the antidual of $\mathcal{H}_n$ (Note: if one identifies $\mathcal{H}_0 = \mathcal{H}_0^{\times}$, by the Riesz lemma, one *cannot* identify the other pairs). On the same pattern, one may replace each space $\mathcal{H}_n$, $n \neq 0$, by a reflexive Banach space V_n, that is, $V_{-n} = V_n^{\times}$ and $V_n = V_{-n}^{\times}$. However, this structure, while mathematically natural and interesting, is of little use so far in quantum mechanics.
- *Lattices of Hilbert or Banach spaces*
  More interesting is the fact that there is in general no reason to require the family $\{\mathcal{H}_n\}$ (or $\{V_n\}$) in (2) to be totally ordered, it suffices to have a notion of inclusion (technically, a partial order). One ends up with a *lattice* of spaces, i.e. every pair of spaces $\mathcal{H}_q, \mathcal{H}_p$ has an infimum $\mathcal{H}_q \cap \mathcal{H}_p$ and a supremum $\mathcal{H}_q + \mathcal{H}_p$. Examples are the nested Hilbert spaces [43, 44]. Again we will consider in the sequel only lattices of Hilbert spaces (LHS).
- *Partial inner product spaces*
  In all the preceding constructions, the basic question for a pair of vectors is whether their inner product exists (then we say that the vectors are *compatible*). This is the case, for instance, if one of them belongs to $\mathcal{H}_n$ and the other to $\mathcal{H}_{-n}$, for some n. But compatibility should not be taken as a condition on individual vectors, but rather on the pair itself. Pushing this idea, one is led to the notion of partial inner product (PIP) space, that unifies all the previous concepts [6, 7, 8].

In the next section, we shall describe in detail these various constructions and supply examples. But, before that, let us see what we have gained.

1.3 What Does One Obtain?

Each one of these formalisms produces what we may call the "traditional" outcome, for which they were conceived in the first place, namely:

— the possibility of justifying the use of singular objects, distributions or generalized functions, very singular operators;
— in particular, the 'eigenvectors' associated to the continuous spectrum of self-adjoint operators (more on that in the next section).

The outcome is a rigorous formulation of the Dirac bra-and-ket formalism of quantum mechanics, which was obtained independently by several authors in the 60's [5, 17, 50, 55, 56] (see also the monograph [33]).

But, in addition, these formalisms have produced a number of results which were to some extent unexpected. For instance:

(1) For the RHS:.

— A nice formulation of symmetries in quantum mechanics (active vs. passive points of view);
— A novel and richer approach to scattering theory, including resonances, Gamow vectors, etc.
— An elegant way of introducing irreversibility in the quantum mechanical measurement problem.

(2) For LHS and PIP*-spaces:.*

— An extended KLMN theorem, which allows to treat very singular Hanilto-nians (e.g. proving their self-adjointness);
— The possibility of describing very singular interactions, for instance the Kronig-Penney model for a one-dimensional crystal with delta potentials (using precisely the extended KLMN theorem);
— An elegant reformulation of Weinberg-van Winter's version of scattering theory, and a proof that the latter is a particular case of the complex scaling method.

In Sections 3 to 5, we will describe a number of these applications to quantum mechanics of the various formalisms (with proper references). In view of these, it is fair to say that the treatment of "Quantum Mechanics beyond Hilbert space" has reaped a rich and diverse harvest, and that it is here to stay, notwithstanding the opinion of some Hilbert space purists.

2. The Mathematical Formalism

2.1 Rigged Hilbert Spaces

Let us begin our survey by describing the rigged Hilbert space. This is the most widely known among the different formalisms we are going to discuss, and also the one wich has led to the largest number of applications in quantum mechanics. It is in fact the natural framework for the latter, as we will see below.

2.1.1 Definition and properties. By a rigged Hilbert space (RHS), we mean a triplet

$$(3) \qquad\qquad \Phi \subset \mathcal{H} \subset \Phi^\times,$$

where $\mathcal{H}$ is a Hilbert space, Φ a dense subspace of $\mathcal{H}$, endowed with a locally convex topology τ_Φ, finer than the norm topology inherited from $\mathcal{H}$ (i.e. a stronger notion of convergence), and $\Phi^\times$ is the space of τ_Φ-continuous antilinear functionals on Φ. By duality [49, 57], each space in (3) is dense in the next one and all embeddings are linear and continuous. Notice that $\Phi^\times$ is determined uniquely by Φ, but that there are two ways of obtaining the couple $\Phi, \mathcal{H}$, namely:

- Either one starts from $\mathcal{H}$ and builds Φ, for instance as the largest domain invariant under a self-adjoint operator $R > 1$; that is, Φ is obtained as $\Phi = \cap_{n>0}\mathcal{H}_n$, where $\{\mathcal{H}_n \equiv D(R^n), n > 0\}$ is a decreasing sequence of Hilbert spaces. We will see an example below.
- Or Φ is supposed to be given and one puts on it an inner product, which makes it into a prehilbert space. Then $\mathcal{H}$ is obtained, either as the completion of Φ with respect to the corresponding norm topology, or as the dual of Φ, with respect to the same norm topology (we use here the Riesz lemma, which says that $\mathcal{H}^\times \sim \mathcal{H}$, so that the two are usually identified).

Standard examples of rigged Hilbert spaces are the Schwartz distribution spaces over $\mathbb{R}$ or $\mathbb{R}^N$, namely $\mathcal{S} \subset L^2 \subset \mathcal{S}^\times$ or $\mathcal{D} \subset L^2 \subset \mathcal{D}^\times$ [40, 60].
In addition, one usually requires that Φ satisfy a number of properties:

- Φ is *complete* with respect to τ_Φ, that is, every Cauchy (generalized) sequence converges to an element of Φ.
- Φ is *reflexive*, that is, $(\Phi^\times)^\times \simeq \Phi$, if $\Phi^\times$ is given its strong dual topology [49, 57].
- In most cases, Φ is obtained as the intersection of a *countable* family of Hilbert spaces, $\Phi = \cap_{n\in\mathbb{N}}\mathcal{H}_n$. It is then a Fréchet space.
- Φ is *nuclear;* in the case where $\Phi = \cap_{n\in\mathbb{N}}\mathcal{H}_n$, this means that, for each n, there is $m > n$ such that the embedding $\mathcal{H}_m \to \mathcal{H}_n$ is a Hilbert-Schmidt operator.

Before proceeding, we have to fix our notation. For $F \in \Phi^\times$, we denote by $F(\phi)$ the value of F on the vector $\phi \in \Phi$. If $F \in \mathcal{H}$, we normalize the duality form by requiring $F(\phi) = \langle\phi|F\rangle$, where $\langle.|.\rangle$ denotes the inner product of $\mathcal{H}$ (remember that F is antilinear). By a slight abuse of notation, we shall use the same notation $\langle\phi|F\rangle$ for *any* $\phi \in \Phi$, $F \in \Phi^\times$, with the obvious convention that $\langle F|\phi\rangle = \overline{\langle\phi|F\rangle}$.

<u>Remark</u>: instead of $\Phi^\times$, one often uses the space Φ' of *linear* functionals over Φ; in that case, the embedding of Φ or $\mathcal{H}$ into Φ' is antilinear, and one writes $F(\phi) = \langle F|\phi \rangle$.

The advantage of the nuclearity property is that it allows one to exploit the *nuclear spectral theorem* of Gel'fand and Maurin [40], which says the following. Let A be a closed operator in $\mathcal{H}$, which maps Φ into itself, continuously for τ_Φ (for simplicity, we denote the operator A and its restriction to Φ by the same symbol). Then A may be transported by duality to a linear operator $A^\times : \Phi^\times \to \Phi^\times$, which is an extension of the usual adjoint A^*, namely:

$$(4) \qquad A^\times F(\phi) = F(A\phi), \ \forall \phi \in \Phi, \ F \in \Phi^\times,$$

which we also write

$$(5) \qquad \langle \phi | A^\times F \rangle = \langle A\phi | F \rangle, \ \forall \phi \in \Phi, \ F \in \Phi^\times.$$

For such an operator, the vector $\xi_\lambda \in \Phi^\times$ is called a *generalized eigenvector* of A, with eigenvalue λ, if one has

$$(6) \qquad A^\times \xi_\lambda(\phi) = \lambda \xi_\lambda(\phi), \ \forall \phi \in \Phi.$$

Assume now that A is self-adjoint and that Φ is nuclear and complete. Then the theorem asserts that A possesses a complete orthonormal set of generalized eigenvectors $\xi_\lambda \in \Phi^\times$, $\lambda \in \mathbb{R}$, which means that, for any $\phi, \psi \in \Phi$, one has

$$
\begin{aligned}
\langle \phi | \psi \rangle &= \int_\mathbb{R} \xi_\lambda(\phi) \, \overline{\xi_\lambda(\psi)} \, d\mu(\lambda) \\
(7) \qquad &\equiv \int_\mathbb{R} \langle \phi | \xi_\lambda \rangle \langle \xi_\lambda | \psi \rangle \, d\mu(\lambda)
\end{aligned}
$$

for some measure μ on $\mathbb{R}$. The latter may further be split as $\mu = \sum_i \delta(\cdot - \lambda_i) + \mu_c$, where $\{\lambda_i\}$ are the eigenvalues of A in $\mathcal{H}$ (in the usual sense) and the support of μ_c contains the continuous spectrum of A in $\mathcal{H}$.

The net result of this theorem is to put on the same footing the eigenvalues and the points of the continuous spectrum of A in $\mathcal{H}$ – exactly what is usually assumed in the Dirac formulation of quantum mechanics. Indeed, using Dirac's notation, (7) may be (formally) rewritten as a decomposition of the identity (I denotes the unit operator):

$$(8) \qquad I = \int_\mathbb{R} |\xi_\lambda\rangle\langle\xi_\lambda| \, d\mu(\lambda),$$

with the proviso that this quantity makes sense only between two vectors of Φ – in other words, I must be interpreted as the (linear) embedding of Φ into $\Phi^\times$, or equivalently as a sesquilinear form on $\Phi \times \Phi$.

Yet a word of warning is necessary here. If one is interested only in the spectral properties of A, one may like that the spectrum of A in $\Phi^\times$ consists exactly of the points of its spectrum in $\mathcal{H}$. If this is the case, one says that (3) is a *tight rigging* for A. But this is by no means guaranteed, and we will see a counterexample in the next section. Sufficient conditions for a tight rigging have been given in [12, 35, 53]. On the other hand, there are cases where one actually needs eigenvalues which do *not* belong to the Hilbert spectrum of A, for instance nonreal values. As we shall see in Section 4.1, scattering theory is a major example, where resonances are associated to nonreal eigenvalues of the Hamiltonian, with the so-called Gamow vectors as generalized eigenvectors.

2.1.2 A simple example. The simplest example of a quantum system is a one-dimensional harmonic oscillator, and it does give the simplest RHS as well. With $\mathcal{H} = L^2(\mathbb{R}, dx)$, the Hamiltonian $H = \frac{1}{2}(p^2 + q^2)$ is self-adjoint, with spectrum $\sigma(H) = \{n + \frac{1}{2}, n = 0, 1, 2, \ldots\}$. Let $R = 1 + H$ and, for $n = 1, 2, \ldots$, denote by $\mathcal{H}_n$ the domain of R^n, equipped with the graph norm $\|f\|_n = \|R^n f\|$. Then each $\mathcal{H}_n$ is complete, hence a Hilbert space, and $\cap_{n \in \mathbb{N}} \mathcal{H}_n = \mathcal{S}(\mathbb{R})$, the Schwartz space of C^∞ functions of fast decrease. Taking duals, one obtains the familiar RHS of tempered distributions

$$(9) \qquad \mathcal{S}(\mathbb{R}) \subset L^2(\mathbb{R}) \subset \mathcal{S}(\mathbb{R})^\times.$$

Note that Fourier transform is an isomorphism of this RHS. Choosing in $L^2(\mathbb{R})$ the orthonormal basis of Hermite functions, which are contained in $\mathcal{S}$, one obtains a RHS of sequences, isomorphic to (9):

$$(10) \qquad s \subset \ell^2 \subset s^\times,$$

where s, resp. $s^\times$, denotes the space of fast decreasing, resp. slowly increasing, sequences. It is immediately seen that, like R, R^{-1} is diagonal, with purely discrete, nondegenerate spectrum $\sigma(R^{-1}) = \{(n + \frac{3}{2})^{-1}, n = 0, 1, 2, \ldots\}$. From this follows that R^{-1} maps $\mathcal{H}_n$ into $\mathcal{H}_{n-1}$, for each n, as a Hilbert-Schmidt operator, so that $\mathcal{S}(\mathbb{R})$ is indeed nuclear.

Consider now the position and momentum operators, q and p. Both are self-adjoint in L^2 and leave $\mathcal{S}$ invariant. Actually $\mathcal{S}$ is the largest domain invariant under both q and p — it is in fact generated by them, since, in addition, the topology of $\mathcal{S}$ is defined by the norms of all powers of q and p. Hence both map $\mathcal{S}$ into itself continuously, and the nuclear spectral theorem applies. For p, we obtain:

$$
\begin{aligned}
\langle \phi | \psi \rangle &= \int_{\mathbb{R}} dk \, \langle \phi | \xi_k \rangle \, \overline{\langle \psi | \xi_k \rangle} \\
(11) \qquad &= \int_{\mathbb{R}} dk \, \overline{\hat{\phi}(k)} \, \psi(k),
\end{aligned}
$$

which is just the Parseval relation. Indeed, the eigenvectors are $\xi_k(x) = e^{ikx} \in \mathcal{S}^\times$, $k \in \mathbb{R}$, so that

$$(12) \qquad \xi_k(\phi) \equiv \langle \phi | \xi_k \rangle = \int dx \, \overline{\phi(x)} \, e^{ikx} = \overline{\hat{\phi}(k)}.$$

Thus the eigenvalue equation for p, written in $\mathcal{S}^\times$, is simply

$$(13) \qquad p\xi_k = k\xi_k, \; k \in \mathbb{R},$$

that is, the RHS (9) is a tight rigging for p. By Fourier transform, the same holds for q, with $\delta_x = \delta(x - \cdot)$, $x \in \mathbb{R}$, as generalized eigenvectors. However, if we replace the space $\mathcal{S}(\mathbb{R})$ by the space $\mathcal{D}(\mathbb{R})$ of C^∞ functions with compact support, which is also nuclear and invariant under p, we obtain for p a much bigger spectrum, $\sigma_{\mathcal{D}^\times}(p) = \mathbb{C}$, that is, every complex number is now an eigenvalue of p. This second rigging is definitely not tight! This is a warning sign that the space Φ must be in some sense 'natural' for the system under consideration, for instance generated by the observables of the system (see the next section).

2.1.3 Applications in quantum mechanics. The RHS approach has many applications in quantum mechanics. Let us mention three of them here. Other ones will come up in later sections.

(i) Rigorous formulation of the Dirac formalism. This was the original motivation for introducing the RHS in quantum mechanics. It was obtained in the 60's, independently by A. Böhm [17], J. Roberts [55, 56] and the present author [5], with many later contributions, e.g. [33, 50]. The idea is to characterize a quantum system, with Hilbert space $\mathcal{H}$, by a family $\mathcal{O}$ of distinguished ('labeled') observables, with are given both a physical interpretation (how does one measure it?) and a mathematical definition (as a self-adjoint operator in $\mathcal{H}$). These are required to have a common dense invariant domain $\mathcal{D}$ (hence $\mathcal{O}$ is an algebra of operators on $\mathcal{D}$). Equip this domain with a suitable ('projective') topology that makes all the operators from $\mathcal{O}$ continuous from the domain into itself, and call the resulting topological vector space (TVS) Φ. Taking duals, one thus obtains the RHS $\Phi \subset \mathcal{H} \subset \Phi^\times$, *defined* by the system. Then, as explained above, assuming that Φ is nuclear, the nuclear spectral theorem yields a rigorous formulation of Dirac's bra-and-ket formalism, which is routinely used by physicists, but cannot be justified solely in Hilbert space.

(ii) Interpretation of quantum measurements and symmetries. Given the RHS just constructed, it seems natural to interpret Φ as the space of *physical* states, i.e. states that can be prepared in actual experiments (notice that, since the Hamiltonian H must certainly be an element of $\mathcal{O}$, all the states in Φ have automatically a finite energy, since they belong to the domain of H). Now an element of $\Phi^\times$ is an antilinear functional on Φ, i.e. a procedure that associates to each state a number, while preserving the linear structure (which results from the superposition principle). This is clealy related to a measurement apparatus or a reference frame – although the precise link has not been made.

This interpretation is supported by an analysis of symmetries. According to Wigner and Bargmann, a (Lie) symmetry group G of the system is realized by a unitary representation U of G in $\mathcal{H}$. Clearly U should transform physical states into physical states, continuously, and similarly for measurement apparatuses. Thus one should have two other realizations of U, in addition to U itself, which acts in $\mathcal{H}$, namely:

- one in Φ, noted U_Φ, which represents the so-called *active* point of view,
- and one in $\Phi^\times$, noted $\check{U}_\Phi$, corresponding to the *passive* point of view.

The equivalence of the two points of view is manifested by the requirement that U_Φ and $\check{U}_\Phi$ are contragredient of each other, that is, $\check{U}_\Phi(g) = U_\Phi(g^{-1})^\times$, i.e.,

$$(14) \qquad \langle \phi | \check{U}_\Phi(g) F \rangle = \langle U_\Phi(g^{-1}) \phi | F \rangle, \, \forall\, g \in G, \phi \in \Phi, F \in \Phi^\times,$$

or, equivalently,

$$(15) \qquad \langle U_\Phi(g) \phi | \check{U}_\Phi(g) F \rangle = \langle \phi | F \rangle, \, \forall\, g \in G, \phi \in \Phi, F \in \Phi^\times,$$

which corresponds to the unitarity of U acting in $\mathcal{H}$:

$$(16) \qquad \langle U(g) f | U(g) h \rangle = \langle f | h \rangle, \, \forall\, g \in G, f, h \in \mathcal{H}.$$

As it is easily verified, this definition implies that $\check{U}_\Phi$ is an extension of both U_Φ and U, as it should in view of (3).

At this point it may be useful to dwell somewhat on the mathematics. As we have said, the space Φ is supposed to be a reflexive Fréchet space. For consistency, we must assume that the representation U_Φ is continuous in Φ, that is, the map $g \mapsto U_\Phi \phi$ is continuous from G to Φ, for every $\phi \in \Phi$. Then the contragredient representation $\check{U}_\Phi$ is automatically continuous in $\Phi^\times$ [28]. Notice that rigged Hilbert spaces of this type have also been used in pure group theory, namely in the decomposition of unitary representations of noncompact groups like $SU(1,1)$ or $SO_o(2,1)$ [51].

If G is a Lie group, one may go further and consider its Lie algebra $\mathfrak{g}$. By deriving $U(g)$, one obtains a representation of symmetric elements of $\mathfrak{g}$ by self-adjoint operators in $\mathcal{H}$, and similarly for the enveloping algebra $\mathfrak{U}(\mathfrak{g})$ – and these are prime candidates for labeled observables, such as energy, momentum, angular momentum, etc. Indeed it is always possible to construct the space Φ in such a way that these operators map it continuously into itself [5], and thus verify all the conditions described so far. Then, proceeding as usual by duality, we obtain a representation of these elements by continuous operators on the large space $\Phi^\times$.

One could also proceed differently and try to derive the representation U_Φ, assuming it is differentiable, in the sense of Bruhat [28]. Under some technical conditions, the result will be the same as before. In that case, indeed, the contragredient representation $\check{U}_\Phi$ is also differentiable on $\Phi^\times$, and the symmetric elements of $\mathfrak{g}$ and $\mathfrak{U}(\mathfrak{g})$ are represented by continuous operators on Φ and, by duality, on $\Phi^\times$ [28]. Notice that here we start from the representations $U, U_\Phi, \check{U}_\Phi$ of G and *derive* them in order to get representations of the Lie algebra $\mathfrak{g}$ and its enveloping algebra $\mathfrak{U}(\mathfrak{g})$. This is quite another problem than trying to *integrate* a representation from $\mathfrak{g}$ to G, where the result depends on the existence of sufficiently many analytic vectors, and domain problems are likely to occur [34]. This would be a step towards the generalization of the theory of integrable representations of Lie algebras in topological vector spaces more general than Hilbert spaces, for which relatively little is known.

(iii) Relativistic scattering states. We also briefly mention here a recent extension of the quantum mechanical RHS formalism to relativistic scattering states [54]. Here again, the starting point is the relevant representation of the Poincaré group.

2.2 Scales and Lattices of Hilbert Spaces

2.2.1 Definition and construction. In most RHS's, the smaller space Φ is *countably Hilbertian*, that is, it is obtained as the intersection of a scale of Hilbert spaces, as in (2)

$$\Phi \equiv \bigcap_n \mathcal{H}_n \subset \ldots \subset \mathcal{H}_2 \subset \mathcal{H}_1 \subset \mathcal{H}_0 \subset \mathcal{H}_{-1} \subset \mathcal{H}_{-2} \ldots \subset \bigcup_n \mathcal{H}_n \equiv \Phi^\times$$

(17)

Notice that this is true both as set and as a TVS, that is, the topology of Φ is given by the family of the norms $\{\|\cdot\|_n, n > 0\}$ of the spaces $\mathcal{H}_n$ (technically, this is called a projective limit [49, 57]). Suppose now that we forget the extreme spaces Φ and $\Phi^\times$. We end up with a scale of Hilbert spaces, but the mathematical properties are exactly the same, including nuclearity, whenever it holds for Φ.

This being understood, one immediately realizes that there is no need for a *total* order in the scale (17), a partial order suffices (a partial order on a set X is a binary relation $\leqslant$ such that $\forall x, y, z \in X$, (i) $x \leqslant x$; (ii) $x \leqslant y$ and $y \leqslant x$ imply $x = y$; and (iii) $x \leqslant y$ and $y \leqslant z$ imply $x \leqslant z$; but two arbitrary elements x, y need not be comparable). Let thus $\mathcal{J} = \{\mathcal{H}_p, p \in J\}$ be a family of Hilbert spaces, partially ordered by inclusion (the index set J has the same order structure). Then $\mathcal{J}$ generates a lattice $\mathcal{I}$, indexed by I, by the operations:

– $\mathcal{H}_{p \wedge q} = \mathcal{H}_p \cap \mathcal{H}_q$, with the projective norm

(18) $$\|f\|_{p \wedge q}^2 = \|f\|_p^2 + \|f\|_q^2,$$

– $\mathcal{H}_{p \vee q} = \mathcal{H}_p + \mathcal{H}_q$, with the inductive norm

(19) $$\|f\|_{p \vee q}^2 = \inf_{f = g + h} \left(\|g\|_p^2 + \|h\|_q^2 \right), \quad g \in \mathcal{H}_p, f \in \mathcal{H}_q.$$

Assume that the original index set J has an involution $q \Leftrightarrow \bar{q}$, with $\mathcal{H}_{\bar{q}} = \mathcal{H}_q^\times$ (by an involution, we mean a one-to-one correspondence such that $p \leqslant q$ implies $\bar{q} \leqslant \bar{p}$ and $\bar{\bar{p}} = p$). Then the lattice $\mathcal{I}$ inherits the same duality structure, with $\mathcal{H}_{p \wedge q} \Leftrightarrow \mathcal{H}_{\bar{p} \vee \bar{q}}$ (it is then called an involutive lattice). For convenience, we will use the notation

$$(20) \qquad V = \sum_{q \in I} \mathcal{H}_q, \quad V^{\#} = \bigcap_{q \in I} \mathcal{H}_q.$$

The resulting structure is called a *lattice of Hilbert spaces* or LHS. This is the framework used throughout this paper. A crucial remark is that this construction is the basic structure of interpolation theory [16]. It immediately extends to a family $\{V_p\}$ of reflexive Banach spaces, but this will not concern us here. Instead, let us give a number of examples.

2.2.2 Examples. All the examples of LHS given here, and some more, may be found in [7, 8].

(i) The canonical scale of quantum mechanics. The first example is, of course, that of Section 2.1.2, which gives Schwartz' space $\mathcal{S}(\mathbb{R})$ as the intersection of the spaces $\mathcal{H}_n = D(R^n)$, $n \in \mathbb{N}$. This may be generalized in two ways. First, the whole construction extends trivially to $\mathbb{R}^N$. Second, one obtains a genuine LHS by taking powers of q and p separately, with norms of the type

$$(21) \qquad \|f\|_{m,n} = \|(1 + q^m p^n)f\|.$$

In this way, one recovers not only the preceding scale, but also all the Sobolev spaces, familiar in PDE theory [62].

(ii) Spaces of sequences. Given a sequence of positive numbers, $r = (r_n)$, $r_n > 0$, we define the weighted Hilbert space $\ell^2(r)$ as

$$\ell^2(r) = \{x = (x_n) \mid \sum_{n=1}^{\infty} |x_n|^2 r_n^{-1} < \infty\}.$$

The lattice operations read:

- involution: $\ell^2(\bar{r}) = \ell^2(r)^\times$, $\bar{r}_n = 1/r_n$.
- infimum: $\ell^2(p) \wedge \ell^2(q) = \ell^2(r)$, $r_n = \min(p_n, q_n)$.
- supremum: $\ell^2(p) \vee \ell^2(q) = \ell^2(s)$, $s_n = \max(p_n, q_n)$.

Indeed one shows easily that the norms of $\ell^2(r)$ and $\ell^2(s)$ are equivalent, respectively, to the projective and inductive norms defined above.
Of course, the central, self-dual, space is $\ell^2(1) = \ell^2(\bar{1}) = \ell^2$. As for the extreme spaces, it is easy to see that the family $\{\ell^2(r)\}$ generates the space ω of *all* complex sequences, while the intersection is the space φ of all *finite* sequences.

(iii) Spaces of locally integrable functions. This is the continuous analogue of
the preceding example. Instead of sequences, we consider locally integrable
(i.e. integrable on bounded sets) functions $f \in L^1_{loc}(\mathbb{R}, dx)$ and define again
weighted spaces:

$$I = \{r \in L^1_{loc}(\mathbb{R}, dx) \mid r(x) > 0, \text{ a.e.}\}$$
$$L^2(r) = \{f \in L^1_{loc}(\mathbb{R}, dx) \mid \int |f(x)|^2 r(x)^{-1} dx < \infty\}, \ r \in I.$$

Then we get exactly the same structure as in (ii):

– involution: $L^2(r) \Leftrightarrow L^2(\bar{r})$, $\bar{r} = 1/r$.
– infimum: $L^2(p) \wedge L^2(q) = L^2(r)$, $r(x) = \min(p(x), q(x))$.
– supremum: $L^2(p) \vee L^2(q) = L^2(s)$, $s(x) = \max(p(x), q(x))$.
– extreme spaces:

$$\bigcup_{r \in I} L^2(r) = L^1_{loc}, \quad \bigcap_{r \in I} L^2(r) = L^\infty_c,$$

where L^∞_c is the space of (essentially) bounded functions of compact support. The central space is, of course, L^2.

The construction extends trivially to $\mathbb{R}^N$, or to any manifold (X, μ). It may
also be done around Fock space, instead of L^2 [6].

(iv) Scales or lattices of Banach spaces. There are many familiar scales
of Banach spaces, such as the sequence spaces ℓ^p, the integration spaces
$L^p([0, 1], dx)$, the Schatten ideals C^p of compact operators on a Hilbert space.
In each case, the spaces with $1 < p < \infty$ are reflexive, whereas the extreme
spaces corresponding to $p = 1$ and $p = \infty$ are not.
If one considers integration spaces L^p over the whole line $\mathbb{R}$, of course they do
not form a scale, since no two of them are comparable, but they do generate
an involutive lattice with the projective and the inductive norms, defined as
above (usually without taking squares: the resulting topologies are the same,
but in the case of Hilbert spaces, the definitions (18) and (19) guarantee that
the resulting norms are Hilbertian).

2.3 Partial Inner Product Spaces

2.3.1 Definition and construction. In the case of a RHS, the construction proceeds "bottom up": one first builds a small space Φ, which generates
the largest space $\Phi^\times$ by duality. Here we are going to do the opposite, and
start from a given vector space V, that will be the largest space of the resulting structure.
As already mentioned, the starting point of this approach is the question:
Given two vectors $f, g \in V$, when does their inner product make sense?
This we abstract into the idea of compatibility. More precisely, a *linear compatibility relation* on the vector space V is a binary relation $f \# g$, which is

symmetric ($f\#g \Leftrightarrow g\#f$) and preserves linearity ($f\#g, f\#h \Rightarrow f\#(\alpha g + \beta h)$). As a consequence, for any subset $S \subset V$, the set

(22) $$S^\# = \{g \in V \mid g\#f, \forall f \in S\},$$

is a vector subspace of V, and also $S^{\#\#} = (S^\#)^\# \supseteq S$. Since the map $S \mapsto S^{\#\#}$ is a closure (in the sense of universal algebra, see [7, 10]), the "closed sets", that is, the subspaces that satisfy the relation $S^{\#\#} = S$, will play the central role. From now on, we call them *assaying subspaces* and denote them by V_r ($r \in F$), where F is the index set. It then follows from general theorems that the family of assaying subspaces $\{V_r\}$, ordered by inclusion, is a *complete involutive lattice* $\mathcal{F}(V, \#)$, that is, the family $\mathcal{F}(V, \#) \equiv \{V_r, r \in F\}$ is stable under the following operations, arbitrarily iterated ($p, q, r \in F$):

– involution : $V_r \Leftrightarrow V_{\bar{r}} \equiv (V_r)^\#$
– infimum : $V_{p \wedge q} \equiv V_r \wedge V_q = V_p \cap V_q$
– supremum : $V_{p \vee q} \equiv V_p \vee V_q = (V_p + V_q)^{\#\#}$.

(the notation is such that the set F of indices is itself a complete lattice; for instance, one has $(V_{p \wedge q})^\# = V_{\overline{p \wedge q}} = V_{\bar{p} \vee \bar{q}} = V_{\bar{p}} \vee V_{\bar{q}}$). Clearly $V^\# = \bigcap_r V_r$ is the smallest element of $\mathcal{F}(V, \#)$, $V = \bigcup_r V_r$ itself the greatest.

Given the linear compatibility relation $\#$ on V, we define a *partial inner product* (PIP) as a hermitian form $\langle \cdot | \cdot \rangle$ defined exactly on compatible pairs of vectors. Then $(V, \#, \langle \cdot | \cdot \rangle)$ is called a partial inner product space or PIP-space. Clearly the PIP defines a notion of orthogonality $\perp$. In order to obtain meaningful results, we will assume that the PIP is nondegenerate, that is, $(V^\#)^\perp = \{0\} : \langle f | g \rangle = 0, \forall f \in V^\#$ implies $g = 0$. From this deceptively simple assumption follows the whole topological structure. Indeed it entails that $(V^\#, V)$, as well as *every* pair $(V_r, V_{\bar{r}})$ of assaying subspaces, is a dual pair, in the sense of topological vector space theory [49, 57]. From this we get for free on each V_r all topologies compatible with the duality, i.e. such that $(V_r[\tau])^\times = V_{\bar{r}}$. Among them, the coarsest one is the weak topology, the finest is the Mackey topology (but *not* in general the strong dual topology). We decide, once and for all, to equip each assaying subspace V_r with its Mackey topology $\tau(V_r, V_{\bar{r}})$. In particular, the latter coincides with the norm topology or the metric topology if there is one. From this, we deduce immediately that the duality $V_r \Leftrightarrow V_{\bar{r}}$ extends to topologies, and moreover, $r < s$ implies $V_r \subset V_s$, where the embedding is continuous and has dense range.

2.3.2 Examples. All the examples of LHS given above are in fact PIP-spaces, namely:

(i) Sequence spaces. Let again $V = \omega$, the space of all complex sequences $x = (x_n)$. Define:

– a compatibility: $x\#y$ iff $\sum_n |x_n y_n| < \infty$.
– a PIP: $\langle x | y \rangle = \sum_n \overline{x_n} y_n$.

Clearly the family $\{\ell^2(r)\}$ of the weighted ℓ^2-spaces is an involutive sublattice of the complete lattice $\mathcal{F}(\omega, \#)$ and $V^\# = \varphi$, the space of finite sequences.

(ii) Locally integrable functions. Similarly, take $V = L^1_{\text{loc}}(\mathbb{R}, dx)$ and define:

$- f\#g$ iff $\int |f(x)g(x)|\, dx < \infty.$

$- \langle f|g \rangle = \int \overline{f(x)}g(x)\, dx.$

Again $\{L^2(r),\ r \in I\}$ is an involutive sublattice of $\mathcal{F}(L^1_{\text{loc}}, \#)$ and $V^\# = L^\infty_c$. It is worth mentioning that this example leads to Köthe's theory of function spaces [49].

(iii) LHS. In general, the complete lattice $\mathcal{F}(V, \#)$ is much too large, and contains all kinds of topological vector spaces. Therefore it is wiser to restrict oneself to a sublattice (not complete in general) of assaying subspaces of the same nature. Thus we define a LHS as a sublattice $V_r, r \in I$ of a PIP-space, such that each space V_r, equipped with its Mackey topology $\tau(V_r, V_{\bar{r}})$ is a Hilbertian space (modulo equivalence of norms, we will say simply that each V_r is a Hilbert space, with dual $V_{\bar{r}} = (V_r)^\times$). As a consequence, one gets:

$$V_{p\vee q} = V_p + V_q = (V_p + V_q)^{\#\#}.$$

Notice that the same would be true for a lattice of reflexive Banach spaces (these two structures were called in [8] *indexed* PIP-*spaces* of type (H), resp. (B)).

Of course, we recover all the previous examples, $\{\ell^2(r)\}$, $\{L^2(r)\}$, the Schwartz scales $\mathcal{H}_n$. Other examples will be discussed in Sections 3 and 4.

(iv) RHS. Obviously, a RHS $\Phi \subset \mathcal{H} \subset \Phi^\times$ is a trivial PIP-space, with compatibility $f\#g$ if, either one of them belongs to Φ, or both to $\mathcal{H}$.

2.4 Operators on PIP-Spaces

The whole idea behind PIP-spaces is that vectors should not be considered individually, but only in terms of the subspaces V_r, which are the building blocks of the theory. For instance,

(23) $f\#g$ iff $\exists r \in F$ such that $f \in V_r, g \in V_{\bar{r}}$.

If one restricts oneself to a LHS, a sublattice of $\mathcal{F}$, this relation will be taken as the definition of compatibility (see [7] for the comparison between I and F in lattice terms). And furthermore, the TVS $\Phi, \Phi^\times$ are replaced by the family of Hilbert spaces $\{V_r,\ r \in I\}$: many spaces, but simple ones.

The same spirit determines the definition of an operator on a PIP-space: only bounded operators between Hilbert spaces are allowed, but an operator is a coherent collection of these. To be more specific, let $V_I = \{V_r,\ r \in I\}$ be a PIP-space. An *operator* on V_I is a map $A : \mathcal{D}(A) \to V_I$, such that:

(i) $\mathcal{D}(A) = \bigcup_{r \in D(A)} V_r$, where $D(A)$ is a nonempty subset of I.

(ii) For every $r \in D(A)$, there is $q \in I$ such that the restriction $A : V_r \to V_q$ is linear and continuous (we denote it by A_{qr}).

(iii) A has no proper extension satisfying (i) and (ii).

The linear continuous operator $A_{qr} : V_r \to V_q$ is called a *representative* of A. Thus A is characterized by two subsets of I :

$$D(A) = \{r \in I \mid \text{there is a } q \text{ such that } A_{qr} \text{ exists}\}$$
$$R(A) = \{q \in I \mid \text{there is a } r \text{ such that } A_{qr} \text{ exists}\}$$

We denote by $J(A)$ the set of all such pairs (r, q) for which A_{qr} exists. Thus the operator A is equivalent to the collection of its representatives

$$(24) \qquad A \simeq \{A_{qr} \mid (r, q) \in J(A)\}.$$

Note that, if $r \in D(A)$ and $s \leq r$, then $s \in D(A)$, and $A_{qs} = A_{qr} E_{rs}$, where E_{rs} is the unit operator. Similarly, if $q \in R(A)$ and $t \geq q$, then $t \in R(A)$ and $A_{tr} = E_{tq} A_{qr}$. This is what we mean by 'coherent'. We denote by $Op(V_I)$ the set of all operators on V_I. Since $V^{\#} \subset V_r$, $\forall r \in I$, an operator may be identified with a separately continuous sesquilinear form on $V^{\#} \times V^{\#}$. But the idea behind the notion of operator is to keep also the algebraic operations on operators, namely:

(i) *Adjoint A^** : every operator $A \in Op(V_I)$ has a unique adjoint $A^* \in Op(V_I)$, defined by:

$$\langle A^* x \mid y \rangle = \langle x \mid A y \rangle,$$
$$\text{for } y \in V_r, r \in D(A), \text{ and } x \in V_{\bar{s}}, s \in R(A),$$

that is, $(A^*)_{\bar{r}\bar{s}} = (A_{sr})^*$ (usual Hilbert space adjoint). This implies that $A^{**} = A$, $\forall A \in Op(V_I)$: no extension is allowed, because of the maximality condition (iii).

(ii) *Multiplication* : AB is defined whenever there is a continuous factorization through some V_q:

$$V_r \overset{B}{\to} V_q \overset{A}{\to} V_s, \quad \text{i.e.} \quad (AB)_{sr} = A_{sq} B_{qr}.$$

This general definition, however, still allows some pathologies (linked to the fact that $V_p + V_q \neq (V_p + V_q)^{\#\#}$ in general), but for a LHS they disappear. Indeed, if V_I is a LHS, $\mathcal{D}(A)$ is always a vector subspace of V and $Op(V_I)$ is a partial *-algebra (which means, in particular, that the usual rule of distributivity is valid).

Among all types of operators, we may single out two useful classes:

(i) A is *symmetric* if $A^* = A$.. Symmetric operators are those for which one may prove a generalized KLMN theorem, which roughly gives conditions under which a symmetric operator has a self-adjoint restriction to the central Hilbert space V_o [6]. A nice application of this theorem is the correct description of a Kronig-Penney crystal [45].

(ii) *Orthogonal projections:* $P^2 = P^* = P$ and $P : V_r \to V_r$, $\forall r$. These correspond one-to-one to PIP-subspaces [9], as in Hilbert space.

Finally, in the case of a LHS, there are two interesting operator *-algebras [52]:

$$\mathcal{A} = \{A \in Op(V_I) \mid A : V_r \to V_r, \, \forall \, r \in I\}$$
$$\mathcal{B} = \{A \in \mathcal{A} \mid \|A\|_{\mathcal{B}} = \sup_r \|A_{rr}\|_r < \infty\},$$

and the latter is a Banach *-algebra for the norm $\|\cdot\|_{\mathcal{B}}$. Notice that projections are symmetric idempotent elements of $\mathcal{A}$.

Since the goal of defining operators on a PIP-space was to be able to catch very singular operators (from the Hilbert space point of view), let us compare the two approaches, and also with the RHS one.

Given $A \in Op(V_I)$, when looked at from $V_o = \mathcal{H}$, there are three possibilities:

- $(o, o) \in J(A)$ $\quad \Leftrightarrow \quad$ $A : \mathcal{H} \to \mathcal{H}$ is bounded.
- there is $r < o$ such that $(r, o) \in J(A)$ but $(o, o) \notin J(A)$ $\quad \Leftrightarrow \quad$ A_{or} is unbounded in $\mathcal{H}$, with domain containing V_r.
- $(r, o) \notin J(A)$, for any $r \leq o$ $\quad \Leftrightarrow \quad$ A is a sesquilinear form on some V_s, $s \leq o$, and, as an operator on $\mathcal{H}$, its domain does not contain any V_r (it may be $\{0\}$).

From the RHS point of view, it suffices to consider the triplet $V^{\#} \subset \mathcal{H} \subset V$, and repeat the argument, replacing V_r ($r < o$) by $V^{\#}$. Then the *-algebra of operators $\mathcal{A}$ defined above may be identified with the *-algebra of operators mapping $V^{\#} \equiv \Phi$ into itself continuously. In the application of RHS to quantum mechanics, the latter is the algebra of labeled observables; here $\mathcal{A}$ plays the same role.

3. Application in Quantum Mechanics: The Fock-Bargmann Representation Revisited

3.1 A RHS of Entire Functions

Besides the familiar position (q) and momentum (p) representations of quantum mechanics, the Fock-Bargmann representation offers an attractive alternative [3, 4, 15]. It is based on the canonical (or oscillator) coherent states and it is characterized by the fact that its wave functions are entire analytic functions of $z = q - ip$. It is therefore a *phase space* representation. As such it is useful for studying the quantum-to-classical transition, quantum optics, path integrals [32], etc.

The Fock-Bargmann Hilbert space is

$$(25) \qquad \mathfrak{F} = \{f(z) \text{ entire} \mid \int_{\mathbb{C}} |f(z)|^2 \, d\mu(z) < \infty\},$$

where $d\mu(z) = \pi^{-1}e^{-|z|^2}d^2z$, with inner product

$$(26) \qquad \langle f|g \rangle = \int_{\mathbb{C}} \overline{f(z)}g(z)\,d\mu(z).$$

The Hilbert space $\mathfrak{F}$ possesses several interesting properties, which are in fact characteristic of all phase space representations [3, 4].

– *Orthonormal basis:*

$$(27) \qquad u_n(z) = \frac{z^n}{\sqrt{n!}},\ n = 0,1,2,\ldots.$$

In this basis, the inner product (26) reads

$$(28) \quad \langle f|g \rangle = \sum_{n=0}^{\infty} n!\,\overline{f_n}\,g_n,\ \text{for } f(z) = \sum_n f_n\,z^n,\ g(z) = \sum_n g_n\,z^n.$$

– *Principal vectors* (coherent states):

$$(29) \qquad e_w(z) = e^{\overline{w}z},\ w,z \in \mathbb{C},$$

so that

$$(30) \qquad f(z) = \langle e_z|f \rangle,$$

that is, e_z is an *evaluation functional* (the equivalent of a delta function, but here e_z is a bona fide vector of the Hilbert space).

– *Reproducing kernel:*

$$(31) \qquad K(w,z) = e^{w\overline{z}} = \langle e_w|e_z \rangle$$

$$(32) \qquad f(w) = \int_{\mathbb{C}} K(w,z)f(z)\,d\mu(z).$$

The question that arises now is, how to build a RHS or a LHS around the Fock-Bargmann space $\mathfrak{F}$. Since $\mathfrak{F}$ consists of entire functions, one may identify immediately the extreme spaces [11]:

– The *maximal* space 3, consisting of *all* entire functions. With uniform convergence on compact sets, the space 3 is a nuclear Fréchet space. Its antidual $3^\times$, the space of *antianalytic functionals*, is thus a nuclear, complete DF-space [62], exactly what is needed for applying the nuclear spectral theorem.

– The *minimal* space Exp, consisting of entire functions of exponential type:

$$(33) \quad \text{Exp} = \{f \in 3 \mid \exists\, a,c > 0 \text{ such that } |f(z)| \le c\,e^{a|z|},\ \forall\, z \in \mathbb{C}\}.$$

It is then a standard result [62] that $3^\times$ is isomorphic to Exp, the correspondence being $\mu \in 3^\times \mapsto \hat{\mu} \in$ Exp, where $\hat{\mu}(w) = \langle \mu, e_w \rangle$, the Fourier-Borel transform. Thus we obtain a natural RHS around $\mathfrak{F}$:

$$(34) \qquad\qquad 3^\times \simeq \text{Exp} \subset \mathfrak{F} \subset 3.$$

The duality between Exp and 3 is given indeed by a natural extension of the inner product of $\mathfrak{F}$:

$$(35) \qquad\qquad \langle \overline{f}, g \rangle = \langle f | g \rangle = \sum_n n! \, \overline{f_n} \, g_n, f \in \text{Exp}, g \in 3.$$

This answers the first half of the question. But how to enrich the RHS (34) into a LHS?

3.2 A LHS of Entire Functions Around $\mathfrak{F}$

On the Hilbert space $\mathfrak{F}$, the two equivalent forms (26) and (28) of the inner product define two natural linear compatibilities:

$$(36) \qquad\qquad f \, \#_1 \, g \quad \Leftrightarrow \quad \int_{\mathbb{C}} |f(z) \, g(z)| \, d\mu(z) < \infty,$$

$$(37) \qquad\qquad f \, \#_2 \, g \quad \Leftrightarrow \quad \sum_n n! |f_n \, g_n| < \infty.$$

Of course, $\#_1$ and $\#_2$ coincide on $\mathfrak{F}$, but they are not comparable on 3 ! Indeed:

- $\#_1$ is too general, and somewhat pathological, since one has $3^{\#_1} = \{0\}$.
- $\#_2$ is more regular, and indeed Exp $\xleftrightarrow{\#_2} 3$, but it applies only to sequences, not analytic functions.

The solution is to restrict the large space 3 to some smaller, more manageable space $\mathfrak{E}$. This may be done in two ways.

(i) A scale of Hilbert spaces. For every $\rho \in \mathbb{R}$, consider the following Hilbert space [15]:

$$(38) \qquad \mathfrak{F}^\rho = \{f \in 3 \mid \int_{\mathbb{C}} |f(z)|^2 \, (1 + |z|^2)^\rho \, d\mu(z) < \infty\},$$

The family $\{\mathfrak{F}^\rho, \, \rho \in \mathbb{R}\}$ is a scale of Hilbert spaces:

$$(39) \qquad\qquad \mathfrak{F}^{\rho_1} \subset \mathfrak{F}^{\rho_2} \quad \Leftrightarrow \quad \rho_1 > \rho_2,$$

and one has

$$(40) \qquad\qquad (\mathfrak{F}^\rho)^\times = \mathfrak{F}^{-\rho} = (\mathfrak{F}^\rho)^{\#_1} = (\mathfrak{F}^\rho)^{\#_2}$$

and

(41) $e_w \in \mathfrak{F}^\rho,\ \forall \rho \in \mathbb{R}, \forall w \in \mathbb{C}.$

Define now the extreme spaces

(42) $\mathfrak{E} = \bigcap_\rho \mathfrak{F}^\rho, \quad \mathfrak{E}^\times = \bigcup_\rho \mathfrak{F}^\rho.$

Then one gets a Hilbert scale, with the structure ($\rho > 0$):

(43) $\mathrm{Exp} \subset \mathfrak{E} \subset \ldots \mathfrak{F}^\rho \ldots \subset \mathfrak{F} \subset \ldots \mathfrak{F}^{-\rho} \ldots \subset \mathfrak{E}^\times \subset \mathfrak{Z},$

and, by restriction, a RHS isomorphic to the Schwartz triplets (9) and (10):

(44) $\mathfrak{E} \subset \mathfrak{F} \subset \mathfrak{E}^\times.$

(ii) A genuine LHS. One may go one step further and try to obtain a LHS by considering more general weights in (38):

(45) $\mathfrak{F}(\rho) = \{ f \in \mathfrak{Z} \mid \int_\mathbb{C} |f(z)|^2\, e^{-\rho(z)}\, d\mu(z) < \infty \},$

where $\rho : \mathbb{C} \to \mathbb{R}$ is a measurable function. Clearly the space $\mathfrak{F}(\rho)$ reduces to $\mathfrak{F}^\rho$ if $\rho(z) = \ln(1 + |z|^2)^{-\rho}$, $\rho \in \mathbb{R}$. In order to obtain a LHS around $\mathfrak{F}$, we must require that the space $\mathfrak{F}(\rho)$ satisfies the following conditions [11], all of them satisfied by every $\mathfrak{F}^\rho$, $\rho \in \mathbb{R}$:
(i) $\mathfrak{F}(\rho)$ is a Hilbert space, i.e. it is complete.
(ii) The set of polynomials is dense in $\mathfrak{F}(\rho)$.
(iii) $\mathfrak{F}(\rho)^\times$ is isomorphic to $\mathfrak{F}(-\rho)$.
(iv) $e_w \in \mathfrak{F}(\rho),$, $\forall w \in \mathbb{C}$.
Each of these conditions imposes some restrictions to the weight function ρ. Let us examine them successively.

– For $\mathfrak{F}(\rho)$ to be complete, it suffices that ρ be locally bounded, that is, bounded on compact sets.
– Condition (ii) is a restriction on the growth of ρ: if ρ grows too fast at infinity, $\mathfrak{F}(\rho)$ may become trivial and $\mathfrak{F}(-\rho)$ too large. When (ii) holds, the monomials $\{u_m(z), m = 1, 2, \ldots\}$ form a basis of $\mathfrak{F}(\rho)$. If ρ is *radial*, i.e. $\rho(z) = \rho(|z|)$, then the functions

$$u_m^{(\rho)}(z) = (\eta_m^{(\rho)})^{-1/2}\, u_m(z), \quad m = 1, 2, \ldots$$

form an orthonormal basis of $\mathfrak{F}(\rho)$. Here we have introduced the coefficients

$$\eta_m^{(\rho)} \;=\; \int |u_m(z)|^2\, e^{-\rho(z)}\, d\mu(z)$$

(46) $\;=\; \dfrac{1}{m!} \displaystyle\int_0^\infty t^m\, e^{-\rho(t)-t}\, dt$

(clearly these coefficients are closely related to a moment problem). In that case, all spaces $\mathfrak{F}(\rho)$ may be realized as weighted ℓ^2 sequence spaces and the compatibility $\#_2$ is easy to handle.

– Condition (iii) is the crucial one, and it is difficult to verify, unless ρ is radial.

– Finally, condition (iv) implies that the principal vectors e_w generate a dense subspace of $\mathfrak{F}(\rho)$, and therefore every operator A in the resulting PIP-space is an integral operator, with kernel $A(w, z) = \langle e_w \mid A\, e_z \rangle$.

Some results may be obtained for general weights ρ, but condition (iii) remains a problem. For radial weights, however, a complete answer may be formulated as follows.

Theorem 3.1 – *Let I be the set of weight functions $\rho(z)$ that satisfy the following three conditions:*
(i) ρ is locally bounded and radial, $\rho(z) = \rho(|z|)$.
(ii) e_w belongs to $\mathfrak{F}(\rho) \cap \mathfrak{F}(-\rho)$, for any $w \in \mathbb{C}$.
(iii) There are positive constants A, B such that

$$(47) \qquad A \leq \eta_m^{(\rho)}\, \eta_m^{(-\rho)} \leq B, \quad \text{for all } m = 0, 1, 2, \ldots,$$

where $\eta_m^{(\rho)}$ is given by (46).
Then the family $\{\mathfrak{F}(\rho),\, \rho \in I\}$ is a LHS with central Hilbert space $\mathfrak{F}(0) \equiv \mathfrak{F}$. The compatibility $\#$ coincides with both $\#_1$ and $\#_2$, and the partial inner product with that induced by $\mathfrak{F}$ and ℓ^2, respectively.

□

For the proof, it suffices to notice that condition (47) is necessary and sufficient for the isomorphism $\mathfrak{F}(\rho)^\times \simeq \mathfrak{F}(-\rho)$ [11]. The resulting LHS, which is *not* a scale, has found interesting applications in the study of Weyl quantization, that is, the setting of a correspondence between functions on phase space $f(q - ip) \equiv \tilde{f}(q, p)$ and operators on $L^2(\mathbb{R}, dx)$ [29, 30, 31].

Remark: An almost identical LHS may be built with functions analytic in the unit disk $\mathcal{D} \subset \mathbb{C}$, with central Hilbert space $L^2(\mathcal{D}, d^2z)$. The corresponding weighted L^2 spaces are standard Bergman spaces [47].

4. Application in Scattering Theory

We turn now to applications of the various "super-Hilbert space" formalisms to quantum scattering theory. Here also, both RHS and LHS have proven to be extremely useful tools.

4.1 RHS: Resonances, Gamow Vectors, Arrow of Time

The notion of *resonance* in quantum scattering theory has a long and controversial history [2, 27]. The conventional definition runs as follows [18, 61]. First one extends the S-matrix analytically to a function $S(E)$ defined on a two-sheeted Riemann surface, associated to the relation $E = p^2$. Then a resonance is defined as a pair of conjugate poles of $S(E)$ on the second sheet (corresponding to a pole p_R in the lower half-plane):

$$(48) \qquad z_R = E_R - i\frac{\Gamma}{2}, \quad z_R^* = E_R + i\frac{\Gamma}{2}$$

(E_R is the resonant energy and Γ the width, according to the Breit-Wigner model). The problem is to find a precise mathematical framework for defining the associated states, the so-called *Gamow states* [19, 22, 39] and correlatively derive the exponential decay law. Since this is known to be impossible within the Hilbert space, it is natural to embed the latter in a RHS. This indeed answers the question, as shown by Böhm and Gadella in a series of papers [19]-[23], [36]-[38] (see also [18]) In addition, it also justifies the intrinsic asymmetry between past and future ("arrow of time" is the fashionable expression for this): a physical state must be prepared *before* its starts to decay [24, 25]. We will summarize this set of applications in this section.

4.1.1 Hardy spaces and Hardy RHS. The basic tool in this approach is the pair of Hardy spaces $H_\pm^2(R)$, that are defined as follows [48]. One starts from the obvious decomposition (the variable is time):

$$(49) \qquad L^2(\mathbb{R}) = L^2(\mathbb{R}_-) \oplus L^2(\mathbb{R}_+).$$

Taking an inverse Fourier transform of $\hat{f}(\xi) \in L^2(\mathbb{R}_+)$, one gets:

$$(50) \qquad f(E) = \int_0^\infty e^{iE\xi}\, \hat{f}(\xi)\, d\xi,$$

and this function may be continued analytically to the whole upper half-plane, $z = E + iy$, $y > 0$. Thus, applying the inverse Fourier transform to (49), one obtains, in the E variable:

$$(51) \qquad L^2(\mathbb{R}) = H_-^2(\mathbb{R}) \oplus H_+^2(\mathbb{R}),$$

where the elements of $H_-^2(\mathbb{R})$, resp. $H_+^2(\mathbb{R})$, have analytic continuation into the lower, resp. upper half-plane. The spaces $H_\pm^2(R)$, which are called *Hardy spaces*, are both Hilbert spaces. The upper space $H_+^2(\mathbb{R})$ may be defined intrinsically as the space of functions $f(z)$ analytic in the upper half-plane, such that

$$(52) \qquad \sup_{y>0} \int_{-\infty}^\infty |f(E+iy)|^2 dE < C,$$

for some constant C. The fact that $H_+^2(\mathbb{R})$ corresponds to $L^2(\mathbb{R}_+)$ by Fourier transform is known as the *Paley-Wiener* theorem. Similar considerations hold true for $H_-^2(\mathbb{R})$.

The spaces $H_\pm^2(\mathbb{R})$ have remarkable properties:

(i) *The Titchmarsh theorem.* If $f \in H_\pm^2(\mathbb{R})$, then

$$(53) \qquad \frac{1}{2\pi i} \int_{-\infty}^{\infty} dE\, \frac{f(E)}{E - \omega} = \pm \vartheta(\pm \mathrm{Im}\,\omega) f(\omega),$$

where ϑ is the usual step function.

(ii) *The van Winter theorems.* For E real, $f \in H_\pm^2(\mathbb{R})$ is uniquely determined by its boundary values on $\mathbb{R}_+$. If we denote by $\theta_\pm : H_\pm^2(\mathbb{R}) \to L^2(\mathbb{R}_+)$ the operator of restriction to $\mathbb{R}_+$, then $\theta_\pm$ is one-to-one and its range is dense in $L^2(\mathbb{R}_+)$.

Combining now the decomposition (49) with the Schwartz triplet (9), and taking Fourier transforms (which yields (51)), we obtain a RHS around $H_+^2(\mathbb{R})$:

$$(54) \qquad \begin{array}{ccccc} \mathcal{S}_+ = \mathcal{S} \cap L^2(\mathbb{R}_+) & \subset & L^2(\mathbb{R}_+) & \subset & (\mathcal{S}_+)^\times \\ \downarrow & & \downarrow & & \downarrow \\ \Psi_+ = \mathcal{S} \cap H_+^2(\mathbb{R}) & \subset & H_+^2(\mathbb{R}) & \subset & (\Psi_+)^\times. \end{array}$$

In the same way one obtains a RHS around $H_-^2(\mathbb{R})$:

$$(55) \qquad \Psi_- = \mathcal{S} \cap H_-^2(\mathbb{R}) \subset H_-^2(\mathbb{R}) \subset (\Psi_-)^\times.$$

Taking now the restriction of all spaces to $\mathbb{R}_+$ and denoting $\Delta_\pm = \theta_\pm(\Psi_\pm)$, we obtain finally

$$(56) \qquad \Delta_\pm \subset L^2(\mathbb{R}_+) \subset (\Delta_\pm)^\times.$$

Notice that one has $\Delta_+ \cap \Delta_- \neq \emptyset$, whereas, of course, $\Psi_+ \cap \Psi_- = \emptyset$, but $\Psi_+ + \Psi_- \neq \mathcal{S}$.

4.1.2 The Hardy RHS in scattering theory. Consider now a simple scattering problem, with total Hamiltonian $H = H_o + V$. Assume that both H and H_o have a nondegenerate, absolutely continuous spectrum equal to $\mathbb{R}_+ = [0, \infty)$. Assume further that the scattering process is asymptotically complete, so that the Møller operators $\Omega^\pm$ exist and are unitary between the respective absolutely continuous subspaces $\mathcal{H}_{ac}(H)$ and $\mathcal{H}_{ac}(H_o)$. Then both H and H_o, restricted to these subspaces, are unitarily equivalent to the operator of multiplication by the variable E on $L^2(\mathbb{R}_+)$:

$$(57) \qquad \mathcal{H}_{ac}(H) \xrightarrow{(\Omega^\pm)^{-1}} \mathcal{H}_{ac}(H_o) \xrightarrow{U} L^2(\mathbb{R}_+).$$

Define the space $\Phi_\pm = \Omega^\mp U^{-1} \Delta_\pm$ (note the change of sign, unfortunate but traditional). Then we get two additional RHS:

(58)
$$\Phi_\pm \subset \mathcal{H}_{ac}(H) \subset (\Phi_\pm)^\times.$$

A state $\psi^+ \in \Phi_-$ tends for $t \to -\infty$ to a free ingoing state ψ^{in} : $\psi^+ = \Omega^+ \psi^{in}$, whereas a state $\psi^- \in \Phi_+$ tends for $t \to +\infty$ to a free outgoing state ψ^{out} : $\psi^- = \Omega^- \psi^{out}$, and of course $\psi^{out} = S\psi^{in}$ [61].

In this framework, one obtains a precise definition of decaying and growing states [23, 36, 37, 38].

- Let $z_R = E_R - i\frac{\Gamma}{2}$ be a pole of $S(E)$ in the lower half of the second sheet. Then a state $|z_R^-\rangle \in (\Phi_+)^\times$ may be associated to this pole, and it is a generalized eigenvector of H:

(59)
$$H|z_R^-\rangle = z_R|z_R^-\rangle.$$

This state is called a *decaying Gamow vector*.

- Similarly, a pole $z_R^* = E_R + i\frac{\Gamma}{2}$ in the upper half of the second sheet corresponds to a generalized eigenvector $|z_R^{*+}\rangle \in (\Phi_-)^\times$:

(60)
$$H|z_R^{*+}\rangle = z_R^*|z_R^{*+}\rangle,$$

called a *growing Gamow vector*.

Actually, these decaying states have long been considered for describing the decay of an unstable system, following the original idea of Gamow [39], but in a rather formal way, within the Hilbert space. Notice that square integrable approximations to decaying Gamow vectors *do* have the required properties, as shown in [58].

Two remarks are in order here:

(i) First, $|z_R^-\rangle$ is a functional over Φ_+ only, which is, so to speak, roughly one half of the space of all physically realizable states (compare (51) and (54)), in the usual RHS interpretation. Correspondingly, $|z_R^{*+}\rangle$ is a functional over Φ_-, i.e. the other half.

(ii) The complex eigenvalues $z_R^{(*)} = E_R \pm i\frac{\Gamma}{2}$ do not belong to the (real) Hilbert space spectrum of H, which is self-adjoint. Hence, the riggings (56) are *not* tight with respect to H. This follows, of course, from the situation described in remark (i).

Using the two RHS (58) simultaneously, one obtains a neat description of the time evolution of the scattering system.

- The time evolution operator e^{-iHt} leaves Φ_+ invariant for $t \geq 0$ only, and Φ_- only for $t \leq 0$. Thus the time evolution splits into two time evolution *semigroups* :

(61)
$$U_-(t): \quad \Phi_- \to \Phi_- \quad \text{for } t \leq 0$$
$$U_+(t): \quad \Phi_+ \to \Phi_+ \quad \text{for } t \geq 0$$

Therefore the time evolution of $|z_R^-\rangle \in (\Phi_+)^\times$ is defined for $t \geq 0$ only, and that of $|z_R^{*+}\rangle \in (\Phi_-)^\times$ for $t \leq 0$ only.

– The scattering process may be described in terms of the evolution semi-groups $U_\pm(t)$, in two steps:

(i) For $t \leq 0$, the state $\psi^+ \in \Phi_-$ is prepared in the remote past from $\psi^{in} = (\Omega^+)^{-1}\psi^+$ and evolves according to $U_-(t)$.

(ii) For $t \geq 0$, the state $\psi^- \in \Phi_+$ decays and is detected in the distant future, where it evolves according to $U_+(t)$ into the free state $\psi^{out} = (\Omega^-)^{-1}\psi^-$. In addition, the decay of ψ^- follows the exponential law.

– Since the preparation of the state ψ^+ must be completed (at time $t = 0$) before the detection begins, there is an intrinsic asymmetry between past and future, that is, one may define an arrow of time.

4.2 LHS: Integral Equations vs. Complex Scaling

In Section 3, we have described a LHS of analytic functions, in which the order parameter was the rate of growth at infinity. In the scale $\{\mathfrak{F}^\rho\}$, one has $\mathfrak{F}^{\rho_1} \subset \mathfrak{F}^{\rho_2}$ iff $\rho_1 > \rho_2$, and similarly for the full lattice $\{\mathfrak{F}(\rho), \rho \in I\}$. In this section, we shall describe another LHS of analytic functions, in which the order parameter is the opening angle of a sector. The spaces are obtained by a Mellin transform from the corresponding ones in a LHS of weighted L^2 spaces, again indexed by a growth factor at infinity. This particular LHS of analytic functions has two advantages. First of all, it simplifies considerably the formulation of scattering theory of van Winter [63, 64] (based on the so-called Weinberg-van Winter or WVW integral equations). Second, it is a crucial tool for proving that the WVW formalism is a particular case of the Complex Scaling or Dilation Analyticity Method (CSM) [1, 13], by now a standard approach to quantum scattering theory.

4.2.1 A LHS for scattering theory.

(i) A Hilbert space of Hardy type. The idea of van Winter [63, 64] is to formulate the WVW integral equations in an appropriate Hilbert space, so that they become Fredholm equations, with a Hilbert-Schmidt kernel, uniquely solvable by standard methods. For simplicity, we restrict ourselves to two-body scattering, but the whole development extends in a straightforward way to many-body situations. We also omit the angular variables, which are totally irrelevant in this context.

Following [63], we define $G(a, b)$ $(-\pi < a < b < \pi)$ as the space of all functions $f(z)$, $z = re^{i\varphi}$, which are analytic in the open sector

$$S_{ab} = \{z = re^{i\varphi}, a < \varphi < b\},$$

and such that the integral

$$\int_0^\infty |f(re^{i\varphi})|^2 \, dr < \infty$$

is uniformly bounded in $\varphi \in (a, b)$. Then it is shown that every function $f(re^{i\varphi}) \in S(a, b)$ possesses well-defined limiting values (in the sense of L^2

limits) $f(re^{ia})$, $f(re^{ib})$ on the boundaries of S_{ab}, and furthermore that $G(a, b)$ is complete, thus a Hilbert space, for the inner product

$$(62) \qquad \langle f \mid g \rangle_{ab} = \int_0^\infty \overline{f(re^{ia})}\, g(re^{ia})\, dr + \int_0^\infty \overline{f(re^{ib})}\, g(re^{ib})\, dr.$$

Notice that the usual Hardy spaces may also be obtained by this construction: $H_+^2(\mathbb{R}) = G(0, \pi)$, $H_-^2(\mathbb{R}) = G(-\pi, 0)$.

Among the bounded linear operators on $G(a, b)$, a distinguished role is played by the class $\mathfrak{A}$ of those operators A for which the quantity

$$(63) \quad \alpha(A, \varphi) = \sup_{f \in G(a,b)} \left[\int_0^\infty |(Af)(re^{i\varphi})|^2\, dr \right] \left[\int_0^\infty |f(re^{i\varphi})|^2\, dr \right]^{-1/2}$$

is uniformly bounded in $\varphi \in (a, b)$. Then it turns out that the class $\mathfrak{A}$ is a Banach *-algebra for the norm $\alpha(A) = \sup_{a<\varphi<b} \alpha(A, \varphi)$. Furthermore, $\mathfrak{A}$ contains a two-sided ideal $\mathfrak{K}$ of integral operators which are all Hilbert-Schmidt.

(ii) Scattering theory in $G(a, b)$. The essence of van Winter's approach [63, 64] is to formulate the entire theory of scattering in the space $G(a, b)$, with the following steps (from now on, we assume that $-\frac{\pi}{2} < a < b < \frac{\pi}{2}$).

- Definition of the interaction, by $v(re^{i\varphi}) \in G(-a, a)$ if it is local, or by $V \in \mathfrak{K}$ if it is nonlocal.
- Definition of the Hamiltonian $H = H_o + V$ and the resolvents $R(z) = (H - z)^{-1}$, $R_o(z) = (H_o - z)^{-1}$ as operators in $G(a, b)$.
- Write the Weinberg-van Winter (WVW) equations in $G(a, b)$:

$$(64) \qquad\qquad R(z) = R_o(z) - R_o(z)V R_o(z).$$

Then it is proven that $R_o(z)V$ belongs to $\mathfrak{K}$, hence is Hilbert-Schmidt, so that the WVW equation is a Fredholm equation in $G(a, b)$, which can be solved uniquely.

(iii) A LHS version of the WVW formalism. We shall now show that the formalism sketched above simplifies considerably if one uses the LHS language [41]. To that effect, we define the Hilbert space

$$(65) \qquad \widetilde{G}(a, b) = \{\widetilde{f}(t) \mid \int_{-\infty}^{+\infty} (e^{2at} + e^{2bt}) |\widetilde{f}(t)|^2\, dt < \infty\}$$

and we introduce the inverse Mellin transform of $\widetilde{f}$:

$$(66) \qquad f(re^{i\varphi}) = (2\pi)^{-1/2} \int_{-\infty}^{+\infty} \widetilde{f}(t)\, (re^{i\varphi})^{it-\frac{1}{2}}\, dt.$$

Then it turns out that $f(re^{i\varphi}) \in G(a, b)$ iff $\widetilde{f}(t) \in \widetilde{G}(a, b)$, and furthermore the Mellin transform $f \mapsto \widetilde{f}$ is a unitary map from $G(a, b)$ onto $\widetilde{G}(a, b)$.

We claim that the family $\{G(a,b), -\frac{\pi}{2} \le a < b \le \frac{\pi}{2}\}$ is a part of a LHS of weighted L^2 spaces. Indeed, for $-\frac{\pi}{2} \le a \le \frac{\pi}{2}$, define the Hilbert space

$$(67) \qquad L^2(a) = \{\widetilde{f} \mid \int_{-\infty}^{+\infty} e^{2at} |\widetilde{f}(t)|^2 \, dt < \infty\}$$

In the notation of Section 2.2.2 (iii), $L^2(a) \equiv L^2(r_a)$, with $r_a(t) = e^{-2at}$. Then consider the lattice generated by the family $\{L^2(a), -\frac{\pi}{2} \le a \le \frac{\pi}{2}\}$:

- Infimum: $L^2(a) \wedge L^2(b) = L^2(a) \cap L^2(b) = L^2(a \wedge b)$, where (we take $a < b$)

$$r_{a \wedge b}(t) = \min(r_a(t), r_b(t)) = \begin{cases} e^{-2at}, & t < 0 \\ e^{-2bt}, & t \ge 0 \end{cases}$$

so that $L^2(a \wedge b) \equiv \widetilde{G}(a,b)$.
- Supremum: $L^2(a) \vee L^2(b) = L^2(a) + L^2(b) = L^2(a \vee b)$, where $r_{a \vee b}(t) = \max(r_a(t), r_b(t))$.
- Duality: $L^2(a \wedge b) \Leftrightarrow L^2(-a \vee -b)$.

Thus we obtain a LHS, with extreme spaces $V^{\#} = L^2(-\frac{\pi}{2}) \cap L^2(\frac{\pi}{2})$, $V = L^2(-\frac{\pi}{2}) + L^2(\frac{\pi}{2})$, which are themselves Hilbert spaces.
In addition,

- All spaces are obtained at the first generation, i.e. they are all of the form $L^2(c \wedge d)$ or $L^2(c \vee d)$. For instance,

$$L^2(a \wedge b) \wedge L^2(-b \vee -a) =$$
$$= \begin{cases} L^2(-|c| \vee |c|), & \text{if } a, b \text{ have the same sign} \\ L^2(-|c| \wedge |c|), & \text{if } a, b \text{ have opposite signs,} \end{cases}$$

where $|c| = \min(|a|, |b|)$.
- All spaces may be obtained from $L^2(-\frac{\pi}{2})$ and $L^2(\frac{\pi}{2})$ by interpolation [16].

In the case $0 < a < b$, for instance, one gets the picture depicted in Figure 1. Duality corresponds to symmetry with respect to the center (i.e., L^2): $a \wedge b \Leftrightarrow -b \vee -a$. In addition, Fourier transform is unitary from $\widetilde{G}(a,b) \equiv L^2(a \wedge b)$ onto $\widetilde{G}(-b, -a) \equiv L^2(-b \wedge -a)$, which is a dense subspace of $L^2(-b \vee -a)$, the dual of $L^2(a \wedge b)$. Notice that one can work with fixed b also, using only the sublattice generated (in the case depicted in Figure 1) generated by $L^2(b), L^2(-b)$.
Since the Mellin transform is unitary, it maps the left hand side of the LHS, that is, $\{\widetilde{G}(a,b), -\frac{\pi}{2} < a < b < \frac{\pi}{2}\}$, onto the family of Hardy spaces $\{G(a,b)\}$, which therefore has the same LHS structure. In addition, the right hand side of the LHS, consisting of the duals of the spaces $G(a,b)$, is now perfectly characterized as well.

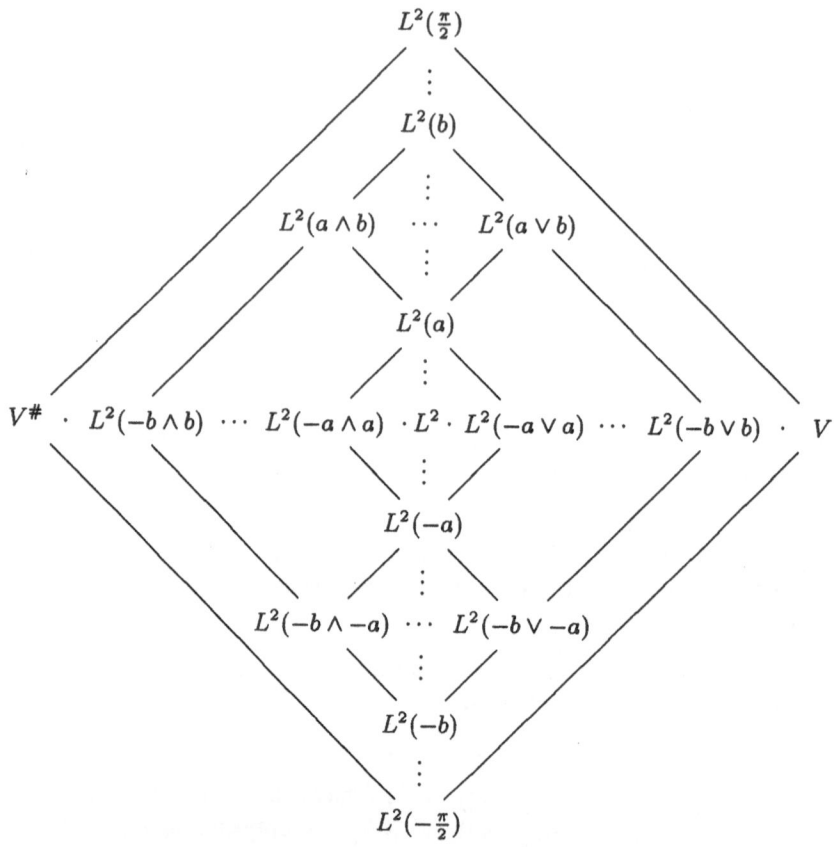

Figure 1: The van Winter LHS

As for operators on this LHS, one shows, using again interpolation theory, that the two operator algebras $\mathcal{A}$ and $\mathcal{B}$ defined in Section 2.4 coincide, i.e., there is a unique Banach *-algebra of bounded operators:

$$\mathcal{A} = \mathcal{B} = \{A \text{ linear } \mid A \text{ maps } L^2(-\frac{\pi}{2}) \text{ and } L^2(\frac{\pi}{2})$$

(68)   into themselves continuously}.

Furthermore, this algebra coincides with the Mellin transform of the van Winter algebra $\mathfrak{A}$, which contains $\mathfrak{K}$ as an ideal. Putting everything together, we obtain a considerable simplification, and clarification, of the WVW formalism.

4.2.2 Connection with the complex scaling method. The dilation analyticity [1, 13] or complex scaling method (CSM) is by now a well established tool in scattering theory, including for numerical work in atomic and molecular systems [2, 27]. Its popularity rests largely on the fact that it turns a resonance into a discrete eigenvalue of a non-self-adjoint operator, thus allowing a complete disentangling of the spectrum of the Hamiltonian. The WVW method is obviously related to this one, but the exact link has not been identified so far. Our aim in this last section is to show, in a precise mathematical way, that the WVW formalism is a particular case of CSM. The starting point of CSM is the familiar unitary representation of the dilation group in $L^2(\mathbb{R}^3)$ (we again omit the angles):

$$(69) \qquad (U(\rho)f)(r) = \rho^{3/2} f(\rho r), \ \rho > 0, \ f \in L^2(\mathbb{R}^3).$$

The basic notion of dilation analyticity applies both to vectors and operators. For $0 \le a < \pi/4$, consider the sector $S_a \equiv S_{--a,a} = \{z = \rho e^{i\varphi} \mid -a < \varphi < a\}$. Then the set of S_a-dilation analytic vectors is defined as

$$D(S_a) = \{f \in L^2 \mid U(\rho)f \text{ has an analytic continuation}$$
$$(70) \qquad\qquad\qquad \text{to } U(\eta)f \text{ for all } \eta = \rho e^{i\varphi} \in S_a\}.$$

Similarly, the potential V is called S_a-dilation analytic if $V(\rho) = U(\rho)VU(\rho)^{-1}$ admits an analytic continuation to $V(\eta)$, $\eta \in S_a$. For such an interaction, one considers the spectrum of the (non-self-adjoint) dilated Hamiltonian $H(\eta) = U(\eta)HU(\eta)^{-1}$. For $\eta = \rho e^{i\varphi}$, it turns out that the bound states remain on the real axis, independently of φ, the absolutely continuous spectrum σ_{ac} is rotated by $-\varphi$, i.e. into the lower half-plane, and the resonances appear as isolated points in the sector $-\varphi < \arg z < 0$, and in fact eigenvalues of $H(\eta)$. Thus the structure of $\sigma(H)$ becomes transparent, and the study of resonances much easier. We note that part of the LHS described above have been considered, for similar purposes, in [46] and in [59].
The key to the comparison between WVW and CSM is the new definition of dilation analytic vectors given in [14], in terms of a new family of Hilbert spaces $D(a, b)$. First, let T_{ab} denote the sector $-b < \arg z < \pi - a$ (since the opening angle is larger than π, all the sectors T_{ab} overlap). Then, for a fixed number $\alpha > -1$, define $D(a, b)$ as the space of all functions $f(z)$ analytic in T_{ab} and such that

$$\int_{\mathbb{C}+} |f(ze^{-i\varphi})|^2 \, y^\alpha \, dx dy < \infty \quad (z = x + iy)$$

is uniformly bounded in $\varphi \in (a, b)$. It turns out that $D(a, b)$ is a Hilbert space, and in addition:

$-$ $D(S_a) = \bigcap_{0 \le c < a} D(-c, c)$.

$-$ For $\psi \in G(a, b)$ and $a < \varphi < b$, consider the function $\hat{V}_\alpha \psi$ given by:

$$(71) \qquad \begin{aligned} (\hat{V}_\alpha \psi)(z) = & [2\pi \Gamma(\alpha + 1)]^{-1/2} \\ & \times \int_0^\infty (re^{i\varphi})^{\alpha + \frac{3}{2}} e^{iz(re^{i\varphi})^2/2} \, \psi(re^{i\varphi}) \, e^{i\varphi} \, dr. \end{aligned}$$

Then this function is analytic in z for $-2\varphi < \arg z < \pi - 2\varphi$. It follows that the restriction of $\hat{V}_\alpha \psi$ to this domain belongs to $D(2a, 2b)$, and moreover, the map $\hat{V}_\alpha$ is unitary from $G(a, b)$ onto $D(2a, 2b)$.

Thus we get *three* unitary equivalent Hilbert spaces:

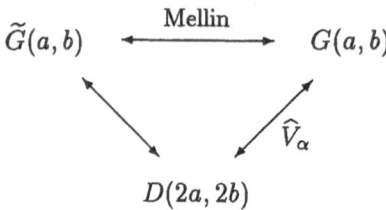

and therefore the LHS structure extends to the three families. Using these three spaces, one may then show [47] that any van Winter interaction that belongs to $G(-a, a)$ (local case) or to $\mathfrak{A}[G(-a, a)]$ (nonlocal case) is necessarily S_a-dilation analytic. In other words, the WVW formalism is a particular case of CSM. In addition, if one lets the parameter α vary, one obtains a scale of Hilbert spaces, that is contained in the Bergman LHS alluded to at the end of Section 3. A detailed account of the results of this section will be presented elsewhere.

In conclusion, the LHS structure makes the WVW approach substantially simpler. In particular, using the reformulation of dilation analyticity given in [14], it allows to embed WVW into CSM, thus settling an old question. Other results, for instance those of [59], point in the same direction. Thus the LHS formalism appears as a natural framework for quantum scattering theory, and we feel that the results reported here are sufficiently convincing to warrant further study.

5. Conclusion

It is fair to say that all the 'super-Hilbert space' formalisms described in this paper have brought substantial advances and new insight into several sectors of quantum mechanics: the Dirac formalism and the measurement process, a generalized Fock-Bargmann representation and its application to Weyl quantization, or various aspects of scattering theory, such as Gamow vectors, the arrow of time and the two corresponding time evolution semigroups, or the van Winter formalism and its connection with the complex scaling method. Other applications, not described here, pertain to decaying systems such as $K^o - \overline{K^o}$ [26], quantum chaos or the accidental degeneracy of resonances (see [25] for a full discussion). Considering all these successes, we expect 'super-Hilbert space' methods to gain wider acceptance and appeal among physicists.

References

1. J. Aguilar and J-M. Combes, A class of analytic perturbations for one-body Schrödinger Hamiltonians, *Commun. Math. Phys.* **22** (1971) 269-279

2. S. Albeverio, L.S. Ferreira and L. Streit (eds.), *Resonances – Models and Phenomena (Proc. Bielefeld 1984)* , Lect. Notes in Phys. Vol. 211, Springer, Berlin, 1984

3. S.T. Ali, J-P. Antoine, J-P. Gazeau and U.A. Mueller, Coherent states and their generalizations: A mathematical overview, *Reviews Math. Phys.* **7** (1995) 1013-1104

4. S.T. Ali, J-P. Antoine and J-P. Gazeau, *Coherent States, Wavelets and Their Generalizations*, Springer, Berlin, 1998 (to appear)

5. J-P. Antoine, Dirac formalism and symmetry problems in Quantum Mechanics. I. General Dirac formalism; II. Dirac formalism and symmetry problems in Quantum Mechanics *J. Math. Phys.* **10** (1969) 53-69, 2276-2290

6. J-P. Antoine and A. Grossmann, Partial inner product spaces. I. General properties. II. Operators, *J. Funct. Anal.* **23** (1976) 369-378, 379-391.

7. J-P. Antoine, Partial inner product spaces. III. Compatibility relations revisited, *J. Math. Phys.* **21** (1980) 268-279.

8. J-P. Antoine, Partial inner product spaces. IV. Topological considerations, *J. Math. Phys.* **21** (1980) 2067-2079.

9. J-P. Antoine and A. Grossmann, Orthocomplemented subspaces of nondegenerate partial inner product spaces, *J. Math. Phys.* **19** (1978) 329-335

10. J-P. Antoine and K. Gustafson, Partial inner product spaces and semi-inner product spaces, *Adv. in Math.* **41** (1981) 281-300

11. J-P. Antoine and M. Vause, Partial inner product spaces of entire functions, *Ann. Inst. H. Poincaré* **35** (1981) 195-224

12. D. Babbitt, Rigged Hilbert spaces and one-particle Schrödinger operators, *Rep. Math. Phys.* **3** (1972) 37-42

13. E. Balslev and J-M. Combes, Spectral properties of many-body Schrödinder operators with dilatation-analtic interactions, *Commun. Math. Phys.* **22** (1971) 280-294

14. E. Balslev, A. Grossmann and T. Paul, A characterization of dilation-analytic operators, *Ann. Inst. H. Poincaré* **45** (1986) 277-292

15. V. Bargmann, On a Hilbert space of analytic functions and an associated integral transform. Part II. A family of related function spaces; application to distribution theory, *Commun. Pure Appl. Math.* **20** (1967) 1-101

16. J. Bergh and J. Löfström, *Interpolation Spaces: An Introduction*, Springer, Berlin, 1976

17. A. Böhm, The rigged Hilbert space in quantum mechanics, in *Boulder Lectures in Theoretical Physics*, **9A** (1966), Interscience, New York, 1966

18. A. Böhm, *Quantum Mechanics*, Springer, Berlin, 1979; 3rd ed., rev. and enlarged, 1993

19. A. Böhm, Gamow state vectors as functionals over subspaces of the nuclear space, *Lett. Math. Phys.* **3** (1979) 455-461

20. A. Böhm, Decaying states in the rigged Hilbert space formulation of quantum mechanics, *J. Math. Phys.* **21** (1980) 1040-1043

21. A. Böhm,The rigged Hilbert space and decaying states, in *Group Theoretical Methods in Physics (Proc. Austin 1978)*, pp. 245-249; Lect. Notes Phys., vol. 94, Springer, Berlin et al., 1979

22. A. Böhm, Resonance poles and Gamow vectors in the rigged Hilbert space formulation of quantum mechanics, *J. Math. Phys.* **22** (1981) 2813-2823

23. A. Böhm and M. Gadella, *Dirac Kets, Gamow Vectors and Gel'fand Triplets*, Lect. Notes in Physics, vol. 348, Springer, Berlin, 1989

24. A. Böhm, Microphysical irreversibility and the time-reversal operation, *Phys. Rev. A* **51** (1995) 1758-1769

25. A. Böhm, S. Maxson, M. Loewe and M. Gadella, Quantum mechanical irreversibility, *Physica A* **236** (1997) 485-549

26. A. Böhm, Irreversible quantum mechanics in the neutral K-system, preprint U. Texas, 1996

27. E. Brändas and N. Elander (eds.), *Resonances (Proc.. Lertorpet 1987)*, Lect. Notes in Physics vol. 325, Springer, Berlin, 1989

28. F. Bruhat, Sur les représentations induites des groupes de Lie, *Bull. Soc. Math. France* **84** (1956) 97-205

29. I. Daubechies, On the distributions corresponding to bounded operators in the Weyl quantization, *Commun. Math. Phys.* **75** (1980) 229-238

30. I. Daubechies and A. Grossmann, An integral transform related to quantization, *J. Math. Phys.* **21** (1980) 2080-2090

31. I. Daubechies, A. Grossmann and J. Reignier, An integral transform related to quantization. II. Some mathematical properties, *J. Math. Phys.* **24** (1983) 239-254

32. I. Daubechies and J.R. Klauder, Constructing measures for path integrals, *J. Math. Phys.* **23** (1982) 1806-1822

33. S.J.L. van Eijndhoven and J. de Graaf, *A Mathematical Introduction to Dirac's Formalism*, North Holland, Amsterdam, 1986

34. M. Flato, J. Simon, H. Snellman and D. Sternheimer, Simple facts about analytic vectors and integrability, *Ann. Sc. Ec. Norm. Sup.* **5** (1972) 423-434

35. D. Fredricks, Tight riggings for a complete set of commuting observables, *Rep. Math. Phys.* **8** (1975) 277-293

36. M. Gadella, A rigged Hilbert space of Hardy-class functions: Applications to resonances, *J. Math. Phys.* **24** (1983) 1462-1469

37. M. Gadella, A description of virtual scattering states in the rigged Hilbert space formulation of quantum mechanics, *J. Math. Phys.* **24** (1983) 2142-2145

38. M. Gadella, On the RHS description of resonances and virtual states, *J. Math. Phys.* **25** (1984) 2481-2485

39. G. Gamow, Zur Quantentheorie des Atomkernes, *Z. Phys.* **51** (1928) 204-212; Zur Quantentheorie der Atomkzertrümmerung, *ibid.***52** (1928) 510-515

40. I.M. Gelfand and N.Ya. Vilenkin, *Generalized Functions. IV*, Academic Press, New York, 1964

41. A. Gollier, *Espaces de fonctions analytiques et théorie de la diffusion*, Mémoire de licence, UCL, 1982 (unpublished)

42. V. Gorini and G. Parravicini, Unstable quantum states and rigged Hilbert spaces, in *Group Theoretical Methods in Physics (Proc. Austin 1978)*, pp. 219-228; Lect. Notes Phys., vol. 94, Springer, Berlin et al., 1979; G. Parravicini, V. Gorini and E.C.G. Sudarshan, Resonances, scattering theory and rigged Hilbert spaces, *J. Math. Phys.* **21** (1980) 2208-2226

43. A. Grossmann, Elementary properties of nested Hilbert spaces, *Commun. Math. Phys.* **2** (1966) 1-30

44. A. Grossmann, Homomorphisms and direct sums of nested Hilbert spaces, *Commun. Math. Phys.* **4** (1967) 190-202

45. A. Grossmann, R. Hoegh-Krohn and M. Mebkhout, A class of explicitly soluble, local, many-center Hamiltonians for one-particle quantum mechanics in two and three dimensions, *J. Math. Phys.* **21** (1980) 2376-2385

46. L.P. Horwitz and E. Katznelson, A partial inner product space of analytic functions for resonances, *J. Math. Phys.* **24** (1983) 848-859

47. M. Klein, *Phénomènes de résonance et espaces de fonctions analytiques*, Mémoire de licence, UCL, 1987 (unpublished)

48. P. Koosis, *Introduction to H_p Spaces*, London Math. Soc. Lect. Notes Series, vol. 40, Cambridge U.P., Cambridge, 1980

49. G. Köthe, *Topological Vector Spaces. I*, Springer, Berlin, 1969.

50. O. Melsheimer, Rigged Hilbert space formalism as an extended mathematical formalism for quantum systems. I. General theory. II. Transformation theory in nonrelativistic quantum mechanics, *J. Math. Phys.* **15** (1974) 902-916, 917-925

51. B. Nagel, Generalized eigenvectors in group representations, in *Studies in mathematical Physics (Proc. Istanbul 1970)*, pp. 135-154; A.O. Barut (ed.), Reidel, Dordrecht and Boston, 1970; G. Lindblad and B. Nagel, Continuous bases for unitary irreducible representations of SU(1,1), *Ann. Inst. H. Poincaré* **13** (1970) 27-56

52. F. Mathot, Some operator algebras in nested Hilbert spaces, *Commun. Math. Phys.* **42** (1975) 183-2193

53. K. Napiorkowski, Good and bad generalized eigenvectors. I II.. *Bull. Acad. Pol. Sc.* **22** (1974) 1215-1218; **23** (1975) 251-252

54. O. Pelc and L.P. Horwitz, Construction of a complete set of states in relativistic scattering theory, *J. Math. Phys.* **38** (1997) 115-138

55. J.E. Roberts, The Dirac bra and ket formalism, *J. Math. Phys.* **7** (1966) 1097-1104;

56. J.E. Roberts, Rigged Hilbert spaces in quantum mechanics, *Commun. Math. Phys.* **3** (1966) 98-119

57. H.H. Schaefer, *Topological Vector Spaces*, Springer, Berlin, 1971

58. E. Skibsted, Truncated Gamow functions, α-decay and the exponential law, *Commun. Math. Phys.* **104** (1986) 591-604

59. E. Skibsted, Resonances eigenfunctions of a dilation-analytic Schrödinger operator, based on the Mellin transform, *J. Math. Anal. Appl.* **117** (1986) 198-219

60. L. Schwartz, *Théorie des Distributions, I-II*, Hermann, Paris, 1957, 1959

61. J.R. Taylor, *Scattering Theory*, Wiley, New York, 1972

62. F. Trèves, *Topological Vector Spaces, Distributions and Kernels*, Academic Press, New York and London, 1967

63. C. van Winter, Fredholm equations on a Hilbert space of analytic functions, *Trans. Amer. Math. Soc.* **162** (1971) 103-139

64. C. van Winter, Complex dynamical variables for multiparticle systems with analytic interactions. I, II, *J. Math. Anal. Appl.* **47** (1974) 633-670, **48** (1974) 368-399

Gamow States in a Rigged Hilbert Space *

C.G. Bollini[1,2], O. Civitarese[1,3], A.L. De Paoli[1,3], and M.C. Rocca[1,3]

[1] Department of Physics, University of La Plata, C.C.67, La Plata (1900), Argentina
[2] fellow of the CIC, Pcia de Bs.As, Argentina
[3] fellow of the CONICET, Argentina

Summary. The space of analytical test fuctions rapidly decreasing on the real axis (i.e: Schwartz test functions on the real axis), is used to construct the Rigged Hilbert Space (RHS) where Resonant Gamow States (GS) are defined starting from Dirac's formula. It is shown that the expectation value of a self-adjoint operator acting on a GS is real.

The treatment of the continuum and the inclusion of decaying states in the definition of the nuclear response is a long-standing problem [1, 2, 3]. The inclusion of resonant states in the one-body Green Function has been studied years ago by Tore Berggren [4, 5]. Lately, the use of these states to calculate one-particle and collective excitations in finite nuclei has been proposed by Liotta et al. [6].

Several methods have been developed in connection with the treatment of GS [7, 8, 9]. The equivalence between some of these methods and the correspondence between Bergreen's and Mittag-Lefler's representations have been explored in dealing with the use of GS in nuclear structure problems [6]. Mathematical properties of GS, in the framework of the Hamiltonian formalism, have been studied by Sudarshan and collaborators [10]. Bohm et al. [11, 12] have shown that the RHS is a suitable framework to describe idealized resonances as generalized eigenvectors of a self-adjoint Hamiltonian with complex eigenvalues. The overlap between GS and wave packets of the Breit-Wigner form has been discussed by Romo [8] by using techniques of analytic continuation. The possibility of defining expectation values of operators in a resonant state has been studied by Tore Berggren in a recent work [13].

In this talk we shall show some results concerning the calculation of expectation values on resonant states [13]. At variance with the usually adopted mathematical formalism [7, 12] we shall use the concepts of tempered ultradistributions and Gelfand's triplets [14]. In the following, only the aspects of the derivation which are relevant to validate Berggren's approximation will be shown. The mathematical details of the formalism are given in [15].

The space of analytical functionals ξ' (tempered ultradistributions) is the minimal space whose Fourier anti-transform accomodates real exponential functions as distributions. This space is the dual of the space of analytical test functions ξ. Together with the Hilbert space $\mathcal{H}$ one can construct the RHS or Gelfand's triplet (GT) [14, 15] $\xi \subset \mathcal{H} \subset \xi'$. In this RHS a linear and

---

* Dedicated to the late Professor Tore Berggren

symmetric operator A acting on ξ, which admits a self-adjoint prolongation $\bar{A}$ acting on $\mathcal{H}$, has a complete set of eigen-functionals on ξ' with real generalized eigenvalues [14]. Let us introduce the GT $(\tilde{\xi}, \mathcal{H}, \tilde{\xi}')$ which is related to $(\xi, \mathcal{H}, \xi')$ by Fourier transforms. The Schwartz Space $\mathcal{S}'$ of tempered distributions is included in ξ' and in $\tilde{\xi}'$ ($\mathcal{S}' \in \xi'$). The extension to ξ' of Dirac's formula is given by [15]

$$(1) \qquad \hat{\psi}_c(z) = \frac{1}{2\pi i} \int\limits_{-\infty}^{+\infty} \frac{1}{t-z} \hat{\psi}(t) \, dt,$$

where $\hat{\psi}(t) = \hat{\psi}_c(t+i0) - \hat{\psi}_c(t-i0)$. In addition to $(\tilde{\xi}, \mathcal{H}, \tilde{\xi}')$ and $(\xi, \mathcal{H}, \xi')$ it exists the GT $(\xi_a, \mathcal{H}_a, \xi'_a)$ which admits the definition of the position operator $\hat{x}$ acting on $\mathcal{H}_a$. If $|x> \in \xi'_a$ then $< x'|x> = \delta(x - x')$ and $< x'|\hat{x}|x> = x\delta(x - x')$.

The relations $|\phi> \in \xi_a \Leftrightarrow < x|\phi> = \phi(x) \in \tilde{\xi}$, $|\varphi> \in \mathcal{H}_a \Leftrightarrow < x|\varphi> = \varphi(x) \in \mathcal{H}$ and $|\psi> \in \xi'_a \Leftrightarrow < x|\psi> = \psi(x) \in \tilde{\xi}'$ represent Dirac's formalism of Quantum Mechanics in a RHS [16]. Let us introduce a self-adjoint operator H, acting on $\mathcal{H}$, with the eigenstates (eigenvalues) given by $H|E_n> = E_n|E_n>$ (for $n \in \mathcal{N}$) and $H|E> = E|E>$ (for $E_0 < E < E_1$). Thus from Eq.(1) one can write

$$(2) \qquad \left(\hat{\psi}(E_G)\right)^* = \frac{1}{2\pi i} \int\limits_{E_0}^{E_1} \frac{1}{E_G^* - E} \left(\hat{\psi}(E)\right)^* dE \, ,$$

with $E_G = E_D + i\Gamma$ $\Gamma > 0$. In Dirac's notation one has

$$(3) \qquad \left(\hat{\psi}(E_G)\right)^* = \frac{1}{2\pi i} \int\limits_{E_0}^{E_1} \frac{1}{E_G^* - E} < \psi|E> \, dE \, .$$

We can now define

$$(4) \qquad |E_G^*> = \frac{1}{2\pi i} \int\limits_{E_0}^{E_1} \frac{1}{E_G^* - E} |E> \, dE \, .$$

Then $\hat{\psi}(E_G) = < E_G|\psi>$ and $\left(\hat{\psi}(E_G)\right)^* = < \psi|E_G^*>$. The state $|E_G^*>$ is by definition a GS eigenstate of H.

The states $|E_G^*>$ are normalizable. The diagonal matrix element of H between GS is given by the expression

(5) $< E_G|H|E_G^* >= E_D + \dfrac{\Gamma}{2} \dfrac{\ln\left[\frac{(E_1-E_D)^2+\Gamma^2}{(E_0-E_D)^2+\Gamma^2}\right]}{\left[\arctan\left(\frac{E_1-E_D}{\Gamma}\right) - \arctan\left(\frac{E_0-E_D}{\Gamma}\right)\right]}$.

The probability distribution associated to a GS is given by

$$P(E) = |< E|E_G^* >|^2$$

(6) $= \dfrac{\Gamma}{(E - E_D)^2 + \Gamma^2} \cdot \dfrac{1}{\left[\arctan\left(\frac{E_1-E_D}{\Gamma}\right) - \arctan\left(\frac{E_0-E_D}{\Gamma}\right)\right]}$.

as proposed in [8, 11].

For a self-adjoint operator A, which is acting on $\mathcal{H}_a$, the expectation value of A between GS

(7) $< E_G|A|E_G^* >= \displaystyle\int_{-\infty}^{+\infty} < E_G|\lambda > \lambda \, d\sigma_a(\lambda) < \lambda|E_G^* > ,$

is real since $< E_G|\lambda >= (< \lambda|E_G^* >)^*$.

Following Berggren's notation [13] the GS state can be defined by

(8) $|E_G^* >= \dfrac{\sqrt{\Gamma}}{i\sqrt{\pi/2}} \displaystyle\int_0^{+\infty} \dfrac{|E(\vec{k}), \hat{k}, l >}{E_G^* - E} \, dE ,$

and in the impulse representation it is written as

(9) $|E_G^* >= \dfrac{\sqrt{\Gamma}}{i\sqrt{\pi/2}} \displaystyle\int_0^{+\infty} \sqrt{\dfrac{k}{m}} \dfrac{|k, \hat{k}, l >}{E_G^* - E(\vec{k})} \, dk .$

Consequently, for the expectation value of A one has the expression

(10)

$< E_G|A|E_G^* >= \dfrac{2\Gamma}{\pi} \displaystyle\sum_{l,l'} \int_0^{+\infty} dk \int_0^{+\infty} dk' \dfrac{\sqrt{kk'}}{m} \dfrac{< k', \hat{k}', l'|A|k, \hat{k}, l >}{(E(\vec{k}') - E_G)(E(\vec{k}) - E_G^*)}$.

We can now compare the result provided by the present method and by Berggren's conjecture, namely: $< A >= Re < E_G^*|A|E_G^* >$, where

(11)

$< E_G^*|A|E_G^* >= \dfrac{2\Gamma}{\pi} \displaystyle\sum_{l,l'} \int_0^{+\infty} dk \int_0^{+\infty} dk' \dfrac{\sqrt{kk'}}{m} \dfrac{< k', \hat{k}', l'|A|k, \hat{k}, l >}{(E(\vec{k}') - E_G^*)(E(\vec{k}) - E_G^*)}$.

The relation between the above equations can be expressed as

$$< A >=< E_G|A|E_G^* >= Re < E_G^*|A|E_G^* > -$$

$$\frac{2i\Gamma^2}{\pi} \sum_{l,l'} \int_0^{+\infty} dk \int_0^{+\infty} dk' \frac{\sqrt{kk'}}{m} \left[E(\vec{k}) - E(\vec{k}') \right] \frac{< k',\hat{k}',l'|A|k,\hat{k},l >}{|E(\vec{k}') - E_G|^2 |E(\vec{k}) - E_G|^2} +$$

$$(12) \qquad \frac{4\Gamma^3}{\pi} \sum_{l,l'} \int_0^{+\infty} dk \int_0^{+\infty} dk' \frac{\sqrt{kk'}}{m} \frac{< k',\hat{k}',l'|A|k,\hat{k},l >}{|E(\vec{k}') - E_G|^2 |E(\vec{k}) - E_G|^2} .$$

It means that the result obtained in [13] is valid at leading order in Γ and that, in general, $< E_G|A|E_G^* > \neq Re < E_G^*|A|E_G^* >$.

References

1. G. Gamow, Z. Phys. **51** (1928) 204; ibid **52** (1928) 510.
2. R. G. Newton, Scattering Theory of Waves and Particles, McGraw-Hill, N.Y. (1966).
3. C. Mahaux and H. A. Weidenmuller, Shell-Model Approach to Nuclear Reactions, North-Holland (1969).
4. T. Berggren, Nucl. Phys. **A 109** (1968) 265.
5. T. Berggren, Nucl. Phys. **A 389** (1982) 261.
6. R. J. Liotta; Proc. Int. Conf. on Giant Resonances. Groningen.1996. (to be published by North Holland) and references therein.
7. W. J. Romo, Nucl. Phys. **A 116** (1968) 617; J. Math. Phys. **21** (1980) 311.
8. W. J. Romo, Nucl. Phys. **A 419** (1984) 333.
9. G. Garcia Calderon, Nucl. Phys. **A 261** (1976) 130.
10. G. Parravicini, V. Gorini and E. C. G. Sudarshan, J. Math. Phys. **21** (1980) 2208.
11. A. Bohm, J. Math. Phys. **22** (1981) 2813; ibid **21** (1980) 1040.
12. A. Bohm, M. Gadella, G. Bruce Mainland, Am. J. Phys. **57** (1989) 1103.
13. T. Berggren, Phys. Lett. **B 373** (1996) 1.
14. I. M. Guelfand and N. Y. Vilenkin: Les distributions, Tome IV, Dunod, Paris (1967).
15. C. G. Bollini, O. Civitarese, A. L. De Paoli and M. C. Rocca J. Math. Phys. **37** (1996) 4235; Phys. Lett. **B 382** (1996) 205.
16. A. Bohm and M. Gadella, Dirac Kets, Gamow Vectors and Gelfand Triplets, Lecture Notes in Physics, ed. by H. Araki, Springer-Verlag (1989).

Some General Properties
of the Liouville Operator

I. Antoniou[1,2], M. Gadella[1,3], and Z. Suchanecki[1,4]

[1] International Solvay Institutes for Physics and Chemistry, C.P. 231, Campus Plaine ULB. Bd. du Triomphe, 1050 Brussels, Belgium
[2] Theoretische Natuurkunde, Free University of Brussels, Belgium
[3] Departamento de Física Teórica, Facultad de Ciencias, E-46011, Valladolid, Spain
[4] Hugo Steinhaus Center and Institute of Mathematics, Wrocław Technical University, ul. Wybrzeże Wyspiańskiego 27, 50-370 Wrocław, Poland

1. Introduction

Usually, textbooks in quantum mechanics define a pure state as the class of vectors which are proportional to a given one. Then, as the representative of this class, one chooses a vector of norm one. This procedure arises naturally as we describe a pure state by a wave function – solution of the Schrödinger equation.

However, the representation of a pure state by a vector is not precise. Properly speaking, a pure state is a subspace of dimension one of the Hilbert space that describes a given situation in quantum mechanics. Therefore, it seems more natural to describe a pure state either as an element of this one dimensional subspace or as the corresponding projection operator. The first approach is more natural when we describe symmetries, as it was realized by Wigner [1]. The second approach is appropriate in quantum statistical mechanics because it allows for a clear mathematical definition of mixed states or mixtures. In this sense mixed states are bounded positive operators with trace one. The spectral theorem for trace class operators [2] allows to express any mixed state as a mixture of pure states [3].

The set of all state operators is a subspace of the space of bounded operators which does not form a Hilbert space. However, there exists a minimal Hilbert space of operators that contains all state operators. This is the Liouville space. The Liouville space has an important property: there exists an one to one mapping between observables on Hilbert space and observables on Liouville space. Therefore, for every self adjoint operator on Hilbert space, there exists one and only one observable on Liouville space – its Liouvillian. However, this mapping is not onto and there exists many self adjoint operators on the Liouville space that do not have a counterpart on Hilbert space. This fact by itself implies that Liouville space has a richer structure than Hilbert space. However, this is not the only reason for which Liouville space has this richer structure as we shall show soon.

There exits a description of the formalism of quantum mechanics on Liouville space that simply lift the formalism on the Hilbert space to the Liouville space [3]. This description does not take account on the peculiarities of the Liouville space, which in fact opens new possibilities. As an example, we should mention that some observables that in Hilbert space have continuous singular spectrum only, have Liouvillians with an absolutely continuous spectrum. This has a very interesting consequence in scattering theory as there are situations in quantum scattering for which do not exist scattering states in Hilbert space and exist in Liouville space.

This means that the study of quantum mechanics in Liouville space has a meaning by itself. In addition, since Liouville space provides a unified representation for mixtures and pure states and deals directly with this representation, it is the adequate framework for the formalization of quantum statistical mechanics [4]. The purpose of this paper is to present a summary of the general properties of Liouville space. This presentation is far from being complete. For instance, we do not include a discussion of scattering theory.

The paper is divided into five sections. In Section 2, we discuss the definitions of the Liouville space and the Liouvillian of an operator and mention some of the properties of the Liouvillian of an Hermitian operator. In Section 3 we present the spectral decomposition of the Liouville operator of a Hamiltonian, being given the spectral decomposition of the latter. Section 4 is devoted to further properties of the spectrum of the Liouvillian of a given Hamiltonian. In particular, we see how a Hamiltonian with purely singular continuous spectrum may have an absolutely continuous spectrum for its Liouvillian. Finally, Section 5 deals with rigged Liouville spaces, which play the same role as Gel'fand triplets for Hilbert spaces.

2. Liouville Spaces and Quantum Liouville Operator

In this section, we want to define some of our fundamental concepts concerning Liouville operator and discuss some of their properties. We define first the Liouville space that corresponds to a given Hilbert space. Then, for any observable H on the Hilbert space, we define the operator which acts on the Liouville space which we call the Liouvillian of H. We shall show that we can define a formal Liouvillian for any Hermitian (symmetric) operator on the Hilbert space. We shall analyze some properties of the Liouvillian of a given Hermitian operator.

To begin with, let us consider a quantum dynamical system in which time evolution is given by a time independent self adjoint Hamiltonian H. We know that any state either pure or mixture is represented by a density operator ρ. Its time evolution is given [3] by the following formula

$$(1) \qquad\qquad \mathcal{U}_t \rho = S_t \rho S_t^\dagger$$

where $S_t = e^{-itH}$ determines, for any value of t, the solution $\psi_t = S_t \psi_0$ of the Schrödinger equation

(2)
$$\partial_t \psi = -iH\psi$$

The operators S_t are defined on the Hilbert space of the pure states $\mathcal{H}$. $S_t^\dagger$ denotes the adjoint of S_t. We shall always denote the adjoint of an operator by attaching to it the symbol $\dagger$.

It is well known that the operators S_t, $-\infty < t < \infty$, form a commutative Lie group of unitary operators with the generator H. In addition, this group is strongly continuous with respect to its parameter t. This means that, being given an arbitrary value of time τ and an arbitrary vector ψ in $\mathcal{H}$, one has that $\lim_{t \to \tau} S_t \psi = S_\tau \psi$. These properties of S_t together with (1) will permit us to characterize $\mathcal{U}_t$. In order to do it, we will work within a formalism that will allow us to define $\mathcal{U}_t$ as a family of unitary operators on certain Hilbert space.

Hilbert-Schmidt class and trace class. We start with the same Hilbert space of pure states of any physical system that we have denoted by $\mathcal{H}$. A bounded linear operator A on $\mathcal{H}$ is called *Hilbert-Schmidt* (HS) if for some orthonormal basis $\{\psi_i\}$

(3)
$$\sum_i \langle A\psi_i | A\psi_i \rangle < \infty.$$

One can prove [2, 3] that this sum does not depend on the chosen orthonormal basis and that (3) defines the norm in the class of all HS operators. In fact HS operators form a Hilbert space with the scalar product

$$(A, B) = \sum_i \langle A\psi_i, B\psi_i \rangle.$$

We denote this space by $\mathcal{B}_\mathcal{H}^2$. The HS operators have the following properties:

i) All HS operators are bounded. Therefore, the adjoint of any HS operator always exist and it is also a HS operator. The product of two HS operators also belongs to $\mathcal{B}_\mathcal{H}^2$. However, the identity operator on $\mathcal{H}$ is not in $\mathcal{B}_\mathcal{H}^2$. As a consequence, the HS class is an involutive algebra without identity. The involution is given by the map $A \longmapsto A^\dagger$.

ii) If $A \in \mathcal{B}_\mathcal{H}^2$ and B is a bounded operator in $\mathcal{H}$, both AB and BA are in $\mathcal{B}_\mathcal{H}^2$. Thus the HS class is an ideal in the algebra of bounded operators in $\mathcal{H}$.

A bounded linear operator A on $\mathcal{H}$ is called *nuclear* (or *trace class*) if

(4)
$$\sum_i |\langle A\psi_i, \psi_i \rangle| < \infty,$$

for any orthonormal sequence $\{\psi_i\}$ in $\mathcal{H}$. One can prove that this sum also does not depend on the choice of ψ_i. Therefore for a given nuclear operator we can correctly define its trace

(5)
$$\mathrm{tr}\, A = \sum_i \langle A\psi_i, \psi_i \rangle$$

We denote by $\mathcal{B}^1_\mathcal{H}$ the trace class which is a Banach space with the norm

$$\|A\|_{\mathrm{tr}} = \sum_i \|A\psi_i\|\,.$$

The trace class has similar algebraic properties as the HS class. Thus, it is a vector space, an algebra and an ideal in the algebra of bounded operators on $\mathcal{H}$. Moreover the trace is a continuous linear functional on $\mathcal{B}^1_\mathcal{H}$. It is easy to prove that any nuclear operator is HS.

Using the notion of trace we can expand the list of properties of HS class

iii)If $A, B \in \mathcal{B}^2_\mathcal{H}$, then $\mathrm{tr}\, AB$ exists and is finite. Moreover the scalar product on $\mathcal{B}^2_\mathcal{H}$ can be expressed through the trace as follows

(6)
$$(A, B) = \mathrm{tr}\,(A^\dagger B)\,.$$

Before going further let us introduce the dyadic notation. We shall denote by kets $|\psi\rangle$ elements of $\mathcal{H}$ and by bra $\langle\varphi|$ elements of the dual $\mathcal{H}^\times$, i.e. continuous linear functionals on $\mathcal{H}$. The action of $\langle\psi|$ onto an arbitrary vector (ket) $|\varphi\rangle$ is naturally given by the scalar product $\langle\psi|\varphi\rangle$. It is well known that $\mathcal{H}^\times$ is a Hilbert space which is isomorphic to $\mathcal{H}$. However since we want to extend the Hamiltonian (Liouvillian) on rigged Hilbert (Liouville) spaces [5], the distinction between kets and bras (states and observables) should be already done on this level.

A bounded operator A is called normal if it commutes with its adjoint: $AA^\dagger = A^\dagger A$. A is positive if $\langle\psi, A\psi\rangle \geq 0$ for all $\psi \in \mathcal{H}$. Positive operators are self adjoint and hence normal. If A is normal and HS, then there exists an orthonormal basis of $\mathcal{H}$ formed by eigenvectors of A. If the eigenvectors are $\{|\psi_i\rangle\}$ and the corresponding eigenvalues are $\{\lambda_i\}$, then we have:

(7)
$$A = \sum_i \lambda_i |\psi_i\rangle\langle\psi_i| \quad \text{with} \quad \sum_i |\lambda_i|^2 < \infty$$

A positive bounded operator has always a unique positive square root. It is clear that $A^\dagger A$ is positive. The modulus $|A|$ is the positive square root of $A^\dagger A$. An operator A belongs to the trace class if and only if $\mathrm{tr}\,|A|$ is finite.

We recall that in Quantum Mechanics a state is represented by a positive trace class operator ρ with trace one. Pure states are characterized by $\rho^2 = \rho$. Each state, as a self adjoint HS operator, has an orthonormal basis formed by its eigenvectors. Its eigenvalues are always real nonnegative numbers with sum equal to one. The set of states is not a subspace of $\mathcal{B}^2_\mathcal{H}$ but a convex subset instead. This means that all sums of the form $\sum_i \lambda_i \rho_i$ where λ_i are nonnegative real numbers with the property $\sum_i \lambda_i = 1$ are also states. The minimal Hilbert space containing all the states is again $\mathcal{B}^2_\mathcal{H}$. This is the reason why it is convenient to use $B^2_{\mathcal{H}}$ in quantum mechanics and quantum statistical mechanics.

As a vector in the Hilbert space $B^2_{\mathcal{H}}$, a state ρ can be written in the form (7) where $|\psi_i\rangle$ are eigenvectors of ρ, λ_i the corresponding eigenvalues and $\operatorname{tr}\rho = \sum_i \lambda_i = 1$. The dyad $|\psi_i\rangle\langle\psi_i|$ represents the orthogonal projection onto the one dimensional subspace spanned by $|\psi_i\rangle$.

Consider now the tensor product $\mathcal{H} \otimes \mathcal{H}^\times$. It is generated by elements of the form $|\psi\rangle \otimes \langle\varphi|$ that we usually write as $|\psi\rangle\langle\varphi|$. Linear combinations of these dyads form a vector space. A scalar product can be introduced for dyads

$$(8) \qquad (|\eta\rangle\langle\xi|, |\psi\rangle\langle\varphi|) = \langle\eta|\psi\rangle\langle\varphi|\xi\rangle$$

We extend this product to linear combinations of dyads by linearity obtaining a vector space with a scalar product. If we now add all limit points we obtain a Hilbert space. This is what we understand by the Hilbert space $\mathcal{H} \otimes \mathcal{H}^\times$. It has the following important property: if $\{|\psi_i\rangle\}$ and $\{|\varphi_j\rangle\}$ are orthonormal basis for $\mathcal{H}$, then, $\{|\psi_i\rangle\langle\varphi_j|\}$ is an orthonormal basis for $\mathcal{H} \otimes \mathcal{H}^\times$. It may happen that both basis coincide. In this case we obtain the basis $\{|\psi_i\rangle\langle\psi_j|\}$. The norm of the dyad $|\psi\rangle\langle\varphi|$ is given by $\||\psi\|\,\|\varphi\|$ as a consequence of the definition of the scalar product.

One can prove that the Hilbert spaces $B^2_{\mathcal{H}}$ and $\mathcal{H} \otimes \mathcal{H}^\times$ can be identified. This is a very important property in our construction, since we shall work either with $B^2_{\mathcal{H}}$ or $\mathcal{H} \otimes \mathcal{H}^\times$ depending in our computational convenience. Observe also that the scalar product given by (8) can be written as a trace:

$$(9) \qquad \operatorname{tr}(|\eta\rangle\langle\xi|)^\dagger (|\psi\rangle\langle\varphi|) = \operatorname{tr}(|\xi\rangle\langle\eta|\psi\rangle\langle\varphi|) = \langle\eta|\psi\rangle\langle\varphi|\xi\rangle$$

We call to $B^2_{\mathcal{H}} \approx \mathcal{H} \otimes \mathcal{H}^\times$ the *Liouville* space associated with the Hilbert space $\mathcal{H}$.

Let us go back to (1). It is not difficult to show that $\mathcal{U}_t$ is a strongly continuous group of operators on $B^2_{\mathcal{H}}$. This property permit us to use the Stone theorem which says there is a self adjoint operator L on $\mathcal{H} \otimes \mathcal{H}^\times$ such that for all real values of t and all ρ in $\mathcal{H} \otimes \mathcal{H}^\times$, one has:

$$(10) \qquad \mathcal{U}_t\rho = e^{-itL}\rho.$$

Now, we are going to recall certain facts concerning unbounded operators. The domain of an operator is the space of vectors in which it acts. In order to be able to define the adjoint, this domain must be dense. An operator A on a Hilbert space $\mathcal{H}$ is Hermitian if for any pair of vectors ψ and φ in the domain $\mathcal{D}(A)$ of A one has $\langle\psi|A\varphi\rangle = \langle A\psi|\varphi\rangle$. Any Hermitian operator always coincides with its adjoint in its domain. However, the domain of the adjoint may be bigger and for that reason Hermitian and self adjoint mean the same if the operator is bounded. This is the case when $\mathcal{D}(A)$ coincides with $\mathcal{H}$. An unbounded operator is self adjoint if it coincides with its adjoint.

An operator A extends another operator B if the domain of A contains the domain of B and both operators are equal in the domain of B. Hermitian operators may have or may have not self adjoint extensions.

Let us come back to the Liouville space $\mathcal{H} \otimes \mathcal{H}^{\times}$. Let H be a Hamiltonian on $\mathcal{H}$ (in the sequel, we always assume that Hamiltonians are self adjoint). We want to define an operator on $\mathcal{H} \otimes \mathcal{H}^{\times}$. Its domain will be the vector space of HS operators ρ such that both $H\rho$ and ρH are HS operators. Note that this is true for all HS operators if and only if H is bounded. We call this domain $\mathcal{D}(\mathcal{L})$. Then, we define the operator, that we call $\mathcal{L}$ as:

$$(11) \qquad \mathcal{L}\rho = [H, \rho] = H\rho - \rho H \ , \quad \forall \rho \in \mathcal{D}(\mathcal{L})$$

The operator $\mathcal{L}$ is essentially self adjoint (it has a unique self adjoint extension) provided H is self adjoint. By definition, the Liouvillian of H is the unique self adjoint extension of the Hermitian operator H and we denote it as L_H. Occasionally, when there is not doubt to which Hamiltonian we refer, we can denote its Liouvillian as L.

Further properties of $\mathcal{L}$. If H is a non self adjoint Hermitian operator, the operator $\mathcal{L}$ can still be defined on every dyad of the form $|\psi\rangle \langle\varphi|$, where $|\psi\rangle$ and $|\varphi\rangle$ are in the domain of H, as follows:

$$(12) \qquad \mathcal{L}|\psi\rangle \langle\varphi| = |H\psi\rangle \langle\varphi| - |\psi\rangle \langle H\varphi|$$

The domain $\mathcal{D}(\mathcal{L})$ is thus a vector subspace of $\mathcal{H} \otimes \mathcal{H}^{\times}$ spanned by these dyads. On this domain, $\mathcal{L}$ is Hermitian as one can check directly. If the domain of H is dense in $\mathcal{H}$, as we always assume, then $\mathcal{D}(\mathcal{L})$ is also dense in $\mathcal{H} \otimes \mathcal{H}^{\times}$. Therefore $\mathcal{L}$ has a well defined adjoint $\mathcal{L}^{\dagger}$. Since $\mathcal{L}$ is Hermitian its adjoint extends it.

This latter case can be studied in some detail. We know that Hermitian operators may or may not have self adjoint extensions. With $\mathcal{L}$ a surprise arises, since it always does have self adjoint extensions. Let us mention briefly a few results in this direction.

Let H be a Hermitian operator. We have four different possibilities for H:

  i) H is self adjoint,
 ii) H has only one self adjoint extension (it is essentially self adjoint),
iii) H has an infinite number of self adjoint extensions,
 iv) H has no self adjoint extensions.

In two first cases we attach to H a unique quantum observable denoted also by H. In the fourth case, H does not describe any quantum observable. The most puzzling situation comes from the third case in which a Hermitian operator produces an infinite number of observables (provided that we assign an observable to each self adjoint operator, a problem we do not want to discuss here). The self adjoint extensions of H can be characterized using certain boundary conditions [6]. One may ask if this situation is somehow relieved in Liouville space in the sense that the Liouvillians corresponding to these different extensions are not different. This would imply that the Liouville space solves this problem by itself by assigning a unique observable to H. Unfortunately, this is not the case. Different self adjoint extensions of H have different Liouvillians as one can immediately deduce from a more general result:

Theorem 1 *Let A and B be two self adjoint operators and let L_A and L_B be their respective Liouvillians. Then, $L_A = L_B$ if and only if $A = B + \alpha I$, where α is a real number and I is the identity operator in $\mathcal{H}$.*

This theorem refers also to the well known fact that physical results must be independent of the choice of the origin of energies. This means that the Hamiltonians H and $H' = H + cI$ where c is a real constant must provide the same dynamics. We see that two Liouvillians are identical if and only if their respective Hamiltonians differ by the choice of the origin of energies. We see that this kind of indeterminacy can be easily removed when we use the Liouville space.

The second result is the surprise we have announced earlier. In the fourth case discussed above we had Hermitian operators H without self adjoint extensions. One may expect that the operator $\mathcal{L}$ defined by (11) does not have self adjoint extensions either. This conjecture is false as it has an infinite number of self adjoint extensions.

We say that an extension $\mathcal{G}$ of $\mathcal{L}$ is factorizable if there exists a self adjoint operator A such that $L_A = \mathcal{G}$, i.e., $\mathcal{G}\rho = [A; \rho]$ for ρ belonging to some dense subspace of $\mathcal{H} \otimes \mathcal{H}^\times$. If H is a non self adjoint Hermitian operator and we construct for this H the Liouvillian, we can expect that $\mathcal{L}$ have some nonfactorizable self adjoint extensions. This has been proven at least when H has finite identical deficiency indices.

We omit here the technical proofs of these results.

3. Resolvent and Spectral Resolution of Liouville Operators

Along the present section, we shall discuss several properties of the Liouvillian L_H, also called the Liouville operator, of a self adjoint operator H, related to its spectral decomposition. Our program is the following: First, we shall construct explicitly the spectral families of the Liouvillian, being given the spectral families of the Hamiltonian. This will allow to define functions of the Liouvillian and, in particular, will provide a formula for the resolvent of L_H in terms of the spectral projections of H. Finally, we shall discuss the problem of the degeneracy of the continuous spectrum of L_H.

We start with the well know fact according to which, for any self adjoint operator H with spectrum $\sigma(H)$, there exists a spectral family of orthogonal projections E_ε on $\mathcal{H}$ labeled by the points ε of $\sigma(H)$ such that for any $|\psi\rangle$ in the domain of H, one has:

$$(13) \qquad \langle\psi|H\psi\rangle = \int_{\sigma(H)} \varepsilon\, d\langle\psi|E_\varepsilon\psi\rangle$$

The properties of the defining spectral family of orthogonal projections show that $d\langle\psi|E_\varepsilon\psi\rangle$ is a well defined positive measure in $\sigma(H)$ for **all** the vectors $|\psi\rangle$ in $\mathcal{H}$. However, the above integral converges if and only if $|\psi\rangle$ is in the domain of H. The spectral family of projections can be extended to all the values of the real line uniquely in such a way that the measure $d\langle\psi|E_\varepsilon\psi\rangle$ is zero outside $\sigma(H)$. Then, the integral limits in (13) can be changed to be from minus infinity to infinity. The family E_ε is called the spectral family of H and (13) gives the spectral decomposition of H. This spectral decomposition is usually written as

$$(14) \qquad H = \int_{\sigma(H)} \varepsilon\, dE_\varepsilon \quad \text{or} \quad H = \int_{-\infty}^{\infty} \varepsilon\, dE_\varepsilon$$

We do not want to discuss here properties of the spectral family E_ε, its relation with the eigenvalues of H, and etc, because this is well known and extensively treated in the standard literature [2, 3]. We are much more interested in the following problem: Since $\boldsymbol{L_H}$ is self adjoint, it has its own spectral family $\boldsymbol{E_\omega}$ so that

$$(15) \qquad \boldsymbol{L_H} = \int_{\sigma(\boldsymbol{L_H})} \omega\, d\boldsymbol{E_\omega} = \int_{-\infty}^{\infty} \omega\, d\boldsymbol{E_\omega},$$

what is the relation between the spectral families E_ε and $\boldsymbol{E_\omega}$? Since we usually start from H to obtain $\boldsymbol{L_H}$, this question can be posed as follows: how can we obtain the projections $\boldsymbol{L_H}$ from E_ε? First of all, we recall that since $\boldsymbol{E_\omega}$ are projection operators, they are bounded and therefore it is enough to obtain their action on dyads of the form $|\psi\rangle\langle\varphi|$. We begin with formulas (1) and (10) written as

$$(16) \qquad e^{i\boldsymbol{L_H}t}\rho = e^{itH}\rho e^{-itH}$$

If we choose as ρ the dyad $|\psi\rangle\langle\varphi|$ and consider the following scalar product in $\mathcal{H}\otimes\mathcal{H}^\times$ then

$$(17) \qquad (e^{it\boldsymbol{L_H}}|\psi\rangle\langle\varphi|\,|\,|\psi\rangle\langle\varphi|) = (e^{itH}|\psi\rangle\langle\varphi|e^{-itH}\,|\,|\psi\rangle\langle\varphi|)$$

Taking into account that the adjoint of $e^{itH}|\psi\rangle\langle\varphi|e^{-itH}$ is $e^{itH}|\varphi\rangle\langle\psi|e^{-itH}$ and the definition of the scalar product in $\mathcal{B}_\mathcal{H}^2$ we have

$$\langle\varphi|e^{itH}\varphi\rangle\langle\psi|e^{-itH}\psi\rangle = \int_{-\infty}^{\infty} e^{it\varepsilon}\,d\langle\varphi|E_\varepsilon\varphi\rangle \int_{-\infty}^{\infty} e^{-it\varepsilon'}\,d\langle\psi|E_{\varepsilon'}\psi\rangle$$

$$(18) \qquad\qquad = \int_{-\infty}^{\infty}\int_{-\infty}^{\infty} e^{it(\varepsilon-\varepsilon')}\,d\langle\varphi|E_\varepsilon\varphi\rangle\,d\langle\psi|E_{\varepsilon'}\psi\rangle$$

The change of variables given by $\varepsilon - \varepsilon' = -\omega$ gives for (18) the following form:

$$(19) \qquad \int_{-\infty}^{\infty} \int_{-\infty}^{\infty} e^{-it\omega} \, d\langle\varphi|E_\varepsilon\varphi\rangle \, d\langle\psi|E_{\omega+\varepsilon}\psi\rangle \cdot$$

Comparing (19) with (15) we obtain

$$(20) \qquad (e^{it\boldsymbol{L}_H}|\psi\rangle\langle\varphi| \, | \, |\psi\rangle\langle\varphi|) = \int_{-\infty}^{\infty} e^{-it\omega} \, d(|\psi\rangle\langle\varphi| \, | \, \boldsymbol{E}_\omega|\psi\rangle\langle\varphi|)$$

If the Fourier transforms of two measures coincide, these two measures must be identical. Therefore,

$$(21) \qquad (|\psi\rangle\langle\varphi| \, | \, \boldsymbol{E}_\omega|\psi\rangle\langle\varphi|) = \int_{-\infty}^{\infty} \langle\psi|E_{\omega+\varepsilon}\psi\rangle \, d\langle\varphi|E_\varepsilon\varphi\rangle$$

which can be written in the following form:

$$(22) \qquad \boldsymbol{E}_\omega|\psi\rangle\langle\varphi| = \int_{-\infty}^{\infty} |E_{\omega+\varepsilon}\psi\rangle \, d\langle E_\varepsilon\varphi|$$

Spectral decompositions of self adjoint operators are useful for constructions of functions of these operators. This is, in particular, true for the Liouville operator $\boldsymbol{L}_H$. Its functions can be defined as follows: let $f(\omega)$ be a measurable complex function on the real line such that:

$$(23) \qquad \int_{-\infty}^{\infty} |f(\omega)|^2 d(\rho|E_\omega\rho) < \infty$$

converges for all ρ in a dense subspace of $\mathcal{H} \otimes \mathcal{H}^\times$. Then, a well know result in functional calculus [2, 7] says that the operator $f(\boldsymbol{L}_H)$ is well defined on all these ρ,

$$(24) \qquad (\rho, f(\boldsymbol{L}_H)\rho) = \int_{-\infty}^{\infty} f(\omega) \, d(\rho, \boldsymbol{E}_\omega\rho)$$

and it is normal. If, in addition, $f(\omega)$ is real, $f(\boldsymbol{L}_H)$ is self adjoint. Equation (24) is usually written as:

$$(25) \qquad f(\boldsymbol{L}_H) = \int_{-\infty}^{\infty} f(\omega) \, d\boldsymbol{E}_\omega$$

The operators $f(\boldsymbol{L}_H)$ commute in their common domain. Because of the properties of the spectral projections associated to a self adjoint operator, the real measure $d(\rho, \boldsymbol{E}_\omega\rho)$ is always finite[1]. Thus, $f(\boldsymbol{L}_H)$ exists and is bounded

---

[1] Because $|(\rho, \boldsymbol{E}_\omega\rho)| \leq ||\rho||^2$. If the norm of ρ is equal to one this measure becomes a probability.

if $f(\omega)$ is bounded. In particular, we can define functions like the following imaginary exponential for any real t:

$$(26) \qquad e^{-itL_H} = \int_{-\infty}^{\infty} e^{-it\omega} \, dE_\omega$$

Another interesting function of L_H is its resolvent:

$$(27) \qquad R_z(L_H) := (L_H - z)^{-1} = \frac{1}{L_h - z}$$

where z is in the resolvent set [2, 3, 7] $\rho(L_H)$ of $L_H{}^2$. Since L_H is self adjoint, $R_z(L_H)$ is well defined for any complex numbers in the resolvent set $\rho(L_H)$. Moreover, the operator valued complex function $R_z(L_H)$ is analytic in the resolvent set of L_H (which is an open set of the complex numbers)[3]. Our next goal is to find a formula for the resolvent $R_z(L_H)$ in terms of the spectral resolution of H. In fact, we shall get a more general result finding this formula for any function $f(\omega)$, for which (23) holds, and then obtaining its particular form for $f(\omega) = \frac{1}{\omega - z}$. This procedure is very simple and it goes as follows:

$$(28) \qquad f(L_H) \, |\psi\rangle \, \langle\varphi| = \int_{-\infty}^{\infty} f(\omega) \, dE_\omega \, |\psi\rangle \, \langle\varphi|$$

If one uses (21), then (28) is equal to:

$$(29) \qquad \int_{-\infty}^{\infty} \int_{-\infty}^{\infty} f(\omega) \, dE_{\omega+\varepsilon} \, |\psi\rangle \, d\langle E_\varepsilon \varphi|$$

The change of variables $\omega + \varepsilon = \varepsilon'$ gives:

$$(30) \qquad \int_{-\infty}^{\infty} f(\omega) \, dE_{\omega+\varepsilon} = \int_{-\infty}^{\infty} f(\varepsilon' - \varepsilon) \, dE_{\varepsilon'} = f(H - \varepsilon)$$

The last identity in (30) is a consequence of the definition of functions of an operator as given before. If we carry (30) into (29), we find:

$$(31) \qquad f(L_H) \, |\psi\rangle \, \langle\varphi| = \int_{-\infty}^{\infty} f(H - \varepsilon) \, |\psi\rangle \, d\langle E_\varepsilon \varphi|$$

or equivalently,

$$(32) \qquad (|\psi\rangle \, \langle\varphi|, f(L_H) \, |\psi\rangle \, \langle\varphi|) = \int_{-\infty}^{\infty} \langle\psi| f(H - \varepsilon) |\psi\rangle \, d\langle E_\varepsilon \varphi|\varphi\rangle$$

---

[2] The resolvent set of an operator is the set of complex numbers that are not in its spectrum.

[3] This is a general property of self adjoint operators. If a self adjoint operator A have continuous spectrum, its resolvent $R_z(A)$ as function of the complex variable z is not even continuous on the continuous spectrum, $\sigma_c(A)$ of A. From the point of view of the theory of complex analytic functions, $\sigma_c(A)$ is a branch cut of $R_z(A)$.

In particular, this formulas give for the resolvent

$$(33) \qquad \frac{1}{\boldsymbol{L}_H - z} |\psi\rangle \langle\varphi| = \int_{-\infty}^{\infty} \frac{1}{H - \varepsilon - z} |\psi\rangle \, d\langle E_\varepsilon \varphi|$$

If we now ignore the last identity in (30), we have

$$(34) \qquad f(\boldsymbol{L}_H) |\psi\rangle \langle\varphi| = \int_{-\infty}^{\infty} \int_{-\infty}^{\infty} f(\varepsilon' - \varepsilon) \, d|E_{\varepsilon'} \psi\rangle \, d\langle E_\varepsilon \varphi|$$

and

$$(35) \qquad \frac{1}{\boldsymbol{L}_H - z} |\psi\rangle \langle\varphi| = \int_{-\infty}^{\infty} \int_{-\infty}^{\infty} \frac{1}{\varepsilon' - \varepsilon - z} \, d|E_{\varepsilon'} \psi\rangle \, d\langle E_\varepsilon \varphi|$$

4. The Spectrum of a Liouville Operator

We want to discuss in this section how to construct the spectrum of the Liouvillian $\boldsymbol{L}_H$ of a given self adjoint Hamiltonian H. This problem has been treated in some publications in the past [3, 8] although not always with the desirable correctness.

The set of the eigenvalues of $\boldsymbol{L}_H$ is easily obtained if we know the eigenvalues of H. In fact, it is easy to check that

$$(36) \qquad \sigma_p(\boldsymbol{L}_H) = \{\varepsilon_i - \varepsilon_j : \varepsilon_i, \varepsilon_j \in \sigma_p(H)\}$$

where $\sigma_p(H)$ denotes the set of eigenvalues of the operator H. If $H|e_i\rangle = \varepsilon_i |e_i\rangle$ for all i, we have the following spectral equation in $\mathcal{H} \otimes \mathcal{H}^\times$:

$$(37) \qquad \boldsymbol{L}_H |e_i\rangle \langle e_j| = (\varepsilon_i - \varepsilon_j) |e_i\rangle \langle e_j|$$

More generally, it is well known that the spectrum $\sigma(H)$ of H determines the spectrum $\sigma(\boldsymbol{L})$ of $\boldsymbol{L}$ [3, 8]:

$$(38) \qquad \sigma(\boldsymbol{L}) = \sigma(H) - \sigma(H)$$

This shows in part the importance of the Liouville space, because we never observe energy values but instead, as spectroscopic measurements show, differences between energy values. Due to this observation, the formulation of Quantum Mechanics in Liouville space seems more natural than the standard one in Hilbert space.

It is interesting to discuss why the above mentioned construction by Spohn of the singular spectrum of the Liouvillian of a given Hamiltonian [4] is not correct.

Denote by $\mathcal{H}_p$ the closed linear hull of all eigenelements of H and write $\mathcal{H}_c = \mathcal{H} \ominus \mathcal{H}_p{}^4$. Recall that the singular continuous subspace $\mathcal{H}_{sc}$ of $\mathcal{H}_c$ consists of all $f \in \mathcal{H}_c$ for which there exists a Borel set B_0 of Lebesgue measure

---

[4] By $\mathcal{H}_c = \mathcal{H} \ominus \mathcal{H}_p$, we mean that $\mathcal{H}_c$ is the orthogonal complement of $\mathcal{H}_p$ in $\mathcal{H}$, i.e., $\mathcal{H} = \mathcal{H}_p \oplus \mathcal{H}_c$

zero such that $\int_{B_0} dE_\lambda f = f$. By $\mathcal{H}_{ac} = \mathcal{H}_c \ominus \mathcal{H}_{sc}$ we shall denote the absolutely continuous subspace of $\mathcal{H}_c$. Recall also that $\mathcal{H}_p$, $\mathcal{H}_c$, $\mathcal{H}_{sc}$ and $\mathcal{H}_{ac}$ are closed linear subspaces of $\mathcal{H}$ which reduce[5] the operator H. The spectra of corresponding reductions will be called respectively point, continuous, singular continuous and absolutely continuous spectrum of H, and will be denoted by $\sigma(H)$ with appropriate subscripts.

Let $\mu = \mu_f$ denotes, for a given $f \in \mathcal{H}$, the spectral measure on $\sigma(H)$ determined by the nondecreasing function $\lambda \longmapsto \langle E_\lambda f | f \rangle$. Let $f = f_p + f_{sc} + f_{ac}$ be the decomposition of f in the direct sum $\mathcal{H}_p \oplus \mathcal{H}_{sc} \oplus \mathcal{H}_{ac}$. Putting $\mu_p = \mu_{f_p}$, $\mu_{sc} = \mu_{f_{sc}}$ and $\mu_{ac} = \mu_{f_{ac}}$ we obtain the decomposition of μ

$$(39) \qquad \mu = \mu_p + \mu_{sc} + \mu_{ac}$$

onto the point, singular continuous and absolutely continuous component. Conversely, having given three finite Borel measures μ_p, μ_{sc} and μ_{ac}, where μ_p is concentrated on a countable set of points, and the other two measures are respectively singular and absolutely continuous, one can always construct a Hilbert space $\mathcal{H}$ and a selfadjoint operator H such that these measures are spectral measures associated with some $f \in \mathcal{H}$. Moreover, the point, singular and absolutely continuous spectrum of H coincides with μ_p, μ_{sc} and μ_{ac} respectively. This can be proved by taking as $\mathcal{H}$ the direct sum $L^2(\mathbf{R}, \mu_{sc}) \oplus L^2(\mathbf{R}, \mu_{sc}) \oplus L^2(\mathbf{R}, \mu_{sc})$ and as H the operator of multiplication by x (see [2]).

In the above mentioned paper [8], it is proved that $\sigma(\mathbf{L}) = \sigma(H) - \sigma(H)$, where $\sigma(A)$ stands for the spectrum of the operator A. This result is not new, since a more general one was knew before [2]. Moreover, along the same paper, it is given a similar characterization for absolutely continuous, singular continuous and point spectrum. What is wrong there, is that $\sigma(\mathbf{L})_{sc} = \sigma(H)_{sc} - \sigma(H)_{sc}$. The argument used in the proof is that if $\sigma(H)_{sc}$ is the set of Lebesgue measure 0, (where "sc" stands for singular continuous) then the complex difference $\sigma(H)_{sc} - \sigma(H)_{sc}$ is also a set of the Lebesgue measure 0. This is, however, not true as we shall show below.

To see it, let us consider a Hamiltonian operator H for which $\sigma(H)$ is a Cantor set. The existence of such H is guaranteed by the the presented above construction of the Hamiltonian with given spectral measure. Let us note, however, that there are more "physical" examples of Hamiltonians with Cantor like spectrum. For example, there is an absolutely summable sequence $\{a_n\}$ such that the spectrum of the Hamiltonian

$$H = -\frac{d^2}{dx^2} + \sum_{n=0}^{\infty} a_n \cos(x2^{-n})$$

---

[5] H reduces $\mathcal{H}_\alpha$ if $H\varphi \in \mathcal{H}_\alpha$ for any vector φ in the intersection of $\mathcal{H}_\alpha$ with the domain of H.

is the Cantor set (see [8, 9] an references therein). Suppose that $\sigma(H) = C$, where C is the Cantor set on the interval $[0,1]$. According to the general property of spectra of functions of the Hamiltonian (see [2]) we have (1).

However, according to a theorem of Steinhaus [10] we have

$$C - C = [-1, 1] \, ,$$

i.e. the complex difference of two Cantor sets covers the whole interval and is thus the set of nonzero Lebesgue measure, contrary to what is claimed in [7]. This property of the Cantor set does not, however, imply that the spectrum of the Liouvillian is absolutely continuous. To show this let us consider the well known "devil's staircase" distribution function $F(x)$ on the Cantor set C which has constant value equal $k/2^n$ (k are such that the fraction $k/2^n$ is non reducible) on each interval which is removed in the n-th step of the construction of the Cantor set [11]. (The "devil staircase" is defined as follows: it is $1/2$ on the interval $(\frac{1}{3}, \frac{2}{3})$, $1/4$ on $(\frac{1}{9}, \frac{2}{9})$, $3/4$ on $(\frac{7}{9}, \frac{8}{9})$, $1/8$ on $(\frac{1}{27}, \frac{2}{27})$, $3/8$ on $(\frac{7}{27}, \frac{8}{27})$, $5/8$ on $(\frac{19}{27}, \frac{20}{27})$, $7/8$ on $(\frac{25}{27}, \frac{26}{27})$, and so on.)

It is well known and not difficult to check that $F(x)$ is a non decreasing continuous function such that $F(x) = 0$ for $x \leq 0$ and $F(x) = 1$ for $x \geq 1$. By above remarks $F(x)$ is a spectral measure corresponding to a Hamiltonian, i.e. there exists a selfadjoint operator $H = \int_{\sigma(H)} \lambda \, dE_\lambda$ on a Hilbert space $\mathcal{H}$ and an element $h \in \mathcal{H}$ such that

(40) $F(x) = \langle E_x h, h \rangle \, , \quad \text{for } x \in \mathbf{R} \, .$

Consider the Liouvillian, corresponding to this H, defined on $\mathcal{H} \otimes \mathcal{H}^\times$ by $\boldsymbol{L} = H \otimes I - I \otimes H$. It follows from (21) that the spectral family $\{\boldsymbol{E}_\lambda\}$ satisfies

(41)

$$(\boldsymbol{E}_\lambda|h\rangle\langle h| \, | \, |h\rangle\langle h|) = \int_{-\infty}^{\infty} \langle E_{\lambda+\mu}h, h \rangle \, d\langle h|E_\mu h \rangle = \int_{-\infty}^{\infty} F(\lambda + \mu) \, dF(\mu) \, .$$

This implies that the function $x \longmapsto G(-x)$, where

(42) $G(x) = (\boldsymbol{E}_x|h\rangle\langle h| \, | \, |h\rangle\langle h|)$

is a convolution of the function $x \longmapsto F(-x)$ with $x \longmapsto F(x)$. By general properties of the convolution the function, $G(x)$ is continuous. We can show, however, that it is not absolutely continuous. In order to do this it is enough to find the explicit expression for the Fourier-Stieltjes transform of $F(x)$ (see [12] for details) which is of the form

(43) $$\widehat{F}(t) = \prod_{k=1}^{\infty} \left[1 + e^{\frac{2it}{3}} \right] \, .$$

Choosing now $t_n = \pi 3^n$ we see that

$$\widehat{F}(t_n) = \prod_{k=1}^{\infty} \frac{1}{2}\left[1 + e^{\frac{2\pi i}{3^k}}\right]$$

does not depend on n. Moreover one can also show that $\widehat{F}(t_n)$ is not zero. This means that the Fourier transform $\widehat{F}(t)$ does not converge to zero as $t \to \infty$. The same is true for the Fourier transform $\widehat{G}(t)$ of $G(x)$ because

(44)
$$|\widehat{G}(t)| = |\widehat{F}(t)|^2.$$

Therefore, in view of the Riemann Lebesgue lemma $F(x)$ as well as $G(x)$ cannot be absolutely continuous.

In fact it can be shown more, namely that $G(t)$ is a strictly monotonically increasing and continuous function on the interval $[-1, 1]$ with the derivative $G'(x) = 0$ at each point.

The above example shows that a Hamiltonian with the singular spectrum can have a Liouvillian with a spectrum which is much bigger. It may even contain an absolutely continuous spectrum. To show it, we use a result in Reference [13] (see also [14]), according to which we can show that it is always possible to find a Hamiltonian which has the purely singular spectral measure concentrated on a set of Hausdorff dimension zero (see [15] for definition of the Hausdorff measure).

It follows from our previous discussion that singular states of the Hamiltonian may have, when transported in the natural way (44) to the Liouville space, a rather complex structure. However, it is easy to see that a state φ decays in $\mathcal{H}$, i.e.

$$\langle e^{itH}\varphi, \varphi \rangle \longrightarrow 0, \quad \text{as } t \to \infty,$$

if and only if the Liouville state $|\varphi\rangle\langle\varphi|$ decays in $\mathcal{H} \otimes \mathcal{H}^{\times}$. This equivalence is a simple consequence of (40), (42) and (44). The states considered in the previous section were non decaying. We can show, however, that similar effects may appear in the class of decaying states, i.e. that starting from the singular spectrum of the hamiltonian we can get a purely absolutely continuous as well as a purely singular continuous spectrum of the Liouvillian [12].

The first possibility follows from Winer and Wintner's paper [16] who constructed a non-negative singular continuous measure μ on $[\pi, \pi)$ such that

(45)
$$\int_{-\pi}^{\pi} e^{-inx}\mu(dx) = O(n^{-\frac{1}{2}+\varepsilon}),$$

for every $\varepsilon > 0$. Then, of course $\mu * \mu$ as well as $\mu\bar{*}\mu$ is absolutely continuous. It is now enough to construct a Hamiltonian corresponding to μ and apply again (40), (42) and (44) to obtain absolute continuity in the Liouville space. The second possibility follows from an Erdös result which shows the existence of a wide class of singular measures with the Fourier transform vanishing at infinity [17]. It is easy to choose a measure μ belonging to this class such that $\mu\bar{*}\mu$ is also singular continuous. In this way we show the existence of singular decaying states $|\varphi\rangle$ in $\mathcal{H}$ with $|\varphi\rangle\langle\varphi|$ decaying in $\mathcal{H} \otimes \mathcal{H}^{\times}$.

5. Rigged Liouville Spaces

The concept of Rigged Liouville Space (RLS) is both an extension and a particular case of the concept of Rigged Hilbert Space (RHS) [5]. It is an extension because we depart from the structure of a RHS to construct a RLS, exactly as we depart from the structure of a Hilbert space to construct its corresponding Liouville space. It is a particular case because as the Liouville space is indeed a Hilbert space, the RLS is a RHS.

We want to recall that the nuclear spectral theorem provides, for every self adjoint operator H on a Hilbert space $\mathcal{H}$, a RHS, $\Phi \subset \mathcal{H} \subset {}^{\times}\Phi$ [6] and a complete set of (generalized) eigenvectors of H, where the set of eigenvalues run out the Hilbert space spectrum of H. If we assume that H does not have singular continuous spectrum and that its absolutely continuous spectrum has uniform multiplicity, then for any "reasonable" function $f(x)$, the operator $f(H)$ can be defined through its spectral decomposition:

$$(46) \qquad f(H) = \sum_{i,j} f(\varepsilon_i)|e_i,j\rangle \langle e_i,j| + \sum_{n=1}^{N} \int_{\sigma(H)} f(\varepsilon)\, |\varepsilon,n\rangle \langle \varepsilon,n|\, d\varepsilon$$

where: $|e_i,j\rangle$ denotes a Hilbert space eigenvector of H with eigenvalue ε_i, the subindex j stands for the degeneracy of the eigenvalue ε_i; $|\varepsilon,n\rangle$ is the (generalized) eigenvalue of H with eigenvector ε and degeneracy N.

In Section 2, we have defined the Liouville space as the Hilbert space $\mathcal{H} \otimes \mathcal{H}^{\times}$. On this space, it has been defined the Liouvillian L_H corresponding to the self adjoint operator H. If the spectrum of H contains only eigenvalues, then L_H has only point spectrum, as we have already seen. Its eigenvalues and corresponding eigenvectors have been shown in equations (36) and (37). If L_H has a continuous spectrum, then $\sigma(L_H)$ is also continuous. Since L_H is also self adjoint, the nuclear spectral theorem assures that L_H must have a complete system of generalized eigenvectors corresponding to its continuous spectrum on the dual of a suitable RHS. This RHS is the **Rigged Liouville Space (RLS)**. We sketch here its construction.

To begin with, pick an arbitrary self adjoint operator H and construct the RHS $\Phi \subset \mathcal{H} \subset {}^{\times}\Phi$ which existence is guaranteed by the nuclear spectral theorem [18]. What is essential here is the space Φ and its topology[7]. From it we can obtain the duals $\Phi^{\times}$ and ${}^{\times}\Phi$. Since the proper eigenvectors, i.e., its eigenvectors in the Hilbert space $\mathcal{H} \otimes \mathcal{H}^{\times}$, of L_H are dyads, it seems natural to look for generalized elements of L_H among the elements of the vector space ${}^{\times}\Phi \otimes \Phi^{\times}$. Now the question is: is such space the topological dual of some vector space satisfying for L_H and $\mathcal{H} \otimes \mathcal{H}^{\times}$ analogous conditions as

---

[6] ${}^{\times}\Phi$ denotes the space of continuous antilinear functionals on Φ while the space $\Phi^{\times}$ is the vectors space of continuous linear functionals on Φ.

[7] The topology on Φ is finer than the topology that Φ has as a subspace of $\mathcal{H}$ and it is constructed in such a way that the spaces Φ and ${}^{\times}\Phi$ form a dual pair [19]

Φ satisfies for H and $\mathcal{H}$? The answer is positive under certain conditions[8]. Then, the triplet

(47) $$\Phi \otimes \Phi \subset \mathcal{H} \otimes \mathcal{H}^{\times} \subset {}^{\times}\Phi \otimes \Phi^{\times}$$

has the desired properties. Here we do not want to discuss certain details like how the topologies on the tensor products of Φ and its duals look like because it is not relevant in our presentation.

Remark concerning notation. Let F be an element of ${}^{\times}\Phi \otimes \Phi^{\times}$. The action of F on $G \in \Phi \otimes \Phi$ is here denoted as $(F|G)$. This allows to avoid confusion with the action of $|F\rangle \in {}^{\times}\Phi$ or $\langle F| \in \Phi^{\times}$ on $\varphi \in \Phi$, denoted here as $\langle \varphi|F\rangle$ or $\langle F|\varphi\rangle$ respectively.

Generalized eigenoperator decomposition of the Liouvillian. Let us assume that we have a Hamiltonian H with continuous spectrum which does not have a singular spectrum although H may have point spectrum. Recall now formula (13) and compare it with (46), with $f(\varepsilon) = \varepsilon$. It follows from it that for ε in the continuous spectrum of H we can, at least formally, write the derivatives:

(48) $$\frac{dE_{\varepsilon}}{d\varepsilon} = \sum_{k} |\varepsilon, k\rangle \langle \varepsilon, k|,$$

which are well defined as operators from the space Φ into ${}^{\times}\Phi$ when acting to the right and into $\Phi^{\times}$ when acting to the left. Henceforth, we shall denote the derivatives (48) by $P(\varepsilon)$ and call them **generalized eigenprojectors** of H. $P(\varepsilon)$ is an RHS analog of the projector on the subspace of $\mathcal{H}$ spanned by all eigenvectors corresponding to the eigenvalue ε.

In the same way we define the generalized eigenprojectors of the Liouvillian:

(49) $$\boldsymbol{P}(\varepsilon) = \frac{d\boldsymbol{E}_{\varepsilon}}{d\varepsilon}$$

which are, due to the properties that we have assigned to $\Phi \otimes \Phi$, well defined operators from $\Phi \otimes \Phi$ into ${}^{\times}\Phi \otimes \Phi^{\times}$.

We can immediately guess the existence of a close connection between the operators $P(\varepsilon)$ and $\boldsymbol{P}(\varepsilon)$. We shall show that this is indeed the case showing this connection explicitly. In order to do it let us consider the following derivatives:

(50) $$\frac{d}{d\varepsilon}(|\psi\rangle \langle \varphi|E_{\varepsilon}) = |\psi\rangle \langle \varphi|P(\varepsilon) \quad ; \quad \frac{d}{d\varepsilon}(E_{\varepsilon}|\psi\rangle \langle \varphi|) = P(\varepsilon)|\psi\rangle \langle \varphi|$$

where $|\psi\rangle$ and $|\varphi\rangle$ are in Φ, and apply the definition of $P(\varepsilon)$. Then,

---

[8] If Φ is nuclear and Fréchet (A topological vector space is a Fréchet space if it is metrizable, i.e., one can define a distance on Φ such that the topology that it produces is identical to the topology we already have on Φ.)

$$P(\varepsilon)|\psi\rangle\langle\varphi| = \frac{dE_\varepsilon}{d\varepsilon}|\psi\rangle\langle\varphi| = \frac{d}{d\varepsilon}\int_{-\infty}^{\infty}|E_{\varepsilon+t}\psi\rangle\,d\langle E_t\,\varphi| =$$

$$(51) \qquad = \int_{-\infty}^{\infty}|\frac{dE_{\varepsilon+t}}{d\varepsilon}\psi\rangle\frac{d}{dt}\langle\varphi|E_t\,dt = \int_{-\infty}^{\infty}P(\varepsilon+t)\,|\psi\rangle\langle\varphi|P(t)\,dt$$

From formulas (25) and (28) giving the spectral resolution of the Liouvillian combined with (49), one easily gets

$$L|\psi\rangle\langle\varphi| = \int_{-\infty}^{\infty}\omega\,dE_\omega\,|\psi\rangle\langle\varphi| = \int_{-\infty}^{\infty}\omega\,\frac{dE_\omega}{d\omega}|\psi\rangle\langle\varphi|\,d\omega$$

$$(52) \qquad = \int_{-\infty}^{\infty}d\omega\int_{-\infty}^{\infty}dt\,\omega\,P(\omega+t)|\psi\rangle\langle\varphi|P(t)$$

For the function $f(\omega)$, we have a similar expression

$$(53) \qquad f(L)|\psi\rangle\langle\varphi| = \int_{-\infty}^{\infty}d\omega\int_{-\infty}^{\infty}dt\,f(\omega)\,P(\omega+t)|\psi\rangle\langle\varphi|P(t)$$

If we perform the following change of variables: $t = t$, $\omega + t = s$, taking account that the Jacobian of this transformation is equal to one, the precedent formulas are changed into

$$(54) \qquad L\,|\psi\rangle\langle\varphi| = \int_{-\infty}^{\infty}ds\int_{-\infty}^{\infty}dt\,(t-s)\,P(s)|\psi\rangle\langle\varphi|P(t)$$

and

$$(55) \qquad f(L)\,|\psi\rangle\langle\varphi| = \int_{-\infty}^{\infty}ds\int_{-\infty}^{\infty}dt\,f(t-s)\,P(s)|\psi\rangle\langle\varphi|P(t)$$

In particular the resolvent of L can be written as

$$(56) \qquad (L-\lambda)^{-1}\,|\psi\rangle\langle\varphi| = \int_{-\infty}^{\infty}ds\int_{-\infty}^{\infty}dt\,\frac{1}{s-t-\lambda}\,P(s)|\psi\rangle\langle\varphi|P(t)$$

In the simplest case, when H as a nondegenerate continuous spectrum only, $P(\varepsilon) = |\varepsilon\rangle\langle\varepsilon|$. Then, formulas (55) and (56) can be written respectively as

$$(57) \qquad f(L)\,|\psi\rangle\langle\varphi| = \int_{-\infty}^{\infty}ds\int_{-\infty}^{\infty}dt\,f(t-s)\,|s\rangle\langle s|\psi\rangle\langle\varphi|t\rangle\langle t|$$

and

$$(58) \qquad \frac{1}{L-\lambda}|\psi\rangle\langle\varphi| = \int_{-\infty}^{\infty}ds\int_{-\infty}^{\infty}dt\,\frac{1}{s-t-\lambda}\,|s\rangle\langle s|\psi\rangle\langle\varphi|t\rangle\langle t|$$

Remark. The generalized eigenprojectors $P(\omega)$ satisfy the following formula:

$$(59) \qquad \int_{-\infty}^{\infty}P(\omega)\,d\omega = I$$

i.e., it is the identity operator (called canonical imbedding because it is not onto) from $\Phi \otimes \Phi$ into $^{\times}\Phi \otimes \Phi^{\times}$.

Generalized eigenvectors of the Liouvillian. The theory of RHS and, in particular the nuclear spectral theorem, permits the construction of generalized eigenvectors of H whose eigenvalues lie on its continuous spectrum. This constructions are based upon the generalized eigenprojections given by $P(\varepsilon) = \sum_\alpha |\varepsilon, \alpha\rangle \langle \varepsilon, \alpha|$, where the label α denotes the degeneracy of the continuous spectrum. Now we would like to have a similar formula for the generalized projectors of $\boldsymbol{P}(\omega)$ as

$$(60) \qquad \boldsymbol{P}(\omega) = \sum_\beta |\omega, \beta) (\omega, \beta|$$

where $|\omega, \beta) \in {}^{\times}\Phi \otimes \Phi^{\times}$ with $\boldsymbol{L}_H |\omega, \beta) = \omega |\omega, \beta)$ so that one has:

$$(61) \qquad \boldsymbol{L}_H = \int_{-\infty}^{\infty} d\omega\, \omega \sum_\beta |\omega, \beta) (\omega, \beta|$$

where the label β stands for the degeneracy of the continuous spectrum of $\boldsymbol{L}_H$. This label cannot be removed even if the continuous spectrum of H is not degenerate. We want to obtain the vectors $|\omega, \beta)$ when H has a nondegenerate continuous spectrum. We present this case because of its simplicity in the notation and because to extend it when H has a continuous degenerate spectrum is easy using (46). We have only to observe that (57) can be written, for $f(\omega) = \omega$, as

$$(62) \qquad \boldsymbol{L}_H |\psi\rangle \langle\varphi| = \int_{-\infty}^{\infty} \omega\, d\omega \int_{-\infty}^{\infty} d\varepsilon\, |\omega + \varepsilon\rangle \langle\omega + \varepsilon|\psi\rangle \langle\varphi|\varepsilon\rangle \langle\varepsilon|$$

After (61), we have

$$(63) \qquad \boldsymbol{L}_H |\psi\rangle \langle\varphi| = \int_{-\infty}^{\infty} \omega\, d\omega \sum_\beta |\omega, \beta) (\omega, \beta| |\psi\rangle \langle\varphi|)$$

Comparison between (62) and (63) suggest the identification of the indices ε and β so that the sum in (63) becomes an integral. After this identification, we conclude that

$$(64) \qquad |\omega, \beta) (\omega, \beta| |\psi\rangle \langle\varphi|) = |\omega + \varepsilon\rangle \langle\omega + \varepsilon|\psi\rangle \langle\varphi|\varepsilon\rangle \langle\varepsilon|$$

so that

$$(65) \qquad (\omega, \beta| |\psi\rangle \langle\varphi|) = \langle\omega + \varepsilon|\psi\rangle \langle\varphi|\varepsilon\rangle$$

and

$$(66) \qquad |\omega, \beta) = |\omega + \varepsilon\rangle \langle\varepsilon| \in {}^{\times}\Phi \otimes \Phi^{\times}$$

We can also show directly that $|\omega, \beta)$ is an eigenvector of $\boldsymbol{L}_H$ with eigenvalue ω. We want to remark that this solution for the eigenvalue problem in RLS is not unique. In fact,

$$(67) \qquad |F_\omega) = \int_{-\infty}^{\infty} |\omega + \varepsilon\rangle \langle \varepsilon| \, d\varepsilon$$

is also a generalized eigenvector of $\boldsymbol{L}_H$ with eigenvalue ω.

References

1. E.P. Wigner, Ann. Math. **40**, 149 (1939).
2. M. Reed and B. Simon, *Methods of Modern Mathematical Physics, Vol I: Functional Analysis*, Academic Press, New York 1980.
3. E. E. Prugovecki *Quantum Mechanics in Hilbert Space*, Academic Press, New York 1981.
4. I. Prigogine *Non-Equilibrium Statistical Mechanics*, Wiley, New York 1962.
5. A. Bohm, Boulder Lectures in Theoretical Physics, vol **19A**, 255 (1965). A. Bohm, *Quantum Mechanics. Foundations and Applications*. Springer Verlag, Berlin (1994). A. Bohm and M. Gadella, *Dirac Kets, Gamow Vectors and Gel'fand Triplets.*, Lecture Notes in Physics, vol **348**, Berlin (1989).
6. M. Reed and B. Simon, *Methods of Modern Mathematical Physics Vol II: Fourier Analysis. Self Adjointness*, Academic Press, New York 1975. B. Pavlov, Russ. Math. Sur., **42**, 127 (1987). See also reference 7.
7. H. Weidmann *Linear Operators in Hilbert space*, Springer Verlag 1980.
8. H. Spohn, The spectrum of the Liouville-von Neumann operator, *J. Math Phys.* **17** 57-60 (1976).
9. J. Bellisard *Schroedinger Operators with Almost Periodic Potential: an Overview* Lecture Notes in Physics **153**, 356-363 (1982).
10. H. Steinhaus *A new property of the Cantor set* (in Polish), Wektor **7**, (1917), 1-3 English translation: H. Steinhaus *Selected Papers*, PWN Warsaw 1985.
11. P. Billingsley *Probability and measure*, Wiley, New York 1985.
12. I. Antoniou, S.A. Shkarin, Z. Suchanecki *Spectrum of the Hamiltonian and spectrum of the Liouvillian*, Preprint 1997.
13. N. Levenberg, Jr., G.J. Martin, A.L. Shields and S. Zdravkovska, Factorizations of Lebesgue measure via convolution, *Proc. Amer. Math. Soc.* **104** 419-430 (1988).
14. L. Bas, B.S. Pavlov *Absolute Continuity of Convolution of Singular Measures and Localization Problems* ULB Preprint (1995)
15. K. Falconer *Fractal geometry: mathematical foundations and applications*, Chichester, Wiley 1990.
16. N. Wiener and A. Wintner, Fourier-Stieltjes transforms and singular infinite convolution, *Amer. J. Math.* **60** 513-522 (1938).
17. P. Erdös, On a family of symmetric Bernoulli convolutions, *Amer. J. Math* **61** 974-976 (1939).
18. I.M. Gel'fand and G.P. Shilov, *Generalized Functions*, vol 4, Academic Press, New York (1964). K. Maurin *Generalized Eigenfunction Expansions and Unitary Representations of Topological Groups*, Polish Scientific Publishers, Warsawa (1968).
19. H.H. Schaeffer, *Topological Vector Spaces*, Springer Veralg, Berlin (1970).

Chapter II

Dynamical Semigroups of Open Systems

On Dynamical Semigroups and Open Systems

A. Kossakowski

Institute of Physics, Nicholas Copernicus University, 87-100 Torun, Poland

1. Introduction

The dynamics of a finite isolated quantum system is represented by a one parameter group of unitary operators in Hilbert space. This formulation makes it difficult to describe irreversible processes like the decay of unstable particles, the approach thermodynamical equilibrium and measurement processes.

It seems that the only possibility of introducing an irreversible behaviour in a finite system is to avoid the unitary time evolution altogether by considering non-Hamiltonian systems.

One way of doing this is to assume an interaction of the considered system S with an external system R (reservoir) like a heat bath, external stochastic field, or a measuring instrument. If the reservoir R is supposed to be finite but large, then the time evolution of the composed system $R + S$ is given by a one parameter group of unitary operators, while the reduced dynamics of the system S has no longer this property.

It seems that this approach to the description of irreversible behaviour of quantum systems has already been recognized by van Hove [1], Zwanzig [2], and has been applied to the theory of lasers [3,4]. The first abstract formulation of irreversible time evolution is due to Sudarshan [5,6].

In the early 70's this problem was rediscovered and initiated the mathematical theory of open systems, quantum stochastic processes, quantum dynamical semigroups, etc. It is difficult to present here a complete list of references concerning the subject, the potential reader is referred to Refs. [7]-[10]. An alternative approach to the irreversibility is presented at this Conference.

2. Positive Dynamical Semigroups

Let $\mathcal{H}$ be a separable complex Hilbert space corresponding to a physical system. Let $T(\mathcal{H})$ and $B(\mathcal{H})$ be the Banach spaces of all trace class operators and all bounded operators on $\mathcal{H}$, respectively. The corresponding norms are denoted by $||\cdot||_1$ and $||\cdot||$. The positive cones in $T(\mathcal{H})$ and $B(\mathcal{H})$, i.e., the sets of all positive operators, will be denoted by $T(\mathcal{H})^+$ and $B(\mathcal{H})^+$, respectively. The set of all density operators will be denoted by $S(\mathcal{H})$.

A *positive dynamical semigroup* (c.f. [11]) (in the Schrödinger picture) is a family of linear maps $\{\Lambda_t^* : T(\mathcal{H}) \to T(\mathcal{H}), t \geq 0\}$ such that

(i) $\Lambda_t^* : T(\mathcal{H})^+ \to T(\mathcal{H})^+$, $t \geq 0$ (positivity)
(ii) $tr(\Lambda_t^* \rho) = tr\rho$, for all $\rho \in T(\mathcal{H})$, $t \geq 0$
(iii) $\Lambda_t^* \Lambda_s^* = \Lambda_{t+s}^*$, $t, s \geq 0$
(iv) $\lim_{t \downarrow 0} \|\Lambda_t^* \rho - \rho\|_1 = 0$, for all $\rho \in T(\mathcal{H})$ (strong continuity).

By (i) and (ii), $\Lambda_t^* : S(\mathcal{H}) \to S(\mathcal{H}), t \geq 0$, so it may be considered as describing the time evolution of an open system, while the time evolution of an isolated system is given in terms of the Hamiltonian dynamics

$$\phi_t \rho = e^{-\frac{i}{\hbar} Ht} \rho \, e^{\frac{i}{\hbar} Ht}, \quad -\infty < t < \infty, \tag{1}$$

where H is the Hamiltonian of the system, and ϕ_t is a one-parameter group. On the other hand (iii) and (iv) allow to write $\Lambda_t^* = \exp(tL^*)$.

The properties (i) - (iv) can be translated to the Heisenberg picture. One defines the dual time evolution as the family $\{\Lambda_t : B(\mathcal{H}) \to B(\mathcal{H}), \ t \geq 0\}$ through the relation

$$tr[a(\Lambda_t^* \rho)] = tr[(\Lambda_t a)\rho], \tag{2}$$

for all $\rho \in T(\mathcal{H})$ and $a \in B(\mathcal{H})$. Then (i) to (iii) translate to

(i′) $\Lambda_t : B(\mathcal{H})^+ \to B(\mathcal{H})^+$, $t \geq 0$
(ii′) $\Lambda_t \mathbb{1} = \mathbb{1}$, $t \geq 0$, $\mathbb{1}$ is the identity operator in $B(\mathcal{H})$,
(iii′) $\Lambda_t \Lambda_s = \Lambda_{t+s}$, $t, s \geq 0$.

The appropriate continuity to require is

(iv′) $\lim_{t \downarrow 0} tr[\rho(\Lambda_t a - a)] = 0$
    for all $\rho \in T(\mathcal{H})$ and $a \in B(\mathcal{H})$.

Furthermore one would like to ensure that Λ_t is the dual of a map on $T(\mathcal{H})$. This is guaranteed by

(v) Λ_t is normal

This means that if $tr(\rho a_n) \to tr(\rho a)$, then also $tr(\rho \Lambda_t a_n) \to tr(\rho \Lambda_t a)$.

From the definitions of positive dynamical semigroups Λ_t^* and Λ_t, $t \geq 0$, it follows that there exist densely defined operators L^* and L on $T(\mathcal{H})$ and $B(\mathcal{H})$, respectively, such that

$$\frac{d}{dt}\Lambda_t^* \rho = L^*(\Lambda_t^* \rho), \quad \rho \in D(L^*) \subset T(\mathcal{H}) \tag{3}$$

and

$$\frac{d}{dt}(\Lambda_t a) = L(\Lambda_t a), \quad a \in D(L) \subset B(\mathcal{H}) \tag{4}$$

The equations (3) and (4) are generalizations of the corresponding von Neumann and Heisenberg equations for isolated system.

There is very little known concerning the structure of the generators of positive dynamical semigroups.

In the case that $L^*(L)$ is a bounded operator, i.e., that the corresponding dynamical semigroups are norm continuous, the following two results can be presented.

Theorem 2.1 (c.f. [12]). Let $P = \{P_1, P_2, \ldots\}$ be a family of projection operators on finite dimensional and mutually orthogonal subspaces of $\mathcal{H}$, such that $P_1 + P_2 + \ldots = \mathbb{1}$. A bounded operator L on $B(\mathcal{H})$ generates a positive dynamical semigroup if and only if for every P the relations

(5) $$a_{ij}(P) \geq 0, \quad i \neq j$$

and

(6) $$\sum_i a_{ij}(P) = 0$$

are satisfied, where

(7) $$a_{ij}(P) = \text{tr}[(LP_i)P_j].$$

Theorem 2.2 (c.f. [13]). Let ϕ be a normal positive bounded map on $B(\mathcal{H})$ and h be a bounded self-adjoint operator on $\mathcal{H}$. Then the bounded operator

(8) $$La = \phi(a) - \frac{1}{2}[\phi(\mathbb{1})a + a\phi(\mathbb{1})] + i[h, a]$$

generates a positive dynamical semigroup on $B(\mathcal{H})$.

In [14] the theorem 2.1 was applied to give the characterization of the generator of spin $-\frac{1}{2}$ system. In this case $\dim \mathcal{H} = 2$ and $B(\mathcal{H}) = M_2(C)$. Let $\sigma_1, \sigma_2, \sigma_3$ be the Pauli matrices. Using the notation $\hat{a} = \frac{1}{2}\vec{a}\vec{\sigma}$, $\vec{a} \in R^3$ one has

Theorem 2.3. A linear operator $L^* : M_2 \to M_2$ is the generator of a positive dynamical semigroup $t \to \Lambda_t^* : M_2 \to M_2$ (in the Schrödinger picture) iff it can be written in the form

(9) $$L^*\rho = -i[\hat{h}, \rho] + \frac{1}{2}\sum_{m,n=1}^{3} (\vec{e}_m, D\vec{e}_n + iA\vec{e}_n)\{[\hat{e}_m, \rho\hat{e}_n] + [\hat{e}_m\rho, \hat{e}_n]\}$$

where $(\vec{e}_m, \vec{e}_n) = \delta_{mn}$, D and A are linear operators on R^3 of the form

(10) $$D = \mathbb{1}\text{tr}\, F - 2F$$

(11) $$A\vec{x} = (F\vec{m}_0 + \vec{m}_0\Lambda\vec{h}) \wedge \vec{x}$$

(12) $$F\vec{x} = \sum_{n=1}^{3} \gamma_n \vec{e}_n (\vec{e}_n, \vec{x}), \quad \gamma_1, \gamma_2, \gamma_3 \geq 0,$$

and $\vec{m}_0 \in S \subset R^3$, where

(13)
$$S = \begin{cases} \{0\} & \text{for } \gamma_1\gamma_2\gamma_3 = 0 \\ \{\vec{z} \in R^3 : \inf_{||\vec{x}||=1}[(F\vec{x}, \vec{x} - \vec{z}) + \vec{x}(\vec{h} \wedge z)] \geq 0\} & \text{for } \gamma_1\gamma_2\gamma_3 > 0 \end{cases}$$

Putting $\rho(t) = \Lambda_t^* \rho$ and $\vec{m}(t) = \frac{1}{2}\text{tr}(\rho(t)\vec{\sigma})$, one obtains the following equations for $\vec{m}(t)$:

(14)
$$\frac{d}{dt}\vec{m}(t) = \vec{h} \times (\vec{m}(t) - \vec{m}_0) - F(\vec{m}(t) - \vec{m}_0)$$

which has the form of the Bloch equation.

3. Complete Positivity

Positive maps on C^*-algebras have been studied in Refs. [15]-[23]. Some results will be presented in the case of $B(\mathcal{H})$.

Let $\phi : B(\mathcal{H}) \to B(\mathcal{H})$ be a linear map, ϕ is said to be *self-adjoint* if $\phi(a^*) = \phi(a)^*$, *normalized* if $\phi(\mathbb{1}) - \mathbb{1}$, and *positive* if $\phi : B(\mathcal{H})^\dagger \to B(\mathcal{H})^\dagger$. A map $\phi : B(\mathcal{H}) \to B(\mathcal{H})$ is said to be *n-positive* if the relation

(15)
$$\sum_{k,\ell=1}^{n} (x_k, \phi(a_{k\ell})x_\ell) \geq 0$$

holds for all $x_1, \ldots, x_n \in \mathcal{H}$, and $a_{k\ell} \in B(\mathcal{H})$ such that

(16)
$$\sum_{k,\ell=1}^{n} (x_k, a_{k\ell}x_\ell) \geq 0.$$

A map ϕ is called to be *completely positive* if it is n-positive for all positive integers n. A map $\phi : B(\mathcal{H}) \to B(\mathcal{H})$ is said to be *n-decomposable* if the relation

(17)
$$\sum_{k,\ell=1}^{n} (x_k, \phi(a_{k\ell})x_\ell) \geq 0$$

is satisfied for all $x_1, \ldots, x_n \in \mathcal{H}$, and $a_{k\ell} \in B(\mathcal{H})$ such that

(18)
$$\sum_{k,\ell=1}^{n} (x_k, a_{k\ell}x_\ell) \geq 0$$

and

(19)
$$\sum_{k,\ell=1}^{n} (x_k, a_{\ell k}x_\ell) \geq 0.$$

A map ϕ is said to be *decomposable* if it is n-decomposable for all positive
integers n. One should note that the notions of positivity 1-positivity and
1-decomposability coincide. The following results are known:

   Theorem 3.1 (c.f. [18]). Every completely positive map $\phi : B(\mathcal{H}) \to B(\mathcal{H})$
has the form

(20)
$$\phi(a) = v^* \pi(a) v$$

where π is a representation of $B(\mathcal{H})$ on some $B(\mathcal{K})$ and v is a bounded
operator from $\mathcal{H}$ to $\mathcal{K}$ (representation is a map $\pi : B(\mathcal{H}) \to B(\mathcal{K})$ such that
$\pi(a^*) = \pi(a)^*, \pi(ab) = \pi(a)\pi(b)$)

   Theorem 3.2 (c.f. [22]). Every decomposable map $\phi : B(\mathcal{H}) \to B(\mathcal{H})$ has
the form

(21)
$$\phi(a) = v^* \alpha(a) v$$

where α is a Jordan homomorphism of $B(\mathcal{H})$ on some $B(\mathcal{K})$, *i.e.* $\alpha(a^2) = \alpha(a)\alpha(a)$ for $a = a^*$, $\alpha(a^*) = \alpha(a)^*$, and v is a bounded operator from $\mathcal{H}$ to
$\mathcal{K}$.

   Let t denote the transposed map on $B(\mathcal{H})$ with respect to some orthogonal
basis. It follows from Theorem 3.2, that the decomposable map ϕ can be
written in the form

(22)
$$\phi(a) = \phi_1(a) + \phi_2(a^t),$$

where ϕ_1 and ϕ_2 are completely positive maps.

   Theorem 3.3 (c.f. [23]). Let $\phi : B(\mathcal{H}) \to B(\mathcal{H})$ be a normal and completely
positive map, then it has the form

(23)
$$\phi(a) = \sum_{k=1}^{\infty} c_k^* a c_k,$$

where $c_k \in B(\mathcal{H})$ and $\sum_{k=1}^{\infty} c_k^* c_k$ is strongly convergent.

   Theorem 3.4 (c.f. [16], [19]). Every positive map $\phi : M_2(C) \to M_2(C)$ is
decomposable.

   One has to point out that the above theorem does not hold for M_n ($n \geq 3$).
It can be shown that there are positive maps of M_n ($n \geq 3$) which are not
decomposable.

   The physical significance of complete positivity has been recognized by
Accardi, Kraus and Lindblad (cf. [13], [23], [25]). Let us consider the model
of the system + reservoir type. If $\mathcal{H}_S$ and $\mathcal{H}_R$ are Hilbert spaces associated
with the system and the reservoir, respectively, then the reduced dynamics
of the system (in the Heisenberg picture) $\Lambda_t : B(\mathcal{H}) \to B(\mathcal{H})$ is defined by
the relation

(24)
$$\mathrm{tr}_S(\rho \Lambda_t a) = \mathrm{tr}_{R,S}[(\rho \otimes \omega) e^{\frac{i}{\hbar} H t}(a \otimes \mathbb{1}_R) e^{-\frac{i}{\hbar} H t}],$$

for all $\rho \in T(\mathcal{H}_S), a \in B(\mathcal{H}_S), H$ is the Hamiltonian of the system + reservoir, and ω is a fixed density operator of the reservoir. One can check that the reduced dynamics Λ_t is completely positive. Complete positivity makes the mathematical theory really going. Physically it is not an additional assumption, since it is always fulfilled if the system together with its surroundings are described by the Hamiltonian dynamics.

4. Completely Positive Dynamical Semigroups

The completely positive dynamical semigroup is defined in the same manner as before except that the positivity of Λ_t is replaced by complete positivity. In what follows the dynamical semigroup is understood as a completely positive one. Two problems have been studied in detail. (i) One knows that $\Lambda_t = \exp(tL)$ what is the form of L. Can one see directly from the form of infinitesimal generator L whether it generates a dynamical semigroup?

(ii) Can one say something about ergodic properties of Λ_t^*, in particular about

$$\lim_{t \to \infty} \Lambda_t^* \rho$$

from the knowlege of the generator L^*. The first problem was answered for dynamical semigroups with a bounded generator (which is equivalent to the norm continuity of the semigroup) and for the case $\dim \mathcal{H} < \infty$.

Theorem 4.1 (c.f. [13]). A bounded operator L^* on $T(\mathcal{H})$, is the generator of the norm-continuous dynamical semigroup $\{\Lambda_t^* : T(\mathcal{H}) \to T(\mathcal{H}), \ t \geq 0\}$ iff it has the form

(25) $\qquad L^* \rho = -i[h, \rho] + \dfrac{1}{2} \sum_{j \in A} \{[V_j, \rho V_j^*] + [V_j \rho, V_j^*]\}, \ \rho \in T(\mathcal{H})$

where h is a bounded self-adjoint operator, $\{V_j\}_{j \in A}$ is a sequence of bounded operators, $\sum_{j \in A} V_j^* V_j$ converges ultraweakly, and the right-hand side converges in the trace norm.

The generator L of the dual semigroup $\{\Lambda_t; t \geq 0\}$ (in the Heisenberg picture) is given by

$$La \quad = \quad i[h, a] + \frac{1}{2} \sum_{j \in A} \{[V_j^*, a]V_j + V_j^*[a, V_j]\}$$

(26) $\qquad\qquad = \quad i[h, a] + \phi(a) - \dfrac{1}{2}[\phi(\mathbb{1})a + a\phi(\mathbb{1})].$

If $\dim \mathcal{H} = n, \ B(\mathcal{H}) = M_n$, one has

Theorem 4.2 (c.f. [26]). A linear operator $L^* : M_n \to M_n$ is the generator of a dynamical semigroup if it has the form

(27) $$L^*\rho = -i[h,\rho] + \frac{1}{2}\sum_{k,\ell=1}^{n^2-1} c_{k\ell}\{[f_k, \rho f_\ell^*] + f_k[\rho, f_\ell^*]\}$$

where $h = h^*, \mathrm{tr}\, h = 0,\ \mathrm{tr}\, f_k = 0, \mathrm{tr}(f_k f_\ell^*) = \delta_{k\ell},\ k, \ell = 1, \ldots, n^2 - 1$, and $||c_{k\ell}||$ is a complex positive definite matrix.

The generator L of the dual semigroup (in the Heisenberg picture) has the form

(28) $$La = i[h,a] + \frac{1}{2}\sum_{k,\ell=1}^{h^2-1} c_{k\ell}\{[f_\ell^*, \rho]f_k + f_\ell^*[\rho, f_k]\}.$$

In the case $\dim \mathcal{H} = n$, the conditions $||c_{k\ell}|| \geq 0$ implicitly express the inequalities satisfied among the physical parameters characterizing the dynamical evolution (such as relaxation times and equilibrium states) which would be weaker or even nonexisting if just positivity were required. In the case $n = 2$ the generator of positive dynamical semigroup is given by (9). In this case, the condition of complete positivity, expressed by the positive definiteness of the matrix $D + iA$, imposes further restrictions on the inverse relaxation times $\gamma_1, \gamma_2, \gamma_3$ and the range of the variation of $\vec{m}_0$. In particular, it implies the inequalities

(29) $$\gamma_1 + \gamma_2 \geq \gamma_3, \quad \gamma_1 + \gamma_3 \geq \gamma_2, \quad \gamma_2 + \gamma_3 \geq \gamma_1$$

The ergodic properties of dynamical semigroups have been investigated in Refs. [27], [28], [29], [30].

A state $\rho \in S(\mathcal{H})$ is said to be invariant (stationary) with respect to the dynamical semigroup $\Lambda_t^* : T(\mathcal{H}) \to T(\mathcal{H})$ if

(30) $$\Lambda_t^*\rho = \rho\ , \quad t \geq 0,$$

and relaxing if there exists an invariant state, $\rho_0 \in S(\mathcal{H})$ such that

(31) $$\lim_{t\to\infty} \mathrm{tr}[a(\Lambda_t^*\rho)] = \mathrm{tr}(a\rho_0)$$

for every $a \in B(\mathcal{H})$ and $\rho \in S(\mathcal{H})$. The state ρ_0 is called an equilibrium state.

Sufficient conditions for the existence of equilibrium states are known, e.g.:

Theorem 4.3(c.f. [27]). Let $\dim \mathcal{H} = n$ and $e^{tL^*} : M_n \to M_n$ be a dynamical semigroup in the Schrödinger picture, where L^* is given by (27), and let the positive definite matrix $||c_{k\ell}||_{k,\ell=1}^{n^2-1}$ have a p-fold degenerate eigenvalue zero. If $p < \frac{n}{2}$, then the dynamical semigroup e^{tL^*} is relaxing and has a unique equilibrium state.

Theorem 4.4 (c.f. [30]). Let $e^{tL^*} : T(\mathcal{H}) \to T(\mathcal{H})$ be a dynamical semigroup with a bounded generator given by (25). If there exists an invariant state ρ_0 with $\overline{\mathrm{Range}\rho_0} = \mathcal{H}$ (i.e., ρ_0 has no zero eigenvalue) the following two properties are equivalent:

(i) ρ_0 is the unique invariant state,

(ii) $M(L^*) = \{h, V_j, V_j^* : j \in A\}' = \{C \cdot \mathbb{1}\}$,

where { }′ denotes the commutant of the set { }. In this article, only an outline of the problems concerning dynamical semigroups has been presented. More details can be found in references [7], [8], [9] and [10].

References

1. Van Hove, L., The approach to equilibrium in quantum statistics, Physica **23**, 441 (1957).
2. Zwanzig, R., Ensemble method in theory of irreversibility, J. Chem. Phys. **33**, 1338 (1960).
3. Haken, H., Laser Theory, Handbuch der Physik, vol. XXV/2c, Springer, Berlin (1970).
4. Haake, F., in Springer Tracts in Modern Physics, **66**, Springer, Berlin (1973).
5. Jordan, T.F.; Sudarshan, E.C.G., Dynamical mappings of density operators in quantum mechanics, J. Math. Phys. **2**, 772 (1961).
6. Jordan, T.F.; Sudarshan, E.C.G. ; Pinsky, M.A., J. Math. Phys. **3**, 848 (1962).
7. Gorini, V.; Frigerio, A.; Veri, M.; Kossakowski, A.; Sudarshan, E.C.G., Properties of quantum markovian master equations, Rep. Math. Phys., B. **73**, 149 (1978).
8. Davies, E.B., Quantum Theory of Open Systems, Academic Press, New York, 1976.
9. Alicki, R.; Lendi, K., Quantum dynamical semigroups and applications, Lecture Notes in Physics **286**, Springer, Berlin (1990).
10. Ingarden, R.S.; Kossakowski, A.; Ahya, M., Information dynamics and open systems, Kluwer Academic Publ. (1997).
11. Kossakowski, A., On quantum statistical mechanics of non-Hamiltonian systems, Rep. Math. Phys. **3**, 247 (1972).
12. Kossakowski, A., On necessary and sufficient conditions for a generator of a quantum dynamical semigroup. Bull. Acad. Polon. Sci. **20**, 1021 (1972).
13. Lindblad, G., On the generators of quantum dynamical semigroup, Comm. Math. Phys. **48**, 119 (1976).
14. Kossakowski, A., On the general form of the generator of a dynamical semigroup for the spin $-\frac{1}{2}$ system, Bull. Acad. Polon. Sci. **21**, 649 (1973).
15. Arveson, W., Subalgebras of C*-algebras, Acta Math. **123**, 141 (1969).
16. Choi, M.D., Positive semidefinite biquadratic forms, Linear Algebra and Its Appl. **12**, 95 (1975).

Covariant Quantum Dynamical Semigroups: Unbounded Generators

A.S. Holevo

Steklov Mathematical Institute, Vavilova 42, 117966 Moscow, Russia

Summary. A survey of probabilistic approaches to quantum dynamical semigroups with unbounded generators is given. An emphasis is made upon recent advances in the structural theory of covariant Markovian master equations. As an example, a complete characterizations of the Galilean covariant irreversible quantum Markovian evolutions is given in terms of the corresponding quantum master and Langevin equations. Important topics for future investigation are outlined.

1. Introduction

Quantum dynamical semigroups are a noncommutative analog of (sub-) Markov semigroups in classical probability: while the latter are semigroups of maps in functional spaces, the former are semigroups of maps in operator algebras, having certain properties of positivity and normalization. In quantum statistical mechanics dynamical semigroups arise when one considers weak or singular coupling or low density limits for open quantum system interacting with surrounding, allowing to neglect the memory effects of the interaction [36]. These semigroups satisfy differential equations that are noncommutative generalization of the Fokker-Planck or Chapman-Kolmogorov equations and represent the general solution of the Cauchy problem for such equations.

Let $\mathcal{B}(\mathcal{H})$ be the algebra of all bounded operators in a Hilbert space $\mathcal{H}$. We denote by I the unit operator in $\mathcal{H}$, and by Id the identity map of $\mathcal{B}(\mathcal{H})$. Since $\mathcal{B}(\mathcal{H})$ is the dual Banach space of the space $\mathcal{T}(\mathcal{H})$ of trace-class operators, it is supplied with the *weak* topology*. On norm-bounded sets this topology coincides with the weak operator topology (see e.g. [6]). A bounded map Φ of $\mathcal{B}(\mathcal{H})$ into itself is *completely positive* (c. p.) if

$$(1) \qquad \sum_{i,j}(\psi_i|\Phi[X_i^* X_j]\psi_j) \geq 0$$

for any finite sets $\{\psi_j\} \in \mathcal{H}, \{X_j\} \in \mathcal{B}(\mathcal{H})$. According to Stinespring's theorem adapted to the case of $\mathcal{B}(\mathcal{H})$ (see [30]), a generic weak*-continuous c. p. map has the representation

$$(2) \qquad \Phi[X] = L^*(X \otimes I_0)L,$$

where L is a bounded operator from $\mathcal{H}$ to $\mathcal{H} \otimes \mathcal{H}_0$ and I_0 is the unit operator in an auxiliary Hilbert space $\mathcal{H}_0$. By using this representation, it is possible to show that such maps satisfying additional normalization condition $\Phi[I] = I$

represent irreversible evolutions of the open quantum system interacting via unitary operator with auxiliary system in a fixed initial state (see [30], [15]).

If $I_0 = \int |\psi_x ><\psi_x|\mu(dx)$ is a resolution of identity in $\mathcal{H}_0$, then (2) implies

$$\Phi[X] = \int L(x)^* X L(x)\mu(x),$$

where $L(x) = (I \otimes \psi_x)^* L$, and $I \otimes \psi_x$ is the operator from $\mathcal{H}$ to $\mathcal{H} \otimes \mathcal{H}_0$, mapping ψ into $\psi \otimes \psi_x$. In particular, if the measure $\mu(dx)$ is discrete, we obtain the familiar representation of a normal c. p. map in $\mathcal{B}(\mathcal{H})$.

By a *dynamical semigroup* in $\mathcal{B}(\mathcal{H})$ (or *quantum dynamical semigroup*) we shall call a semigroup Φ_t; $t \geq 0$, of weak* continuous completely positive maps in $\mathcal{B}(\mathcal{H})$, satisfying $\Phi_0 = \mathrm{Id}$, and $\Phi_t[I] \leq I$. Moreover, for any X the function $t \to \Phi_t[X]$ is required to be weak*-continuous. Φ_t is called *unital* if $\Phi_t[I] = I$.

In the case of finite-dimensional $\mathcal{H}$ the weak* continuity is equivalent to the norm continuity; every quantum dynamical semigroup then has the form $\Phi_t = \exp t\mathcal{L}$, where $\mathcal{L}$ is the *generator* of the semigroup. The generator is *conditionally completely positive* map, which means that inequality of the type (1) holds provided $\sum_j X_j \psi_j = 0$, and satisfies the normalization condition $\mathcal{L}[I] \leq 0$ (or $\mathcal{L}[I] = 0$ for unital semigroup). The semigroup is the unique solution of the *backward* and the *forward Markovian master equations* (M. m. e.)

$$(3) \qquad \frac{d}{dt}\Phi_t = \mathcal{L} \circ \Phi_t; \qquad \frac{d}{dt}\Phi_t = \Phi_t \circ \mathcal{L},$$

satisfying $\Phi_0 = \mathrm{Id}$.

The conditional complete positivity of $\mathcal{L}$ is equivalent to the *standard representation*

$$(4) \qquad \mathcal{L}[X] = \Phi[X] - K^* X - XK,$$

where Φ is a c. p. map of the form (2), and the normalization condition is equivalent to $L^* L \leq K^* + K$ (with equality for unital semigroups). Similar results hold for *norm-continuous* semigroups in infinite-dimensional $\mathcal{H}$ with all operators in question being *bounded*. The representation (4) for this case was established by Lindblad [31], and independently an equivalent representation was obtained by Gorini, Kossakowski and Sudarshan [18] for $\dim \mathcal{H} < \infty$. A physical interpretation for the standard representation can be seen from the Dyson expansion of the solution of the forward M. m. e.

$$(5) \ \Phi_t = \hat{\Phi}_t + \sum_{n=1}^{\infty} \int \cdots \int_{0 \leq t_1 \leq \cdots \leq t_n \leq t} \hat{\Phi}_{t_1} \circ \Phi \circ \hat{\Phi}_{t_2 - t_1} \cdots \Phi \circ \hat{\Phi}_{t - t_n} dt_1 \ldots dt_n$$

described as sequence of "spontaneous jumps" of the magnitude Φ occuring at times $t_1 \leq \ldots \leq t_n$ on the background of the "relaxing evolution" given by the semigroup $\hat{\Phi}_t[X] = e^{-K^* t} X e^{-Kt}$.

Let $g \to V_g$ be a unitary representation of a group G in $\mathcal{H}$. The dynamical semigroup Φ_t is called *covariant* if

$$(6) \qquad \Phi_t[V_g^* X V_g] = V_g^* \Phi_t[X] V_g.$$

The property of covariance reflects presence of certain symmetries in the interacting open quantum system and is important for applications. For example, covariance with respect to various subgroups of the orthogonal group is characteristic for Bloch type equations and is relevant for optical or magnetic resonance spectroscopy [1] or theory of anisotropic relaxation of spin systems [2]. The combination of covariance and complete positivity imposes strong restrictions on the generator, in some cases defining practically uniquely the form of the corresponding M. m. e.. It can be shown [20] that the generator of a covariant norm continuous semigroup admits the representation (4) in which Φ is covariant and K commutes with V_g (provided G is amenable). This in general is not true for non-norm continuous case, see Section 4.

Already in the mid-seventies when the structure of norm-continuous dynamical semigroups was well understood, it became clear that the non-norm continuous case, while being interesting from both physical and mathematical points of views, poses difficult problems. The generator of such a semigroup may be unbounded with domain not even necessarily being a * - algebra. Therefore it was difficult to formulate a generalization of conditional complete positivity (or equivalent property) useful enough to obtain a kind of standard representation for $\mathcal{L}$ with all its important consequences. The very formulation of the standard representation needed clarification and it was not obvious whether there are "non-standard" unbounded generators. Among very few papers on the subject, Davies [12] established a standard representation for semigroups on $\mathcal{B}(\mathcal{H})$ having invariant pure state. Bratteli et al. [6] studied semigroups on rather general C*-algebra covariant with respect to a compact Abelian group and satisfying rather strong restriction that $\mathcal{L}$ vanishes on the fixed point subalgebra, and showed a kind of Levy-Khinchin formula for $\mathcal{L}$. There were few papers on quasi-free dynamical semigroups on CCR algebra, generators of which are certainly unbounded and standard, e. g. [14], [37]. Unbounded generators arise when the semigroup is covariant with respect to a non-compact symmetry group (such as translations or Galilei group, [3]). While enormous attention was paid to the study of reversible evolutions generated by Schrödinger operators, much less is known about their irreversible Markovian counterparts.

A substantial progress in this direction was achieved in the past few years by making use of profound analogies from the classical theory of Markov semigroups, as developed by Feller, Dynkin, Ito and McKeane, see e. g. [16], [28], or by direct use of classical probabilistic methods. This development concerns the following topics:

– The minimal dynamical semigroup [11], [9], [10], [21], [33]. Existence and uniqueness of solutions of Markovian master equations [24], [26];

- The structure of (covariant) Markovian master equations [21], [22], [24], [25];
- Noncommutative excessive functions and arrival times [4], [23]. Nonstandard generators [23], [27];
- Stochastic representations and hyperdissipativity [26]. Relations to continuous measurement processes and nonlinear stochastic Schrödinger equations [17], [26], [29]. Dilations to quantum Langevin equations [25].

In what follows we shall concentrate on the second topic, restricting to brief comments concerning other topics; further details can be found in the references given above.

2. The Quantum Markovian Master Equations

The starting point of our approach, just as in the classical probability theory, is not a semigroup itself, but the differential equation it satisfies. This is also more natural for physical applications. A quantum M. m. e. must be an equation for matrix elements of the semigroup; thus we assume that there is a dense domain $\mathcal{D} \subset \mathcal{H}$, such that the following derivative

$$(7) \qquad \frac{d}{dt} < \psi | \Phi_t[X] \phi >|_{t=0} = \mathcal{L}(\psi; X; \phi)$$

exists for $\phi, \psi \in \mathcal{D}$, $X \in \mathcal{B}(\mathcal{H})$. The form $\mathcal{L}(\psi; X; \phi)$ is called *form-generator*. It can be characterized by a number of nice properties including conditional complete positivity [21], [24]. These properties turn out to be equivalent to the *standard representation*

$$(8) \quad \mathcal{L}(\psi; X; \phi) = < L\psi|(X \otimes I_0)L\phi > - < K\psi|X\phi > - < \psi|XK\phi >,$$

where L, K are (unbounded) operators defined on $\mathcal{D}$ and satisfying the *dissipativity condition*

$$\|L\psi\|^2 \leq 2 \, \mathrm{Re} < \psi|K\psi >, \ \psi \in \mathcal{D}.$$

In particular, K is accretive: $\mathrm{Re} < \psi|K\psi >> 0, \psi \in \mathcal{D}$. The (backward) M. m. e. takes the form

$$(9) \qquad \frac{d}{dt} < \psi | \Phi_t[X] \phi >= \mathcal{L}(\psi; \Phi_t[X]; \phi); \ \phi, \psi \in \mathcal{D}.$$

The relation between the form-generator and the generator resembles relation between a formal differential operator and its closed extensions determined by certain boundary conditions. To see this let $\Psi_t = (\Phi_t)_*$ be the strongly continuous *preadjoint semigroup* in $\mathcal{T}(\mathcal{H})$, such that $\Psi_t^* = \Phi_t$. Denoting its generator $\mathcal{L}_*$, one has

$$\mathcal{L}(\psi; X; \phi) = \mathrm{Tr} \mathcal{L}_*[|\phi > < \psi|]X$$

for $\phi, \psi \in \mathcal{D}$, $X \in \mathcal{B}(\mathcal{H})$. The assumption that the derivative (7) exists for all X is equivalent to $\mathrm{dom}\mathcal{L}_* \supset$ D, where

$$(10) \qquad\qquad \mathrm{D} = \mathrm{lin}\{|\phi><\psi| : \phi, \psi \in \mathcal{D}\}$$

is a dense domain in $\mathcal{T}(\mathcal{H})$. The M. m. e. (9) takes the form

$$(11) \qquad \frac{d}{dt}\,\mathrm{Tr}\rho\Phi_t[X] = \mathrm{Tr}\mathcal{L}_*[\rho]\Phi_t[X], \qquad \rho \in \mathrm{D}, X \in \mathcal{B}(\mathcal{H}).$$

If D is a core for $\mathcal{L}_*$, then this equation determines Φ_t uniquely, otherwise it may have non-unique solution. Under the condition that the closure of K is maximal accretive, one can show that there exists a dynamical semigroup $\Phi_t{}^\infty$ giving the *minimal solution* of the equation (9) in the sense that for any other solution Φ_t the difference $\Phi_t - \Phi_t{}^\infty$ is completely positive. Of special interest is the case of a *unital* generator, satisfying $\mathrm{Tr}\mathcal{L}_*[\rho] \equiv 0, \rho \in \mathrm{D}$, or

$$\sum_j \|L_j\psi\|^2 = 2\,\mathrm{Re} < \psi|K\psi >, \ \psi \in \mathcal{D}.$$

In general $\Phi_t{}^\infty$ may not be unital; however if it is, then $\Phi_t{}^\infty$ is the unique solution of (9).

The method of construction of the minimal dynamical semigroup developed in [11] for resolvents, in [9], [10] for associated integral equation, and in [21], [24] for the backward M. m. e., is the noncommutative extension of the Feller's method [16]. It is based on a standard representation, i. e. on a decomposition of the relevant object into completely positive and relaxing parts. The starting point is the relaxing semigroup $\hat{\Phi}_t[X] = e^{-K^*t}Xe^{-Kt}$ providing the unique solution of the equation (9) with

$$\mathcal{L}(\psi; X; \phi) = - < K\psi|X\phi > - < \psi|XK\phi >,$$

which is then perturbed with the completely positive form $< L\psi|(X \otimes I_0)L\phi >$ introducing spontaneous jumps on the background of the relaxing evolution $\hat{\Phi}_t$. It may be viewed upon as a generalization of the expansion (5) to the case of unbounded but completely positive perturbations. Just as in the classical case, "explosion" may occur if the infinite number of jumps happens during finite interval and the process reaches "boundary" in a finite time (this can never happen for a bounded generator). If Φ_t^∞ is not unital, then there is a positive probability of explosion, and additional "boundary conditions" are required to specify the solution, which amounts to certain maximal extension of $\mathcal{L}_*$ from D.

Under the additional assumption that operator L^* satisfies

$$\sum_j \|L^*(\psi \otimes e_j)\|^2 < \infty, \ \psi \in \mathcal{D}^*,$$

where $\{e_j\}$ is an orthonormal basis in $\mathcal{H}_0$, and $\mathcal{D}^* \subset \mathrm{dom}K^*$ is a dense domain in $\mathcal{H}_0$, one can write also the *forward* Markovian master equation for the preadjoint semigroup Ψ_t:

$$(12) \qquad \frac{d}{dt} < \phi | \Psi_t[\rho] \psi > = \mathcal{L}_*(\phi; \Psi_t[\rho]; \psi); \quad \phi, \psi \in \mathcal{D}^*,$$

where

$$\mathcal{L}_*(\phi; \rho; \psi) = \mathrm{Tr}\rho\mathcal{L}[|\psi><\phi|]$$
$$= \sum_j < L^*(\phi \otimes e_j) | \rho L^*(\psi \otimes e_j) > - < K^*\phi | \rho\psi > - < \phi | \rho K^* \psi >,$$

and $\mathcal{L}$ is the generator of Φ_t defined on $\mathbf{D}^* = \{|\psi><\phi| : \phi, \psi \in \mathcal{D}^*\}$. Assuming K^* to be maximal accretive one can prove that $\Psi_t^\infty = (\Phi_t^\infty)_*$ is the minimal solution of the forward equation [25], [26]. However in general the forward and the backward equations are no longer equivalent. Thus the situation is similar to that for the Kolmogorov-Feller differential equations in the theory of Markov processes [16].

Going back to the problem of standard representation, we can make the following remarks. The fact that a form-generator has the standard representation (8) implies the possibility of decomposing the generator $\mathcal{L}_*$ into completely positive and relaxing parts only on the subspace $\mathbf{D}$ which need not be a core for $\mathcal{L}_*$. If explosion occurs, these two parts need not be separately extendable onto a core for $\mathcal{L}_*$. On the other hand, generators of different dynamical semigroups restricted to $\mathbf{D}$ can give rise to one and the same standard expression (8). One may formalize the notion of standard representation by saying that a dynamical semigroup is *standard* if it can be constructed as the minimal semigroup for some M. m. e., that is by a completely positive perturbation of a relaxing semigroup. In [23] a possible noncommutative extension of "boundary conditions" for conservative form-generator was proposed as very singular completely positive perturbations vanishing on the dense domain $\mathbf{D}$. By using such a perturbation the author gave a construction of non-standard dynamical semigroup on $\mathcal{B}(\mathcal{H})$ [23], [27].

3. An Example

Let $\xi_t, t \geq 0$ be stochastic process with stationary independent increments [16]. Roughly speaking, the (generalized) time derivative of ξ_t is a continuous analog of a sequence of independent identically distributed random variables, that is a classical "noise" process. One of the beautiful results of probability theory is the Levy-Khinchin formula describing the possible form of the characteristic function of such process:

$$(13) \quad \mathrm{M}\exp{i\lambda\xi_t} = \exp{t[i\beta\lambda - \frac{\alpha}{2}\lambda^2 + \int_{0<|y|} (e^{iy\lambda} - 1 - iy\lambda 1_h(y))\mu(dy)],}$$

where β is real number, $\alpha \geq 0$, h is arbitrary but fixed positive number, $1_h(y)$ is the indicator of the set $|y| \leq h$, and $\mu(dy)$ is a positive measure on the set $\mathbf{R} \setminus \{0\}$, satisfying the condition

$$(14) \qquad \int_{0<|y|} [y^2 1_h(y) + (1 - 1_h(y))]\mu(dy) < \infty.$$

In (13) the term $i\beta\lambda - \frac{\alpha}{2}\lambda^2$ corresponds to the Gaussian component of the process ξ_t, which is a continuous process. If in the integral term we take $\mu(dy) = \mu\, \delta(y - y_0)dy$ with $\mu > 0$, then for $|y_0| > h$ we obtain logarithm of the characteristic function of the Poisson process with the jumps of the magnitude y_0. Therefore for arbitrary measure $\mu(dy)$ the integral $\int_{h<|y|}(e^{iy\lambda} - 1)\mu(dy)$ describes the mixture of independent Poisson processes with various magnitudes y, $|y| > h$. It corresponds to the discontinuous (pure jump) component of the process ξ_t (with magnitudes of jumps $|y| > h$). The value of h is arbitrary but fixed, so the name "big jumps" is only conventional. The term related to "small jumps" (of magnitudes $|y| \leq h$) corresponds to the situation when infinitely many small jumps can accumulate during finite time, and one must include portions of linear drift between jumps in order that the total increment will remain finite. The process ξ_t itself can be decomposed into three components – continuous Gaussian, Poisson "big jumps" and "small jumps", according to Ito's formula (see e. g. [32]):

$$(15) \qquad d\xi_t = \beta dt + \sqrt{\alpha}dW_t + \int_{h<|y|} y\Pi(dy\ dt) + \int_{0<|y|\leq h} \tilde{\Pi}(dy\ dt),$$

where W_t is the standard Wiener process, $\Pi(dy\ dt)$ is the Poisson random measure on $\mathbf{R}^2$ with the compensator $\mu(dy)dt$, so that

$$(16) \qquad \mathsf{M}dW_t = 0, \quad \mathsf{M}\Pi(dy\ dt) = \mu(dy)\ dt,$$

and $\tilde{\Pi}(dy\ dt) = \Pi(dy\ dt) - \mu(dy)\ dt$ is the compensated random measure. Note that $\Pi([y_1, y_2], [t_1, t_2])$ is just the number of jumps of the process ξ_t on the time interval $[t_1, t_2]$, which have magnitudes $y \in [y_1, y_2]$.

Now consider the Hilbert space $\mathcal{H} = L^2(\mathbf{R})$, and let $Q = x, P = i^{-1}\frac{d}{dx}$ be, respectively, the self-adjoint position and momentum operators for one-dimensional quantum system, so that $V_y = \exp(iyQ), y \in \mathbf{R}$, and $U_x = \exp(-ixP), x \in \mathbf{R}$, are the unitary groups in $\mathcal{H}$ satisfying the Weyl canonical commutation relation (CCR):

$$(17) \qquad U_x V_y = \exp(-ixy)V_y U_x.$$

Defining

$$(18) \qquad \Phi_t[X] = \mathsf{M}U^*_{\xi_t} X U_{\xi_t}, \quad t \geq 0,$$

one easily sees that Φ_t is a unital dynamical semigroup in $\mathcal{H}$. Indeed, operators Φ_t are manifestly completely positive; the semigroup property follows from the fact that ξ_t has stationary independent increments; the weak* continuity properties follow from the continuity properties of U_x and of the expectation. The semigroup (18) represents the dynamics of quantum system in $\mathcal{H}$ interacting with the classical noise via unitary operators $\exp(-i\xi_t P)$, averaged with respect to the distribution of the noise. To find the generator of this semigroup, one can use the Ito formula for $\exp(i\xi_t P)$ (cf. [19]):

$$d\exp(i\xi_t P) = \exp(i\xi_t P)\{[i\beta P - \frac{\alpha}{2}P^2 + \int_{0<|y|\leq h}(\exp(iyP) - 1 - iyP)\mu(dy)]dt$$

$$+i\sqrt{\alpha}PdW_t + \int_{h<|y|}[\exp(iyP) - 1]\Pi(dy\ dt) + \int_{0<|y|\leq h}[\exp(iyP) - 1]\tilde{\Pi}(dy\ dt)\},$$

and the Ito product rule

(19) $$dW_t^2 = dt, \quad \Pi(dy\ dt)^2 = \Pi(dy\ dt),$$

with all other products of stochastic differentials (including dt) equal to zero. Taking into account (16), one can obtain both backward and forward M. m. e. (9), (12) with $\mathcal{D} = \mathcal{D}^* = C_0^2(\mathbf{R})$, the subspace of twice continuously differential functions with compact support, where the form-generators correspond to the expression

(20)
$$\mathcal{L}[X] = i\beta[P, X] - \frac{\alpha}{2}[P, [P, X]] + \int_{0<|y|}(U_y^* X U_y - X - iy[P, X]1_h(y))\mu(dy),$$

defined for $X \in D$. Here the first term is the Hamiltonian "drift", the second term corresponds to the interaction with the Gaussian "white" noise and is typical for diffusion approximations, while the last term reflects the influence of the Poisson "shot" noises arising in low density limits. The generator (20) is bounded if and only if $\alpha = 0, \beta = 0$ and $\mu(dy)$ is a finite measure on $\mathbf{R} \setminus 0$.

The Gaussian noise gives rise to the diffusive generator

(21) $$\mathcal{L}[X] = -\frac{\alpha}{2}[P, [P, X]] = \frac{\alpha}{2}(2PXP - P^2X - XP^2),$$

with the obvious standard representation on D. A standard representation for the last term in (20) can be obtained by taking

$$L = \int_{0<|y|}(I \otimes |y>)(U_y - I)\mu(dy), \quad K = \int_{0<|y|}(I - U_y - iyP1_h(y))\mu(dy),$$

where $\{|y>\}$ is the canonical family of "ket" vectors in $\mathcal{H}_0 = L^2(\mathbf{R}, \mu)$.

From the CCR it follows that the semigroup is covariant with respect to the representation $y \to V_y$ describing translations in the momentum space. As shown in [21], for covariant M. m. e. the non-explosion in $\mathcal{B}(\mathcal{H})$ is equivalent to the non-explosion in the fixed-point subalgebra $\mathcal{A}_V = \{X : V_g^* X V_g = X, g \in G\}$ of the representation $g \to V_g$. If this subalgebra is Abelian then the problem is reduced to the well-studied problem of non-explosion for a classical Markov process. In our example the fixed point algebra is the maximal Abelian subalgebra $\mathcal{A}_Q$ of operators of the form $X = f(Q)$; by the CCR

$$\Phi_t[f(Q)] = \mathsf{M}f(Q + \xi_t),$$

and

$$\mathcal{L}f(x) = \beta\frac{df(x)}{dx} + \frac{\alpha}{2}\frac{d^2f(x)}{dx^2} + \int_{0<|y|}[f(x+y) - f(x) - yf'(x)1_h(y)]\mu(dy)$$

is the generator of the semigroup corresponding to the process ξ_t with stationary independent increments, for which explosion can never occur [16]. This is also strictly related to the additional property of covariance with respect to the space translations $x \to U_x$, shared by the semigroup (18). However the situation is different for more general momentum translation covariant M. m. e..

To see this, following [24], consider the Hilbert space $\mathcal{H} = L^2(l, \infty)$, the domain $\mathcal{D} = C_0^2(l, \infty)$ consisting of continuously twice differentiable functions with compact support, vanishing at l, and the form-generator

(22) $\mathcal{L}(\phi, X, \psi)$
$$= < (P + L(Q))\phi|X(P + L(Q))\psi > - < K\phi|X\psi > - < \phi|XK\psi >,$$

defined for $\phi, \psi \in \mathcal{D}$, where $K = \frac{P^2}{2} + PL(Q) + \frac{|L(Q)|^2}{2}$, and $L(Q)$ is a continuously differentiable complex function. This form-generator is covariant with respect to the representation $y \to V_y = \exp(iyQ)$, and hence the corresponding minimal dynamical semigroup is also covariant [21]. The restriction to the fixed point algebra $\mathcal{A}_Q$ corresponds to the classical diffusion on (l, ∞) with the generator

$$\mathcal{L}f(x) = 2\mathrm{Im}L(x)\frac{df(x)}{dx} + \frac{1}{2}\frac{d^2f(x)}{dx^2}.$$

Non-explosion means that both l and ∞ are non-absorbing boundaries for this diffusion. The necessary and sufficient condition for this is Feller's test [28], saying that the function

$$\int_{x_0}^x \left[\exp\int_x^y 4\mathrm{Im}L(z)dz\right]dy,$$

where $x_0 \in (l, \infty)$, must be non-integrable in the neighbourhoods of both l and ∞. In particular, if $L(x) \equiv 0$ (pure diffusion with no drift), then the probability of absorption at l is positive, hence the minimal semigroup is non-unital and the solution of the M. m. e. is not unique. This minimal semigroup is the extension onto $\mathcal{B}(\mathcal{H})$ of the Markov semigroup corresponding to the Brownian motion on (l, ∞) killed at the boundary l. Other solutions of the backward M. m. e. are obtained by taking perturbations corresponding to various boundary conditions at l. An example of non-standard dynamical semigroup on $\mathcal{B}(\mathcal{H})$ is constructed as a singular perturbation of this minimal semigroup corresponding to rebounding from l to a fixed quantum state ρ_0 [23], [27].

4. Covariant Evolutions and the Group Cohomology

Consider a backward M. m. e. given by a form-generator $\mathcal{L}(\phi, X, \psi)$. The standard representation (8) of the form-generator is not unique even if it is subjected to further condition of minimality [24]. If D is a unitary operator in $\mathcal{H}_0$, $a \in \mathcal{H}_0$, and b is a real number, then the operators

$$L' = (I \otimes D)L + I \otimes a, \quad K' = K + (I \otimes a)^*(I \otimes D)L + [\frac{1}{2}\|a\|^2 - ib]I,$$

where $(I \otimes a)$ is the operator from $\mathcal{H}$ to $\mathcal{H} \otimes \mathcal{H}_0$ acting as $(I \otimes a)\psi = \psi \otimes a$, give another standard representation for $\mathcal{L}(\phi, X, \psi)$ satisfying the minimality condition. The transformations $(D, a, b) : (L, K) \to (L', K')$ form a kind of a "gauge group" (cf. also [34]) under the multiplication law

(23) $(D', a', b')(D, a, b) = (D'D, D'a + a', b + b' - \text{Im} < a'|D'a >)$.

We denote this group by $G(\mathcal{L})$. It is endowed with the natural topology as a subset of the product $\mathcal{U}(\mathcal{H}_0) \times \mathcal{H}_0 \times \mathbf{R}$, where $\mathcal{U}(\mathcal{H}_0)$ is the group of unitary operators in $\mathcal{H}_0$ with the weak operator topology.

Let now the form-generator be *covariant* under a (projective) unitary representation $g \to V_g$ of a symmetry group G, namely, the domain $\mathcal{D}$ be invariant under V_g and

$$\mathcal{L}(\phi, V_g^* X V_g, \psi) = \mathcal{L}(V_g \phi, X, V_g \psi), \quad \phi, \psi \in \mathcal{D}.$$

Theorem 1. *There is a representation* $g \to (D_g, a_g, b_g)$ *of* G *in* $G(\mathcal{L})$ *such that*

(24) $(V_g^* L V_g, V_g^* K V_g) = (D_g, a_g, b_g)(L, K)$.

If V_g *is a continuous representation of a topological group* G *then the representation* (D_g, a_g, b_g) *is continuous if and only if the scalar function* $g \to \mathcal{L}(\phi, X V_g^*, V_g \psi)$ *is continuous for all* $\phi, \psi \in \mathcal{D}, X \in \mathcal{B}(\mathcal{H})$.

The proof of this theorem may be found in [24], and here we discuss briefly the way it can be applied to find the form of a covariant generator for concrete symmetry groups G. From this theorem taking into account (24) it follows that D_g is a unitary representation of G and a_g is a first order cocycle for this representation in $\mathcal{H}_0$: $a_{g'g} = D_{g'}a_g + a_{g'}$. Moreover, the real function b_g satisfies the coboundary equation $b(g') + b(g) - b(g'g) = \text{Im} < a_{g'}|D_{g'}a_g >$. Thus the structure of the covariant form-generator is determined by the low order cohomology of the group G, which was studied in detail for many interesting groups (see, e. g. [35]). In particular, it is well known that the low order cohomology is trivial for compact groups, that is every cocycle is a coboundary for such groups, $a_g = (D_g - I)a$ for some $a \in \mathcal{H}_0$. It follows that in this case, similarly to the case of bounded generators, the covariant form-generator has the standard representation (8) where the c. p. component and the relaxing terms are separately covariant. That this is not the case for non-compact groups, can be easily seen from the example of the diffusive generator (21).

5. Galilean Covariant Markovian Evolutions

Let $(\xi, \tau) \in \mathbf{R}^2$ be a point in the 2-dimensional non-relativistic space-time, and let $(x, v, t) : (\xi, \tau) \to (\xi', \tau')$ be the Galilei transformation

$$(25) \qquad \xi' = \xi + x + v\tau, \qquad \tau' = \tau + t,$$

where $x \in \mathbf{R}$ is the space shift, $v \in \mathbf{R}$ the Galilean boost. For simplicity we consider zero-spin unit mass elementary system characterized by the Weyl operators

$$W_{x,v} = \exp i(vQ - xP) = V_v U_x \exp(\frac{i}{2}vx),$$

constituting irreducible representation of the CCR.

A dynamical semigroup Φ_t is *Galilean covariant* [3], [25], if

$$\Phi_t[W_{x,v}^* X W_{x,v}] = W_{x-vt,v}^* \Phi_t[X] W_{x-vt,v}.$$

Let $\mathcal{D} \subset \mathcal{H}$ be the dense domain

$$\mathcal{D} = \bigcap_{x,v\in\mathbf{R}} \mathrm{dom}(vQ - xP)^2,$$

and let $\mathrm{D} \subset \mathcal{B}(\mathcal{H})$ be the domain defined by the relation (10). We remark that $\mathcal{D}$ is invariant under $W_{x,v}$, and that D is norm-dense in $\mathcal{T}(\mathcal{H})$ and weakly* - dense in $\mathcal{B}(\mathcal{H})$. We make the following assumption

(A) The domain D is contained both in $\mathrm{dom}\mathcal{L}_*$ and $\mathrm{dom}\mathcal{L}$.

Theorem 2. *A unital Galilean covariant dynamical semigroup satisfying the condition (A) has the generator $\mathcal{L}$ given by the following expression on* D

$$\mathcal{L}[X] = i[\frac{P^2}{2}, X] + i[\beta_P P + \beta_Q Q, X]$$

$$(26) \qquad -\frac{1}{2}\{\alpha_{PP}[P, [P, X]] + \alpha_{PQ}[P, [Q, X]] + \alpha_{QQ}[Q, [Q, X]]\}$$

$$+ \int \int_{x^2+v^2>0} \{W_{x,v}^* X W_{x,v} - X - i[xP - vQ, X]1_h(x, v)\}\nu(dx \, dv),$$

where $\beta_P, \beta_Q \in \mathbf{R}$, the real matrix

$$\begin{bmatrix} \alpha_{PP} & \alpha_{PQ} \\ \alpha_{PQ} & \alpha_{QQ} \end{bmatrix}$$

is positive definite, $1_h(x, v)$ is the indicator of the set $x^2+v^2 \le h$ and $\nu(dx \, dv)$ is a positive measure on $\mathbf{R}^2\backslash\{0\}$ satisfying the Levy condition

$$\int \int_{x^2+v^2>0} \{(x^2 + v^2)1_h(x, v) + [1 - 1_h(x, v)]\}\nu(dx \, dv) < \infty.$$

Moreover, the domain D is a core for both $\mathcal{L}_*$ and $\mathcal{L}$ and the corresponding M. m. e. (9), (12) have Φ_t (resp. Ψ_t) as the unique solution.

The last statement of the Theorem applies to particular Galilean covariant M. m. e. arising in various physical applications, such as quantum optics [7], precision experiments [5], nonlinear quantum mechanics [13] etc. The uniqueness of the solution of the M. m. e. is related to the fact that the fixed point algebra of the representation $(x, v) \to W_{x,v}$ is trivial, that is consists of multiples of the identity operator (cf. [21]).

A derivation of (26) can be based on Theorem 1. By subtracting from $\mathcal{L}$ the Hamiltonian term corresponding to the free motion we obtain a generator $\mathcal{L}_0$ satisfying the condition of Weyl covariance

$$\mathcal{L}_0[W_{x,v}^* X W_{x,v}] = W_{x,v}^* \mathcal{L}_0[X]W_{x,v}.$$

Let L, K be the components of the standard representation of the corresponding form-generator. According to Theorem 1 there is a unitary representation $(x, v) \to D_{x,v}$ of the Abelian group $\mathbf{R}^2$ and the cocycle $a_{x,v}$ in the Hilbert space $\mathcal{H}_0$ such that L, K satisfy the covariance equations

$$W_{x,v}^* L W_{x,v} = (I \otimes D_{x,v})L - I \otimes a_{-x,-v},$$

$$W_{x,v}^* K W_{x,v} = K - (I \otimes a_{-x,-v})^* L + [\tfrac{1}{2}\|a_{x,v}\|^2 - ib_{x,v}]I.$$

These equations can be solved by diagonalizing the representation $D_{x,v}$ and by using the structure of cocycles for representations of Abelian locally compact groups [22]. The "Gaussian" part of the generator $\mathcal{L}_0$ arises from the identity subrepresentation of $D_{x,v}$ while the orthogonal complement gives the "jump" part. We conjecture that the assumption (A) can be deduced from the Galilean covariance itself, as we were able to deduce it from the Weyl covariance (see [25], where an alternative proof of Theorem 2 is given).

The generator (20) considered in Section 3 is a particular case of (26), provided we exclude the free Hamiltonian term. That generator arose from the semigroup (18) describing interaction of quantum system with the classical noise. It turns out to be possible to give a similar explicit description of the

Galilean covariant quantum open systems, as systems interacting with specific classical noises. Let ξ_t, η_t be a classical stochastic process with stationary independent increments in $\mathbf{R}^2$, defined by the characteristic function of the Levy-Khinchin form

$$\text{Mexp}i(\mu\xi_t - \lambda\eta_t) = \text{exp}t\{i(\mu\beta_P - \lambda\beta_Q) - \frac{1}{2}(\alpha_{PP}\mu^2 + 2\alpha_{PQ}\mu\lambda + \alpha_{QQ}\lambda^2)$$

$$(27) \quad + \int\int_{x^2+v^2>0} [e^{i(\mu x - \lambda v)} - 1 - i(\mu x - \lambda v)1_h(x,v)]\nu(dx\ dv)\},$$

where $\beta_P, \beta_Q; \alpha_{PP}, \alpha_{PQ}, \alpha_{QQ}$ and $\nu(dx\ dv)$ are taken from (26). Consider the stochastic differential equations

$$(28) \qquad dQ_t = \frac{P_t}{m}dt + d\xi_t,\ dP_t = d\eta_t,$$

with the initial conditions $Q_0 = Q, P_0 = P$. These will be the Heisenberg equations for our open quantum system. They correspond to the infinitesimal canonical transformation with the Hamiltonian

$$dH_t = \frac{P^2}{2}dt + Pd\xi_t - Qd\eta_t.$$

Defining the chronologically ordered exponential

$$U_t(\xi, \eta) = \mathcal{T}\text{exp}(-i\int_0^t dH_s)$$

as the solution of the corresponding stochastic differential equation, we can prove (see [25]) that

$$\Phi_t[X] = MU_t(\xi, \eta)^* X U_t(\xi, \eta).$$

This relation is a generalization of the representation (18) and the proof proceeds along similar lines by using the stochastic differential equation for $U_t(\xi, \eta)$ and the distribution of ξ_t, η_t defined by (27). Equations (28) are the Langevin equation giving the dilation of the dynamical semigroup Φ_t with the classical stationary independent increment processes as the driving noises.

6. Discussion

The results described in the previous Section are due to the very restrictive nature of the full Galilean covariance. We obtain much broader and physically interesting class of quantum Markovian evolutions by omitting space translations and restricting only to Galilean boosts, that is to the fundamental symmetry of a non-relativistic particle in a potential field. The class of resulting evolutions is described in detail in [25] for the case where the position space is the whole $\mathbf{R}^3$. Discussion at the end of Section 3 suggests that contrary to the case of full Galilean covariance, there is no automatic non-explosion, and boundary conditions should play an important role, especially for systems with restricted position domains. This case deserves much more detailed study. Other interesting problems are related to introducing spin degrees of freedom along with spatial ones and to gauge covariance.

Another important distinction of the boost covariant evolutions is that they describe open systems interacting with quantum rather than classical noises. This means that the corresponding M. m. e. at least formally can be dilated to the Langevin equations (see [25]) which are quantum stochastic differential equations driven by quantum Brownian motion or Poisson-type processes in the sense of [34], but in general, with unbounded operator coefficients. For example, the Langevin equation dilating the diffusive M. m. e. defined by the form-generator (22) supplemented with the Hamiltonian term has the form

$$df(Q_t) = i[\frac{P_t^2}{2}, f(Q_t)]dt + f'(Q_t)i[(dA_t + L(Q_t)dt)^\dagger - \text{h. c.}] + \frac{1}{2}f''(Q_t)dt,$$

$$dP_t = U'(Q_t)dt + i[\bar{L}'(Q_t)(dA_t + \frac{1}{2}L(Q_t)dt) - \text{h.c.}],$$

where U is the potential, $A_t^\dagger, A_t$ are creation-annihilation processes representing quantum Brownian motion and h. c. denotes hermitean conjugated terms.

Remarkably, at least for the minimal solution of M. m. e. there always exists a representation via solutions of certain *classical* dissipative stochastic equation in the Hilbert space of the system. It provides a powerful probabilistic tool for study of the problem of non-explosion for quantum dynamical semigroups [26], and of the nonlinear stochastic Schrödinger equation arising in the theory of continuous quantum measurement processes [17], [26], [29].

Acknowledgements. The author acknowledges support from Arnold Sommerfeld Institute for Mathematical Physics, Technical University Clausthal, during the XXI International Colloquium on Group Theoretical Methods in Physics. The work was partially supported by RFBR grant no. 96-01-01709.

References

1. R. Alicki, K. Lendi, Quantum Dynamical Semigroups and Applications, Lect. Notes. Phys. **286**, Springer-Verlag, Berlin-Heidelberg-NY 1987.
2. A. Yu. Artem'ev, Teor. Mat. Fiz. **87**, 34-39 (1991); Application of Methods of Completely Positive Semigroups in Study of Dynamics of Open Spin Systems, Ph. D. Thesis, Moscow 1991.
3. A. Barchielli, L. Lanz, Il Nuovo Cimento **44B**, 241-264 (1978).

4. B. V. Bhat, K. R. Parthasarathy, Ann. Inst. H. Poincare, ser. B **31**, 601-652 (1995).
5. V. B. Braginsky, F. Ya. Khalili, Quantum Measurements, Cambridge University Press 1992.
6. O. Bratteli, D. W. Robinson, Operator Algebras and Quantum Statistical Mechanics I, Springer-Verlag, Berlin 1981.
7. H. J. Carmichael, An Open System Approach to Quantum Optics, Lect. Notes Phys. m18, Springer-Verlag, Berlin-Heidelberg-NY 1993.
8. O. Bratteli, P. E. T. Jørgensen, A. Kishimoto, D. W. Robinson, Ann. Inst. Fourier **34**, 155-187 (1984).
9. A. M. Chebotarev, J. of Soviet Math. **56**, 2697–2719 (1991).
10. A. M. Chebotarev, F. Fagnola, J. Funct. Anal. **118**, 131-153 (1993).
11. E. B. Davies, Rep. Math. Phys. **11**, 169–188 (1977).
12. E. B. Davies, J. Funct. Anal. **34**, 421-432 (1979).
13. H.-D. Doebner, J. D. Hennig, P. Nattermann, Lecture at the XX International Colloquium on Group Theoretical Physics, Osaka 1994.
14. G. G. Emch, S. Albeverio, J.-P. Eckmann, Rept. Math. Phys. **13**, 73-85 (1978).
15. D. E. Evans, J. T. Lewis, Dilations of Irreversible Evolutions in Algebraic Quantum Theory, Commun. Dublin Inst. Adv. Studies, Ser. A, **24** Dublin 1977.
16. W. Feller, An Introduction to Probability Theory and Its Applications II, J. Wiley, NY (1966).
17. D. Gatarek, N. Gisin, J. Math. Phys. **32**, 2152–2157 (1991).
18. V. Gorini, A. Kossakowski, E. Sudarshan, Journ. Math. Phys. **17**, 821-825 (1976).
19. A. S. Holevo, Lect. Notes. Math. **1442**, 211-215 (1990).
20. A. S. Holevo, Rept. Math. Phys. **32**, 211-216 (1993).
21. A. S. Holevo, Rept. Math. Phys. **33**, 95-110 (1993).
22. A. S. Holevo, Izvestiya: Mathematics **59**:2, 427-443 (1995).
23. A. S. Holevo, Izvestiya: Mathematics **59**:6, 1311-1325 (1995).
24. A. S. Holevo, J. Funct. Anal. **131**, 255-278 (1995).
25. A. S. Holevo, J. Math. Phys. **37**(4), 1812-1832 (1996).
26. A. S. Holevo, Probab. Theory Rel. Fields **104**, 483-500 (1996).
27. A. S. Holevo, Uspekhi Mat. Nauk. **51**(6), 225-226 (1996).
28. K. Ito, H. P. McKeane, Diffusion Processes and Their Sample Paths, Springer-Verlag, Berlin-Heidelberg-NY 1965.
29. V. N. Kolokoltsov, J. Math. Phys. **36**, 2741-2760 (1995).
30. K. Kraus, States, Effects and Operations, Lect. Notes Phys. **190**, Springer-Verlag, Berlin-Heidelberg-NY 1983.
31. G. Lindblad, Commun. Math. Phys. **48**, 119–130 (1976).
32. R. Sh. Liptser, A. N. Shiryaev, Theory of Martingales, Kluwer, Dodrecht 1986.
33. A. Mohari, K. B. Sinha, Proc. Indian Acad. Sci., **102**, 159-173 (1992).
34. K. R. Parthasarathy, An Introduction to Quantum Stochastic Calculus, Birkhäuser Verlag, Basel-Boston-Berlin 1992.
35. K. R. Parthasarathy, K. Schmidt, Positive Definite Kernels, Continuous Tensor Products and the Central Limit Theorem of Probability Theory, Lect. Notes Math. **272**, Springer-Verlag, Berlin-Heidelberg 1972.
36. H. Spohn, Rev. Mod. Phys. **53**, 569-615 (1980).
37. P. Vanheuverzwijn, Ann. Inst. H. Poincare, ser. A **19**, 123-138 (1978).

Quantum Stochastic Semigroups and Their Generators

V.P. Belavkin

Mathematics Department, University of Nottingham NG7 2RD, UK.

Summary. A rigged space characterisation of the unbounded generators of quantum completely positive (CP) stochastic semigroups is given. The general form and the dilation of the stochastic completely dissipative (CD) equation over the algebra $\mathcal{L}(\mathcal{H})$ is described, as well as the unitary quantum stochastic dilation of the subfiltering and contractive flows with unbounded generators is constructed.

1. Introduction

Quantum stochastic dynamics gives beautiful solvable models for the interaction of a quantum system with the quantum noise, which is produced by a heat bath, measurement apparatus, or any other environment with infinite number of freedom. It can be defined by a weakly continuous evolution semigroup on a rigged Hilbert space with a special unbounded form-generator, corresponding to the singular boundary-type interaction. The Heisenberg picture of such interaction is described by the quantum stochastic Langevin equation.

In quantum theory of open systems there is a well known Lindblad's form [1] of quantum Markovian master equation, satisfied by the one-parameter semigroup of completely positive (CP) maps over the algebra $\mathcal{L}(\mathcal{H})$ of bounded operators on the system Hilbert space $\mathcal{H}$. This is nonstochastical equation, which can be obtained by averaging stochastic Langevin equation for quantum flow [2] over the driving noises, represented in a Fock space $\mathcal{F}$. On the other hand the quantum EH-flow corresponds to the interaction representation for a one parametric group of dynamical authomorphisms over $\mathcal{L}(\mathcal{H} \otimes \mathcal{F})$, which are obviously completely positive due to *-multiplicativity of these representations. The authomorphisms (representations) give the examples of pure, i.e. extreme point CP maps, but among the extreme points of the convex cone of all CP maps over $\mathcal{L}(\mathcal{H})$ there are not only the representations. This means a possibility to construct the stochastic representations of dynamical CP semigroups as averagings of pure, i.e. non-mixing irreversible quantum stochastic CP dynamics, which can not be driven by a Langevin equation. Such irreversible dynamics, corresponding to the interaction representation for a dynamical CP semigroup over $\mathcal{L}(\mathcal{H} \otimes \mathcal{F})$, are described by quantum stochastic flows of CP maps, which should satisfy a generalized form of Lindblad equation with quantum stochastic unbounded generators.

The examples of such dynamics having recently been found many physical applications, will be considered in the first section. The rest of the paper will be devoted to the mathematical derivation of the general structure for the unbounded generators of the dynamical CP semigroups, corresponding to the quantum stochastic CP flows over $\mathcal{L}(\mathcal{H})$ with the noises represented in $\mathcal{F}$. The results of the paper not only generalize the Evans-Hudson (EH) flows [2] from the representations to the general CP maps, but also prove the existence of the homomorphic dilations for the subfiltering and contractive CP flows. This gives the subfiltering CP flows as conditional expectations of EH flows, generalizing the similar representation for contractive CP semigroups. Here in the introduction we would like to outline the generalized structure of the generators on the formal level.

As was proved in [3], every stationary quantum stochastic processes $t \in \mathbb{R}_+ \mapsto \Lambda(t, a)$ parametrized by $a \in \mathfrak{a}$ with $\Lambda(0, a) = 0$ and independent increments $d\Lambda(t, a) = \Lambda(t + dt, a) - \Lambda(t, a)$, forming an Itô $\star$-algebra

$$(1) \quad d\Lambda(a)^* \, d\Lambda(a) = d\Lambda(a^*a), \quad \sum \lambda_i d\Lambda(a_i) = d\Lambda\left(\sum \lambda_i a_i\right),$$
$$d\Lambda(a)^* = d\Lambda(a^*),$$

can be represented in the Fock space $\mathfrak{F}$ over the space of $\mathcal{E}$-valued square-integrable functions on $\mathbb{R}_+$ as $\Lambda(t, a) = a_\nu^\mu \Lambda_\mu^\nu(t)$ with respect to the vacuum state $\delta_\emptyset \in \mathfrak{F}$. Here

$$(2) \qquad a_\nu^\mu \Lambda_\mu^\nu(t) = a_\bullet^\bullet \Lambda_\bullet^\bullet(t) + a_+^\bullet \Lambda_\bullet^+(t) + a_\bullet^- \Lambda_-^\bullet(t) + a_+^- \Lambda_-^+(t),$$

is the canonical decomposition of Λ into the exchange $\Lambda_\bullet^\bullet$, creation $\Lambda_\bullet^+$, annihilation $\Lambda_-^\bullet$ and preservation (time) $\Lambda_-^+ = tI$ processes of quantum stochastic calculus [4], [5] having the mean values $\langle \Lambda_\mu^\nu(t) \rangle = t \delta_+^\nu \delta_\mu^-$ with respect to the vacuum state in $\mathfrak{F}$, and $\mathcal{E}$ is a pre-Hilbert space of the quantum noise in $\mathfrak{F}$. Thus the parametrizing algebra $\mathfrak{a}$ can be always identified with a $\star$-subalgebra of the algebra $\mathcal{Q}(\mathcal{E})$ of all quadruples $a = (a_\nu^\mu)_{\nu=+,\bullet}^{\mu=-,\bullet}$, where $a_\nu^\mu : \mathcal{E}_\nu \to \mathcal{E}_\mu$ are the linear operators on $\mathcal{E}_\bullet = \mathcal{E}, \mathcal{E}_+ = \mathbb{C} = \mathcal{E}_-$, having the adjoints $a_\nu^{\mu*} \mathcal{E}_\mu \subseteq \mathcal{E}_\nu$, with the Hudson–Parthasarathy (HP) multiplication table [6]

$$(3) \qquad a \bullet b = (a_\bullet^\mu b_\nu^\bullet)_{\nu=+,\bullet}^{\mu=-,\bullet},$$

and the involution $a_{-\nu}^{*\mu} = a_{-\mu}^{\nu*}$, where $-(-) = +, -\bullet = \bullet, -(+) = -$.

The stochastic differential of a CP flow $\phi = (\phi_t)_{t>0}$ over an operator algebra $\mathcal{B} \subseteq \mathcal{L}(\mathcal{H})$ is written in terms of the quantum canonical differentials as $d\phi = \phi \circ \lambda_\nu^\mu d\Lambda_\mu^\nu$ with $\phi_0 = \imath$ at $t = 0$, where $\imath(B) = B$ is the identical representation of $\mathcal{B}$. The main result of this paper is the derivation and the dilation of the linear quantum stochastic evolution equation

$$d\phi_t(B) + \phi_t(K^*B + BK - L^*\jmath(B)L) \, dt = \phi_t(L^\bullet\jmath(B)L_\bullet - B \otimes \delta_\bullet^\bullet) \, d\Lambda_\bullet^\bullet$$

$$(4) \qquad + \phi_t(L^\bullet\jmath(B)L - K^\bullet B) \, d\Lambda_\bullet^+ + \phi_t(L^*\jmath(B)L_\bullet - BK_\bullet) \, d\Lambda_-^\bullet,$$

where $\jmath$ is an operator representation of $\mathcal{B}$, $\delta_\bullet^\bullet$ is the identity operator in $\mathcal{E}$, and the operator K satisfies the dissipativity condition $K + K^\dagger \geq L^*L$ with the Hamiltonian part $H = \mathrm{Im} K$. This differential form for the CP flows was discovered in [7] as the general completely dissipative (CD) structure of the bounded quantum stochastic generators $\lambda_\nu^\mu : \mathcal{B} \to \mathcal{B}$ over a von Neumann algebra $\mathcal{B}$ even in the nonlinear case. In the matrix form $\boldsymbol{\lambda} = (\lambda_\nu^\mu)_{\nu=+,\bullet}^{\mu=-,\bullet}$ this can be written similar to the Lindblad form for the nonstochastic generator $\lambda = \lambda_+^-$ as

(5) $$\boldsymbol{\lambda}(B) = \boldsymbol{L}^* \jmath(B) \boldsymbol{L} - \boldsymbol{K}^* B - B\boldsymbol{K}.$$

The dilation of the stochastic differentials for CP processes over arbitrary $*$-algebras $\mathfrak{a}$ and $\mathcal{B}$, giving this structure for the bounded generators over a von Neumann algebra $\mathcal{B}$ as a consequence of the Christensen-Evans theorem [8], was obtained in [7, 9].

Here we shall prove that the quantum stochastic extension (5) of the Lindblad's structure $\lambda(B) = L^* \jmath(B) L - K^* B - BK$, can always be used for the construction and the dilation of the CP flows also in the case of the unbounded maps λ_ν^μ over the algebra $\mathcal{B} = \mathcal{L}(\mathcal{H})$. The existence of minimal CP solution which has been recently constructed under certain continuity conditions in [10] proves that this structure is also sufficient for the CP property of any solution to this stochastic equation. The construction of the differential dilations and the CP solutions of such quantum stochastic differential equations with the bounded generators over the simple finite-dimensional Itô algebra $\mathfrak{a} = \mathcal{Q}(\mathcal{E})$ and the arbitrary $\mathcal{B} \subseteq \mathcal{L}(\mathcal{H})$ was also recently discussed in [11, 12].

The nonstochastic case $\Lambda(t, a) = \alpha t \mathrm{I}$ is described by the simplest, one-dimensional Itô algebra $\mathfrak{a} = \mathbb{C}d$ with $l(a) = \alpha \in \mathbb{C}$ and the nilpotent multiplication $\alpha^\star \alpha = 0$ corresponding to the non-stochastic (Newton) calculus $(\mathrm{d}t)^2 = 0$ in $\mathcal{E} = 0$. The standard Wiener process $Q = \Lambda_\bullet^- + \Lambda_\bullet^+$ in Fock space is described by the second order nilpotent algebra $\mathfrak{a}$ of pairs $a = (\alpha, \xi)$ with $d = (1, 0)$, $\xi \in \mathbb{C}$, represented by the quadruples $a_+^- = \alpha$, $a_\bullet^- = \xi = a_+^\bullet$, $a_\bullet^\bullet = 0$ in $\mathcal{E} = \mathbb{C}$, corresponding to $\Lambda(t, a) = \alpha t \mathrm{I} + \xi Q(\mathrm{t})$. The unital $\star$-algebra $\mathbb{C}$ with the usual multiplication $\zeta^\star \zeta = |\zeta|^2$ can be embedded into the two-dimensional Itô algebra $\mathfrak{a}$ of $a = (\alpha, \zeta)$, $\alpha = l(a)$, $\zeta \in \mathbb{C}$ as $a_\bullet^\bullet = \zeta$, $a_+^\bullet = +i\zeta$, $a_\bullet^- = -i\zeta$, $a_+^- = \zeta$. It corresponds to $\Lambda(t, a) = \alpha t \mathrm{I} + \zeta \mathrm{P}(\mathrm{t})$, where $\mathrm{P} = \Lambda_\bullet^\bullet + i(\Lambda_\bullet^+ - \Lambda_\bullet^-)$ is the representation of the standard Poisson process, compensated by its mean value t. These two commutative cases exhaust the possible types of two-dimensional Itô algebras. Thus, our results [9, 10] are also applicable to the classical stochastic differentials of completely positive processes, corresponding to the commutative Itô algebras, which are always decomposable into the Wiener, Poisson and Newton orthogonal components.

2. Quantum Sub-Filtering Dynamics

The quantum filtering theory, which was outlined in [13, 14] and developed then since [15], provides the derivations for new types of irreversible stochastic equations for quantum states, giving the dynamical solution for the well-known quantum measurement problem. Some particular types of such equations have been considered also in the phenomenological theories of quantum permanent reduction [16, 17], continuous measurement collapse [18, 19], spontaneous jumps [26, 20], diffusions and localizations [21, 22]. The main feature of such dynamics is that the reduced irreversible evolution can be described in terms of a linear dissipative stochastic wave equation, the solution to which is normalized only in the mean square sense.

The simplest dynamics of this kind is described by the continuous filtering wave propagators $V_t(\omega)$, defined on the space Ω of all Brownian trajectories as an adapted operator-valued stochastic process in the system Hilbert space $\mathcal{H}$, satisfying the stochastic diffusion equation

$$(6) \qquad dV_t + KV_t dt = LV_t dQ, \quad V_0 = I$$

in the Itô sense. Here $Q(t, \omega)$ is the standard Wiener process, which is described by the independent increments $dQ(t) = Q(t + dt) - Q(t)$, having the zero mean values $\langle dQ \rangle = 0$ and the multiplication property $(dQ)^2 = dt$, K is an accretive operator, $K + K^\dagger \geq L^*L$, defined on a dense domain $\mathcal{D} \subseteq \mathcal{H}$, with $K^\dagger = K^*|\mathcal{D}$, and L is a linear operator $\mathcal{D} \to \mathcal{H}$. This stochastic wave equation with $K + K^\dagger = L^*L$ was first derived [24] from a unitary cocycle evolution by a quantum filtering procedure. A sufficient analyticity condition, under which it has the unique solution in the form of stochastic multiple integral even in the case of unbounded K and L is given in [10]. Using the Itô formula

$$(7) \qquad d(V_t^* V_t) = dV_t^* V_t + V_t^* dV_t + dV_t^* dV_t,$$

and averaging $\langle \cdot \rangle$ over the trajectories of Q, one obtains $\langle V_t^* V_t \rangle \leq I$ as a consequence of $d\langle V_t^* V_t \rangle \leq 0$. Note that the process V_t is not necessarily unitary if the filtering condition $K^\dagger + K = L^*L$ holds, and even if $L^\dagger = -L$, it might be only isometric, $V_t^* V_t = I$, in the unbounded case.

Another type of the filtering wave propagator $V_t(\omega) : \psi_0 \in \mathcal{H} \mapsto \psi_t(\omega)$ in $\mathcal{H}$ is given by the stochastic jump equation

$$(8) \qquad dV_t + KV_t dt = LV_t dP, \quad V_0 = I.$$

at the random time instants $\omega = \{t_1, t_2, ...\}$. Here $L = J - I$ is the jump operator, corresponding to the stationary discontinuous evolutions $\psi_{t+} = J\psi$ at $t \in \omega$, and $P(t, \omega) = N(t, \omega) - t$ is the standard Poisson process, counting the number $N(t, \omega) = |\omega \cap [0, t)|$ compensated by its mean value t. It is described as the process with independent increments $dP(t) = P(t + dt) - P(t)$, having the values $\{0, 1\}$ at $dt \to 0$, with zero mean $\langle dP \rangle = 0$, and the multiplication property $(dP)^2 = dP + dt$. This stochastic wave equation was first derived in [23] under the filtering condition $L^*L = K + K^\dagger$ by the conditioning with respect to the spontaneous reductions $J : \psi_t \mapsto \psi_{t+}$. An analyticity condition under which it has the unique solution in the form of the multiple stochastic integral even in the case of unbounded K and L is also given in [10]. Using the Itô formula (7) with $dV_t^*dV_t = V_t^*L^*LV_t(dP+dt)$, one can obtain

$$d\left(V_t^*V_t\right) = V_t^*\left(L^*L - K - K^\dagger\right)V_tdt + V_t^*\left(L^\dagger + L + L^*L\right)V_tdP.$$

Averaging $\langle\cdot\rangle$ over the trajectories of P, one can easily find that $d\langle V_t^*V_t\rangle \leq 0$ under the sub-filtering condition $L^*L \leq K+K^\dagger$. Such evolution is not needed to be unitary, but in the filtering case it might be isometric, $V_t^*V_t = I$ if the jumps are isometric, $J^*J = I$.

This proves in both cases that the stochastic wave function $\psi_t\left(\omega\right) = V_t\left(\omega\right)\psi_0$ is not normalized for each ω, but it is normalized in the mean square sense to the survival probability $\langle\|\psi_t\|^2\rangle \leq \|\psi_0\|^2 = 1$, a decreasing probability for a quantum unstable system not to be demolished during its observation up to the time t. In the stable case $\langle\|\psi_t\|^2\rangle = 1$ the positive stochastic function $p_t\left(\omega\right) = \|\psi_t\left(\omega\right)\|^2$ is the probability density of a diffusive $\widehat{Q}$ or counting $\widehat{P}$ output process up to the given t with respect to the standard Wiener Q or Poisson P input processes correspondingly, in the general case this is given by the conditional probability density $\|\psi_t\left(\omega\right)\|^2/\langle\|\psi_t\|^2\rangle$.

Using the Itô formula for $\rho_t\left(\omega\right) = V_t\left(\omega\right)\rho_0V_t\left(\omega\right)^*$, one can obtain the stochastic equations

(9) $$d\rho_t + \left(K\rho_t + \rho_tK^* - L\rho_tL^*\right)dt = \left(L\rho_t + \rho_tL^*\right)dQ,$$

(10) $$d\rho_t + \left(K\rho_t + \rho_tK^* - L\rho_tL^*\right)dt = \left(J\rho_tJ^* - \rho_t\right)dP,$$

describing the stochastic evolution $\Phi_t : \rho_0 \mapsto \rho_t$ of an initially normalized density operator $\rho_0 \geq 0$, $\mathrm{tr}\rho_0 = 1$ as the stochastic density operator $\rho_t\left(\omega\right) = \Phi_t\left(\omega,\rho_0\right)$, normalized to the probability density $p_t\left(\omega\right) = \mathrm{tr}\rho_t\left(\omega\right)$. The stochastic dynamical maps $\Phi_t\left(\rho\right) = V_t\rho V_t^*$ are obviously positive but in general irreversible if $V_t\left(\omega\right)$ are not unitary, although they preserve the pure states in this particular case.

Although the filtering equations (8), (6) look very different, they can be unified in the form of quantum stochastic equation

(11) $$dV_t + KV_tdt + K^-V_td\Lambda_- = \left(J - I\right)V_td\Lambda + L_+V_td\Lambda^+$$

where $\Lambda^+\left(t\right)$ is the creation process, corresponding to the annihilation $\Lambda_-\left(t\right)$ on the interval $[0,t)$, and $\Lambda\left(t\right)$ is the number of quanta on this interval. Indeed, the standard Poisson process P as well as the Wiener process Q can be represented in $\mathfrak{F}$ by the linear combinations [6]

(12) $$P\left(t\right) = \Lambda\left(t\right) + i\left(\Lambda^+\left(t\right) - \Lambda_-\left(t\right)\right), \quad Q\left(t\right) = \Lambda^+\left(t\right) + \Lambda_-\left(t\right),$$

so the equation (11) corresponds to the stochastic diffusion equation (6) if $J = I$, $L_+ = L = -K^-$, and it corresponds to the stochastic jump equation (8) if $J = I + L$, $L_+ = iL = K^-$. These canonical quantum stochastic processes, representing the quantum noise with respect to the vacuum state $|0\rangle$ of the Fock space $\mathcal{F}$ over the single-quantum Hilbert space $L^2(\mathbb{R}_+)$ of square-integrable functions of $t \in [0, \infty)$, are formally given in [25] by the integrals

$$\Lambda_-(t) = \int_0^t \Lambda_-^r \, dr, \quad \Lambda^+(t) = \int_0^t \Lambda_r^+ \, dr, \quad \Lambda(t) = \int_0^t \Lambda_r^+ \Lambda_-^r \, dr,$$

where Λ_-^r, Λ_r^+ are the generalized quantum one-dimensional fields in $\mathcal{F}$, satisfying the canonical commutation relations

$$[\Lambda_-^r, \Lambda_s^+] = \delta(s-r) I, \quad [\Lambda_-^r, \Lambda_-^s] = 0 = [\Lambda_r^+, \Lambda_s^+].$$

They can be defined by the independent increments with

(13) $\qquad \langle 0|d\Lambda_-|0\rangle = 0, \quad \langle 0|d\Lambda^+|0\rangle = 0, \quad \langle 0|d\Lambda|0\rangle = 0$

and the noncommutative multiplication table

(14) $\quad d\Lambda d\Lambda = d\Lambda, \quad d\Lambda_- d\Lambda = d\Lambda_-, \quad d\Lambda d\Lambda^+ = d\Lambda^+, \quad d\Lambda_- d\Lambda^+ = dtI$

with all other products being zero: $d\Lambda d\Lambda_- = d\Lambda^+ d\Lambda = d\Lambda^+ d\Lambda_- = 0$.

The corresponding quantum stochastic equation for the density operator $\rho_t = V_t \rho_0 V_t^*$ has the following form

$$d\rho_t + (K\rho_t + \rho_t K^* - L\rho_t L^*) \, dt = (J\rho_t J^* - \rho_t) \, d\Lambda$$

(15) $$+ (J\rho_t L^- - K^- \rho_t) \, d\Lambda_- + (L_+ \rho_t J^* - \rho_t K_+) \, d\Lambda^+,$$

where $L^- = L_+^*, K_+^* = K^-$. The equation (15), coinciding with either (9) or with (10) in the particular cases, is obtained from (11) by using the Itô formula (7) with the multiplication table (14). In the another particular case

$$J = S, \quad K^- = L^- S, \quad L_+ = SK_+, \quad S^*S = I,$$

it corresponds to the Hudson–Evans quantum stochastic flow [2] if $S^* = S^{-1}$. Such evolution is isometric, and identity preserving, $V_t V_t^* = I$, i.e. unitary at least in the case of the bounded K and L.

In the Heisenberg picture the stochastic dynamics is described by the dual transformations $\phi_t(\omega) = \Phi_t'(\omega)$, such that for any density operator ρ_0 and for any bounded observable B on $\mathcal{H}$

$$\operatorname{tr}[\Phi_t'(\omega, B) \rho_0] = \operatorname{tr}[B\Phi_t(\omega, \rho_0)].$$

The linear stochastic maps $B \mapsto Y_t = \phi_t(B)$ are obviously Hermitian in the sense that $Y_t^* = Y_t$ if $B^* = B$ and completely positive, but in contrast to the usual Hamiltonian dynamics, they are multiplicative, $\phi_t(B^*C) = \phi_t(B)^* \phi_t(C)$ only in the case, corresponding to the HE flow, even if they are not averaged with respect to ω. Moreover, they are usually not normalized,

$R_t(\omega) := \phi_t(\omega, I) \neq I$, although the stochastic positive operators $R_t = V_t^* V_t$ under the filtering condition are usually normalized in the mean, $\langle R_t \rangle = I$, and satisfy the martingale property $\epsilon_t[R_s] = R_t$ for all $s > t$, where ϵ_t is the conditional expectation with respect to the history of the processes P or Q up to time t. The sub-filtering condition $K + K^\dagger \geq L^- L_+$ for the equation (11) defines in both cases the positive operator-valued stochastic process $R_t = \phi_t(I)$ as a sub-martingale with $R_0 = I$, or a martingale in the case $K + K^\dagger = L^- L_+$.

Although the filtering dynamics with unbounded coefficients of the particular types has been studied elsewhere [27] by means of the classical stochastic differential equations, the general structure of such equations has not been discovered, and the general filtering CP flows have not been constructed. In the next sections we define a multidimensional analog of the quantum stochastic equation (15), and will show that the general structure of its generator indeed follows just from the property of complete positivity of the dual stochastic maps $\phi_t = \Phi_t'$ for all $t > 0$ and the normalization condition $\phi_t(I) = R_t$ to a form-valued sub-martingale with respect to the natural filtration of the quantum noise in the Fock space $\mathfrak{F}$.

3. Quantum Completely Positive Flows

Throughout the complex pre-Hilbert space $\mathcal{D} \subseteq \mathcal{H}$ is a Fréchet (i.e. metrizable complete) space with respect to a stronger topology, $\mathcal{E} \otimes \mathcal{D}$ denotes the projective tensor product (π-product) with another such space $\mathcal{E}$, $\mathcal{D}' \supseteq \mathcal{H}$ denotes the dual space of continuous antilinear functionals $\eta' : \eta \in \mathcal{D} \mapsto \langle \eta | \eta' \rangle$, with respect to the canonical pairing $\langle \eta | \eta' \rangle$ given by $\|\eta\|^2$ for $\eta' = \eta \in \mathcal{H}$, $\mathcal{B}(\mathcal{D})$ denotes the linear space of all continuous sesquilinear forms $\langle \eta | B\eta \rangle$ on $\mathcal{D}$, identified with the continuous linear operators $B : \mathcal{D} \to \mathcal{D}'$ (kernels), $B^\dagger \in \mathcal{B}(\mathcal{D})$ is the Hermit conjugated form (kernel) $\langle \eta | B^\dagger \eta \rangle = \langle \eta | B\eta \rangle^*$, and $\mathcal{L}(\mathcal{D}) \subseteq \mathcal{B}(\mathcal{D})$ denotes the algebra of all strongly continuous operators $B : \mathcal{D} \to \mathcal{D}$. Any such space $\mathcal{D}$ can be considered as a projective limit with respect to an increasing sequence of norms $\|\cdot\|_p > \|\cdot\|$ on $\mathcal{D}$; for the definitions and properties of this standard topological notions see for example [28]. The spaces $\mathcal{D}'$ and $\mathcal{B}(\mathcal{D})$ will be equipped with w*- topologies induced by their preduals $\mathcal{D}$ and $\mathcal{D} \otimes \mathcal{D}$, and coinciding with the weak topology on each bounded subset with respect to a norm $\|\cdot\|_p$. Any operator $A \in \mathcal{L}(\mathcal{D})$ with $A^\dagger \in \mathcal{L}(\mathcal{D})$ can be uniquely extended to a weakly continuous operator onto $\mathcal{D}'$ as $A^{\dagger *}$, denoted again as A, where A^* is the dual operator $\mathcal{D}' \to \mathcal{D}'$, $\langle \eta | A^* \eta' \rangle = \langle A\eta | \eta' \rangle$, defining the involution $A \mapsto A^*$ for the continuations $A : \mathcal{D}' \to \mathcal{D}'$. We say that the operator A commutes with a sesquilinear form, $BA = AB$ if $\langle \eta | BA\eta \rangle = \langle A^\dagger \eta | B\eta \rangle$ for all $\eta \in \mathcal{D}$. The commutant $\mathcal{A}^c = \{B \in \mathcal{B}(\mathcal{D}) : [A, B] = 0, \forall A \in \mathcal{A}\}$ of an operator $*$-algebra $\mathcal{A} \subseteq \mathcal{L}(\mathcal{D})$ is weakly closed in $\mathcal{B}(\mathcal{D})$, so that the weak closure $\overline{\mathcal{B}} \subseteq \mathcal{B}(\mathcal{D})$ of any $\mathcal{B} \subseteq \mathcal{A}^c$ also commutes with $\mathcal{A}$.

Let us denote $\mathcal{B} = \mathcal{L}(\mathcal{H})$ the algebra of all bounded operators $B : \mathcal{H} \to \mathcal{H}$, $\|B\| < \infty$, $\overline{\mathcal{B}} = \mathcal{B}(\mathcal{D})$ means the weak closure of $\mathcal{B} \subseteq \mathcal{B}(\mathcal{D})$, and let $(\Omega, \mathfrak{A}, P)$ be a probability space with a filtration $(\mathfrak{A}_t)_{t>0}$, $\mathfrak{A}_t \subseteq \mathfrak{A}$ of σ-algebras on Ω. One can assume that the filtration $\mathfrak{A}_t \subseteq \mathfrak{A}_s, \forall t < s$ is generated by the pieces $x_{t]} = \{r \mapsto x(r) : r \leq t\}$ of a stochastic process $x(t, \omega)$ with independent increments $dx(t) = x(t + \Delta) - x(t)$, and the probability measure P is invariant under the measurable representations $\omega \mapsto \omega_s \in \Omega$, $A_s^{-1} = \{\omega : \omega_s \in A\} \in \mathfrak{A}$, $\forall A \in \mathfrak{A}$ of the time shifts $t \mapsto t + s, s > 0$ on $\Omega \ni \omega$, corresponding to the shifts of the random increments

$$dx(t, \omega_s) = dx(t + s, \omega), \quad \forall \omega \in \Omega, t \in \mathbb{R}_+.$$

The *stochastic dynamics* over $\mathcal{B}$ with respect to the process $x(t)$ is described by a cocycle flow $\phi = (\phi_t)_{t>0}$ of linear completely positive [29] w*-continuous stochastic adapted maps $\phi_t(\omega) : \mathcal{B} \to \overline{\mathcal{B}}$, $\omega \in \Omega$ such that the stochastic process $y_t(\omega) = \langle \eta | \phi_t(\omega, B) \eta \rangle$ is causally measurable for each $\eta \in \mathcal{D}$, $B \in \mathcal{B}$ in the sense that $y_t^{-1}(B) \in \mathfrak{A}_t, \forall t > 0$ and any Borel $B \subseteq \mathbb{C}$. The maps ϕ_t can be extended on the $\mathfrak{A}$-measurable functions $Y : \omega \mapsto Y(\omega)$ with values $Y(\omega) \in \overline{\mathcal{B}}$ as the normal maps $\phi_t[Y](\omega) = \overline{\phi}_t(\omega, Y(\omega_t))$ for almost all $\omega \in \Omega$, where the linear maps $\overline{\phi}_t : \overline{\mathcal{B}} \to \overline{\mathcal{B}}$ are defined by the normal extensions of ϕ_t from the positive cone $\mathcal{B}_+$ onto $\overline{\mathcal{B}}_+$, so that the cocycle condition $\phi_r(\omega) \circ \phi_s(\omega_r) = \phi_{r+s}(\omega), \forall r, s > 0$ reads as the semigroup condition $\phi_r[\phi_s[Y]] = \phi_{r+s}[Y]$ of the extended maps. As it was noted in the previous section, the maps $\phi_t(\omega)$ are not considered to be normalized to the identity, and can be even unbounded, but in the case of filtering dynamics they are supposed to be normalized, $\phi_t(\omega, I) = R_t(\omega)$, to an operator-valued martingale $R_t = \epsilon_t[R_s] \geq 0$ with $R_0(\omega) = I$, or to a positive submartingale, $R_t \geq \epsilon_t[R_s], \forall s > t$ in the subfiltering case, where ϵ_t is the conditional expectation over ω with respect to $\mathfrak{A}_t$.

Now we give an algebraic generalization and a Fock space representation of the filtering (or subfiltering) CP flows for a commutative Itô algebra $\mathfrak{a}$, which was suggested in [10] even in the noncommutative case.

The role of the classical process $x(t)$ will play the quantum stochastic process

$$X(t) = A \otimes I + I \otimes \Lambda(t, a), \quad A \in \mathcal{A}, a \in \mathfrak{a}$$

parametrized by an Abelian *-subalgebra $\mathcal{A} \subset \mathcal{L}(\mathcal{D})$ and a commutative Itô algebra $\mathfrak{a}$. Here $\Lambda(t, a)$ is the process with independent increment on a dense subspace $\mathfrak{F} \subset \Gamma(\mathfrak{C})$ of the Fock space $\Gamma(\mathfrak{C})$ over the space $\mathfrak{C} = L_{\mathcal{E}}^2(\mathbb{R}_+)$ of all square-norm integrable $\mathcal{E}$-valued functions on $\mathbb{R}_+$, where $\mathcal{E}$ is a pre-Hilbert space of the representation $a \in \mathfrak{a} \mapsto (a_\nu^\mu)_{\nu=+,\bullet}^{\mu=-,\bullet}$ for the Itô *-algebra $\mathfrak{a}$. Every commutative Itô *-algebra is the sum $\mathfrak{a} = \mathbb{C}d + \mathfrak{a}_0 + \mathfrak{a}_1$ of the Wiener and Poisson algebras $\mathbb{C}d + \mathfrak{a}_0$, $\mathbb{C}d + \mathfrak{a}_1$, such that each $a \in \mathfrak{a}$ has the unique decomposition $a = l(a)d + b + c$, where d is the death of $\mathfrak{a}$, $b \in \mathfrak{a}_0$ is defined by the conditions

$$b_+^- = l\,(b) = 0, \qquad b_\bullet^\bullet = j\,(b) = 0,$$

and $c \in \mathfrak{a}_1$ is orthogonal to b: $bc = 0 = cb$, defined by the condition $c_\bullet^\bullet = 0 \Rightarrow c_+^\bullet = 0 = c_\bullet^-$. Thus the space $\mathcal{E}$ is decomposed into the orthogonal sum $\mathcal{E}_0 \oplus \mathcal{E}_1 \oplus \mathcal{E}_\perp$ with $\mathcal{E}_0$ generated by $k\,(\mathfrak{a}_0)$, $\mathcal{E}_1$ generated by $k\,(\mathfrak{a}_1)$, and $\mathcal{E}_\perp$ is the orthogonal complement, which is zero if $\mathcal{E}$ is the minimal space of the representation of $\mathfrak{a}$.

Assuming that $\mathcal{E}$ is a Fréchet space, given by an increasing sequence of Hilbertian norms $\|e^\bullet\|\,(\xi) > \|e^\bullet\|$, $\xi \in \mathbb{N}$, we define $\mathfrak{F}$ as the projective limit $\cap_\xi \Gamma\,(\mathfrak{E}, \xi)$ of the Fock spaces $\Gamma\,(\mathfrak{E}, \xi) \subseteq \Gamma\,(\mathfrak{E})$, generated by coherent vectors $f^\otimes$, with respect to the norms

$$(16) \quad \left\| f^\otimes \right\|^2 (\xi) = \int_\Gamma \left\| f^\otimes\,(\tau) \right\|^2 (\xi)\,\mathrm{d}\tau$$

$$:= \sum_{n=0}^\infty \frac{1}{n!} \left(\int_0^\infty \|f^\bullet\,(t)\|^2\,(\xi)\,\mathrm{d}t \right)^n = e^{\|f^\bullet\|^2(\xi)}.$$

Here $f^\otimes\,(\tau) = \bigotimes_{t \in \tau} f^\bullet\,(t)$ for each $f^\bullet \in \mathfrak{E}$ is represented by tensor-functions on the space Γ of all finite subsets $\tau = \{t_1, ..., t_n\} \subseteq \mathbb{R}_+$. Moreover, we shall assume that the Itô algebra $\mathfrak{a}$ is realized as a $\star$-subalgebra of Hudson-Parthasarathy (HP) algebra $\mathcal{Q}\,(\mathcal{E})$ of all quadruples $a = (a_\nu^\mu)_{\nu=+,\bullet}^{\mu=-,\bullet}$ with $a_\bullet^\bullet \in \mathcal{L}\,(\mathcal{E})$, strongly representing the $\star$-semigroup $1 + \mathfrak{a}$ on the Fréchet space $\mathcal{E}$ by projective contractions $\delta_\bullet^\bullet + a_\bullet^\bullet \in \mathcal{L}\,(\mathcal{E})$ in the sense that for each $\zeta \in \mathbb{N}$ there exists ξ such that $\|e^\bullet + a_\bullet^\bullet e^\bullet\|\,(\zeta) \le \|e^\bullet\|\,(\xi)$ for all $e^\bullet \in \mathcal{E}$. The following theorem proves that these are natural assumptions (which are not restrictive in the simple Fock scale for a finite dimensional $\mathfrak{a}$.)

Proposition 1 *The exponential operators* $W\,(t, a) =: \exp\,[\Lambda\,(t, a)]$: *defined as the solutions to the quantum Itô equation*

$$(17) \qquad \mathrm{dW}_t\,(g) = \mathrm{W}_t\,(g)\,\mathrm{d}\Lambda\,(t, g\,(t)), \quad \mathrm{W}_0\,(g) = \mathrm{I}, g\,(t) \in \mathfrak{a}$$

with $g\,(t) = a$, *are strongly continuous,* $W\,(t, a) \in \mathcal{L}\,(\mathfrak{F})$, *iff all* $\widehat{a}_\bullet^\bullet = \delta_\bullet^\bullet + a_\bullet^\bullet$ *are projective contructions on* $\mathcal{E}$. *They give an analytic representation*

$$(18) \quad W\,(t, a \star a) = W\,(t, a)^* W\,(t, a), \quad W\,(t, 0) = I, \quad W\,(t, d) = e^t I$$

of the unital $\star$-*semigroup* $1 + \mathfrak{a}$ *for the Itô* $\star$-*algebra* $\mathfrak{a}$ *with respect to the* $\star$-*product* $a \star a = a + a^\star a + a^\star$.

Proof. – The solutions $W\,(t, a)$ are uniquely defined on the coherent vectors as analytic functions

$$(19) \quad W\,(t, a)\,f^\otimes\,(\tau)$$

$$= \otimes_{r \in \tau}^{r \le t} \left(\widehat{a}_\bullet^\bullet f^\bullet\,(r) + a_+^\bullet \right) \exp \left[\int_0^t \left(a_\bullet^- f^\bullet\,(r) + a_+^- \right)\mathrm{d}r \right] \otimes_{r \in \tau}^{r \ge t} f^\bullet\,(r),$$

which obey the properties (18), see for example [3]. Thus the span of coherent vectors is invariant, and it is also invariant under $W(t,a)^* = W(t,a^*)$. They can be extended on $\mathfrak{F}$ by continuity which follows from the continuity of Wick exponentials $\otimes \hat{a}^\bullet_\bullet$ for the projective contractions $\hat{a}^\bullet_\bullet \in \mathcal{E}$, and boundedness of $a^\bullet_+ \in \mathcal{E}$, $a^-_\bullet \in \mathcal{E}'$.

Let $\mathfrak{D}$ denote the Fréchet space $\mathcal{D} \otimes \mathfrak{F}$, generated by $\psi = \eta \otimes f^\otimes$, $\eta \in \mathcal{D}$, $f^\bullet \in \mathfrak{E}$. Assuming for simplicity the separability of the Itô algebra in the sense $\mathcal{E} \subseteq \ell^2$ such that $f^\bullet = (f^m)^{m \in \mathbb{N}}$, one can identify each $\psi' \in \mathfrak{D}'$ with a sequence of $\mathcal{D}'$-valued symmetric tensor-functions $\psi'_{m_1,\ldots,m_n}(t_1,\ldots t_n)$, $n = 0,1,2,\ldots$. Let $(\mathfrak{D}_t)_{t>0}$ be the natural filtration and $(\mathfrak{D}_{[t})_{t>0}$ be the backward filtration of the subspaces $\mathfrak{D}_t = \mathcal{D} \otimes \mathfrak{F}_t$, $\mathfrak{D}_{[t} = \mathcal{D} \otimes \mathfrak{F}_{[t}$ generated by $\eta \otimes f^\otimes$ with $f^\bullet \in \mathfrak{E}_t$ and $f^\bullet \in \mathfrak{E}_{[t}$ respectively, where $\mathfrak{E}_t = L^2_\mathcal{E}[0,t)$, $\mathfrak{E}_{[t} = L^2_\mathcal{E}[t,\infty)$ are embedded into $\mathfrak{E}$. The spaces $\mathfrak{D}_t$, $\mathfrak{D}_{[t}$ of the restrictions $E_t \psi = \psi | \Gamma_t$, $E_{[t} \psi = \psi | \Gamma_{[t}$ onto $\Gamma_t = \{\tau_t = \tau \cap [0,t)\}$, $\Gamma_{[t} = \{\eta_t = \tau \cap [t,\infty)\}$ are embedded into $\mathfrak{D}$ by the isometries $E^\dagger_t : \psi \mapsto \psi_t$, $E^\dagger_{[t} : \psi \mapsto \psi_{[t}$ as $\psi_t(\tau) = \psi(\tau_t)\delta_\emptyset(\eta_t)$, $\psi_{[t}(\tau) = \delta_\emptyset(\tau_t)\psi(\eta_t)$, where $\delta_\emptyset(\tau) = 1$ if $\tau = \emptyset$, otherwise $\delta_\emptyset(\tau) = 0$. The projectors $E_t, E_{[t}$ onto $\mathfrak{D}_t, \mathfrak{D}^t$ are extended onto $\mathfrak{D}'$ as the adjoints to $E^\dagger_t, E^\dagger_{[t}$. The time shift on $\mathfrak{D}'$ is defined by the semigroup $(T^t)_{t>0}$ of adjoint operators $T^t = T^*_t$ to $T_t \psi(\tau) = \psi(\tau+t)$, where $\tau+t = \{t_1+t,\ldots,t_n+t\}$, $\emptyset+t = \emptyset$, such that $T^t \psi(\tau) = \delta_\emptyset(\tau_t)\psi(\eta_t - t)$ are isometries for $\psi \in \mathfrak{D}$ onto $\mathfrak{D}_{[t}$. A family $(Z_t)_{t>0}$ of sesquilinear forms $\langle \psi | Z_t \psi \rangle$ given by linear operators $Z_t : \mathfrak{D} \to \mathfrak{D}'$ is called $adapted$ (and $(Z^t)_{t>0}$ is called $backward\ adapted$) if

(20)
$$Z_t(\eta \otimes f^\otimes) = \psi' \otimes E_{[t}f^\otimes \quad (Z^t(\eta \otimes f^\otimes) = \psi' \otimes E_t f^\otimes), \quad \forall \eta \in \mathcal{D}, f^\bullet \in \mathfrak{E},$$

where $\psi' \in \mathfrak{D}'_t$ $(\mathfrak{D}'_{[t})$ and $E_{[t}$ (E_t) are the projectors onto $\mathfrak{F}_{[t}$ $(\mathfrak{F}_t)$ correspondingly.

The (vacuum) $conditional\ expectation$ on $\mathcal{B}(\mathfrak{D})$ with respect to the past up to a time $t \in \mathbb{R}_+$ is defined as a positive projector, $\epsilon_t(Z) \geq 0$, if $Z \geq 0$, $\epsilon_t = \epsilon_t \circ \epsilon_s$, $\forall s > t$, giving an adapted sesquilinear form $Z_t = \epsilon_t(Z)$ in (20) for each $Z \in \mathcal{B}(\mathfrak{D})$ by $\psi' = E_t Z E^\dagger_t \psi$, where $\psi = \eta \otimes E_t f^\otimes$. The time shift $(\theta^t)_{t>0}$ on $\mathcal{B}(\mathfrak{D})$ is uniquely defined by the covariance condition $\theta^t(Z) T^t = T^t Z$ as a backward adapted family $Z^t = \theta^t(Z)$, $t > 0$ for each $Z \in \mathcal{B}(\mathfrak{D})$. As in the bounded case [5] between the maps ϵ_t and θ^t we have the relation $\theta^r \circ \epsilon_s = \epsilon_{r+s} \circ \theta^r$ which follows from the operator relation $T^r E_s = E_{r+s} T^r$. An adapted family $(M_t)_{t>0}$ of positive $\langle \psi | M_t \psi \rangle \geq 0, \forall \psi \in \mathfrak{D}$ Hermitian $M^\dagger_t = M_t$ forms $M_t \in \mathcal{B}(\mathfrak{D})$ is called $martingale\ (submartingale)$ if $\epsilon_t(M_s) = M_t$ $(\epsilon_t(M_s) \leq M_t)$ for all $s \geq t \geq 0$.

Let $\mathfrak{B}$ denote the space of all $Y \in \mathcal{B}(\mathfrak{D})$, commuting with all $X = \{X(t)\}$ in the sense

$$AY = YA, \quad \forall A \in \mathcal{A}, \quad YW(t,a) = W(t,a)Y, \quad \forall t > 0, a \in \mathfrak{a},$$

where $A\left(\eta\otimes\varphi\right)=A\eta\otimes\varphi$, $W\left(\eta\otimes\varphi\right)=\eta\otimes W\varphi$, and the unital $*$-algebra $\mathcal{B}\subseteq\mathcal{L}\left(\mathcal{H}\right)$ be weakly dense in the commutant $\mathcal{A}^c$ (we can take $\mathcal{B}=\mathcal{L}\left(\mathcal{H}\right)$ only if $\mathcal{A}=0$, corresponding to $X\left(0\right)=0$.) The quantum filtration $(\mathfrak{B}_t)_{t>0}$ is defined as the increasing family of subspaces $\mathfrak{B}_t\subseteq\mathfrak{B}_s, t\leq s$ of the adapted sesquilinear forms $Y_t\in\mathfrak{B}$. The covariant shifts $\theta^t: Y\mapsto Y^t$ leave the space $\mathfrak{B}$ invariant, mapping it onto the subspaces of backward adapted sesquilinear forms $Y^t=\theta^t\left(Y\right)$.

The *quantum stochastic positive flow* over $\mathcal{B}$ is described by a one-parameter family $\phi=(\phi_t)_{t>0}$ of linear w*-continuous maps $\phi_t:\mathcal{B}\to\mathfrak{B}$ satisfying

1. the causality condition $\phi_t\left(B\right)\subseteq\mathfrak{B}_t, \quad\forall B\in\mathcal{B}, t\in\mathbb{R}_+$,
2. the complete positivity condition $[\phi_t\left(B_{kl}\right)]\geq 0$ for each $t>0$ and for any positive definite matrix $[B_{kl}]\geq 0$ with $B_{kl}\in\mathcal{B}$,
3. the cocycle condition $\phi_r\circ\phi_s^r=\phi_{r+s}, \forall t,s>0$ with respect to the covariant shift $\phi_s^r=\theta^r\circ\phi_s$.

Here the composition $\circ$ is understood as $\phi_r\left[\phi_s\left(B\right)\right]=\phi_{r+s}\left(B\right)$ in terms of the linear normal extensions of $\phi_t\left[B\otimes Z\right]=\overline{\phi}_t\left(B\right)Z^t$ to the CP maps $\mathfrak{B}\to\mathfrak{B}$, forming a one-parameter semigroup, where $B\in\overline{\mathcal{B}}, \overline{\phi}_t:\overline{\mathcal{B}}\to\mathfrak{B}_t$ are the normal extensions of ϕ_t, $Z^t=\theta^t\left(Z\right)$, $Z\in\mathcal{B}\left(\mathfrak{F}\right)$. These can be defined like in classical case as $\phi_t\left[Y\right]\left(\bar{f}^\bullet,f^\bullet\right)=\overline{\phi}_t\left(\bar{f}^\bullet,Y\left(\bar{f}_t^\bullet,f_t^\bullet\right),f^\bullet\right)$ with $f_t^\bullet\left(r\right)=f^\bullet\left(t+r\right)$ by the coherent matrix elements $Y\left(\bar{f}^\bullet,f^\bullet\right)=F^*YF$ for $Y\in\mathfrak{B}$ given by the continuous operators $F:\eta\mapsto\psi_f=\eta\otimes f^\otimes$, $\eta\in\mathcal{D}$ for each $f^\bullet\in\mathfrak{E}_t$ with the adjoints $F^*\psi'=\int_{\tau<t}f^\otimes\left(\tau\right)^*\psi'\left(\tau\right)\mathrm{d}\tau$ for $\psi'\in\mathcal{D}'$.

The flow is called *(sub)-filtering*, if $R_t=\phi_t\left(I\right)$ is a (sub)-martingale with $R_0=I$, and is called *contractive*, if $I\geq R_t\geq R_s$ for all $0\leq t\leq s\in\mathbb{R}_+$.

Proposition 2 *The complete positivity for adapted linear maps $\phi_t:\mathcal{B}\to\mathcal{B}\left(\mathfrak{D}\right)$ can be written as*

$$(21)\quad\sum_{f,h\in\mathfrak{E}_t}\sum_{B,C\in\mathcal{B}}\langle\xi_B^f|\phi_t\left(\bar{f}^\bullet,B^*C,h^\bullet\right)\xi_C^h\rangle$$

$$:=\langle\eta^k|\phi_t\left(\bar{f}_k^\bullet,B_k^*B_l,h_l^\bullet\right)\eta^l\rangle\geq 0,\quad\forall t>0$$

(the usual summation rule over repeated cross-level indices is understood), where $\xi_B^f=\eta^k$ if $f^\bullet=f_k^\bullet$ and $B=B_k$ with $f_k^\bullet\in\mathfrak{E}_t, B_k\in\mathcal{B}, k=1,2,...$, otherwise $\xi_B^f=0$, and $\phi_t\left(B,f^\bullet\right)=\phi_t\left(B\right)F, \phi_t\left(\bar{f}^\bullet,B\right)=F^*\phi_t\left(B\right)$.

Proof. – By definition the map ϕ into the sesquilinear forms is completely positive on $\mathcal{B}$ if $\langle\psi^k|\phi\left(B_{kl}\right)\psi^l\rangle\geq 0$ whenever $\langle\eta^k|B_{kl}\eta^l\rangle\geq 0$, where η^k,ψ^k are arbitrary finite sequences. Approximating from below the latter positive forms by sums of the forms $\sum_{kl}\langle\eta^k|B_{ik}^*B_{il}\eta^l\rangle\geq 0$, the complete positivity can be tested only for the forms $\sum_{kl}\langle\eta^k|B_k^*B_l\eta^l\rangle\geq 0$ due to the additivity $\phi\left(\sum_i B_{ik}^*B_{il}\right)=\sum_i\phi\left(B_{ik}^*B_{il}\right)$. If ϕ_t is adapted, this can be written as

$$\sum_{B,C\in\mathcal{B}}\langle\chi_B|\phi\left(B^*C\right)\chi_C\rangle=\langle\psi^k|\phi\left(B_k^*B_l\right)\psi^l\rangle:=\sum_{k,l}\langle\psi^k|\phi\left(B_k^*B_l\right)\psi^l\rangle\geq 0,$$

where $\chi_B = \psi^k \in \mathfrak{D}_t$ if $B = B_k \in \mathcal{B}$, otherwise $\chi_B = 0$. Because any $\psi \in \mathfrak{D}_t$ can be approximated by a $\mathcal{D}$-span $\sum_f \eta^f \otimes f^\otimes$ of coherent vectors over $f_k^\bullet \in \mathfrak{C}_t$, it is sufficient to define the CP property only for such spans as

$$0 \le \sum_{f,h} \sum_{B,C} \left\langle \xi_B^f \otimes f^\otimes | \phi \left(B^* C \right) \left(\xi_C^h \otimes h^\otimes \right) \right\rangle$$

$$= \sum_{f,h} \sum_{B,C} \left\langle \xi_B^f | \phi \left(\bar{f}^\bullet, B^* C, h^\bullet \right) \xi_C^h \right\rangle .$$

Note that the subfiltering (filtering) flows can be considered as a quantum stochastic CP dilations of the quantum sub-Markov (Markov) semigroups $\theta = (\theta_t)_{t>0}$, $\theta_r \circ \theta_s = \theta_{r+s}$ in the sense $\theta_t = \epsilon \circ \phi_t$, where $\epsilon(Y) \eta = EY\psi_0$, $E\psi' = \psi'(\emptyset)$, $\forall \psi' \in \mathfrak{D}'$, with $\theta_s(I) \le \theta_t(I) \le I (\theta_t(I) = I)$, $\forall t \le s$. The contraction $C_t = \theta_t(I)$ with $C_0 = I$ defines the probability $\langle \eta | C_t \eta \rangle \le 1$, $\forall \eta \in \mathcal{H}, \|\eta\| = 1$ for an unstable system not to be demolished by a time $t \in \mathbb{R}_+$, and the conditional expectations $\langle \eta | AC_t \eta \rangle / \langle \eta | C_t \eta \rangle$ of the initial nondemolition observables $A \in \mathcal{A}$ in any state $\eta \in \mathcal{D}$, and thus in any initial state $\psi_0 \in \eta \otimes \delta_\emptyset$. The following theorem shows that the submartingale (or the contraction) $R_t = \phi_t(I)$ is the density operator with respect to $\psi_0 = \eta \otimes \delta_\emptyset$, $\eta \in \mathcal{H}$ (or with respect to any $\psi \in \mathcal{H} \otimes \mathfrak{F}$) also for the conditional state of the restricted nondemolition process $X_{t]} = \{r \mapsto X(r) : r \le t\}$.

Theorem 3 *Let $t \mapsto R_t \in \mathfrak{B}_t$ be a positive (sub)-martingale and $(\mathfrak{g}_t)_{t>0}$ be the increasing family of $\star$-semigroups $\mathfrak{g}_t$ of step functions $g : \mathbb{R}_+ \to \mathfrak{a}$, $g(s) = 0$, $\forall s \ge t$ under the $\star$-product*

$$(22) \qquad (g_k \star g_l)(t) = g_l(t) + g_k(t)^* g_l(t) + g_k(t)^*$$

of $g_k^\star = g_k \star 0$ and $g_l = 0 \star g_l$. The generating function $\vartheta_t(g) = \epsilon[R_t W_t(g)]$ of the output state for the process $\Lambda(t)$, defined for any $g \in \mathfrak{g}_t$ and each $t > 0$ as

$$(23) \qquad \langle \eta | \vartheta_t(g) \eta \rangle = \langle \psi_0 | R_t W_t(g) \psi_0 \rangle, \quad \psi_0 = \eta \otimes \delta_\emptyset,$$

is $\mathcal{B}^c$-valued, positive, $\vartheta_t \ge 0$ in the sense of positive definiteness of the kernel

$$(24) \qquad \langle \eta^k | \vartheta_t (g_k \star g_l) \eta^l \rangle \ge 0, \quad \forall g_k \in \mathfrak{g}_t; \eta^k \in \mathcal{D},$$

and $\vartheta_t \ge \vartheta_s | \mathfrak{g}_t$ in this sense for any $s \ge t$. If $R_0 = I$, then $\vartheta_0(0) = I \ge \vartheta_t(0)$, and if R_t is a martingale, then $\vartheta_t = \vartheta_s | \mathfrak{g}_t$ for any $s \ge t$, and $\vartheta_t(0) = I$ for all $t \in \mathbb{R}_+$. Any family $\vartheta = (\vartheta_t)_{t>0}$ of positive-definite functions $\vartheta_t : \mathfrak{g}_t \to \mathcal{B}^c$, satisfying the above consistency and normalization properties, is the state generating function of the form (23) iff it is absolutely continuous in the following sense

$$(25) \qquad \lim_{n \to \infty} \sum_{g \in \mathfrak{g}_t} \eta_n^g \otimes g_+^\otimes = 0 \Rightarrow \lim_{n \to \infty} \sum_{g,h \in \mathfrak{g}_t} \langle \eta_n^g | \vartheta_t (g \star h) \eta_n^h \rangle = 0,$$

where $g_+^\otimes(\tau) = \otimes_{t\in\tau} g_+^\bullet(t)$ and $\eta_n^g = 0$ for almost all g (i.e. except for a finite number of $g \in \mathfrak{g}_t$).

The proof is given in [10] even for the general (noncommutative) algebras $\mathcal{A}$ and $\mathfrak{a}$.

4. Generators of Quantum CP Dynamics

The quantum stochastically differentiable positive flow ϕ is defined as a weakly continuous function $t \mapsto \phi_t$ with CP values $\phi_t : \mathcal{B} \to \mathfrak{B}_t$, $\phi_0(B) = B \otimes I, \forall B \in \mathcal{B}$ such that for any product-vector $\psi_f = \eta \otimes f^\otimes$ given by $\eta \in \mathcal{D}$ and $f^\bullet \in \mathfrak{E}$

(26) $\quad \dfrac{d}{dt} \langle \psi_f | \phi_t(B) \psi_f \rangle = \langle \psi_f | \phi_t \left(\lambda \left(\bar{f}^\bullet(t), B, f^\bullet(t) \right) \right) \psi_f \rangle, \qquad B \in \mathcal{B},$

where $\lambda(\bar{e}^\bullet, B, e^\bullet) = \lambda(B) + e_\bullet \lambda^\bullet(B) + \lambda_\bullet(B) e^\bullet + e_\bullet \lambda_\bullet^\bullet(B) e^\bullet$, $e_\bullet = \bar{e}^\bullet$ is the linear form on $\mathcal{E}$ with $e_\bullet^* = e^\bullet \in \mathcal{E}$ and $\langle \psi_f | \phi_0(B) \psi_f \rangle = \langle \eta | B\eta \rangle \exp \|f^\bullet\|^2$. The generator $\lambda(B) = \lambda(0, B, 0)$ of the quantum dynamical semigroup $\theta_t = \epsilon \circ \phi_t$ is a linear w*-continuous map $B \mapsto \lambda(B) \in \mathcal{A}^c$, $\lambda^\bullet = \lambda_\bullet^\dagger$ is a linear w*-continuous map given by the Hermitian adjoint values $\lambda_\bullet(B^*) = \lambda^\bullet(B)^\dagger$ in the continuous operators $\mathcal{E} \to \mathcal{A}^c$, and $\lambda_\bullet^\bullet : \mathcal{B} \to \mathcal{B}(\mathcal{D} \otimes \mathcal{E})$is a w*-continuous map with the values $\lambda_\bullet^\bullet(B)$ given by continuous operators $\mathcal{E} \otimes \mathcal{E} \to \mathcal{A}^c$. The differential evolution equation (26) for the coherent vector matrix elements $\langle \psi_f | \phi_t(B) \psi_f \rangle$ corresponds to the Itô form [6] of the quantum stochastic equation

(27) $\quad d\phi_t(B) = \phi_t \circ \lambda_\nu^\mu(B) d\Lambda_\mu^\nu := \sum_{\mu,\nu} \phi_t(\lambda_\nu^\mu(B)) d\Lambda_\mu^\nu, \qquad B \in \mathcal{B}$

with the initial condition $\phi_0(B) = B$, for all $B \in \mathcal{B}$. Here λ_ν^μ are the flow generators $\lambda_+^- = \lambda$, $\lambda_+^\bullet = \lambda^\bullet$, $\lambda_-^- = \lambda_\bullet$, $\lambda_\bullet^\bullet$, called the structural maps, and the summation is taken over the indices $\mu = -, \bullet$, $\nu = +, \bullet$ of the standard quantum stochastic integrators Λ_μ^ν. For simplicity we shall assume that the pre-Hilbert Fréchet space $\mathcal{E}$ is separable, $\mathcal{E} \subseteq \ell^2$. Then the index $\bullet$ can take any value in $\{1, 2, ...\}$ and $\Lambda_\mu^\nu(t)$ are indexed with $\mu \in \{-, 1, 2, ...\}$, $\nu \in \{+, 1, 2, ...\}$ as the standard time $\Lambda_-^+(t) = tI$, annihilation $\Lambda_-^m(t)$, creation $\Lambda_n^+(t)$ and exchange-number $\Lambda_n^m(t)$ operator integrators with $m, n \in \mathbb{N}$. The infinitesimal increments $d\Lambda_\nu^\mu(t) = \Lambda_\nu^{t\mu}(dt)$ are formally defined by the HP multiplication table [6] and the $\star$-property [15],

(28) $\qquad\qquad\qquad d\Lambda_\mu^\alpha d\Lambda_\beta^\nu = \delta_\beta^\alpha d\Lambda_\mu^\nu, \qquad \Lambda^* = \Lambda,$

where δ_β^α is the usual Kronecker delta restricted to the indices $\alpha \in \{-, 1, 2, ...\}$, $\beta \in \{+, 1, 2, ...\}$ and $\Lambda_{-\nu}^{*\mu} = \Lambda_{-\mu}^{\nu*}$ with respect to the reflection $-(-) = +$, $-(+) = -$ of the indices $(-, +)$ only.

The linear equation (27) of a particular type, (quantum Langevin equation) with bounded finite-dimensional structural maps λ_ν^μ was introduced by Evans and Hudson [2] in order to describe the $*$-homomorphic quantum stochastic evolutions. The constructed quantum stochastic $*$-homomorphic flow (EH-flow) is identity preserving and is obviously completely positive, but it is hard to prove these algebraic properties for the unbounded case. However the typical quantum filtering dynamics is not homomorphic or identity preserving, but it is completely positive and in the most interesting cases is described by unbounded generators λ_ν^μ. In the general content the equation (22) was studied in [31], and the correspondent quantum stochastic, not necessarily homomorphic and normalized flow was constructed even for the infinitely-dimensional non-adapted case under the natural integrability condition for the chronological products of the generators λ_ν^μ in the norm scale (16). The EH flows with unbounded λ_ν^μ, satisfying certain analyticity conditions, have been recently constructed in strong sense by Fagnola-Sinha in [30] for the non-Hilbert class L^∞ of test functions $f^\bullet$. Here we will formulate the necessary differential conditions which follow from the complete positivity, causality, and martingale properties of the filtering flows, and which are sufficient for the construction of the quantum stochastic flows obeying these properties in the case of the bounded λ_ν^μ. As we showed in [7], the found properties are sufficient to define the general structure of the bounded generators, and this structure will help us in construction of the minimal completely positive weak solutions for the quantum filtering equations also with unbounded λ_ν^μ.

Obviously the linear w*-continuous generators $\lambda_\nu^\mu : \mathcal{B} \to \mathcal{A}^c$ for CP flows $\phi_t^* = \phi_t$, where $\phi_t^*(B) = \phi_t(B^*)^\dagger$, must satisfy the $\star$-property $\lambda^\star = \lambda$, where $\lambda_{-\mu}^{\star\nu} = \lambda_{-\nu}^{\mu*}$, $\lambda_\nu^{\mu*}(B) = \lambda_\nu^\mu(B^*)^*$ and are independent of t, corresponding to cocycle property $\phi_s \circ \phi_r^s = \phi_{s+r}$, where ϕ_t^s is the solution to (27) with $\Lambda_\nu^\mu(t)$ replaced by $\Lambda_\nu^{s,\mu}(t)$, and $\lambda_+^-(I) = 0$ if ϕ is a filtering flow, $\phi_t(I) = I$, as it is in the multiplicative case [2]. We shall assume that $\boldsymbol{\lambda} = (\lambda_\nu^\mu)_{\nu=+,\bullet}^{\mu=-,\bullet}$ for each $B^* = B$ defines a continuous Hermitian form $\boldsymbol{b} = \boldsymbol{\lambda}(B)$ on the Fréchet space $\mathcal{D} \oplus \mathcal{D}_\bullet$,

$$\langle \eta | \boldsymbol{b}\, \eta \rangle = \sum_{m,n} \langle \eta^m | b_n^m \eta^n \rangle + \sum_m \langle \eta^m | b_+^m \eta \rangle + \sum_n \langle \eta | b_n^- \eta^n \rangle + \langle \eta | b_+^- \eta \rangle,$$

where $\eta \in \mathcal{D}$, $\eta^\bullet = (\eta^m)^{m \in \mathbb{N}} \in \mathcal{D}_\bullet = \mathcal{D} \otimes \mathcal{E}$. We say that an Itô algebra $\mathfrak{a}$, represented on $\mathcal{E}$, commutes in HP sense with a $\boldsymbol{b}$, given by the form-generator $\boldsymbol{\lambda}$ if $(I \otimes a_\bullet^\mu) b_\bullet^\nu = b_\bullet^\mu (I \otimes a_\nu^\bullet)$ (For simplicity the ampliation $I \otimes a_\nu^\mu$ will be written again as a_ν^μ.) Note that if we define the matrix elements a_ν^μ, b_ν^μ also for $\mu = +$ and $\nu = -$, by the extension

$$a_\nu^+ = 0 = a_-^\mu, \qquad \lambda_\nu^+(B) = 0 = \lambda_-^\mu(B), \quad \forall a \in \mathfrak{a}, B \in \mathcal{B},$$

the HP product (3) of $\boldsymbol{a}$ and $\boldsymbol{b}$ can be written in terms of the usual matrix product $\mathbf{ab} = [\mathbf{a}_\lambda^\mu \mathbf{b}_\nu^\lambda]$ of the extended quadratic matrices $\mathbf{a} = [\mathbf{a}_\nu^\mu]_{\nu=-,\bullet,+}^{\mu=-,\bullet,+}$

and $\mathbf{b} = b\mathbf{g}$, where $\mathbf{g} = [\delta^\mu_{-\nu}]$. Then one can extend the summation in (27) so it is also over $\mu = +$, and $\nu = -$, such that $b^\mu_\nu d\Lambda^\nu_\mu$ is written as the trace $\mathbf{b} \cdot d\mathbf{\Lambda}$ over all μ, ν. By such an extension the multiplication table for $d\Lambda\,(\mathbf{a}) = \mathbf{a} \cdot d\mathbf{\Lambda}$, $d\Lambda\,(\mathbf{b}) = \mathbf{b} \cdot d\mathbf{\Lambda}$ can be represented as $d\Lambda\,(\mathbf{a})\,d\Lambda\,(\mathbf{b}) = \mathbf{ab} \cdot d\mathbf{\Lambda}$, and the involution $\mathbf{b} \mapsto \mathbf{b}^\star$, defining $d\Lambda\,(\mathbf{b})^\dagger = \mathbf{b}^\star \cdot d\mathbf{\Lambda}$, can be obtained by the pseudo-Hermitian conjugation $b^{\star\nu}_\alpha = g_{\alpha\mu} b^{\mu*}_\beta g^{\beta\nu}$ respectively to the indefinite Minkowski metric tensor $\mathbf{g} = [g_{\mu\nu}]$ and its inverse $\mathbf{g}^{-1} = [g^{\mu\nu}]$, given by $g^{\mu\nu} = \delta^\mu_{-\nu} I = g_{\mu\nu}$.

Now let us find the differential form of the normalization and causality conditions with respect to the quantum stationary process, with independent increments $dX\,(t) = X\,(t + \Delta) - X\,(s)$ generated by an Itô algebra $\mathfrak{a}$ on the separable space $\mathcal{E}$.

Proposition 4 *Let ϕ be a flow, satisfying the quantum stochastic equation (27), and $[W_t\,(g), \phi_t\,(B)] = 0$ for all $g \in \mathfrak{g}, B \in \mathcal{B}$. Then the coefficients $b^\mu_\nu = \lambda^\mu_\nu\,(B)$, $\mu = -, \bullet$, $\nu = +, \bullet$, where $\bullet = 1, 2, ...$, written in the matrix form $b = (b^\mu_\nu)^{\mu=-,\bullet}_{\nu=+,\bullet}$, commute in the sense of the HP product with $a = (a^\mu_\nu)^{\mu=-,\bullet}_{\nu=+,\bullet}$ for all $a \in \mathfrak{a}$ and $B \in \mathcal{B}$:*

(29) $$[a, b] := (a^\mu_\bullet b^\bullet_\nu - b^\mu_\bullet a^\bullet_\nu)^{\mu=-,\bullet}_{\nu=+,\bullet} = 0.$$

Proof. – Since $\epsilon_t\,(\phi_s\,(I) - \phi_t\,(I))$ is a negative Hermitian form,

$$\epsilon_t\,(d\phi_t\,(I)) = \epsilon_t\,(\phi_t\,(\lambda^\mu_\nu\,(I))\,d\Lambda^\nu_\mu) = \phi_t\,(\lambda^-_+\,(I))\,dt \le 0.$$

Since $Y_t = \phi_t\,(B)$ commutes with $W_t\,(g)$ for all B and $g\,(t) = a$, we have by virtue of quantum Itô's formula

$$d\,[Y_t, W_t] = [dY_t, W_t] + [Y_t, dW_t] + [dY_t, dW_t] = 0.$$

The equations (17), (27) and commutativity of a^μ_ν with Y_t and W_t imply

$$([\phi_t\,(b^\mu_\nu), W_t] + [Y_t, a^\mu_\nu W_t] + \phi_t\,(b^\mu_\bullet)\,a^\bullet_\nu W_t - a^\mu_\bullet W_t \phi_t\,(b^\bullet_\nu))\,d\Lambda^\nu_\mu$$
$$= W_t\,(\phi_t\,(b^\mu_\bullet)\,a^\bullet_\nu - a^\mu_\bullet \phi_t\,(b^\bullet_\nu))\,d\Lambda^\mu_\nu = W_t \phi_t\,(b^\mu_\bullet a^\bullet_\nu - a^\mu_\bullet b^\bullet_\nu)\,d\Lambda^\nu_\mu = 0.$$

Thus $a \bullet b = b \bullet a$ by the argument [4] of independence of the integrators $d\Lambda^\nu_\mu$.

In order to formulate the CP differential condition we need the notion of *quantum stochastic germ* for the CP flow ϕ at $t = 0$. It was defined in [31, 9], for a quantum stochastic differential (27) with $\phi_0\,(B) = B, \forall B \in \mathcal{B}$ as $\gamma^\mu_\nu = \lambda^\mu_\nu + \imath^\mu_\nu$, where λ^μ_ν are the structural maps $B \mapsto \lambda^\mu_\nu\,(B)$ given by the generators of the quantum Itô equation (27) and $\imath^\mu_\nu : B \mapsto B\delta^\mu_\nu$ is the ampliation of $\mathcal{B}$. Let us prove that the germ-maps γ^μ_ν of a CP flow ϕ must be conditionally completely positive (CCP) in a degenerated sense as it was found for the finite-dimensional bounded case in [7, 11]. Another, equivalent, but not so explicit characterization was suggested for this particular case in [12].

Theorem 5 *If ϕ is a completely positive flow satisfying the quantum stochastic equation (27) with $\phi_0(B) = B$, then the germ-matrix $\gamma = (\lambda_\nu^\mu + \iota_\nu^\mu)_{\nu=+,\bullet}^{\mu=-,\bullet}$ is conditionally completely positive in the sense*

$$\sum_{B\in\mathcal{B}} \iota(B)\,\zeta_B = 0 \Rightarrow \sum_{B,C\in\mathcal{B}} \langle \zeta_B | \gamma(B^*C)\,\zeta_C \rangle \ge 0.$$

Here $\zeta \in \mathcal{D} \oplus \mathcal{D}_\bullet$, $\mathcal{D}_\bullet = \mathcal{D} \otimes \mathcal{E}$, and $\iota = (\iota_\nu^\mu)_{\nu=+,\bullet}^{\mu=-,\bullet}$ is the degenerate representation $\iota_\nu^\mu(B) = B\delta_\nu^+\delta_-^\mu$, written both with γ in the matrix form as

$$(30) \qquad \gamma = \begin{pmatrix} \gamma & \gamma_\bullet \\ \gamma^\bullet & \gamma_\bullet^\bullet \end{pmatrix}, \qquad \iota(B) = \begin{pmatrix} B & 0 \\ 0 & 0 \end{pmatrix},$$

where $\gamma = \lambda_+^-$, $\quad \gamma^m = \lambda_+^m$, $\quad \gamma_n = \lambda_n^-$, $\quad \gamma_n^m = \iota_n^m + \lambda_n^m$ with $\iota_n^m(B) = B\delta_n^m$ such that

$$(31)\, \gamma(B^*) = \gamma_\bullet(B)^*, \qquad \gamma^n(B^*) = \gamma_n(B)^*, \qquad \gamma_n^m(B^*) = \gamma_m^n(B)^*.$$

If ϕ is subfiltering, then $D = -\lambda_+^-(I)$ is a positive Hermitian form, $\langle \eta | D\eta \rangle \ge 0$, for all $\eta \in \mathcal{D}$, and if ϕ is contractive, then $\boldsymbol{D} = -\boldsymbol{\lambda}(I)$ is positive in the sense $\langle \boldsymbol{\eta} | \boldsymbol{D\eta} \rangle \ge 0$ for all $\boldsymbol{\eta} \in \mathcal{D} \oplus \mathcal{D}_\bullet$.

The proof is given in [9, 10] even for the general (noncommutative) algebras $\mathfrak{a}$ and $\mathcal{A}$.

Obviously the CCP property for the germ-matrix γ is invariant under the transformation $\gamma \mapsto \varphi$ given by

$$(32) \qquad \varphi(B) = \gamma(B) + \iota(B)\boldsymbol{K} + \boldsymbol{K}^*\iota(B),$$

where $\boldsymbol{K} = (K_\nu^\mu)_{\nu=+,\bullet}^{\mu=-,\bullet}$ is an arbitrary matrix of $K_\nu^\mu \in \mathcal{L}(\mathcal{D})$ with $K_{-\nu}^{*\mu} = K_{-\mu}^{\nu*}$. As was proven in [7, 11] for the case of finite-dimensional matrix γ of bounded γ_ν^μ, see also [12], the matrix elements K_ν^- can be chosen in such way that the matrix map $\varphi = (\varphi_\nu^\mu)_{\nu=+,\bullet}^{\mu=-,\bullet}$ becomes CP from $\mathcal{B}$ into the quadratic matrices of $\varphi_\nu^\mu(B)$. (The other elements can be chosen arbitrarily, say as $K_+^\bullet = 0$, $K_\bullet^\bullet = \frac{1}{2}I_\bullet^\bullet$, because (32) does not depend on $K_+^\bullet, K_\bullet^\bullet$.) Thus the generator $\boldsymbol{\lambda} = \gamma - \iota$ for a quantum stochastic CP flow ϕ can be written (at least in the bounded case) as $\varphi - \iota\boldsymbol{K} - \boldsymbol{K}^*\iota$:

$$(33) \qquad \lambda_\nu^\mu(B) = \varphi_\nu^\mu(B) - B\left(\tfrac{1}{2}\delta_\nu^\mu I + \delta_-^\mu K_\nu\right) - \left(\tfrac{1}{2}\delta_\nu^\mu I + K^\mu\delta_\nu^+\right)B,$$

where $\varphi_\nu^\mu : \mathcal{B} \to \mathcal{B}(\mathcal{D})$ are matrix elements of the CP map φ and $K_\nu \in \mathcal{L}(\mathcal{D})$, $K^- = K_+^*$, $K^m = K_m^*$. Now we show that the germ-matrix of this form obeys the CCP property even in the general case of unbounded K_ν^-, $\varphi_\nu^\mu(B) \in \mathcal{B}(\mathcal{D})$.

Proposition 6 *The matrix map $\gamma = (\gamma_\nu^\mu)_{\nu=+,\bullet}^{\mu=-,\bullet}$ given in (32) by*

$$(34)$$

$$\varphi = \begin{pmatrix} \varphi & \varphi_\bullet \\ \varphi^\bullet & \varphi_\bullet^\bullet \end{pmatrix}, \quad \text{and} \quad \boldsymbol{K} = \begin{pmatrix} K & K_\bullet \\ 0 & \frac{1}{2}I_\bullet^\bullet \end{pmatrix}, \quad \boldsymbol{K}^* = \begin{pmatrix} K^* & 0 \\ K_\bullet^* & \frac{1}{2}I_\bullet^\bullet \end{pmatrix},$$

with $\varphi = \varphi_+^-$, $\quad \varphi^m = \varphi_+^m$, $\quad \varphi_n = \varphi_n^-$ and $\varphi_n^m = \gamma_n^m$ is CCP with respect to the degenerate representation $\iota = \left(\delta_-^\mu \delta_\nu^+ \iota\right)_{\nu=+,\bullet}^{\mu=-,\bullet}$, where $\iota\left(B\right) = B$, if φ is a CP map.

Proof. – If $\iota\left(B_k\right)\boldsymbol{\eta}^k = 0$, then

$$\left\langle \boldsymbol{\eta}^k \middle| \iota\left(B_k^* B_l\right) \boldsymbol{K} + \boldsymbol{K}^* \iota\left(B_k^* B_l\right) \boldsymbol{\eta}^l \right\rangle$$

$$= 2\mathrm{Re}\left\langle \iota\left(B_k\right)\boldsymbol{\eta}^k \middle| \iota\left(B_l\right)\boldsymbol{K}\boldsymbol{\eta}^l \right\rangle = 0.$$

Hence the CCP for γ is equivalent to the CCP property for (32) and follows from its CP property:

$$\left\langle \boldsymbol{\eta}^k \middle| \gamma\left(B_k^* B_l\right)\boldsymbol{\eta}^l \right\rangle = \left\langle \boldsymbol{\eta}^k \middle| \varphi\left(B_k^* B_l\right)\boldsymbol{\eta}^l \right\rangle \geq 0$$

for such sequences $\boldsymbol{\eta}^k \in \mathcal{D} \oplus \mathcal{D}_\bullet$.

5. Construction of Quantum CP Flows

The necessary conditions for the stochastic generator $\boldsymbol{\lambda} = (\lambda_\nu^\mu)_{\nu=+,\bullet}^{\mu=-,\bullet}$ of a CP flow ϕ at $t = 0$ are found in the previous section in the form of a CCP property for the corresponding germ $\boldsymbol{\gamma} = (\gamma_\nu^\mu)_{\nu=+,\bullet}^{\mu=-,\bullet}$. In the next section we shall show, these conditions are essentially equivalent to the assumption (33), corresponding to

(35)
$$\gamma^m\left(B\right) = \varphi^m\left(B\right) - K_m^* B = \gamma_m^*\left(B\right), \quad \gamma\left(B\right) = \varphi\left(B\right) - K^* B - B K,$$

where $\boldsymbol{\varphi} = (\varphi_\nu^\mu)_{\nu=+,\bullet}^{\mu=-,\bullet}$ is a CP map with $\varphi_n^m = \gamma_n^m$. Here we are going to prove under the following conditions for the operators $K, K_\bullet$ and the maps φ_ν^μ that this general form is also sufficient for the existence of the CP solutions to the quantum stochastic equation (27). We are going to construct the minimal quantum stochastic positive flow $B \mapsto \phi_t\left(B\right)$ for a given w*-continuous unbounded germ-matrix map of the above form, satisfying the following conditions.

1. First, we suppose that the operator $K \in \mathcal{B}\left(\mathcal{D}\right)$ generates the one parametric semigroup $\left(e^{-Kt}\right)_{t>0}$, $e^{-Kr} e^{-Ks} = e^{-K(r+s)}$ of continuous operators $e^{-Kt} \in \mathcal{L}\left(\mathcal{D}\right)$ in the strong sense

$$\lim_{t\searrow 0} \frac{1}{t}\left(I - e^{-Kt}\right)\eta = K\eta, \quad \forall \eta \in \mathcal{D}.$$

(A contraction semigroup on the Hilbert space $\mathcal{H}$ if K defines an accretive $K + K^\dagger \geq 0$ and so maximal accretive form.)

2. Second, we suppose that the solution $S_t^n, n \in \mathbb{N}$ to the recurrence

$$S_t^{n+1} = S_t^o - \int_0^t S_{t-r}^o \sum_{m=1}^{\infty} K_m S_r^n d\Lambda_-^m, \quad S_t^0 = S_t^o,$$

where $S_t^o = e^{-Kt} \otimes T_t \in \mathcal{L}(\mathfrak{D})$ is the contraction given by the shift co-isometries $T_t : \mathfrak{F} \to \mathfrak{F}$, strongly converges to a continuous operator $S_t \in \mathcal{L}(\mathfrak{D})$ at $n \longrightarrow \infty$ for each $t > 0$.

3. Third, we suppose that the solution $R_t^n, n \in \mathbb{N}$ to the recurrence

$$R_t^{n+1} = S_t^* S_t + \int_0^t d\Lambda_\mu^\nu \left(r, S_r^* \varphi_\nu^\mu \left(R_{t-r}^n \right) S_r \right), \quad R_t^0 = S_t^* S_t,$$

where the quantum stochastic non-adapted integral is understood in the sense [31], weakly converges to a continuous form $R_t \in \mathcal{B}(\mathfrak{D})$ at $n \longrightarrow \infty$ for each $t > 0$.

The first and second assumptions are necessary to define the existence of free evolution semigroup $S^o = (S_t^o)_{t>0}$ and its perturbation $S = (S_t)_{t>0}$ on the product space $\mathfrak{D} = D \otimes \mathfrak{F}$ in the form of multiple quantum stochastic integral

(36) $\quad S_t = S_t^o$

$$+ \sum_{n=1}^{\infty} (-1)^n \int \cdots \int_{0<t_1<\ldots<t_n<t} K_{m_n}(t-t_n) \cdots K_{m_1}(t_2-t_1) S_{t_1}^o d\Lambda_-^{m_1} \cdots d\Lambda_-^{m_n},$$

iterating the quantum stochastic integral equation

(37) $\qquad S_t = S_t^o - \int_0^t \sum_{m=1}^{\infty} K_m(t-r) S_r d\Lambda_-^m, \quad S_0 = I,$

where $K_m(t) = S_t^o(K_m \otimes I)$. The third assumption supplies the weak convergence for the series

(38) $\quad R_t = S_t^* S_t$

$$+ \sum_{n=1}^{\infty} \int \cdots \int_{0<t_1<\ldots<t_n<t} d\Lambda_{\mu_1 \ldots \mu_n}^{\nu_1 \ldots \nu_n}\left(t_1, \ldots, t_n, \varphi_{\nu_1 \ldots \nu_n}^{\mu_1 \ldots \mu_n}\left(t_1, \ldots, t_n, S_{t-t_n}^* S_{t-t_n} \right) \right)$$

of non-adapted n-tuple integrals, i.e. for the multiple quantum stochastic integral [31] with

(39) $\quad \varphi_{\nu_1 \ldots \nu_n}^{\mu_1 \ldots \mu_n}(t_1, \ldots, t_n) = \varphi_{\nu_1 \ldots \nu_{n-1}}^{\mu_1 \ldots \mu_{n-1}}(t_1, \ldots, t_{n-1}) \circ \varphi_{\nu_n}^{\mu_n}(t_n - t_{n-1}),$

where $\quad \varphi_\nu^\mu(t, B) = S_t^* \varphi_\nu^\mu(B) S_t.$

The following theorem gives a characterization of the evolution semigroup S in terms of cocycles with unbounded coefficients, characterized by Fagnola [32] in the isometric and unitary case.

Proposition 7 *Let the family $V^\circ = (V_t^\circ)_{t>0}$ be a quantum stochastic adapted cocycle, $V_r^\circ T_s V_s^\circ = T_s V_{r+s}^\circ$, satisfying the HP differential equation*

$$(40) \quad dV_t^\circ + KV_t^\circ dt + \sum_{m=1}^{\infty} K_m V_t^\circ d\Lambda_-^m + \sum_{n=1}^{\infty} V_t^\circ d\Lambda_n^n = 0, \quad V_0^\circ = I.$$

Then $S_t = T_t V_t^\circ$ is a semigroup solution, $S_r S_s = S_{r+s}$ to the non-adapted integral equation (37) such that $S_t \psi_f = S_t(f^\bullet)\, \eta \otimes \delta_\varnothing, \forall \eta \in \mathcal{D}$ on $\psi_f = \eta \otimes f^\otimes$ with $f^\bullet \in \mathfrak{E}_t$. Conversely, if $S = (S_t)_{t>0}$ is the non-adapted solution (36) to the integral equation (37), then $V_t^\circ = T_t^* S_t$ is the adapted solution to (40), defined as $V_t^\circ \psi_f = S_t(f^\bullet)\, \eta, \forall \eta \in \mathcal{D}$, where $S_t(f^\bullet) = F^* S_t F$ is given by $F\eta = \eta \otimes f^\otimes$ with $f^\bullet \in \mathfrak{E}_t$.

Proof. – First let us show that the equation (40) is equivalent to the integral one

$$V_t^\circ = e^{-Kt} \otimes I_t - \int_0^t \sum_{m=1}^{\infty} \left(e^{-K(t-r)} K_m \otimes I_{t-r}^r \right) V_r^\circ d\Lambda_-^m, \quad V_0^\circ = I,$$

where $I_t = T_t^\dagger T_t = 0^{\Lambda_\bullet^\bullet(t)}$ is the decreasing family of orthoprojectors onto $\mathfrak{F}_{[t}$, and $I_s^r = \theta^r(I_s)$. Indeed, multiplying both parts of the integral equation from the left by $\left(e^{K(t-s)} \otimes I\right)$ and differentiating the product $e^{K(t-s)} V_t^\circ$ at $t = s$, we obtain (40) by taking into account that $dI_t + \sum_{n=1}^{\infty} I_t d\Lambda_n^n = 0$ and $d\Lambda_n^n d\Lambda_-^m = 0$. Conversely, the integral equation can be obtained from (40) by the integration:

$$
\begin{aligned}
V_t^\circ - e^{-Kt} \otimes I_t &= \int_0^t d\left(\left(e^{-K(t-r)} \otimes I_{t-r}^r \right) V_r^\circ \right) \\
&= \int_0^t \left(e^{-K(t-r)} \otimes I_{t-r}^r \right) (dV_r^\circ + KV_r^\circ dr + V_r^\circ d\Lambda_\bullet^\bullet) \\
&= -\int_0^t e^{-K(t-r)} \left(K_\bullet \otimes I_{t-r}^r \right) V_r^\circ d\Lambda_-^\bullet,
\end{aligned}
$$

where we used that $dI(r) = I(r) d\Lambda_\bullet^\bullet$ and $d(I(r) V_r^\circ) = dI(r) V_r^\circ + I(r) dV_r^\circ$ for the backward-adapted process $I(r) = I_{t-r}^r, \forall r \leq t$. The non-adapted equation (37) is obtained by applying the operator $T_t = T_{t-r} T_r$ to both parts of this integral equation and taking into account the commutativity of $e^{K(r-t)} K_m$ with T_r. Moreover, due to the adaptiveness of V_t°, $S_t \psi_f = T_t \left(E_t V_t^\circ \psi_f \otimes E_{[t} f^\otimes \right) = S_t(f^\bullet) \eta \otimes f_t^\otimes$, where $f_t^\otimes = T_t f^\otimes$, and $S_t(f^\bullet) = E V_t^\circ F$ is the solution to the equation

$$S_t(f^\bullet) = e^{-Kt} + \int_0^t e^{-K(t-r)} K_\bullet f^\bullet(r) S_r(f^\bullet) dr, \quad S_0(f^\bullet) = I.$$

Hence $S_t F = E^* S_t(f^\bullet)$ if $f^\bullet \in \mathfrak{E}_t$, and $F^* S_t F = S_t(f^\bullet)$ as $EF = I$. Since this equation is equivalent to the differential one

(41)

$$\frac{d}{dt} S_t \left(f^\bullet\right) \eta + \left(K_\bullet f^\bullet \left(t\right) + K\right) S_t \left(f^\bullet\right) \eta = 0, \quad S_0 \left(f^\bullet\right) \eta = \eta, \qquad \forall \eta \in \mathcal{D},$$

the function $t \mapsto S_t \left(f^\bullet\right)$, $f^\bullet \in \mathfrak{E}$ is a strongly continuous cocycle,

$$S_r \left(f_s^\bullet\right) S_s \left(f^\bullet\right) = S_{r+s} \left(f^\bullet\right), \ \forall r, s > 0, \qquad f_s^\bullet \left(t\right) = f^\bullet \left(t + s\right), \quad S_0 \left(f^\bullet\right) = I.$$

As was proved in [31], the multiple integral (36) gives a solution to the integral equation (37), and so the multiple integral for $V_t^\circ \psi_f = S_t \left(f^\bullet\right) \eta \otimes f^\otimes$,

$$S_t \left(f^\bullet\right)$$

$$= e^{-Kt} + \sum_{n=1}^{\infty} \left(-1\right)^n \int \cdots \int_{0 < t_1 < \ldots < t_n < t} K \left(t, t_n\right) \cdots K \left(t_2, t_1\right) e^{-Kt_1} dt_1 \cdots dt_n,$$

where $K \left(t, r\right) = e^{-K\left(t-r\right)} K_\bullet f^\bullet \left(r\right)$, corresponding to the iteration of the integral equation for V_t° on ψ_f, satisfies the HP equation (40).

The following theorem reduces the problem of solving of differential evolution equations to the problem of iteration of integral equations similar to the nonstochastic case [33, 34].

Proposition 8 *Let $S_t = T_t V_t^\circ$, where $V_t^\circ \in \mathcal{L} \left(\mathfrak{D}\right)$ are continuous operators defining the adapted cocycle solution to the equation (40). Then the linear stochastic evolution equation (27) is equivalent to the quantum non-adapted (in the sense of [31]) integral equation*

$$(42) \qquad \phi_t \left(B\right) = S_t^* B S_t + \int_0^t d\Lambda_\mu^\nu \left(\mathrm{r}, \phi_\mathrm{r} \left[\varphi_\nu^\mu \left(S_{t-\mathrm{r}}^* B S_{t-\mathrm{r}}\right)\right]\right)$$

with $\phi_0 \left(B\right) = B \in \mathcal{B}$, where φ_ν^μ are extended onto $\mathfrak{B}$ in the normal way by w^*-continuity and linearity as $\varphi_\nu^\mu \left(B \otimes Z\right) = \overline{\varphi}_\nu^\mu \left(B\right) \otimes Z$ for $B \in \overline{\mathcal{B}}$, $Z \in \mathcal{B} \left(\mathfrak{F}\right)$.

The proof is given in [35, 10].

Theorem 9 *Let φ be a w^*-continuous CP-map, and $S_t = T_t V_t^\circ$ be given by the solution to the quantum stochastic equation (40). Then the solutions to the evolution equation (27) with the generators, corresponding to (35), have the CP property, and satisfy the submartingale (contractivity) condition $\phi_t \left(I\right) \le \epsilon_t \left[\phi_s \left(I\right)\right]$ for all $t < s$ if $\varphi \left(I\right) \le K + K^\dagger$ ($\phi_t(I) \le \phi_s(I)$ if $\varphi(I) \le K + K^\dagger$). The minimal solution can be constructed in the form of multiple quantum stochastic integral in the sense [31] as the series*

$$(43) \quad \phi_t \left(B\right)$$

$$= \sum_{n=0}^{\infty} \int \cdots \int_{0 < t_1 < \ldots < t_n < t} d\Lambda_{\mu_1 \ldots \mu_n}^{\nu_1 \ldots \nu_n} \left(t_1, \ldots, t_n, \varphi_{\nu_1 \ldots \nu_n}^{\mu_1 \ldots \mu_n} \left(t_1, \ldots, t_n, S_{t-t_n}^* B S_{t-t_n}\right)\right)$$

of non-adapted n-tuple CP integrals with $S_t^* B S_t$ at $n = 0$ and

$$\varphi_{\nu_1 \ldots \nu_n}^{\mu_1 \ldots \mu_n}(t_1, \ldots, t_n) = \varphi_{\nu_1}^{\mu_1}(t_1) \circ \varphi_{\nu_2}^{\mu_2}(t_2 - t_1) \circ \ldots \circ \varphi_{\nu_n}^{\mu_n}(t_n - t_{n-1}),$$

where $\varphi_\nu^\mu(t, B) = S_t^* \varphi_\nu^\mu(B) S_t$. If φ is bounded, then the solution to the equation is unique, and $\phi_t(I) = \epsilon_t[\phi_s(I)]$ for all $t < s$ if $K + K^\dagger = \varphi(I)$ ($\phi_t(I) = I$ if $K + K^\dagger = \varphi(I)$).

The proof is given in [35, 10].

6. The Structure of the Generators and Flows

First, let us prove the structure (33) for the (unbounded) form-generator of CP flows over the algebra $\mathcal{B} = \mathcal{L}(\mathcal{H})$ of all bounded operators, assuming that $\mathcal{A} = 0$. This algebra contains the one-dimensional operators $|\eta'\rangle\langle\eta^0| : \eta \mapsto \langle\eta^0|\eta\rangle \eta'$ given by the vectors $\eta^0, \eta' \in \mathcal{H}$.

Let us fix a vector $\eta^0 \in \mathcal{D} \oplus \mathcal{D}_\bullet$ with the unit projection $\eta^0 \in \mathcal{D}$, $\|\eta^0\| = 1$, and make the following assumption of the weak continuity for the linear operator $\eta' \mapsto \gamma(|\eta'\rangle\langle\eta^0|)\eta^0$.

0) The sequence $\eta_n' = \gamma(|\eta_n'\rangle\langle\eta^0|)\eta^0 \in \mathcal{D}' \oplus \mathcal{D}_\bullet'$ of anti-linear forms

$$\eta \in \mathcal{D} \oplus \mathcal{D}_\bullet \mapsto \langle\eta|\eta_n'\rangle := \langle\eta|\gamma(|\eta_n'\rangle\langle\eta^0|)\eta^0\rangle$$

converges for each sequence $\eta_n' \in \mathcal{H}$ converging in $\mathcal{D}' \supseteq \mathcal{H}$.

Proposition 10 *Let the CCP germ-matrix γ satisfy the above continuity condition for a given η^0. Then there exist strongly continuous operators $K \in \mathcal{L}(\mathcal{D})$, $K_\bullet : \mathcal{D}_\bullet \to \mathcal{D}$ defining the matrix operator K in (33), such that the matrix map (32) is CP, and there exists a Hilbert space $\mathcal{K}$, a *-representation $\jmath : \mathcal{B} \mapsto \mathcal{B} \otimes J$ of $\mathcal{B} = \mathcal{L}(\mathcal{H})$ on the Hilbert product $\mathcal{G} = \mathcal{H} \otimes \mathcal{K}$, given by an orthoprojector J in $\mathcal{K}$, such that*

$$(44) \qquad \varphi(B) = (L^\mu \jmath(B) L_\nu)_{\nu=+,\bullet}^{\mu=-,\bullet} = L^* \jmath(B) L.$$

Here $L = (L, L_\bullet)$ is a strongly continuous operator $\mathcal{D} \oplus \mathcal{D}_\bullet \to \mathcal{G}$ with $L = L_+$, $L^- = L^*$, $L^\bullet = L_\bullet^*$ which is always possible to make

$$(45) \qquad \langle\eta^0 \otimes e|L\eta^0\rangle = 0, \qquad \forall e \in \mathcal{K}_1,$$

where $\mathcal{K}_1 = J\mathcal{K}$. If $D = -\lambda(I) \geq 0$, then one can make $L^* L = K + K^\dagger$ in a canonical way $L = L^\circ$, and in addition one can make $L^* L_\bullet = K_\bullet$, $L_\bullet^* L_\bullet = I_\bullet^\bullet$, where $I_\bullet^\bullet = I\delta_\bullet^\bullet$ for a canonical $L_\bullet = L_\bullet^\circ$ if $D = -\lambda(I) \geq 0$.

The proof is given in [10].

Thus we have proved that the equation (27) for a completely positive quantum stochastic flows over $\mathcal{B} = \mathcal{L}(\mathcal{H})$ has the following general form

$$d\phi_t(B) + \phi_t(K^*B + BK - L^* \jmath(B)L) dt$$

$$= \sum_{m,n=1}^{\infty} \phi_t(L_m^* \jmath(B)L_n - B\delta_n^m) d\Lambda_m^n + \sum_{m=1}^{\infty} \phi_t(L_m^* \jmath(B)L - K_m^*B) d\Lambda_m^+$$

$$+ \sum_{n=1}^{\infty} \phi_t(L^* \jmath(B)L_n - BK_n) d\Lambda_-^n,$$

generalizing the Lindblad form [1] for the semigroups of completely positive maps. This can be written in the tensor notation form as

(46)
$$d\phi_t(B) = \phi_t\left(L_\alpha^\mu \jmath_\beta^\alpha(B)L_\nu^\beta - \imath_\nu^\mu(B)\right) d\Lambda_\mu^\nu = \phi_t(\mathbf{L}^* \mathbf{\jmath}(B)\mathbf{L} - \imath(B)) \cdot d\Lambda,$$

where the summation is taken over all $\alpha, \beta = -, \circ, +$ and $\mu, \nu = -, \bullet, +$, $\jmath_-^-(B) = B = \jmath_+^+(B)$, $\jmath_\circ^\circ(B) = \jmath(B)$, $\jmath_\beta^\alpha(B) = 0$ if $\alpha \neq \beta$, $\imath_\nu^\mu(B) = B\delta_\nu^\mu$, and $\mathbf{L}^* = \left[L_\beta^{*\mu}\right]_{\beta=-,\circ,+}^{\mu=-,\bullet,+}$ with $L_{-\alpha}^{*\mu} = L_{-\mu}^{\alpha*}$ is the triangular matrix, pseudoadjoint to $\mathbf{L} = [L_\nu^\alpha]_{\nu=-,\bullet,+}^{\alpha=-,\circ,+}$ with $L_-^- = I = L_+^+$,

$$L_\bullet^\circ = L_\bullet, \qquad L_+^\circ = L, \qquad L_\bullet^- = -K_\bullet, \qquad L_+^- = -K.$$

(All other L_ν^α are zero.) If the Hilbert space $\mathcal{K}$ is separable, $\mathcal{K}_1 = \ell^2(\mathbb{N}_1)$ for a subset $\mathbb{N}_1 \subseteq \mathbb{N}$. Then the equation (46) can be resolved as $\phi_t(B) = V_t^*(B \otimes I_t) V_t$, where $V = (V_t)_{t>0}$ is an (unbounded) cocycle on the product $\mathcal{D} \otimes \mathfrak{F}$ with Fock space $\mathfrak{F}$ over the Hilbert space $L^2(\mathbb{N} \times \mathbb{R}_+)$ of the quantum noise, and I_t is a decreasing family of orthoprojectors in $\mathfrak{F}$, satisfying the stochastic equation $dI_t + \sum_{n \notin \mathbb{N}_1} I_t d\Lambda_n^n = 0$ with $I_0 = I$. The cocycle V can be found from the quantum stochastic equation $dV_t = (L_\nu^\mu - I\delta_\nu^\mu) V_t d\Lambda_\mu^\nu$ with $V_0 = I \otimes I$ of the form

(47)
$$dV_t + KV_t dt + \sum_{n=1}^{\infty} K_n V_t d\Lambda_-^n = \sum_{m,n=1}^{\infty} (L_n^m - I\delta_n^m) V_t d\Lambda_m^n + \sum_{m=1}^{\infty} L^m V_t d\Lambda_m^+,$$

where L_n^m and L^m are the operators in $\mathcal{D}$, defining

(48) $\varphi_n^m(B) = \sum_{k \in \mathbb{N}_1} L_m^{k*} B L_n^k, \qquad \varphi(B) = \sum_{k \in \mathbb{N}_1} L^{k*} B L^k$

$\varphi^m(B) = \sum_{k \in \mathbb{N}_1} L_m^{k*} B L^k, \qquad \varphi_n(B) = \sum_{k \in \mathbb{N}_1} L^{k*} B L_n^k$

with $\sum_{k=1}^{\infty} L^{k*} L^k = K + K^\dagger$ if $D \geq 0$, and in addition $\sum_{k=1}^{\infty} L^{k*} L_n^k = K_n$, $\sum_{k=1}^{\infty} L_m^{k*} L_n^k = I\delta_n^m$ if $D \geq 0$. The formal derivation of the equation (47) from

(46) is obtained by a simple application of the HP Itô formula. The martingale M_t, describing the density operator for the output state of $\Lambda(t, a)$, is then defined as $M_t = V_t^* V_t$.

The following theorem ensures the existence of a $*$-representation ι : $\Lambda(t, a) \mapsto \Lambda(t, i(a)) := i_\beta^\alpha(a)\Lambda_\alpha^\beta(t)$ of the quantum stochastic process (2), commuting with $Y_t = \phi_t(B)$ for all $a \in \mathfrak{a}, B \in \mathcal{L}(\mathcal{H})$, with $A = X(0) = 0$, in the form

$$\Lambda(t, i(a)) = i_\circ^\circ(a)\Lambda_\circ^\circ(t) + i_+^\circ(a)\Lambda_\circ^+(t) + i_\circ^-(a)\Lambda_-^\circ(t) + i_+^-(a)\Lambda_-^+(t).$$

Here $i = \left(i_\beta^\alpha\right)_{\beta=+,\circ}^{\alpha=-,\circ}$ is a $\star$-representation

$$i_\beta^\alpha(a^* a) = i_\circ^\alpha(a^*) i_\beta^\circ(a), \quad i_{-\beta}^\alpha(a^*) = i_{-\alpha}^\beta(a)^*$$

of the Itô algebra $\mathfrak{a}$ in the operators $i_\beta^\alpha(a) : \mathcal{K}_\beta \to \mathcal{K}_\alpha$, with a domain $\mathcal{K}_\circ \subseteq \mathcal{K}$, $\mathcal{K}_- = \mathbb{C} = \mathcal{K}_+$, and $\Lambda_\alpha^\beta(t)$ are the canonical quantum stochastic integrators in the Fock space $\Gamma(\mathfrak{K})$ over $\mathfrak{K} = L_\mathcal{K}^2(\mathbb{R}_+)$, the space of $\mathcal{K}$-valued square-integrable functions on $\mathbb{R}_+$. We shall extend i to the triangular matrix representation $\mathbf{i} = \left[i_\beta^\alpha\right]_{\beta=-,\circ,+}^{\alpha=-,\circ,+}$ on the pseudo-Hilbert space $\mathbb{C} \oplus \mathcal{K} \oplus \mathbb{C}$ with the Minkowski metrics tensor $\mathbf{g} = \left[\delta_{-\beta}^\alpha\right] = \mathbf{g}^{-1}$, by $i_\beta^+(a) = 0 = i_-^\alpha(a)$, for all $a \in \mathfrak{a}$, as it was done for $\mathbf{a} = [a_\nu^\mu]_{\nu=-,\bullet,+}^{\mu=-,\bullet,+}$, and denote the ampliation $I \otimes i_\beta^\alpha(a)$ again as $i_\beta^\alpha(a)$. Note that if the stochastic generator of the form (33) is restricted onto an operator algebra $\mathcal{B} \subseteq \mathcal{L}(\mathcal{H})$ with the weak closure $\bar{\mathcal{B}} = \mathcal{A}^c$, and all the sesquilinear forms $\gamma_\nu^\mu(B)$, $B \in \mathcal{B}$ commute with the $*$-algebra $\mathcal{A} \subset \mathcal{L}(\mathcal{D})$, then $\lambda_\nu^\mu(B) \in \bar{\mathcal{B}}$.

Proposition 11 *Let* $b = \gamma(B) - \imath(B)$ *satisfy the commutativity conditions (29) for all* $a \in \mathfrak{a}$, $B \in \mathcal{L}(\mathcal{H})$. *Then there exists a $\star$-representation* $a \mapsto i(a)$ *of the Itô algebra* $\mathfrak{a}$, *defining the operators* $i_\beta^\alpha(a) : \mathcal{K}_\beta \to \mathcal{K}_\alpha$, *with* $i_\beta^\alpha(a)^* \mathcal{K}_\alpha \subseteq \mathcal{K}_\beta$, *where* $\mathcal{K}_- = \mathbb{C} = \mathcal{K}_+$, *such that* $L_\mu^\alpha(I \otimes a_\nu^\mu) = \left(I \otimes i_\beta^\alpha(a)\right) L_\nu^\beta$ *for all* $a \in \mathfrak{a}$. *By omitting* $I \otimes$ *this can be written as*

$$
\begin{array}{ll}
(49) \qquad L_\bullet a_\bullet^\bullet = i(a) L_\bullet, & a_+^- - K_\bullet a_+^\bullet = i^-(a) L + i_+^-(a), \\
\qquad L_\bullet a_+^\bullet = i(a) L + i_+(a), & a_\bullet^- - K_\bullet a_\bullet^\bullet = i^-(a) L_\bullet,
\end{array}
$$

where we take the convention $i^- = i_\circ^-$, $i_+ = i_+^\circ$ *and* $i = i_\circ^\circ$. *If* $[A, \gamma_\nu^\mu(B)] = 0$ *for all* $A \in \mathcal{A}$ *and* $B \in \mathcal{B}$, *where* $\mathcal{B} \subseteq \mathcal{L}(\mathcal{H})$ *is a $*$-subalgebra, and* $\bar{\mathcal{B}} = \mathcal{A}^c$, *then there exists a triangular $\star$-representation* $\mathbf{j} = \left[j_\beta^\alpha\right]_{\beta=-,\circ,+}^{\alpha=-,\circ,+}$ *of the operator algebra* $\mathcal{A}$ *with* $j_\circ^\circ(I) = J$ *such that*

$$(50) \quad \mathbf{JLA} = \mathbf{j}(A)\mathbf{L}, \quad [\mathbf{j}(A), \mathbf{i}(a)] = 0,$$

$$[\mathbf{j}(A), \mathbf{j}(B)] = 0, \quad \forall A \in \mathcal{A}, a \in \mathfrak{a}, B \in \mathcal{B}.$$

The proof is given in [10] even for the general (noncommutative) Itô algebra $\mathfrak{a}$.

Now we are going to construct the quantum stochastic dilation for the flow $\phi_t(B)$ and the quantum state generating function $\vartheta_t^a = \epsilon[R_t W(t, a)]$ of the output process $\Lambda(t, a)$ in the form

$$\phi_t(B) = V_t^*(I_t \otimes B) V_t, \quad \vartheta_t(g) = \epsilon[V_t^*(W_t^a \otimes I) V_t], \quad \forall B \in \mathcal{L}(\mathcal{H}), a \in \mathfrak{a},$$

where V_t is an operator on $\mathfrak{D}$ into $\Gamma(\mathfrak{K}) \otimes \mathfrak{D}$, intertwining the Weyl operators $W(t, a)$ with the operators $W_t^a = W(t, i(a)) I_t$ in the Fock space $\Gamma(\mathfrak{K})$,

$$dW(t, i(a)) = W(t, i(a)) d\Lambda(t, i(a)), \quad W(0, i(a)) = I,$$

and $I_t \geq I_s, \forall t \leq s$ is a decreasing family of orthoprojectors.

In order to prove the existence of the Fock space dilation, we need the following assumptions in addition to the continuity assumptions of this and previous sections.

1) The minimal quantum stochastic *-flow [2] over the operator algebra $\mathcal{A}$, resolving the quantum Langevin equation

$$d\tau_t(A) = \tau_t(\mathbf{j}(A) - \iota(A)) \cdot d\Lambda, \quad A \in \mathcal{A},$$

where $\mathbf{j}(I) = \mathbf{J} \otimes I$, $\iota(A) = I \otimes A$, constructed by its iterations with $\tau_0(A) = I \otimes A$ as it was done in the Sec 4 for the flow ϕ, is the multiplicative flow, satisfying the condition $\tau_t(I) = I_t \otimes I$, where I_t is the solution to $dI_t = (\mathbf{J} - I)_\circ^\circ I_t d\Lambda_\circ^\circ$ with $I_0 = I$.

2) The operators $L_\nu(\bar{e}) = (I \otimes e^*) L_\nu$, given for all $e \in \mathcal{G}$ as $\langle L_\nu(\bar{e}) \eta | \eta' \rangle = \langle L_\nu \eta | \eta' \otimes e \rangle \quad \forall \eta \in \mathcal{D}, \eta' \in \mathcal{D}'$, are strongly continuous onto $\mathcal{D}$. This is necessary for the weak definition of the operators $V_t(\sigma) : \mathfrak{D} \to \mathcal{K}^{\otimes|\sigma|} \otimes \mathfrak{D}_{[t}$ on finite subsets $\sigma \subset [0, t)$ by the recurrence

$$V_t(\sigma) \psi = \left(I^{\otimes|\sigma|} \otimes V_t^\circ(s) \right) \left(L V_s(\sigma \backslash s) \psi + \sum_{n=1}^{\infty} L_n V_s(\sigma \backslash s) \psi^n(s) \right),$$

with $V_t(\emptyset) = V_t^\circ(0)$, $s = \max \sigma$. Here $V_t^\circ(s) = T_s^* V_{t-s}^\circ T_s$, V_t° is the solution to the equation (37), the operators $L_\nu : \mathcal{D} \to \mathcal{K} \otimes \mathcal{D}$ act on $\mathcal{K}^{\otimes|\sigma\backslash s|} \otimes \mathfrak{D}_{[s}$ as $I^{\otimes|\sigma\backslash s|} \otimes L_\nu \otimes I_{[s}$, and $\psi^n(s)$ are the components of $\psi^\bullet(\tau, s) = \psi(\tau \sqcup s)$, where $\tau \sqcup s$ is defined for almost all s ($s \notin \sigma$) as the disjoint union of the single point $\{s\}$ with a finite subset $\tau \in \mathbb{R}_+$.

3) The operator-valued function $\sigma \mapsto V_t(\sigma)$, defined for all such $\sigma \in \Gamma_t$, is weakly square integrable for each t with respect to the measure $d\sigma = \prod_{s \in \sigma} ds$ in the sense

$$\int_{\Gamma_t} \|V_t(\sigma) \psi\|^2 d\sigma := \sum_{n=0}^{\infty} \int \cdots \int_{0 < s_1 \ldots s_n < t} \|V_t(s_1, \ldots, s_n) \psi\|^2 ds_1 \ldots ds_1 < \infty,$$

for all $\psi \in \mathfrak{D}$. Thus the operators $V_t (\cdot)$ define a Fock space one $V_t :$ $\mathfrak{D} \to \Gamma(\mathfrak{K}_t) \otimes \mathfrak{D}_{[t}$. They form a cocycle, $V_{t-r}^r (\sigma) V_r (\sigma) = V_t (\sigma)$, where $V_s^r (\sigma) = I^{\otimes |\sigma_r|} \otimes T_r^* V_s (\sigma_{[r} - r) T_r$.

Theorem 12 *Under the given assumptions 0)–3) there exist:*

(i) *A cocycle dilation $V_t : \mathfrak{D} \to \Gamma(\mathfrak{K}_t) \otimes \mathfrak{D}$ of the minimal CP flow ϕ, intertwining the Weyl operator $W(t,a)$ with W_t^a:*

$$(51) \quad V_t (I \otimes W(t,a)) = (W_t^a \otimes I) V_t,$$
$$\phi_t (B) = V_t^* (I_t \otimes B) V_t , \quad \forall a \in \mathfrak{a}, B \in \mathcal{L}(\mathcal{H}),$$

where $I_t \leq I_s,$ $\forall t < s$ are orthoprojectors in $\Gamma(\mathfrak{K})$;

(ii) *A $*$-multiplicative flow $\tau = (\tau_t)$ over $\mathcal{A}$ in $\Gamma(\mathfrak{K}) \otimes \mathcal{H}$ with the properties $\tau_t (I) = I_t,$*

$$(52) \quad V_t A = \tau_t (A) V_t, \quad [\tau_t (A), W_t^a] = 0,$$
$$[\tau_t (A), I \otimes B] = 0, \quad \forall A \in \mathcal{A}, a \in \mathfrak{a}, B \in \mathcal{B}.$$

(iii) *If $\lambda(I) \leq 0$, then one can make $M_t = V_t^* V_t$ martingale, and, if $\lambda(I) \leq 0$, one can make V_t isometric, $V_t^* V_t = I$.*

(iv) *Moreover, let $U = (U_t)_{t \geq 0}$ be a one-parametric weakly continuous cocycle of unitary operators on $\Gamma(\mathfrak{K}) \otimes \mathcal{H} \otimes \Gamma(\mathfrak{E})$, satisfying the quantum stochastic equation*

$$(53) \quad dU_t + \left(K dt + K_\bullet^- d\Lambda_\bullet^\bullet + K_\circ^- d\Lambda_\bullet^\circ \right) U_t$$
$$= \left(L_+^\circ d\Lambda_\circ^+ - I_\bullet^\circ d\Lambda_\bullet^\bullet + J_\bullet^\circ d\Lambda_\bullet^\bullet + J_\circ^\circ d\Lambda_\bullet^\circ + (J_\circ^\circ - I_\circ^\circ) d\Lambda_\circ^\circ \right) U_t$$

with $U_0 = I$ and the necessary differential unitarity conditions

$$K + K^\dagger = L_\circ^- L_+^\circ, \quad K_\bullet^- = L_\circ^- J_\bullet^\circ, \quad J_\bullet^\circ J_\bullet^\circ = I_\bullet^\bullet,$$
$$K_\circ^- = L_\circ^- J_\circ^\circ, \quad J_\circ^\circ = I_\circ^\circ - J_\bullet^\circ J_\bullet^\circ,$$

where $L_\circ^- = L_+^{\circ*}, J_\circ^\bullet = J_\bullet^{\circ*}$. If $\lambda(I) \leq 0$ and $L_+^\circ = L^\circ$ is the canonical operator in the dilation (44), then

$$(54)$$
$$\langle \psi | (A \otimes I) \phi_t^a (B) \psi \rangle = \langle U_t (\delta_\emptyset \otimes \psi) | (\tau_t^a (A) (I \otimes B)) U_t (\delta_\emptyset \otimes \psi) \rangle$$

for all $A \in \mathcal{A}, a \in \mathfrak{a}, B \in \mathcal{B}$ and any initial $\psi = \eta \otimes \delta_\emptyset, \eta \in \mathcal{D}$, where

$$\phi_t^a (B) = (I \otimes W(t,a)) \phi_t (B), \quad \tau_t^a (A) = (W_t^a \otimes I) \tau_t (A).$$

This unitary cocycle dilation is valied for any state $\psi \in \mathcal{D} \otimes \mathfrak{F}$ if in addition $J_\bullet^\circ = L_\bullet^\circ$ is the canonical isometry in (44) for the case $\lambda(I) \leq 0$.

Proof. – (Sketch). The cocycle $V = (V_t)_{t>0}$ is recurrently constructed due to the above assumptions (1)–(3). It obviously intertwines the Weyl operators (19) with the operators W_t^a, acting in the same way in $\Gamma(\mathfrak{K})$, by virtue of the property (49).

Let us denote by $\mathfrak{K}_1 = L_{\mathcal{K}}^2(\mathbb{R}_+)$ the functional Hilbert space corresponding to the minimal dilation (44) sub-space $\mathcal{K} = \mathcal{K}_1$ for the CP map φ, given by the orthoprojector $J = J_1$ in the space $\mathcal{K}_\circ$ of the canonical dilation, and $\mathfrak{K}_0$ its orthogonal compliment, corresponding to $\mathcal{K}_0 = J_0 \mathcal{K}_\circ$, where $J_0 = I - J_1$. Representing $\Gamma(\mathfrak{K}_0 \oplus \mathfrak{K}_1)$ as $\Gamma(\mathfrak{K}_0) \otimes \Gamma(\mathfrak{K}_1)$, let us denote by I_t the survival orthoprojectors

$$I_t \chi\left(\sigma^0, \sigma^1\right) = \delta_\emptyset\left(\sigma_t^0\right) \chi\left(\sigma^0, \sigma^1\right), \qquad \sigma_t = \sigma \cap [0, t),$$

where $\chi\left(\sigma^0, \sigma^1\right) = \chi\left(\sigma^0 \sqcup \sigma^1\right) \in \mathcal{K}^{\otimes|\sigma^0|} \otimes \mathcal{K}^{\otimes|\sigma^1|}$ is the set function, representing a $\chi \in \Gamma(\mathfrak{K}_0 \oplus \mathfrak{K}_1)$. The decreasing family $(I_t)_{t>0}$ defines the decay orthoprojectors $E_t = I - I_t$ in $\Gamma(\mathfrak{K}_\circ)$ satisfying the quantum stochastic equation $dE_t = E_t J_0 \cdot d\Lambda_\circ^\circ$ with $E_0 = 0$, and $\Lambda_\circ^\circ$ is the number integrator in the Fock space $\Gamma(\mathfrak{K}_\circ)$ over $\mathfrak{K}_\circ = \mathfrak{K}_0 \oplus \mathfrak{K}_1$. Then one easily find that the minimal CP flow (43) can be represented as $\phi_t(B) = V_t^*(I_t \otimes B) V_t$.

We may also construct by iteration the minimal quantum stochastic *-flow $\tau = (\tau_t)$ over the operator algebra $\mathcal{A}$, resolving the quantum Langevin equation by its iteration. It is unique if normalised $\tau_t(I) = I_t$ to the solution of the Langevin equation for $A = I$, and it is *-multiplicative as in [30] due to the differential *-multiplicativity of $\mathbf{j}$. Then the properties (52) follow from the definition of the operators $\widehat{V}_t$, and can be checked recurrently by use of (49) and (50).

The cocycle $U = (U_t)$ is constructed to give the unitary solution to the HP equation (15), which always exists due to differential unitarity relations. If the solution is unique as in the case of all bounded coefficients K and L, it can be represented in the form of the stochastic multiple integral of the chronologically ordered products of the coefficients of the quantum differential equation under the integrability conditions given in [10].

If $K + K^\dagger \geq \varphi(I)$, the HP unitarity condition [6] is satisfied for the canonical choice $L_+^\circ = L^\circ$ and arbitrary isometric operator $J_\bullet^\circ$, $J_\bullet^\circ J_\bullet^\circ = I_\bullet^\bullet$ with $K_\bullet^- = L_\circ J_\bullet^\circ$, $K_\circ^- = L_\circ J_\circ^\circ$, $J_\circ^\circ = I_\circ^\circ - J_\bullet^\circ J_\bullet^\bullet$; if $K + K^\dagger \geq \varphi(I)$, in addition we make the choice $J_\bullet^\circ = L_\bullet^\circ$ from the canonical dilation, and so $K_\bullet^- = L_\circ L_\bullet^\circ = K_\bullet$, where $L_\bullet^* = L^\circ$, $J_\bullet^\circ = J_\bullet^{\circ*}$. In the first, subfiltering case $\lambda(I) \leq 0$ such a choice gives the coincidence $U_t(\delta_\emptyset \otimes \psi_0) = V_t \psi_0$ of the stochastic multiple integrals for any initial vacuum $\psi_0 = \eta \otimes \delta_\emptyset$, $\eta \in \mathcal{D}$, and therefore $\|V_t \psi_0\| = \|\psi_0\|$. Thus $M_t = V_t^* V_t$ is a martingale, and the equation (54) is satisfied for any initial ψ_0. In the second, contractive case $\lambda(I) \leq 0$ the canonical choice gives $U_t(\delta_\emptyset \otimes \psi) = V_t \psi$ and therefore $\|V_t \psi\| = \|\psi\|$ for any $\psi \in \mathcal{D} \otimes \mathfrak{F}$. Thus $V_t^* V_t = I$, and the equation (54) is satisfied for any state ψ.

References

1. Lindblad, G. On the Generators of Quantum Dynamical Semigroups. Comm. Math. Phys., **48**, pp. 119–130, 1976.
2. Evans, M. P. and Hudson, R. L. Multidimensional Quantum Diffusions. Lect. Notes Math., **1303**, pp. 69–88, 1988.
3. Belavkin, V. P. Chaotic States and Stochastic Integration in Quantum Systems. Russian Math. Survey, **47**, (1), pp. 47–106, 1992.
4. Parthasarathy, K. R. An Introduction to Quantum Stochastic Calculus. Birkhäuser, Basel, 1992.
5. Meyer, P. A. Quantum Probability for Probabilists, Lecture Notes in Mathematics, 1538, Springer-Verlag, Heidelberg, 1993.
6. Hudson, R. L. and Parthasarathy, K. R. Quantum Itô's Formula and Stochastic Evolution. Comm. Math. Phys., **93**, pp. 301–323, 1984.
7. Belavkin, V. P. On Stochastic Generators of Completely Positive Cocycles. Russ. J. Math. Phys., **3**, pp. 523–528, 1995.
8. Christensen, E. and Evans, D. E., Cohomology of Operator Algebras and Quantum Dynamical Semigroups. J. London Math. Soc., **20**, pp. 358–368, 1979.
9. Belavkin, V. P. Positive Definite Germs of Quantum Stochastic Processes. Comptes Rendus, **322**, 1, pp385–390. 1996.
10. Belavkin, V. P. Quantum Stochastic Positive Evolutions: Characterization, Construction, Dilation. Commun. Math. Phys. **184**, pp. 533–566, 1997.
11. Belavkin, V. P. On the General Form of Quantum Stochastic Evolution Equation. Stochastic Analysis and Applications, Proc. of Fifth Gregynog Symposium, World Scientific, Singapore 1996, pp. 91–106.
12. Lindsay, J. M. and Parthasarathy, K. R. Positivity and Contractivity of Quantum Stochastic Flows. Stochastic Analysis and Applications, Proc. of Fifth Gregynog Symposium, World Scientific, Singapore 1996, pp. 315–329.
13. Belavkin, V. P. Nondemolition Measurements and Nonlinear Filtering of Quantum Stochastic Processes. Lecture Notes in Control and Information Sciences, **121**, pp. 245–266, Springer-Verlag, 1988.
14. Belavkin, V. P. Nondemolition Calculus and Nonlinear Filtering in Quantum Systems. In: Stochastic Methods in Mathematics and Physics, pp. 310–324, World Scientific, 1989.
15. Belavkin, V. P. Quantum Stochastic Calculus and Quantum Nonlinear Filtering. J. Multivariate Analysis, **42** (2), pp. 171–201, 1992.
16. Gisin, N. Phys. Rev. Lett., **52**, pp. 1657–60, 1984.
17. Diosi, L. Phys. Rev. A **40**, pp. 1165–74, 1988.
18. Barchielli, A. and Belavkin, V.P. Measurement Continuous in Time and a Posteriori States in Quantum Mechanics. J. Phys. A: Math. Gen., **24**, pp. 1495–1514, 1991.
19. Belavkin, V. P. Quantum Continual Measurements and a Posteriori Collapse on CCR. Commun. Math. Phys., **146**, pp. 611–635, 1992.
20. Milburn, G. Phys. Rev. A, **36**, p. 744, 1987.
21. Pearle, P. Phys. Rev. D, **29**, p. 235, 1984.
22. Ghirardi, G. C., Pearle, P. and Rimini, A. Markov Processes in Hilbert Space and Continuous Spontaneous Localization of Systems of Identical Particles. Phys. Rev. A, **42**, pp. 78–89, 1990.
23. Belavkin, V. P. A Continuous Counting Observation and Posterior Quantum Dynamics. J. Phys. A: Math. Gen. **22**, pp. L1109–14, 1989.
24. Belavkin, V. P. A Posterior Schrödinger Equation for Continuous Nondemolition Measurement. J. Math. Phys., **31**, pp. 2930–34, 1990.

25. Collett, M. J. and Gardiner C. W. Input and output in Damped Quantum System. Phys. Rev., A, **31**, pp. 3761–74, 1985.
26. Carmichael, H. Open Systems in Quantum Optics. Lecture Notes in Physics, **18**, Springer-Verlag, 1993.
27. Holevo, A. S. Prob. Theory and Related Fields, **104**, 483-500, 1996.
28. Obata, N. White Noise Calculus and Fock Space. Lecture Notes in Mathematics, 1577, Springer-Verlag, Heidelberg, 1994.
29. Stinespring, W. F. Positive Functions on C*-algebras, Proc. Amer. Math. Soc. **6**, pp. 242–247, 1955.
30. Fagnola, F. and Sinha, K. B. Quantum Flows with Unbounded Structure Maps and Finite Degrees of Freedom. J. London Math. Soc., **48**, pp. 537-551, 1993.
31. Belavkin, V. P. A Quantum Nonadapted Itô Formula and Stochastic Analysis in Fock Scale. J. Funct. Anal., **102**, No. 2, pp. 414–447, 1991.
32. Fagnola, F. Characterization of Isometric and Unitary Weakly Differentiable Cocycles in Fock Space, Quantum Probability and Related Topics, 8, World Scientific, Singapore 1993, pp. 143–164.
33. Belavkin, V. P. Multiquantum Systems and Point Processes I, Rep. in Math. Phys., **28**, No. 1, pp. 57–90, 1989.
34. Chebotarev, A. M. The Theory of Conservative Dynamical Semigroup and its Applications. J. of Soviet Math., **56**, pp. 2697-2719,1991.
35. Belavkin, V. P. Stochastic Positive Flows and Their Generators. In: Probability Theory Towards 2000, ed L. Accardy and C. C. Heyde, Springer, 1996.

From Stochastic Semigroups
to Chaotic Dynamics

I. Antoniou[1,2], K. Gustafson[1,3], and Z. Suchanecki[1,4]

[1] International Solvay Institutes for Physics and Chemistry, C.P. 231, Campus Plaine ULB, Bd. du Triomphe, 1050 Brussels, Belgium
[2] Theoretische Natuurkunde, Free University of Brussels, Belgium
[3] Department of Mathematics, CB 395, University of Colorado, Boulder, CO 80309-0395, USA
[4] Hugo Steinhaus Center and Institute of Mathematics, Wrocław Technical University, ul. Wybrzeże Wyspiańskiego 27, 50-370 Wrocław, Poland

Summary. We answer qualitatively the inverse coarse-graining problem of statistical physics, namely: which microscopic dynamics give rise to a given physically observed Markov semigroup as a result of exact coarse graining? We show in particular that all measure preserving stationary Markov processes arise as projections of Kolmogorov dynamical systems. This result shows moreover the physical significance of the Misra-Prigogine-Courbage theory of irreversibility. Because we want positivity preserving transformations, our procedure although analogous to the Sz-Nagy-Foias Dilation theory has a different viewpoint, that of positive dilations.

1. Introduction and Background

Some time ago one of us (K.G.) conjectured in [1] the possibility of dilating any given stochastic Markov semigroup evolution into a unitary group evolution arising from a point dynamics. This question is not only essential for the foundations of statistical physics, but more generally sheds light into the intrinsic nature of probabilistic processes because a positive answer means that the physically observed Markov processes such as kinetic or diffusive processes may arise as projections of corresponding dynamical systems in larger spaces without any approximations. The question is also relevant for the physical significance of the Misra-Prigogine-Courbage theory of irreversibility.

According to the Misra-Prigogine-Courbage theory [2]–[15], originally formulated in L^2 spaces and later in L^1 spaces [16], Kolmogorov dynamical systems can be projected (coarse grained) onto or intertwined with irreversible Markov semigroups without any approximation or weak coupling assumptions. In the other direction, the following are known:

1. If a reversible dynamical system can be projected onto an irreversible Markov semigroup, then the original system necessarily has the Kolmogorov property [5, 7].

2. If a reversible dynamical system is similar to an irreversible Markov semigroup, then the original system necessarily has absolutely continuous spectrum [8], i.e. it is between mixing and Kolmogorov systems in the ergodic hierarchy [17]–[19].

3. A class of Markov semigroups, which we have called exact Markov semigroups and which includes all semigroups in the coarse grained Misra-Prigogine-Courbage theory of irreversibility, possesses positive dilations to Kolmogorov dynamical systems [20].

4. A larger class of Markov semigroups, but restricted to the discrete parameter case, possesses dilations to measurable but not necessarily measure preserving dynamical systems. When the semigroups are measure preserving, the positive dilations are Kolmogorov systems [21].

At this point some background comment on dilations theories should be provided for the reader. We may describe the situation as composed of three such dilation theories, those which we shall call for brevity the Sz.-Nagy–Foias (also Halmos, Schaeffer, others) theory, the Kolmogorov–Rokhlin (and others) theory, and the Akcoglu–Sucheston (also Vershik, others) theory. For brevity we shall just call these the Sz.-Nagy approach, the Kolmogorov approach, and the Akcoglu approach. It is convenient to classify the situation as follows. Suppose one has linear operators T on functions f on a region Ω carrying a measure μ and a point transformation S. Then the Sz.-Nagy theory dilates by extending the operators T, the Kolmogorov theory dilates by extending the transformation S, and the Akcoglu theory dilates by extending the measure μ.

This overly simplified but usefully clarifying picture of the three dilation approaches will be amplified in Section 3 where we present our main result [22]. The Sz.-Nagy theory does not provide meaningful positivity (in our opinion) in the dilations. Therefore in [20] we instead used the Kolmogorov theory to dilate with meaningful positivity the Misra-Prigogine-Courbage semigroups. In [21] we employed the Akcoglu theory to dilate with meaningful positivity more general Markov semigroups. However, that theory restricted us to the discrete case (i.e. Markov chains).

In the present paper we consider the general case, with parameter t either discrete (system, chain) or continuous (flow, process). Our approach here depends on extending conditional probabilities on cylindrical algebras using the Ionescu-Tulcea Theorem [23]. In section 2 we recall the salient facts about dynamical systems and Markov semigroups that we will need in the analysis. Key roles are played by the Koopman and Frobenius-Perron operators of dynamical systems and by the Markov and Stochastic operators of Markov processes. Our dilation results are presented in section 3. Section 4 contains a summary of the state of these investigations at the present time.

2. Dynamical Systems and Markov Processes

A one parameter evolution semigroup $\{S_t\}$ of transformations of the space Ω of states defines a dynamical system. The variable t signifies time and is continuous for flows and discrete for cascades. For reversible systems $\{S_t\}$ is a group of automorphisms of Ω. The space Ω is equipped with a topology in order to formulate nearness and stability, and with a measure structure, compatible with the topology, for the description of the statistical properties. The measure structure includes an algebra $\mathcal{A}$ of measurable sets and an S_t invariant measure μ representing an equilibrium distribution.

The phase functions f evolve according to the Koopman operators

$$V_t f(\omega) = f(S_t \omega) , \quad \omega \in \Omega ,$$

induced by the given dynamical system. The Koopman operators are isometries on the Banach space $L^p = L^p(\Omega, \mathcal{A}, \mu)$, $p \geq 1$ of p-integrable functions and unitary operators when restricted to the Hilbert space L^2 and the transformations S_t are automorphisms. Koopman introduced in 1931 [24] these operators in order to study the ergodic properties of dynamical systems using the powerful tools of operator theory. One year later, Koopman and von Neumann [25] observed that for dynamical systems with continuous spectra, "the states of motion corresponding to any set become more and more spread out into an amorphous everywhere dense *chaos*. Periodic orbits, and such like, appear only as very special possibilities of negligible probability". This result was refined later into the ergodic hierarchy of mixing, Kolmogorov and exact systems [17]–[19] which is the basis of the modern theory of "chaos". To our knowledge, the paper of Koopman and von Neumann was the first use of the term chaos in the context of dynamical systems.

The relation of the point dynamics with the Koopman operators is clarified by asking the converse question, namely: what types of isometries on L^p spaces are implementable by point transformation? Such theorems on the implementability of isometries on L^p spaces , $p \neq 2$, are known as Banach-Lamperti theorems [1]. It is perhaps interesting to briefly comment on the history of these results. Banach essentially answered the converse question for $1 \leq p < \infty$ in his famous book [26]. Lamperti [27] somewhat later extended the answer to the converse question to $0 < p < 1$. The converse to Koopman's lemma in the case $p = 2$ came later yet, in Goodrich, Misra and one of us (K.G.) [9]. An interpretation for this rather strange situation in which the non Hilbert space theorems are found before the Hilbert space case is resolved is important to interject here: there are a great many isometries in Hilbert space, but not many in the other spaces. Thus it takes very little to find that the latter all have underlying point transformations. However, in the Hilbert space case, we do not want all of the isometries: only those which represent the physical situation of interest to us. In the present context, we need to preserve the positivity of probability densities. The result is not only simple but also physically appealing: any isometry V on the space $L^2(\Omega, \mathcal{A}, \mu)$ which preserves the constants and the positivity of phase functions is implementable by a necessarily measure preserving transformation S

$$V f(\omega) = f(S\omega) , \quad \omega \in \Omega .$$

We will use this result later in order to point out the significance of positive dilations.

After Koopman's idea to apply operator methods to dynamical systems, operators were also used for the study of Markov processes [28]–[33]. Markov processes provide the best known prototype for the description of irreversible evolutions such as kinetic or diffusion processes. A stationary Markov process on the phase space $(\Omega, \mathcal{A}, \mu)$ is described by the transition probabilities $Q_t(A, \omega)$ from the point ω in Ω into the measurable set A in $\mathcal{A}$, in time $t \geq 0$. Precisely, we assume that $Q(\cdot, \cdot)$ is a function on $\Omega \times \mathcal{A}$ such that

a) for every $\omega \in \Omega$ $Q(\cdot, \omega)$ is a probability measure on $(\Omega, \mathcal{A})$
b) for every $A \in \mathcal{A}$ $Q(A, \cdot)$ is a measurable function on $(\Omega, \mathcal{A})$.

The probability densities f evolve according to the Markov semigroup $\{M_t\}$ on $L^p(\Omega, \mathcal{A}, \mu)$ defined by the condition:

$$(1) \qquad \int_A (M_t f)(\omega)\, \mu(d\omega) = \int_\Omega Q_t(A, \omega) f(\omega)\, \mu(d\omega), \quad f \in L^p.$$

The Markov operators M_t have the following properties:

$\alpha)$ M_t are contractions on L^p

$$\|M_t f\| \leq \|f\|$$

$\beta)$ M_t preserve the probability densities

$$M_t f \geq 0 \quad \text{if} \quad f \geq 0$$

$\gamma)$ M_t preserve the probability normalization, i.e.

$$\int_\Omega M_t f\, d\mu = \int_\Omega f d\mu \quad \text{for} \quad f \geq 0$$

In the following we will mainly consider the *measure preserving Markov semigroups*, i.e. those with the additional property

$\delta)$ M_t preserves constants

$$M_t 1 = 1$$

We remark that generally an abstract linear operator M on L^p which satisfies properties $\beta)$, $\gamma)$ and $\delta)$ is called a *doubly stochastic operator* and that in such a case the property $\alpha)$ follows from the others. We also note that for L^2 spaces, $\alpha)$ and $\gamma)$ implies $\delta)$. Under additional rather standard assumptions on the space $(\Omega, \mathcal{A})$, for example if Ω is a Polish space and $\mathcal{A}$ is a Borel σ-algebra, see [34], Prop.V.4.4, such M are induced by transition probabilities. For applications there is an additional interesting condition

$\varepsilon)$ Irreversible approach to equilibrium

$$\|M^n f - 1\|_{L^p} \longrightarrow 0, \quad \text{as } n \to \infty \text{ for each probability density } f \in L^p$$

Semigroups with the properties $\alpha) - \varepsilon)$ have been called strong Markov semigroups in [2], monotonic Markov semigroups in [5], strongly irreversible Markov semigroups in [5]–[7] and exact Markov semigroups in [12]. In this paper however we shall adopt the term *irreversible Markov semigroup* for those satisfying the conditions $\alpha) - \varepsilon)$ and reserve the term *exact Markov semigroup* for semigroups which are the Frobenius-Perron semigroups of exact dynamical systems [17]–[19]. We have shown [20] that Misra, Prigogine, Courbage semigroups are exact Markov semigroups and arise as projections of Kolmogorov systems, thus providing a converse to the Misra, Prigogine and Courbage theory of irreversibility [2]–[15].

3. Positive Dilations of Probabilistic Processes

The natural framework from which to study the question whether a Markov probabilistic process may arise as a projection of a deterministic dynamical system is the theory of operator dilations. A semigroup of isometries $\tilde{V}^n$ on a Banach space $\tilde{X}$ is a dilation of the contractive semigroup V^n on the Banach space X if there exist two linear operators $I : X \longrightarrow \tilde{X}$ and $E : \tilde{X} \longrightarrow X$ such that

$$V^n = E\tilde{V}^n I , \quad \text{for each } n \in \mathbf{N} .$$

In other words the following diagram commutes

$$
\begin{array}{ccc}
X & \xrightarrow{V^n} & X \\
I \downarrow & & \uparrow E \\
\tilde{X} & \xrightarrow{\tilde{V}^n} & \tilde{X}
\end{array}
$$

Usually, a dilation is realized in such a way that X is isomorphic to a subspace of $\tilde{X}$, I is the canonical injection of X into $\tilde{X}$, and E is the projection of $\tilde{X}$ onto X.

If X is a Hilbert space and V is a unitary operator we may speak about unitary dilations of contractions. Dilations of contractive semigroups were constructed by Sz.-Nagy [35]–[37] following the Halmos [38, 39] construction of a unitary dilation of a contraction and Naimark's idea [40] of extending operators beyond a given Hilbert Space. The motivation of dilation theory was to find a way to study general bounded operators through isometries or unitary operators on larger spaces. An excellent treatment of the dilation theory and in particular of the structure of the minimal isometric and unitary dilations may found in the book Sz-Nagy and Foias [35].

We are interested here however, in more specific dilations of a Markov semigroup $\alpha) - \delta)$ acting on the Banach space $X = L^p(\Omega, \mathcal{A}, \mu)$, to a group acting on a larger Banach space $\tilde{X} = L^p(\tilde{\Omega}, \tilde{\mathcal{A}}_0, \tilde{\mu})$ and which corresponds to a dynamical system. Therefore we want to show that

a) The dilated evolution $\tilde{V}$ is implementable by a measure preserving transformation S on $\tilde{A}$

b) The projection E onto the space X is the conditional expectation.

In particular we are looking for a dilation such that both $\tilde{V}^n$, $n = 0, \pm 1, \pm 2, \ldots$ and E are positivity preserving.

Let us now see if one can define positivity in Sz-Nagy-Foias construction summarized [35], Ch.I], as follows:

1. Consider the Hilbert space

$$\mathcal{H} = \bigoplus_{n=-\infty}^{\infty} H_n \; , \quad H_n = H \; , \quad n = 1, 2, \ldots$$

of square summable bilateral sequences $(\ldots, f_{-2}, f_{-1}, f_0, f_1, f_2, \ldots)$

2. Define the unitary operator U on the space $\mathcal{H}$ as follows:

$$U(\ldots, f_{-2}, f_{-1}, f_0, f_1, \ldots) = (\ldots, f_{-1}, \sqrt{I - V^*V} f_0 - V^* f_{-1}, f_0 + \sqrt{I - VV^*} f_1, f_2, \ldots)$$
$$= (\ldots, (Uf)_{-2}, (Uf)_{-1}, (Uf)_0, (Uf_1), \ldots) \; .$$

U is a unitary Dilation of V but not a minimal one.

3. The space $\tilde{H}$ of the minimal dilation is the following U-reducing subspace of the space $\mathcal{H}$

$$\tilde{H} = \bigvee_{n=-\infty}^{\infty} U^n H \; .$$

4. The minimal unitary dilation $\tilde{V}$ is the restriction of the operator U onto the space $\tilde{H}$.

It is clear from the explicit construction of the Sz-Nagy-Foias Dilation that the positivity of the sequence space is physically meaningless, because componentwise positivity is an artifact of the construction which carries no special positivity properties from the original Markov semigroup. Apart from that, one has to show moreover that the terms $(Uf)_{-1}$ and $(Uf)_0$ are positive, a property which puts severe restrictions to the Markov semigroup $\tilde{V}^n$. Attempts to apply the Sz-Nagy dilation construction to Markov processes associated with Brownian motion and thus "make a heat bath" [41] did not clarify the positivity question either. They eliminated instead the concept of positivity from the definition of Kolmogorov systems and they gave the name "Kolmogorov structure" to unitary shifts [35].

Faced with the difficulty to obtain meaningful positive dilations of Markov semigroups using the Sz-Nagy dilation theory, we constructed [20] a positive dilation based on the natural extensions of dynamical systems [17]–[18]. There the positivity is clearly related to that of the inducing semigroup. We applied our construction to the Misra-Prigogine-Courbage semigroups which were shown to satisfy the properties of exact Markov semigroups. The result of [20] is that any Misra-Prigogine-Courbage semigroup arises as a projection of a Kolmogorov dynamical system. The projection E is the conditional expectation over the cells of the K-partition. In this way we proved the converse statement to the Misra-Prigogine-Courbage construction, that shows that those constructions must fall within the confines of a theory of exact Markov semigroups which may be regarded as projections of Kolmogorov systems.

We then turned to the case of Markov chains (t the discrete integer parameter system) and were able to use the Akcoglu dilation theory to show that every Markov chain satisfying conditions $\alpha) - \gamma)$ possesses an extension to a deterministic dynamical system on a larger space. The Akcoglu et al theory [42]–[45] for L^p spaces is spread over several papers and moreover depends on previous results of Rokhlin [46] and Maharam [47] on structures of measure spaces. Kern, Nagel and Palm [48] published a simpler construction of these dilations for Banach lattices using only analytic methods. During this period there were similar dilations results obtained by Vershik and others in Russia [49]. Similar versions were explored earlier by Douglas, see eg. [50], and others in America. Thus what we are referring to as the Akcoglu dilation theory existed as much in the folklore as in the published literature during the 1970's. To our knowledge the rigorous published theory does not cover the continuous parameter case.

It is now our task to extend the dilation approach to any arbitrary Markov Semigroup. Let us consider first the case of a single doubly stochastic operator M on L^1 i.e. M satisfies $\beta) - \delta)$. Our aim is to construct a positive dilation $\tilde{M}$ of M which is implemented by a measure preserving transformation S of some measure space $(\tilde{\Omega}, \tilde{\mathcal{A}}, \tilde{\mu})$. Moreover, we will show that if M satisfies additionally $\varepsilon)$ then the dynamical system $(\tilde{\Omega}, \tilde{\mathcal{A}}, \tilde{\mu}; S)$ is a Kolmogorov system. The case of dilations of an ordinary Markov chain, which satisfies only $\alpha) - \gamma)$ discussed earlier using Akcoglu et al methods [21] will also follow from these considerations.

Let us recall that M corresponds to a transition probability $Q(A, \omega)$ through (1) and let us denote by W the predual stochastic operator of M on L^∞. Then the formula (1) can be equivalently expressed for W as follows

$$ W f(\omega) = \int_\Omega f(\omega') \, Q(d\omega', \omega) \ . $$

Now we can start the construction of the dilation. First we consider the discrete case and define the product space

$$ \tilde{\Omega} = \prod_{n=-\infty}^{\infty} \Omega_n \ , \quad \text{where } \Omega_n = \Omega \ , $$

and the σ-algebra $\tilde{\mathcal{A}}$ – generated by the cylindrical sets

$$ (2) \qquad C_{A_1,\ldots,A_k} = \ldots \times \Omega \times A_1 \times A_2 \times \ldots \times A_k \times \Omega \times \ldots \ , $$

where $k \in \mathbf{N}$, $A_i \in \mathcal{A}$.

On $(\tilde{\Omega}, \tilde{\mathcal{A}})$ we define the cylindrical measure $\tilde{\mu}$

$$\tilde{\mu}(C_{A_1,\ldots,A_k}) = \int_{A_1} \int_{A_2} \cdots \int_{A_k} Q(d\omega_k, \omega_{k-1})\ldots Q(d\omega_2, \omega_1)\, \mu(d\omega_1)\,.$$

It can be shown [22] that $\tilde{\mu}$ is correctly defined, i.e. the value of $\tilde{\mu}$ does not depend on either right or left shifts of the base sets of the cylinder (2). Now, using the Kolmogorov theorem [51] we can extend $\tilde{\mu}$ on the σ-algebra $\tilde{\mathcal{A}}$. Denote by S the left shift on $\tilde{\Omega}$, i.e.

$$S\tilde{\omega} = S(\omega_n)_{-\infty}^{\infty} \overset{\text{def}}{=} (\omega_{n+1})_{-\infty}^{\infty}\,,\ \tilde{\omega} \in \tilde{\Omega}\,.$$

Then as was shown above $\tilde{\mu}$ is invariant with respect to S. Let $\tilde{M}$ be the operator on $L^1(\tilde{\Omega}, \tilde{\mu})$ associated with S

$$\tilde{M}f(\tilde{\omega}) = f(S^{-1}\tilde{\omega})\,,$$

We have shown [22] that $\tilde{M}$ is a dilation of M. Precisely, we have

$$M^n = E\tilde{M}^n I\,,\ n \in \mathbf{N}\,,$$

where I is the canonical injection $L^1(\Omega)$ into $L^1(\tilde{\Omega})$,

$$(If)(\ldots, \omega_{-1}, \omega_0, \omega_1, \ldots) \overset{\text{def}}{=} f(\omega_0)$$

and E is the conditional expectation with respect to the σ-algebra generated by the cylinders

$$C_A^0 = \{\tilde{\omega}\ :\ \omega_0 \in A\}\,,\ A \in \mathcal{A}\,.$$

Consequently the function $E\tilde{M}^n If$ depends only on the 0–coordinate and can be uniquely identified with a function on Ω.

Let us next consider the case in which M is an irreversible Markov operator, i.e. it satisfies additionally the condition ε). This additional assumption implies that S is a K-automorphism of $(\tilde{\Omega}, \tilde{\mathcal{A}})$. To be more precise, we choose as a distinguished σ-algebra $\tilde{\mathcal{A}}_0$ the σ-algebra generated by those cylinders $C_{A_0,\ldots,A_k}^{0,\ldots,k}$ ($k = 0, 1, \ldots, A_0, \ldots, A_k \in \mathcal{A}$) for which the set A_0 is placed on the 0th coordinate and note that the σ-algebra $S^{-n}\tilde{\mathcal{A}}_0$, $n \in \mathbf{Z}$, is the σ-algebra generated by $C_{A_0,\ldots,A_k}^{n,\ldots,n+k}$ with A_0 placed on the nth coordinate. Then we can prove (see [22] for details) that

(i) $S\tilde{\mathcal{A}}_0 \supset \tilde{\mathcal{A}}_0$

(ii) $\sigma(\bigcup_{n=-\infty}^{\infty} S^n \tilde{\mathcal{A}}_0) = \tilde{\mathcal{A}}.$

(iii) $\bigcap_{n=-\infty}^{\infty} S^n \tilde{\mathcal{A}}_0$ is the trivial σ-algebra.

The above construction for a single operator M extends to the case of continuous parameter semigroup of doubly stochastic operators $\{M_t\}_{t \geq 0}$ on the space $L^1(\Omega, \mu)$ which satisfies ε). As before we assume that each M_t corresponds through (1) to the transition probability $Q_t(A, \omega)$. Now the dilation space $\tilde{\Omega}$ will be the product

$$\tilde{\Omega} = \prod_{t \in \mathbf{R}} \Omega_t \, , \quad \text{where } \Omega_t = \Omega \, .$$

The σ-algebra $\tilde{\mathcal{A}}$ is generated by the cylinders

$$C_{A_0, \ldots, A_k}^{t_0, \ldots, t_k} \, , \quad \text{where } t_0, \ldots, t_k \in \mathbf{R}, \quad A_0, \ldots, A_k \in \mathcal{A} \, .$$

For such cylinders we define conditional probabilities

$$\tilde{P}_\omega(C_{A_0, \ldots, A_k}^{t_0, \ldots, t_k}) = \mathbf{1}_{A_0}(\omega) \int_{A_1} \cdots \int_{A_k} Q_{t_k - t_{k-1}}(d\omega_k, \omega_{k-1},) \ldots Q_{t_1 - t_0}(d\omega_1, \omega)$$

(see [51] section 43.2) which can be extended by the Ionescu–Tulcea Theorem [23] on $\tilde{\mathcal{A}}$. Then using the P_t – invariant measure μ we define the invariant measure $\tilde{\mu}$.

$$\tilde{\mu}(C_{A_0, \ldots, A_k}^{t_0, \ldots, t_k}) = \int_\Omega \tilde{P}_\omega(C_{A_0, \ldots, A_k}^{t_0, \ldots, t_k}) \, \mu(d\omega) \, .$$

By double stochasticity of $\{M_t\}$ $\tilde{\mu}$ is automatically invariant, i.e.

$$\tilde{\mu}(C_{A_0, \ldots, A_k}^{t_0+s, \ldots, t_k+s}) = \tilde{\mu}(C_{A_0, \ldots, A_k}^{t_0, \ldots, t_k}) \, , \quad \text{for each } s \, .$$

Then we may define the group of shifts transformations of $\tilde{\Omega}$

$$S_t \tilde{\omega} = \omega(\cdot + t) \, , \quad \text{for each function } \tilde{\omega} = \omega(\cdot) \in \tilde{\Omega} \, .$$

The distinguished σ-algebra $\tilde{\mathcal{A}}_0$ is defined in the same way as in the discrete case. Also the conditions (i), (ii) and (iii) for the flow $\{S_t\}$ are satisfied [22]. Finally, we define the operators

$$\tilde{M}_t f(\tilde{\omega}) = f(S_{-t}) \, , \quad f \in L^1(\tilde{\Omega}, \tilde{\mu}) \, ,$$

and check as in [22] that $\tilde{M}_t$ are the dilations of M_t.

4. Concluding Remarks

1. The question of how one may embed probabilistic processes into larger deterministic dynamics has been investigated. We showed in [20] a natural converse to the Misra-Prigogine-Courbage theory of the reduction from deterministic dynamics to probabilistic predictions. In [21] we extended our theory to Markov chains. Here we have extended the theory to arbitrary Markov semigroups. If the Markov semigroup is also measure preserving, the dilated

evolution turns out to be a Kolmogorov system. However the projection to the Markov evolution is an averaging over the generating partition and not over the partition which Misra, Prigogine and Courbage considered [2]–[8]. This means that only exact Markov semigroups arise as projections of Kolmogorov systems according to the Misra-Prigogine-Courbage construction, while the general measure preserving Markov processes arise from Kolmogorov systems as projections over the generating partition.

2. We realized also that although one may obtain abstract dilations by the Sz.-Nagy approach [35], the positivity in them is not clearly related to that of the inducing semigroup. The Akcoglu approach [42] which we used in [21] gives a meaningful positive dilation but is specific to the discrete case. Abstract positive dilations on Banach lattices by the Akcoglu approach were constructed by Kern, Nagel and Palm [48]. They showed moreover that the positive dilations preserve, in most cases, the ergodicity, mixing and irreducibility of the original Markov semigroup. However Kern, Nagel and Palm [48] do not address the Kolmogorov property because their arguments are based on the theory of Banach lattices and not on the specific measure theoretic constructions of Akcoglu et al [42]–[45].

3. Even if the dilation of an irreversible measure preserving Markov semigroups is a reversible K-system, the dynamical instability of K-systems necessitates manifestly irreversible realizations. Such realizations correspond for example to the Misra-Prigogine-Courbage construction [2]–[7] or to the recently obtained time-asymmetric spectral decomposition of the Frobenius-Perron operator [52, 53].

4. We have not touched the question of intertwining by a similarity Λ an arbitrary Markov semigroup with some deterministic dynamics. The answer to this question would provide another converse to the Misra-Prigogine-Courbage construction of non-unitary transformations mentioned in the introduction. One reason we have not considered this converse question is because the original situation (to which the converse would apply) is not yet well enough understood. For example, the correct physical properties of such Λ transformations is not yet clarified.

5. The space L^1 on which the doubly stochastic operator M of section 3 acts can be replaced by any other L^p-space, $1 \leq p \leq \infty$. Then if M satisfies on L^p properties $\beta) - \delta)$ it extends to a doubly stochastic operator on L^1. Indeed, in such a case M satisfies conditions $\beta) - \delta)$ on the "smallest" space L^∞, which is equivalent to the double stochasticity of the adjoint M^* on L^1 [54] Prop.1.1, p.6. Restricting M^* to L^∞ and taking once more its adjoint we obtain the required result. Moreover, the L^1-norm is the weakest (for bounded spaces Ω) among all L^p-norms, $p \geq 1$. Thus condition $\varepsilon)$ in any L^p norm is enough. Recall as we pointed out in Section 2, that isometries are much more prevalent in L^2, so there the positivity requirement is more interesting physically.

6. The dilation $\tilde{M}$ obtained in Section 3 is minimal in the following sense. Suppose that there exists another positive dilation Q of M which is implemented by a measure preserving transformation T on some intermediate space Ω'

$$\Omega \subset \Omega' \subset \tilde{\Omega} .$$

Then $\Omega' = \tilde{\Omega}$ and $T = S$. To prove this let us restrict our doubly stochastic operator M and its dilation to L^2-spaces. Then the Sz.-Nagy–Foias minimal (but not necessary positive) dilation V is on the Hilbert space

$$\mathcal{H} = \bigotimes_{n=-\infty}^{\infty} H_n , \quad \text{where} \quad H_n = L^2(\Omega) .$$

Therefore the space $L^2(\Omega')$ of the dilation Q must be between $\mathcal{H}$ and $L^2(\tilde{\Omega})$. However, the elements (functions) from $\mathcal{H}$ are dense in $L^2(\tilde{\Omega})$, thus $L^2(\Omega')$ must coincide with $L^2(\tilde{\Omega})$.

7. Some further details of this theory were presented in the lectures [55]. As pointed out by Professor Ojima's appended comments there, one may take a more general view which encompasses both classical systems and also quantum systems. Then the differences with different p choices between L^p spaces become more significant. In particular the L^∞, L^1 and L^2 settings carry quite different meanings. However, actually in [55], Remark 1.2.4, p.62 we already pointed out that we preferred the positivity preserving, more physical, L^2 point of view, but that one could as well obtain results from a multiplication preserving $L^1 \cap L^2$ point of view, or from a von Neumann algebra, L^∞ point of view. How these other approaches apply is discussed in [1].

8. There is a further way to dilate, beyond the three philosophies we have delineated here, that could prove interesting for treating quantum systems. That is the dilation theory for operator algebras. Then the natural positivity is that of completely positive maps [56]. The potential of this approach to yield physically meaningful underlying chaotic dynamics from given stochastic semigroups will be investigated elsewhere.

9. In summary, our result provides a qualitative answer to the inverse problem of statistical physics: which microscopic dynamics give rise to a given physically observed irreversible Markov processes as an exact coarse graining projection? The answer is that all such microscopic dynamics are chaotic and they are isomorphic to the Kolmogorov system constructed through the positive dilation of the Markov process. This Kolmogorov system therefore serves as a prototype for the inverse problem of statistical physics.

Acknowledgments: We thank Professor Ilya Prigogine for his interest, encouragement and support of this work. This work was partly supported by the Belgian Interuniversity Attraction Poles, the NATO CR Grant 920977R, and the ESPRIT Contract 21042 CTIAC.

References

1. K. Gustafson, R. Goodrich, A Banach-Lamperti theorem and Similarity transformations in Statistical Mechanics, *Coll. Math. Soc. Ianos Bolyai* **35**, 567-579 (1980).
2. B. Misra, I. Prigogine, M. Courbage, From deterministic dynamics to probabilistic descriptions, *Physica* **98A**, 1-26 (1979).
3. I. Prigogine, *From being to becoming*, Freeman, San Francisco, 1980.
4. S. Goldstein, B. Misra, M. Courbage, On intrinsic randomness of dynamical systems, *J. Stat. Phys.* **25**, 111-126 (1981).
5. B. Misra, I. Prigogine, On the foundations of kinetic theory, *Suppl. Progress of Theor. Phys.* **69**, 101-110 (1980).
6. B. Misra, I. Prigogine, Time Probability and Dynamics in *Long Time Predictions in Dynamical Systems*, ed. C. Horton, L. Reichl, V. Szebehely, Wiley, N.Y., 21-43, 1983.
7. R. Goodrich, K. Gustafson, B. Misra, On K-flows and irreversibility, *J. Stat. Phys.* **43**, 317-320 (1986).
8. B. Misra, Nonequilibrium entropy, Lyapunov variables, and ergodic properties of classical systems, *Proc. Natl. Acad. USA* **75**, 1627-1631 (1978).
9. K. Goodrich, K. Gustafson, B. Misra, On converse to Koopman's Lemma, *Physica* **102A**, 379-388 (1980).
10. I. Prigogine, C. George, F. Henin, I.Rosenfeld, A unified formulation of dynamics and thermodynamics, *Chem. Scr.* **4**, 5-32 (1973).
11. I. Antoniou, B. Misra, The relativistic internal time algebra, *Nuclear Physics, Proceed. Suppl. Sect.* **6**, 240-242 (1989).
12. I. Antoniou, B. Misra, Characterization of semi-direct sum Lie algebras, *J. Math. Phys.* **32**, 864-868 (1991).
13. I. Antoniou, B. Misra, Non-unitary transformation of conservative to dissipative evolutions, *J. Phys.* A: Math. Gen. **24**, 2723-2729 (1991).
14. Z. Suchanecki, I. Antoniou, S. Tasaki, Nonlocality of the Misra-Prigogine-Courbage semigroup, *J. Stat. Physics* **75**, 919-928 (1994).
15. Z. Suchanecki, A. Weron, Applications of an operator stochastic integral in Prigogine's theory of irreversible dynamical systems, *Expo. Math.* **8**, 67-79 (1990).
16. Z. Suchanecki, An L^1 extension of stochastic dynamics for irreversible systems, *Lecture Notes in Math.* No.**1391**, 367-374 (1989).
17. I. Cornfeld, S. Fomin, Ya. B. Sinai, *Ergodic Theory*, Springer, Berlin, 1982.
18. Ya. B. Sinai (ed.), *Dynamical Systems . II: Ergodic Theory with applications to Dynamical Systems and Statistical Mechanics*, Springer, Berlin, 1989.
19. A. Lasota, M. Mackey, *Probabilistic Properties of Deterministic Systems*, Cambridge University Press, Cambridge, 1985.
20. I. Antoniou, K. Gustafson, From probabilistic description to deterministic dynamics, *Physica* **A 197**, 153-166 (1993).
21. I. Antoniou, K. Gustafson, From irreversible Markov semigroups to chaotic dynamics, *Physica* A (1997), to appear.
22. I. Antoniou, K. Gustafson, Z. Suchanecki, Dilations of stationary Markov processes, *Math. Nachr.* (submitted)
23. C.T. Ionescu Tulcea, Mesures dans les espaces produits, *Lincei – Rend. Sc. fis. mat. e nat.* **7**, 208-211 (1949).
24. B. Koopman, Hamiltonian systems and transformations in Hibert space, *Proc. Nat. Acad. Sci. USA* **17**, 315-318 (1931).
25. B. Koopman, J. von Neumann, Dynamical systems of continuous spectra, *Proc. Nat. Acad. Sci. USA* **18**, 255-266 (1932).

26. S. Banach, *Théorie des Opérations Lineaires*, Monografje Matematyczne, War-szawa 1932.
27. J. Lamperti, On the isometries of certain function spaces, *Pacific J. Math.* **8** 459-466 (1958).
28. K. Yosida, S. Kakutani, Operator–theoretical treatment of Markoff's process and mean ergodic theorem, *Annals of Mathematics* **42**, 188-229 (1941).
29. E. Hopf, The general temporally discrete Markov process, *J. Rational. Mech. and Anal.* **3**, 13-44 (1954).
30. K. Yosida, *Functional Analysis*, 6th ed. Springer, Berlin, 1980.
31. E. Dynkin, *Markov Processes*, Springer, Berlin, 1965.
32. S. Foguel, *The Ergodic Theory of Markov Processes*, Van Nostrand, New York, 1969.
33. M. Lin, Mixing for Markov operators, *Z. Wahrscheinlichkeitstheorie Verw. Geb.* **19**, 231-242 (1971).
34. J. Neveu, *Mathematical Foundations of the Calculus of Probability*, Holden–Day Inc. 1965.
35. B. Sz-Nagy, C Foias, *Harmonic Analysis of Operators in Hilbert Space*, North-Holland, Amsterdam, 1970.
36. B. Sz-Nagy, Sur les contractions de l'espace de Hilbert, *Acta Sci. Math.* **15**, 87-92 (1953).
37. B. Sz-Nagy, Extensions of Linear Transformations in Hilbert space which ex-tend beyond this space, Appendix to the classical treatise. F. Riesz, B. Sz-Nagy, *Functional Analysis*, F. Ungar, 1955, Dover Reprint, 1990.
38. P. Halmos, Normal dilations and extensions of operator, *Summa Brasiliensis Math.* **2**, 125-134 (1950).
39. J. Schaeffer, On unitary dilations of contractions, *Proc. Am. Math. Soc.* **6**, 332 (1955).
40. M. Naimark, Selfadjoint extensions of the second kind of a symmetric operator *Izv. Acad. Sci. USSR Math. Ser.* **4**, 53-104 (1940).
41. J. Lewis, L. Thomas, How to make a heat bath, in *Functional Integration*, A. Arthurs (ed.), Oxford, Clarendon Press, U.K., 1975.
42. M. Akcoglu, Positive contractions of L_1-spaces, *Math. Zeit.* **143**, 5-13 (1975).
43. M. Akcoglu, L. Sucheston, On convergence of iterates of positive contractions in L_p spaces, *J. Approx. Theory* **13**, 348-362 (1975).
44. M. Akcoglu, L. Sucheston, On positive dilations to isometries in L_p spaces, *Springer Lect. Notes Math* **541**, 389-401 (1976).
45. M. Akcoglu, P.E. Kopp, Construction of dilations of positive L_p-contractions, *Math. Zeit.* **155**, 119-127 (1977).
46. V. Rokhlin, On the fundamental ideas of measure theory, *Am. Math. Soc. Transl.* **39**, 1-36 (1964).
47. D. Maharam, Decompositions of measure algebras and spaces, *Trans. Amer. Math. Soc.* **69**, 142-160 (1950).
48. M. Kern, R. Nagel, G. Palm, Dilations of positive operators: construction and ergodic theory, *Math. Z.* **156**, 265-277 (1977).
49. A. Vershik, Many valued measure preserving mappings (polymorphisms), *J. Sov. Math.* **23**, 2243-2266 (1983).
50. R.G. Douglas, *Pacific J. Math.* **15**, pp. 443-462, 1965
51. M. Loève, *Probability Theory*, vol.I,II, Springer – Verlag 1977,78.
52. I. Antoniou, S. Tasaki, Generalized spectral decomposition of mixing dynamical systems, *Int. J. Quantum Chemistry* **46**, 425-474 (1993).
53. I. Antoniou, S. Tasaki, Generalized spectral decomposition of the β-adic baker's transformation and intrinsic irreversibility, *Physica* **A 190**, 303-329 (1992).

54. J.R. Brown, *Ergodic Theory and Topological Dynamics*, Academic Press, New York 1976.
55. K. Gustafson, *Lectures on Computational Fluid Dynamics, Mathematical Physics, and Linear Algebra*, Kaigai Publications, Tokyo 1996.
56. M. Takesaki, *Theory of Operator Algebras I*, Springer, Berlin 1979.

Complete Positivity and Neutral Kaon Decay

F. Benatti

Dipartimento di Fisica Teorica, Università di Trieste, Strada Costiera 11, I-34100 Trieste, Italy

1. Introduction

Ever since the discovery of the $\tau - \theta$ puzzle the K-mesons have been providing a most fertile ground to experimental tests of fundamental symmetries: invariance under space-reflection (P), charge conjugation and space reflection (CP), time-reversal (T) and, finally, invariance under CPT [1, 2, 3]. It now appears that neutral kaons K^0, $\overline{K^0}$ might be used to probe the standard formulation of quantum dynamics itself.

Indeed, CPT-invariance is necessary in Lorentz-covariant, local quantum field theories, like the standard model, whose success seemingly dooms any attempt at non-local approaches. Thus, recently, it has been proposed that CPT-violating effects possibly arise from an effective dynamics of the neutral kaon system differing from the Weisskopf-Wigner one by the presence of a mechanism that transforms pure into mixed states [4, 5].

The motivations behind such an approach are based on quantum gravity arguments that predict loss of coherence at the Planck's length due to fluctuations of the gravitational background [6] and a derivation from string theory was also attempted [7]. The idea [4] is that the evolution of kaon states represented by density matrices ρ is generated by the Weisskopf-Wigner term $-iH\,\rho + i\,\rho\,H^\dagger$ to which a linear map $T[\,\rho\,]$ is added. The latter is required to transform initial pure states $\rho = |\psi\rangle\langle\psi|$ into mixtures $T[\rho] = \sum_j \lambda_j\,|\phi_j\rangle\langle\phi_j|$, and to preserve positivity, that is $T[\rho]$ (like ρ) must have positive spectrum, and probability, that is $\mathrm{Tr}\,T[\rho] = \mathrm{Tr}\,\rho$.

Phenomenological considerations [4] fix its form and introduce three parameters α, β and γ, beside the CP-violating ones $\epsilon_{S,L}$. These new phenomenological parameters must obey $\alpha\gamma > \beta^2$ in order to guarantee the positivity of the linear map $T[\cdot]$. They have dimensions of energy and are expected to be small (of the order of $m_K^2/m_{Pl} \simeq 10^{-19}\ GeV$) so that, together with $\epsilon_{S,L}$, allow a perturbative solution of the non-standard quantum mechanical effective kaon dynamics. As a consequence their value might in principle be deduced by comparing certain theoretical quantities (asymmetries) with available experimental data [8, 9, 10].

However, the condition of simple positivity is not the most natural one when dealing with subsystems in interaction with some background (not necessarily gravitational) and tries to deduce an open system effective dynamics for the subsystem alone. A natural condition arises, more stringent than asking that the dynamics be simply positive, known as complete positivity, which is common to several subsystem dynamics [11, 12, 13, 14, 15].

Roughly, if a system S evolves under a positive dynamics and it is coupled to some n-level system E_n (whatever n be) without any disturbance on its evolution, one would like that not only the positivity of states of S be preserved in time, but also the positivity of states of the compound system $S + E_n$ [16, 17]. This is guaranteed if and only if the dynamics of the system S is not only positive, but completely positive.

Interestingly, the phenomenological Weisskopf-Wigner evolution of the kaon system is completely positive, but not the non-standard ones with an extra-piece $T[\,\cdot\,]$ added, unless $T[\,\cdot\,]$ has a specific form depending on 6 new phenomenological parameters α, β, γ, a, b, c that must necessarily satisfy certain inequalities [18, 19] which, apart for rather particular choices, are note satisfied by the simply positive non-standard extra-piece in [4].

Sufficiently accurate experimental data should put bounds on the parameters and one might in principle check whether the hypothesis of a non-standard completely positive evolution for the kaon system is physically tenable or not. Because of their precision, it is doubtful that from the present data a clear cut indication might emerge, but it is not excluded that future results will prove sufficiently accurate. Thus, quite interestingly, the neutral kaon dynamics appears to provide a physically promising testing ground for complete positivity.

Moreover, having complete positivity deep relations with the entanglement of coupled systems [20, 21], experiments on correlated kaons might be of importance in discriminating between the presence of a simply positive or a completely positive time-evolution [22].

The purpose of this review is to examine the decay of the neutral kaon system in the light of non-standard quantum mechanical theories with the additional request of complete positivity [18, 19, 22] instead of simple positivity [4, 8, 10, 23, 24].

The second section will be devoted to a self-contained presentation of the technical notion of complete positivity and to its relevance for physics. The third section examines the standard phenomenological description of the neutral kaon system, while the fourth one deals with the non-standard completely positive approach to the system decay. Finally, the fifth section considers the connections between the non-standard phenomenological parameters of the theory and available or forthcoming experimental results, with particular emphasis on the case of entangled kaon dynamics in ϕ-factories.

2. Positivity and Complete Positivity

Positivity and complete positivity are properties of certain linear maps on the observables of quantum systems. More precisely, we shall consider the Hilbert space $\mathcal{H}_S$ associated with a quantum system S and the bounded operators on it. They form an (C*) algebra $B(\mathcal{H}_S)$ and we shall deal with linear maps $\Lambda : B(\mathcal{H}_S) \to B(\mathcal{H}_S)$. An observable X is positive if $\langle \psi|X|\psi \rangle \geq 0$ for all $\psi \in \mathcal{H}_S$.

• Positivity of Λ

   Λ is positive if and only if $X \geq 0$ implies $\Lambda[X] \geq 0$.

Common instances of positive linear transformations on $B(\mathcal{H}_S)$ come from the time-evolution of either the states (Schrödinger picture) of quantum systems, or of their observables (Heisenberg picture). In what follows, quantum states will be described by density matrices ρ, that is by positive operators of $B(\mathcal{H}_S)$ with $\text{Tr}\rho \leq 1$. Strictly speaking, the eigenvalues of ρ represent the probabilities for the system being found in the corresponding eigenvectors of ρ and the sum of all the eigenvalues, that is $\text{Tr}\rho$, should equal 1. However, we shall be interested in decaying quantum systems and hence be dealing with probability losses.

If $L_t : \rho \mapsto \rho(t) := L_t[\rho]$ are linear maps giving the time-evolution of an initial state ρ of the system S, the evolution of an observable $X \in B(\mathcal{H}_S)$ will correspond to the so-called dual linear map $\Lambda_t : X \mapsto X(t) := \Lambda_t[X]$. In full generality, given a linear map Λ on $B(\mathcal{H}_S)$, its dual linear map $\rho \mapsto L[\rho]$ on the states of S is determined by looking at the mean values:

$$\text{(1)} \qquad \text{Tr}\left(L[\rho]\,X\right) = \text{Tr}\left(\rho\,\Lambda[X]\right) ,$$

and using that $\text{Tr}X\,Y = \text{Tr}Y\,X$. We notice that dual maps L from positive Λ's transform states into states (probability is preserved if $\Lambda[1] = 1$).

   a) Probability preserving time-evolutions

Given any initial state ρ and a self-adjoint time-independent Hamiltonian H,

$$\text{(2)} \qquad \rho \mapsto L_t[\rho] = \exp\left(-iHt\right)\rho \exp\left(iHt\right) ,$$

solves the Liouville von-Neumann evolution equation

$$\text{(3)} \qquad \frac{\partial}{\partial t}\rho(t) = -i\left[H, \rho(t)\right] .$$

The duals of the linear maps $L_t : \rho \mapsto L_t[\rho]$ are given by

$$\text{(4)} \qquad X \mapsto \Lambda_t[X] = \exp\left(iHt\right)X \exp\left(-iHt\right) .$$

The Λ_t are positive linear transformations on $B(\mathcal{H}_S)$ and form a group.

   b) Decaying quantum systems

The time reversal symmetry of (3) is broken in decaying quantum systems. In the Schrödinger picture, these are conventionally described by phenomenological evolution equations

$$\text{(5)} \qquad \frac{\partial}{\partial t}\rho(t) = -iH\,\rho(t) + i\rho(t)\,H^\dagger ,$$

with Hamiltonians of Weisskopf-Wigner type, $H = M - \frac{i}{2}\Gamma$, M and Γ hermitian, $\Gamma \geq 0$. The solutions are semi-groups $\{L_t\}_{t\geq 0}$ of linear maps

(6) $$\rho \mapsto L_t[\rho] = \exp(-iHt)\,\rho\,\exp(iH^\dagger t)\,.$$

Probability is not conserved as $\partial \mathrm{Tr} L_t[\rho]/\partial_t = -\mathrm{Tr}\rho\Gamma \leq 0$. The dual maps

(7) $$\Lambda_t[X] = \exp(iH^\dagger t)\,X\,\exp(-iHt)\,,$$

form semi-groups of positive linear transformations on $B(\mathcal{H}_S)$.

  c) Time-reversal

Let us consider the time-reversal linear $*$-anti-automorphism $\mathcal{V}$ on $B(\mathcal{H}_S)$, with $\mathcal{V}(XY) = \mathcal{V}(Y)\mathcal{V}(X)$, $\mathcal{V}(X^\dagger) = \mathcal{V}(X)^\dagger$, which is implemented by

(8) $$\mathcal{V}(X) = V X^\dagger V\,,$$

where $V = V^\dagger$ is a basis-dependent conjugation (anti-unitary operator) on $\mathcal{H}_S$. The map $\mathcal{V}$ is a positive linear transformation on $B(\mathcal{H}_S)$ acting as a transposition with respect to the chosen basis.

Complete positivity concerns the extension of positive linear maps on $B(\mathcal{H}_S)$ when the system S is coupled with an n-level system E_n. The algebra of observables of the latter is the algebra of $n \times n$ matrices. The bounded observables of the global system $S + E_n$ are $n \times n$ matrices $[X_{ij}]$, with entries $X_{ij} \in B(\mathcal{H}_S)$. Any linear positive map $\Lambda : B(\mathcal{H}_S) \to B(\mathcal{H}_S)$ extends to a linear map

(9) $$\tilde{\Lambda}_n = \Lambda \otimes \mathbf{1}_n : [X_{ij}] \mapsto [\Lambda[X_{ij}]]\,,$$

on $B(\mathcal{H}_{S+E_n})$, $\mathbf{1}_n$ being the $n \times n$ unit matrix.

• Complete Positivity of Λ

  If the extended linear map $\tilde{\Lambda}_n$ is positive on $B(\mathcal{H}_{S+E_n})$, then Λ is called n-positive. If n-positivity holds for all n, Λ is called completely positive [16, 17].

  In the following, we shall call completely positive also the linear maps L on states which are the dual of completely positive maps Λ on the algebra of observables.

Complete positivity is a more stringent request on a linear transformation Λ than simple positivity. As an example of a positive linear map Λ which is not even 2-positive is the following [11]. Let E_n be a two-level system and $\mathcal{H}_S$ itself a bidimensional Hilbert space, so that $\mathcal{H}_{S+E_2}$ is 4-dimensional. The following operator $[X_{ij}]$ of $B(\mathcal{H}_{S+E_2})$ is positive, with eigenvalues 0 and 1:

(10) $$[X_{ij}] = \frac{1}{2}\begin{pmatrix} 0 & 0 & 0 & 0 \\ 0 & 1 & -1 & 0 \\ 0 & -1 & 1 & 0 \\ 0 & 0 & 0 & 0 \end{pmatrix}\,.$$

The extended action of the time-reversal automorphism $\tilde{\mathcal{V}}_2 = \mathcal{V} \otimes 1_2$ acts as follows [19]

$$(11) \qquad \tilde{\mathcal{V}}_2\left([X_{ij}]\right) = \frac{1}{2} \begin{pmatrix} 0 & 0 & 0 & -1 \\ 0 & 1 & 0 & 0 \\ 0 & 0 & 1 & 0 \\ -1 & 0 & 0 & 0 \end{pmatrix} ,$$

is no more a positive operator having eigenvalues 1 and -1.

Interestingly, the map (8) also indicates the possible consequences of the absence of complete positivity when dealing with linear transformations on states. Indeed, if the two quantum systems S and E_2 are identified with two spin $\frac{1}{2}$ particles, then the operator $[X_{ij}]$ is also the density matrix corresponding to the singlet state of the compound system $S + E_2$. By means of (1) one finds that the dual of $\mathcal{V}$ acts on the states ρ_S of the spin 1/2 system S as $\mathcal{V}$. Thus it transforms them into states again, conserving their positivity and their trace. One would expect global states ρ_{S+E_2} of the compound system to be turned by the dual $\tilde{\mathcal{V}} = \mathcal{V} \otimes 1_2$ into global states as well. However, this is not true as the singlet state (10) gets mapped into (11), develops a negative eigenvalue and thus, if still interpreted as a quantum state, would assign negative probability to the corresponding eigenstate.

It is important to notice that the singlet state in (10) is an entangled state of $S + E_2$, namely it cannot be written neither as a factor state $\rho_S \otimes \rho_{E_2}$ nor as a general uncorrelated state

$$(12) \qquad \rho_{S+E_2} = \sum_{ij} \lambda_{ij}\, \rho_S^i \otimes \rho_{E_2}^j ,$$

where $\lambda_{ij} \geq 0$ and $\sum_{ij} \lambda_{ij} = 1$.

2.1 Completely Positive Maps

A technical advantage of completely positive maps Λ on $B(\mathcal{H}_S)$ with respect to simply positive ones is that, unlike the latter, which are far from being given a general characterization, the former have necessarily the form [16, 17]:

$$(13) \qquad \Lambda[X] = \sum_j W_j^\dagger X W_j ,$$

where W_j are suitable bounded operators such that $\sum_j W_j^\dagger W$ is again a bounded operator on $\mathcal{H}_S$. Their dual action L on quantum states of S are easily obtained from (1):

$$(14) \qquad L[\rho] = \sum_j W_j\, \rho\, W_j^\dagger .$$

Consequently, the time-evolutions (2), (4), respectively (6), (7) are instances of completely positive maps, while (8) is not.

More in general, if we ask that the dynamics of a quantum system S be given by a one-parameter semi-group of completely positive maps Λ_t, $t \geq 0$, on $B(\mathcal{H}_S)$, such that the norm $\|(\Lambda_t - 1)/t\| \to 0$ when $t \to 0^+$, then the time-evolution $\rho \to \rho(t) := L_t[\rho]$ of the states of S is generated by [11, 25]:

(15)
$$\frac{\partial}{\partial t}\rho(t) = -i\,[H, \rho(t)] - \frac{1}{2}\sum_j \left\{ A_j^\dagger A_j\, \rho(t) + \rho(t)\, A_j^\dagger A_j \right\} + \sum_j A_j\, \rho(t)\, A_j^\dagger \ ,$$

where $H = H^\dagger$, A_j and $R = \sum_j A_j^\dagger A_j$ belong to $B(\mathcal{H}_S)$. This is also a sufficient condition for the semi-group $\{L_t\}_{t\geq 0}$ to consist of completely positive maps. These families of dynamical maps are called quantum dynamical semi-groups.

Notice that the above evolution equation has $\partial \mathrm{Tr}\rho(t)/\partial t = 0$. If we allow for probability losses as in decaying quantum systems, the phenomenological Hamiltonian H is not hermitian and the commutator in (15) is to be replaced by $H\,\rho(t) - \rho(t)\,H^\dagger$ (compare (5)).

2.2 Physical Motivations

In the preceding section some technical advantages of complete positivity vs simple positivity have been outlined. More important is the physics behind the request of complete positivity. We have already seen that the time-evolutions (2) and (6) are automatically completely positive because they are of the form (14).

More in general, complete positivity quite naturally emerges when dealing with systems in interaction with suitable environments E. They are then considered as (open) subsystems S of global (closed) systems $S + E$ [12, 13, 14]. Given a global Hamiltonian

(16) $$H_{S+E} = H_S \otimes 1_E + 1_S \otimes H_E + \lambda H_{SE} \ ,$$

which is the sum of the Hamiltonian H_S of the subsystem, of the Hamiltonian H_E of the environment and of the interaction Hamiltonian H_{S+E} between them, the states ρ_{S+E} of the compound system evolve according to (2). A reduced dynamics for the subsystem S is consistently provided when S and E are initially uncorrelated and the state of the total system $S+E$ is $\rho_{S+E} = \rho_S \otimes \rho_E$. In such a case, by tracing over the environment degrees of freedom at time t one obtains a state of the subsystem S at time t, $\rho_S(t)$, which is the result of the action of a linear operation $L_t[\,\cdot\,]$ on the subsystem initial state ρ_S:

(17) $$\rho_S(t) := L_t[\rho_S] = \mathrm{Tr}_E \left[U(t)\, (\rho_S \otimes \rho_E)\, U(t)^\dagger \right] \ ,$$

where $U(t) = \exp(-i\, t\, H_{S+E})$. Let then X_S be an observable of the system S only and use (1). The dual linear transformation Λ_t on the observables of S is:

$$(18) \qquad \Lambda_t[X_S] = \mathrm{Tr}_E \left[\rho_E\, U(t)^\dagger \, (X_S \otimes \mathbf{1}_E)\, U(t) \right] \ ,$$

where $\mathbf{1}_E$ represents the identity operator over the Hilbert space $\mathcal{H}_E$ of the environment. Consequently, if $X_S \in B(\mathcal{H}_S)$ is positive, the same is true of $\Lambda_t[X_S]$. Actually, Λ_t are not only positive, but also completely positive [13].

In general, $\Lambda_{t+s} \neq \Lambda_t \circ \Lambda_s$, not even for $t, s \geq 0$. Nevertheless, semi-groups of completely positive dynamical maps naturally emerge in the description of the evolution of subsystems [12, 14]. In fact, let the environment E be a heat bath in equilibrium at a certain temperature with a small coupling constant λ that justifies the factorization of the initial state $\rho_{S+E} = \rho_S \otimes \rho_E$ in (17), meaning that the S and E never become too correlated. If the characteristic relaxation time of the heat bath is much smaller than the characteristic time of S, one might plausibly expect the reduced dynamics of S to have a dissipative character due to the presence of the heat bath, but not to develop memory effects (Markov property). Indeed, a semi-group of completely positive maps emerges as reduced dynamics for S after rescaling time, $t \to t/\lambda^{-2}$, and sending $\lambda \to 0$ (van Hove limit) [26].

Quantum dynamical semi-groups arise as effective dynamics also in situations involving environments quite different from heat baths. A typical example is the one describing the interaction of a micro-system with a macroscopic measuring apparatus; in this case the operators A_j in (15) are typically the projection operators onto the eigenstates of the measured observable [27, 26]. More in general, the time evolution of any open quantum system can be conveniently described by a dynamical semi-group. It has also been used to construct a consistent dynamical description of the wave-packet reduction in ordinary Quantum Mechanics [28].

We conclude this section by considering lack of complete positivity in the light of the entanglement of coupled systems. The consequences might have testable physical consequences as shall be discussed in section 6.2.

We have already seen that the time-reversal anti-automorphism (8) is not 2-positive and transforms the entangled singlet state of two spin $1/2$ particles into an observable with negative eigenvalues that, as such, cannot be considered as a reasonable physical state anymore. Let $\Lambda : B(\mathcal{H}_S) \to B(\mathcal{H}_S)$ be a linear positive map, $\Lambda_n = \Lambda \otimes \mathbf{1}_n$ its lifting to the compound system $S + E_n$ and ρ_{S+E_n} any global state of the coupled systems.

If ρ_{S+E_n} is of the uncorrelated form (12) and L is the dual action of Λ on the states of S, then the dual of the extension $\tilde{\Lambda}_n = \Lambda \otimes \mathbf{1}_n$ is

$$(19) \qquad \tilde{L}_n[\rho_{S+E_n}] = \sum_{ij} \lambda_{ij}\, L[\rho_S^i] \otimes \rho_{E_n}^j \ .$$

Since Λ is a positive linear map, its dual L preserves the positivity of the states S. Moreover, the absence of correlations in ρ_{S+E_n} guarantees that this is true also of the global action $\tilde{L}_n := L \otimes 1_n$. Therefore, for any coupling $S + E_n$, uncorrelated global states ρ_{S+E_n} are mapped into states by the dual actions of the extensions of both simply positive and completely positive maps Λ on $B(\mathcal{H}_S)$.

Assume now that $\Lambda : B(\mathcal{H}_S) \to B(\mathcal{H}_S)$ be simply positive, but fail to be, say, n-positive. This means that there exists some positive observable $[X_{ij}]$ of the system $S + E_n$ that does not remain positive under the action of the extension $\tilde{\Lambda}_n = \Lambda \otimes 1_n$. This means that there must exist a state ρ_{S+E_n} of the system $S + E_n$ such that, by using the dual map $\tilde{L}_n = L \otimes 1_n$ and (1),

$$(20) \qquad \mathrm{Tr}\left(\rho_{S+E_n} \, \tilde{\Lambda}_n[[X_{ij}]] \right) = \mathrm{Tr}\left(\tilde{L}_n[\rho_{S+E_n}] \, [X_{ij}] \right) \leq 0 \, .$$

Being $[X_{ij}]$ a positive observable of $S + E_n$, $\tilde{L}_n[\rho_{S+E_n}]$ must present some negative eigenvalue in order to give rise to a negative mean value of $[X_{ij}]$, hence it is no more interpretable as a global state of $S + E_n$.

On the other hand, if Λ is completely positive, it is by definition n-positive, therefore any positive observable $[X_{ij}]$ of $S + E_n$ is transformed into a positive observable by $\tilde{\Lambda}_n = \Lambda \otimes 1_n$ so that $\tilde{L}_n = L \otimes 1_n$ preserves the positivity of the spectrum of whatever state ρ_{S+E_n}, be it entangled or uncorrelated, whence $\tilde{L}_n[\rho_{S+E_n}]$ is also interpretable as a state of the compound system.

3. Neutral Kaon System

The K^0-$\overline{K^0}$ system can be effectively described in a two-dimensional Hilbert space [1, 2, 3], where we fix the orthonormal states

$$(21) \qquad |K_1\rangle = \frac{1}{\sqrt{2}}\left[|K^0\rangle + |\overline{K^0}\rangle \right] \, , \quad |K_2\rangle = \frac{1}{\sqrt{2}}\left[|K^0\rangle - |\overline{K^0}\rangle \right] \, .$$

K^0 and $\overline{K^0}$ are eigenvectors of the strangeness (S) operator, transform one into the other under CP and, in their rest frame, are mass-degenerate eigenvectors of the (S-invariant) electro-magnetic plus strong Hamiltonian:

$$(22) \qquad S|K^0\rangle = |K^0\rangle \qquad S|\overline{K^0}\rangle = -|\overline{K^0}\rangle$$

$$(23) \qquad CP|K^0\rangle = |\overline{K^0}\rangle \qquad CP|\overline{K^0}\rangle = |K^0\rangle \, .$$

Then, $CP|K_1\rangle = |K_1\rangle$ and $CP|K_2\rangle = -|K_2\rangle$.

Weak interaction (not S-preserving) terms H_{wk} in the Hamiltonian may originate (hadronic) decays into two and three pions

$$(24) \qquad K \to \pi^+ \pi^- \, , \ K \to \pi^0 \pi^0 \, , \ K \to \pi^+ \pi^- \pi^0 \, ,$$

or the semileptonic decays ($\ell = e \, , \, \mu$)

$$(25) \qquad K^0 \to \pi^- \ell^+ \nu \, , \ \overline{K^0} \to \pi^+ \ell^- \overline{\nu} \, .$$

The above last two decays obey the so-called $\Delta S = \Delta Q$ rule, where ΔS is the strangeness and ΔQ the non-leptonic charge variation.

In the standard quantum mechanical description, the decay properties of the K^0-$\overline{K^0}$ system are described by an effective Hamiltonian (the Weisskopf-Wigner Hamiltonian) which includes a non-hermitian part, characterizing the natural width of the states. It is conventionally written as [1, 2, 3]

$$(26) \qquad H = M - \frac{i}{2}\Gamma \ , \ M = M^\dagger \ , \ \Gamma = \Gamma^\dagger \geq 0 \ .$$

Let us consider eigenvectors and eigenvalues of the effective Hamiltonian,

$$(27) \qquad H|K_{S,L}\rangle = \lambda_{S,L}|K_{S,L}\rangle \ , \ \lambda_{S,L} = m_{S,L} - \frac{i}{2}\gamma_{S,L} \ ,$$

where $\gamma_S^{-1} = \tau_S \simeq 0.8 \times 10^{-10}$ sec, $\gamma_L^{-1} = \tau_L \simeq 5 \times 10^{-8}$ sec are the lifetimes of the short and long lived kaons, m_S, m_L their masses that satisfy $\Delta m := m_L - m_S \simeq 3.5 \times 10^{-6}$ eV $\simeq \gamma_S/2$.

If CP were conserved, only $|K_1\rangle$ would decay into two pions ($CP = 1$), whereas $|K_2\rangle$ would be allowed to decay into three pions or semileptonically ($CP = -1$). The phase-space at disposal for the 3π and $\pi l\nu$ decays is much less than for the 2π decay, hence K_2 would be expected to live longer than K_1 and identifiable with K_S. Instead, the experimental evidence that $K_2 \to 2\pi$ established CP-violation and implied the mixing

$$(28) \qquad |K_S\rangle = N_S \left[|K_1\rangle + \epsilon_S|K_2\rangle\right] \ , \quad |K_L\rangle = N_L \left[\epsilon_L|K_1\rangle + |K_2\rangle\right] \ ,$$

where $\epsilon_{S,L}$ are complex CP-violating parameters (of order 10^{-3}) and $N_{S,L}$ are normalization factors. Furthermore, we have [3]

$$(29) \qquad \epsilon_S = \epsilon_L = \epsilon \neq 0 \ \Leftrightarrow \ CPT \text{ conserved} \ , \ T \text{ violated}$$

$$(30) \qquad \epsilon_S = -\epsilon_L = \epsilon \neq 0 \ \Leftrightarrow \ CPT \text{ violated} \ , \ T \text{ conserved} \ .$$

4. Neutral Kaons Effective Dynamics

With respect to the CP eigenstates (21), any (mixed) kaon state $\rho(t)$ will be represented as a positive matrix:

$$(31) \qquad \rho(t) = \begin{pmatrix} \rho_1(t) & \rho_3(t) \\ \rho_4(t) & \rho_2(t) \end{pmatrix} \ , \quad \rho_4(t) \equiv \rho_3(t)^* \ .$$

In spite of the fact that, due to kaon decay, probability is not conserved (see (5)) and $\text{Tr}\rho(t) \leq 1$, we shall refer to these $\rho(t)$ as to quantum states or density matrices as long as their positivity is preserved by the time-evolution.

It is useful to introduce a vector notation by rewriting the matrix ρ in (31) as a four-dimensional vector $|\rho\rangle$, with entries $(\rho_1, \rho_2, \rho_3, \rho_4)$. Then, the evolution equation (5) takes the form of a Schrödinger-like equation:

(32)
$$\frac{d}{dt}|\rho(t)\rangle = \mathcal{H}\,|\rho(t)\rangle \;,$$

where $\mathcal{H}$ is a 4×4 matrix that describes the action of the Weisskopf-Wigner Hamiltonian H on the four-component vector $|\rho(t)\rangle$.

In view of the smallness of the parameters ϵ_S and ϵ_L ($\simeq 10^{-3}$), it is reasonable to seek a solution of (32) by means of perturbation theory. In the following, use will be made also of the following positive combinations:

(33)
$$\Delta\Gamma = \gamma_S - \gamma_L \simeq 10^{-16}\;GeV\;,\quad \Gamma_\pm = \Gamma \pm i\Delta m\;,$$

(34)
$$\Delta\Gamma_\pm = \Delta\Gamma \pm 2i\Delta m\;,\quad \Gamma = \frac{\gamma_S + \gamma_L}{2}\;.$$

corresponding to the differences between decay widths and masses of the states K_S and K_L. The matrix $\mathcal{H}$ can be expanded in the small parameters ϵ_S and ϵ_L,

(35)
$$\mathcal{H} = \mathcal{H}_0 + \mathcal{H}_1 + \mathcal{H}_2 + \cdots \;.$$

Explicitly, for the first terms in the expansion (35) one finds:

(36)
$$\mathcal{H}_0 = \begin{pmatrix} -\gamma_S & 0 & 0 & 0 \\ 0 & -\gamma_L & 0 & 0 \\ 0 & 0 & -\Gamma_- & 0 \\ 0 & 0 & 0 & -\Gamma_+ \end{pmatrix}$$

(37)
$$\mathcal{H}_1 = \frac{1}{2}\begin{pmatrix} 0 & 0 & \epsilon_L^*\Delta\Gamma_+ & \epsilon_L\Delta\Gamma_- \\ 0 & 0 & -\epsilon_S\Delta\Gamma_- & -\epsilon_S^*\Delta\Gamma_+ \\ -\epsilon_S^*\Delta\Gamma_+ & \epsilon_L\Delta\Gamma_- & 0 & 0 \\ -\epsilon_S\Delta\Gamma_- & \epsilon_L^*\Delta\Gamma_+ & 0 & 0 \end{pmatrix}$$

(38)
$\mathcal{H}_2$
$$= \begin{pmatrix} -\mathcal{R}e(\epsilon_S\epsilon_L\Delta\Gamma_-) & 0 & 0 & 0 \\ 0 & \mathcal{R}e(\epsilon_S\epsilon_L\Delta\Gamma_-) & 0 & 0 \\ 0 & 0 & -i\mathcal{I}m(\epsilon_S\epsilon_L\Delta\Gamma_-) & 0 \\ 0 & 0 & 0 & i\mathcal{I}m(\epsilon_S\epsilon_L\Delta\Gamma_-) \end{pmatrix}\;.$$

Since $\mathcal{H}_0$ is diagonal, it is natural to separate the "free" dynamics that it generates by writing:

(39)
$$|\rho(t)\rangle = e^{\mathcal{H}_0 t}|\sigma(t)\rangle\;,\quad |\rho(0)\rangle = |\sigma(0)\rangle\;,$$

so that (32) reduces to:

(40)
$$\frac{d}{dt}|\sigma(t)\rangle = e^{-\mathcal{H}_0 t}\mathcal{K}\,e^{\mathcal{H}_0 t}|\sigma(t)\rangle\;,\quad \mathcal{K} = \mathcal{H}_1 + \mathcal{H}_2 + \cdots\;.$$

One can now solve this equation by iteration, to any order in the small parameters. For our purposes, it will be sufficient to stop at second order. Simple manipulations show that the time evolution of $|\rho(t)\rangle$ is then given by the following expression:

$$|\rho(t)\rangle = e^{\mathcal{H}_0 t} |\rho(0)\rangle$$
$$+ \int_0^t ds\, e^{\mathcal{H}_0(t-s)}\, \mathcal{H}_1\, e^{\mathcal{H}_0 s} |\rho(0)\rangle$$
$$+ \int_0^t ds\, e^{\mathcal{H}_0(t-s)}\, \mathcal{H}_2\, e^{\mathcal{H}_0 s} |\rho(0)\rangle$$
$$+ \int_0^t ds_1 \int_0^{s_1} ds_2\, e^{\mathcal{H}_0(t-s_1)}\, \mathcal{H}_1\, e^{\mathcal{H}_0(s_1-s_2)}\, \mathcal{H}_1\, e^{\mathcal{H}_0 s_2} |\rho(0)\rangle \ .$$

(41)

In the r.h.s., the first two lines correspond to the zero-th and first order perturbation expansion, while the last two give the second order contributions.

4.1 Observables

The time-evolution of any physical property of the K^0-$\overline{K^0}$ system can be obtained by studying the behavior of suitable combinations of traces of the density matrix $\rho(t)$ with some hermitian operators $\mathcal{O} = \begin{pmatrix} O_1 & O_3 \\ O_4 & O_2 \end{pmatrix}$. For instance, in the basis of the CP eigenstates $K_{1,2}$ in (21), the probability for the system S in the state $\rho(t)$ being a K^0, respectively a $\overline{K^0}$, will be obtained by computing the traces $\mathrm{Tr}\big(\rho(t)\, \mathcal{O}_{\ell\pm}\big)$, where

(42)

$$\mathcal{O}_{\ell+} := |K^0\rangle\langle K^0| = \frac{1}{2}\begin{pmatrix} 1 & 1 \\ 1 & 1 \end{pmatrix} \ , \quad \mathcal{O}_{\ell-} := |\overline{K^0}\rangle\langle\overline{K^0}| = \frac{1}{2}\begin{pmatrix} 1 & -1 \\ -1 & 1 \end{pmatrix} \ .$$

The notation indicates that, by the $\Delta S = \Delta Q$ rule, the presence of a K^0 is signaled by the first of the semi-leptonic decays (25) and that of a $\overline{K^0}$ by the second one.

We quote some of the measurable quantities that are relevant to our discussion because they seemingly provide a connection between experiments on neutral kaons and the technical notion of completely positive non-standard quantum mechanical evolution. In order to better discuss them, it is useful to introduce the matrices [8]:

(43)
$$\Delta(t) = \rho_{\overline{K^0}}(t) - \rho_{K^0}(t) = \begin{pmatrix} \Delta_1(t) & \Delta_3(t) \\ \Delta_4(t) & \Delta_2(t) \end{pmatrix} ,$$

(44)
$$\Sigma(t) = \rho_{\overline{K^0}}(t) + \rho_{K^0}(t) = \begin{pmatrix} \Sigma_1(t) & \Sigma_3(t) \\ \Sigma_4(t) & \Sigma_2(t) \end{pmatrix} ,$$

where, $\rho_{K^0}(t = 0) = |K^0\rangle\langle K^0|$ and similarly for $|\overline{K^0}\rangle\langle\overline{K^0}|$.

• $R_{2\pi}$

The decay rate of the neutral kaons into two (neutral or charged) pions can be reasonably associated with the presence of a K_1 [10] (this would be the only possibility in case of conserved CP). Thus, the significant observable is the projection $\mathcal{O}_{2\pi} := |K_1\rangle\langle K_1| = \begin{pmatrix} 1 & 0 \\ 0 & 0 \end{pmatrix}$ and the experimentally testable quantity is

$$(45) \qquad R_{2\pi}(t) = \frac{\mathrm{Tr}\big(\rho(t)\,\mathcal{O}_{2\pi}\big)}{\mathrm{Tr}\big(\rho(0)\,\mathcal{O}_{2\pi}\big)} = \frac{\rho_1(t)}{\rho_1(0)} \ .$$

• $A_{2\pi}$

The simplest asymmetry measures the difference between the decays of the K^0 and the $\overline{K^0}$ into two pions. Using (43), (44) and $\mathcal{O}_{2\pi}$, the experimentally accessible quantity is

$$(46) \qquad A_{2\pi}(t) = \frac{\mathrm{Tr}\Big(\big(\rho_{\overline{K^0}}(t) - \rho_{K^0}(t)\big)\,\mathcal{O}_{2\pi}\Big)}{\mathrm{Tr}\Big(\big(\rho_{\overline{K^0}}(t) + \rho_{K^0}(t)\big)\,\mathcal{O}_{2\pi}\Big)} = \frac{\Delta_1(t)}{\Sigma_1(t)} \ .$$

• A_T

The asymmetry A_T is connected with the semileptonic decays of the K^0. It signals the violation of time-reversal by measuring the rate difference between the process $K^0 \to \overline{K^0}$ and its time-conjugate process $\overline{K^0} \to K^0$. Within the $\Delta S = \Delta Q$ rule (see (25)), the former corresponds to starting off with a K^0 and observing a final decay $\overline{K^0} \to \pi^+ \, \ell^- \, \overline{\nu}$. Hence, the associated observable is $\mathcal{O}_{\ell-}$ The latter process corresponds to starting off with $\overline{K}^0$ and observing the final decay $K^0 \to \pi^- \, \ell^+ \, \nu$. The associated observable is thus $\mathcal{O}_{\ell+}$ and, correspondingly, the measurable quantity is:

$$A_T(t) \quad = \quad \frac{\mathrm{Tr}\Big(\big(\rho_{\overline{K^0}}(t)\,\mathcal{O}_{\ell+}\big) - \big(\rho_{K^0}(t)\,\mathcal{O}_{\ell-}\big)\Big)}{\mathrm{Tr}\Big(\big(\rho_{\overline{K^0}}(t)\,\mathcal{O}_{\ell+}\big) + \big(\rho_{K^0}(t)\,\mathcal{O}_{\ell-}\big)\Big)}$$

$$(47) \qquad\qquad = \quad \frac{\Delta_1(t) + \Delta_2(t) + \Sigma_3(t) + \Sigma_4(t)}{\Sigma_1(t) + \Sigma_2(t) + \Delta_3(t) + \Delta_4(t)} \ .$$

• A_{CPT}

Also this asymmetry is connected with the semileptonic decays of the neutral kaons. In this case one measures the rate difference between the process $K^0 \to K^0$ and its CPT-conjugate process $\overline{K^0} \to \overline{K^0}$. Analogously to the previous case, one works out the measurable quantity:

$$A_{CPT}(t) = \frac{\text{Tr}\Big(\big(\rho_{\overline{K^0}}(t)\,\mathcal{O}_{\ell-}\big) - \big(\rho_{K^0}(t)\,\mathcal{O}_{\ell+}\big)\Big)}{\text{Tr}\Big(\big(\rho_{\overline{K^0}}(t)\,\mathcal{O}_{\ell-}\big) + \big(\rho_{K^0}(t)\,\mathcal{O}_{\ell+}\big)\Big)}$$

$$(48) \qquad = \frac{\Delta_1(t) + \Delta_2(t) - \Sigma_3(t) - \Sigma_4(t)}{\Sigma_1(t) + \Sigma_2(t) - \Delta_3(t) - \Delta_4(t)} \ .$$

• δ

A different observable is the CP violating charge asymmetry $\delta(t)$. This is obtained by observing the semileptonic decays. Within the $\Delta S = \Delta Q$ rule, the experimentally accessible quantity is:

$$(49) \qquad \delta(t) = \frac{\text{Tr}\Big(\rho_{K^0}\big(\mathcal{O}_{\ell+} - \mathcal{O}_{\ell-}\big)\Big)}{\text{Tr}\Big(\rho_{K^0}\big(\mathcal{O}_{\ell+} + \mathcal{O}_{\ell-}\big)\Big)} = \frac{\rho_3(t) + \rho_4(t)}{\rho_1(t) + \rho_2(t)} \ .$$

5. Non-Standard Quantum Description

The time-evolution L_t in (6) transforms initial pure states $\rho = |\psi\rangle\langle\psi|$ into pure states $\rho(t) = |\psi(t)\rangle\langle\psi(t)|$, in spite of the fact that the quantum system decays. There is no mixing enhancing mechanism at work. Thus, if the effective dynamics is thought to transform pure states into mixtures one has to ask for more than for a non-hermitian Hamiltonian.

Following the ideas of [4], but motivated by the discussion of section 2., we will rely on the dissipative dynamics on the states ρ of a 2-level system generated by [18, 19]:

$$(50) \frac{\partial}{\partial t}\rho(t) = -iH\rho(t) + i\rho(t)H^\dagger - \frac{1}{2}\big(R\,\rho(t) + \rho(t)\,R\big) + \sum_j A_j\,\rho(t)\,A_j \ ,$$

where H is the phenomenological non-hermitian Weisskopf-Wigner Hamiltonian. The operators A_j are hermitian, this being a sufficient condition for the increase in time of the von Neumann entropy [29].

The evolution $\rho \mapsto \rho(t) := L_t[\rho]$ generated by (50) is a so-called quantum dynamical semi-group. As we argued in section 2., they provide a natural and physically motivated scenario (quantum open systems) for the mixing enhancing dynamical maps we are looking for. The non-standard term

$$(51) \qquad \rho \mapsto T[\rho] := -\frac{1}{2}\big(R\,\rho + \rho\,R\big) + \sum_j A_j\,\rho\,A_j^\dagger \ ,$$

guarantees that the linear maps L_t be not simply positive, but completely positive [12, 13, 14, 25] (see also section 2.2).

Let us remark that evolution equations of the type (50) do not respect conservation laws even in presence of symmetries [30, 31]. This comes as no surprise since these equations give an effective, phenomenological description of a reduced dissipative dynamics: symmetries correspond to quantities that are conserved by the global fundamental time-evolution.

In two-dimensions one can more explicitly work out (51), by writing ρ and the matrices A_j's in terms of Pauli matrices σ_i and the identity $\sigma_0 = 1_2$:

$$(52) \qquad \rho = \rho_\mu \sigma_\mu \,, \quad A_j = a_\mu^j \sigma_\mu \,, \qquad \mu = 0, 1, 2, 3 \,.$$

For notational convenience, we introduce the following three vectors with components labeled by the index j

$$(53) \qquad \vec{a}_1 = (a_1^j) \,, \quad \vec{a}_2 = (a_2^j) \,, \quad \vec{a}_3 = (a_3^j) \,.$$

Thus, the map (51) can be represented by a 4×4 matrix $[T_{\mu\nu}]$, μ, $\nu = 0, 1, 2, 3$, acting on the column vector with components $(\rho_0, \rho_1, \rho_2, \rho_3)$. The entries $T_{\mu\nu}$ of this matrix are then expressed in terms of the scalar products of the vectors (53):

$$(54)\ T_{0\nu} = T_{\mu 0} = 0 \,, \quad T_{rs} = 2\,\vec{a}_r \cdot \vec{a}_s - 2\,\delta_{rs} \sum_{k=1}^{3} \vec{a}_k \cdot \vec{a}_k \,, \quad r, s = 1, 2, 3 \,.$$

Notice that the a_0^j components of the matrices A_j do not appear and the number of free parameters is six: the lengths of the vectors $\vec{a}_1$, $\vec{a}_2$, $\vec{a}_3$, and the corresponding angles between them. Therefore, there is a minimal choice for the matrices A_j, the one for which: $j = 1, 2, 3$, the components $a_0^j = 0$, and $\vec{a}_1$, $\vec{a}_2$ and $\vec{a}_3$ are linearly independent three-dimensional real vectors.

The matrix $[T_{\mu\nu}]$ is therefore symmetric and can be parameterized by the real constants a, b, c, α, β, and γ:

$$(55) \qquad [T_{\mu\nu}] = -2 \begin{pmatrix} 0 & 0 & 0 & 0 \\ 0 & a & b & c \\ 0 & b & \alpha & \beta \\ 0 & c & \beta & \gamma \end{pmatrix} \,,$$

with a, α and γ non-negative (cfr. (54)). These parameters are not all independent since they satisfy the following inequalities:

$$(56) \qquad a \le \alpha + \gamma \qquad 4b^2 \le \gamma^2 - (a - \alpha)^2$$

$$(57) \qquad \alpha \le a + \gamma \qquad 4c^2 \le \alpha^2 - (a - \gamma)^2$$

$$(58) \qquad \gamma \le a + \alpha \qquad 4\beta^2 \le a^2 - (\alpha - \gamma)^2 \,.$$

These follow from the positivity of the norms and from the Schwartz inequality applied to the scalar products of the three vectors $\vec{a}_1$, $\vec{a}_2$ and $\vec{a}_3$.

Note that if one of the diagonal entries a, α, γ is zero, then the other two must be equal and the off-diagonal terms must vanish. For example, taking $a = 0$, the matrix $[T_{\mu\nu}]$ reduces to:

$$(59) \qquad [T_{\mu\nu}] = -2 \begin{pmatrix} 0 & 0 & 0 & 0 \\ 0 & 0 & 0 & 0 \\ 0 & 0 & \alpha & 0 \\ 0 & 0 & 0 & \alpha \end{pmatrix}.$$

Therefore, the non-standard term in the treatment of [4, 8, 10], which corresponds to having a matrix

$$(60) \qquad [T_{\mu\nu}^{EHNS}] = -2 \begin{pmatrix} 0 & 0 & 0 & 0 \\ 0 & 0 & 0 & 0 \\ 0 & 0 & \alpha & \beta \\ 0 & 0 & \beta & \gamma \end{pmatrix},$$

is recovered by setting $a = b = c = 0$ in our approach, but the conditions $\alpha \neq \gamma$, $\beta \neq 0$, render the corresponding linear map $T[\cdot]$ only positive and not completely positive.

It comes out as a surprise that our phenomenological description of the neutral kaon systems requires three more parameters despite the more stringent requirement of complete positivity. The reason is that the choice made in [4, 8, 10] for the map $T[\rho]$ is a particular one: there, it is argued that modifications of Quantum Mechanics coming from quantum gravity might violate strangeness conservation only as a second order effect. In our approach, such an attitude would force us to take $a = 0$, and thus to reduce to (59). (According to the analysis in [9], such a possibility seems incompatible with actual experimental data.)

We would like to stress that our approach is completely phenomenological in nature. In particular, we do not make any assumption on the microscopic phenomena that would give rise to the dynamical semi-group that drives the decay of the K^0-$\overline{K^0}$ system according to the evolution (50). Indeed, as discussed in the previous Section, very general considerations indicate that the evolution of quantum irreversible processes may be effectively modeled using completely positive maps. The larger number of parameters in $[T_{\mu\nu}]$ thus available need to be fixed from the experimental data.

5.1 The Formalism

We now follow the technique developed in section 3. and apply it to the non-standard quantum mechanical evolution equation (50) whose Schrödinger like form (32) is now:

$$(61) \qquad \frac{d}{dt}|\rho(t)\rangle = \mathcal{H}|\rho(t)\rangle + \mathcal{T}|\rho(t)\rangle .$$

The non-standard term $T[\rho]$, represented as a 4×4 matrix acting on the 4-component vector $|\rho(t)\rangle$, (cfr $\mathcal{H}$ in section 3.) reads as:

$$(62) \qquad \mathcal{T} = \begin{pmatrix} -\gamma & \gamma & -C & -C^* \\ \gamma & -\gamma & C & C^* \\ -C^* & C^* & -A & B \\ -C & C & B^* & -A \end{pmatrix},$$

where we have introduced the convenient notation [18, 19]:

$$(63) \qquad A = \alpha + a, \quad B = \alpha - a + 2ib, \quad C = c + i\beta.$$

Notice that, unlike in section 3. the basis is now given by the Pauli matrices, whence the difference between the 4×4 matrix representations (55) and (62). The parameters entering the non-standard term of the evolution equation are expected to be small, roughly of the order of $m_K^2/m_{Pl} \simeq 10^{-19}$, that is of the same order of the standard CP-violating contribution $|\epsilon_{S,L}| \, \Delta\Gamma$. Then, as was for the standard Weisskopf-Wigner one in section 3., it is reasonable to seek a perturbative solution of (50) .

Since all the entries in $\mathcal{T}$ are of first order in the other small parameters a, b, c, α, β and γ, what we have to do is to replace in the perturbation expansion (41) the standard quantum mechanical first order term $\mathcal{H}_1$ by the non-standard one $\mathcal{H}_1 + \mathcal{T}$.

Up to first order in the small parameters, the time-dependence of the four components of density matrix $\rho(t)$ is explicitly given in formulae (A.1)-(A.4) of Appendix A (for the second order contributions see [19]). The explicit solutions will be used to compute, in the given approximation, the time-dependence of various observables like (45)–(49) that relevant to the analysis of the neutral kaon system and have been measured experimentally. From the available data it is in principle possible to obtain bounds on the values of the parameters a, b, c, α, β and γ.

We conclude by emphasizing that putting the non-standard phenomenological parameters $A = B = C = 0$, we recover the usual Weisskopf-Wigner approximated solution depending on the standard phenomenological parameters $\epsilon_{S,L}$. On the other hand, the conditions $a = b = c = 0$, $\alpha \neq \gamma$, $\beta \neq 0$, would reproduce the analysis of [4, 8, 10], but violate the condition of complete positivity that we put upon the non-standard quantum mechanical evolution (compare (60)).

6. Complete Positivity and Experiments

In what follows, we shall first consider how tests of CP-violation might be used also as tests of complete positivity and then we shall examine the possibility that experiments at ϕ-factories might be influenced by the presence of a non-standard evolution of the neutral kaons that be simply, but not completely positive.

6.1 Single Kaon Decay

Using the perturbative solution $\rho(t)$ of the non-standard quantum mechanical time-evolution (50), explicit expressions can be given to the entries of the matrices $\Delta(t)$ and $\Sigma(t)$ in (43) and (44) up to the required order of approximation (in Appendix B, formulae (B.1)–(B.8) are first order contributions). With the entries $\Delta_i(t)$, $\Sigma_j(t)$, one has then at disposal the theoretical non-standard expressions for the measurable quantities (45)–(49) up to desired order in the small standard ($\epsilon_{S,L}$) and non-standard ($A = a + \alpha$, $B = \alpha - a + 2\,i\,b$, $C = c + i\beta$) phenomenological parameters.

• $R_{2\pi}$

In the case of (45) the comparison with experimental results requires second order contributions and some working assumption on the non-standard phenomenological parameters [19]. Starting with an initial pure K^0 one then finds the large time limit

$$(64) \qquad R_{2\pi}^L = \left| \epsilon_L + \frac{2C^*}{\Delta\Gamma_-} \right|^2 + \frac{\gamma}{\Delta\Gamma} - 8 \left| \frac{C}{\Delta\Gamma_+} \right|^2 - 4\,\mathcal{R}e\left(\frac{\epsilon_L C}{\Delta\Gamma} \right).$$

• $A_{2\pi}$

As before, second order contributions are needed and the resulting approximated expression is

$$(65)$$

$$A_{2\pi}(t) = \frac{2\,\mathcal{R}e\left(\epsilon_L - \frac{2C^*}{\Delta\Gamma_-} \right) - 2\left(1 - At \right)e^{\Delta\Gamma t/2}\,\mathcal{R}e\left[\left(\epsilon_L - \frac{2C^*}{\Delta\Gamma_-} \right) e^{-i\Delta m\,t} \right]}{1 + e^{\Delta\Gamma t}\left[\left| \epsilon_L + \frac{2C^*}{\Delta\Gamma_-} \right|^2 + \frac{\gamma}{\Delta\Gamma} - 8\left| \frac{C}{\Delta\Gamma_+} \right|^2 - 4\,\mathcal{R}e\left(\frac{\epsilon_L C}{\Delta\Gamma} \right) \right]}.$$

The above expression is useful for checking whether the ordinary quantum mechanical CP and CPT violating effects in the K^0-$\overline{K}^0$ system can be mimicked by the extra piece $T[\rho]$ in the time-evolution equation (50). Dimensional considerations indicate that CP- and CPT-violating effects can not be completely ascribed to $T[\rho]$ [19].

• A_T

The first order expressions of the entries of the matrices (43)–(44) suffice to compute the non-standard large time limit

$$(66) \qquad A_T = \frac{2\,\mathcal{R}e(\epsilon_S + \epsilon_L) + 4\,\mathcal{R}e\left(\frac{C}{\Delta\Gamma_+} - \frac{C}{\Delta\Gamma_-} \right)}{1 + 2\frac{\gamma}{\Delta\Gamma}},$$

which can be directly compared with the experimental data.

• A_{CPT}

Analogously, the asymptotic value of the CPT-asymmetry (48) is:

$$(67) \qquad A_{CPT} = \frac{2\mathcal{R}e(\epsilon_S - \epsilon_L) - 4\mathcal{R}e\left(\frac{C}{\Delta\Gamma_+} + \frac{C}{\Delta\Gamma_-}\right)}{1 + 2\frac{\gamma}{\Delta\Gamma}} .$$

Assuming CPT invariance, *i.e.* $\epsilon_S = \epsilon_L$ (cfr. (29)) standard Quantum Mechanics predicts $A_{CPT} = 0$, whereas here $A_{CPT} \neq 0$.

• δ

The large and short time contributions to CP-charge asymmetry (49) that can be directly compared with the available data are:

$$(68) \qquad \delta_L = 2\mathcal{R}e\left(\epsilon_L + \frac{2C^*}{\Delta\Gamma_-}\right) , \quad \delta_S = 2\mathcal{R}e\left(\epsilon_S + \frac{2C}{\Delta\Gamma_-}\right) .$$

By using the above expressions and the available experimental values for $\delta_{L,S}$, $A_{T,CPT}$, $R_{2\pi}$ and other observables (see [19]), one can obtain estimates on the parameters entering the matrix (55) responsible for the complete positivity of the evolution equation (50). The actual derivation of the numerical values for these parameters requires careful fits and appropriate estimates of the errors. The accuracy of the available experimental results is not expected to be sufficient to test the inequalities (56)–(58) that have to be satisfied by the phenomenological parameters a, b, c, α, β and γ. However, a precise experimental test of these formulae and hence of the complete positivity of the time-evolution of the K^0-$\overline{K^0}$ system is foreseeable in the next generation of experiments. In particular, the conditions met in the future planned experiments at ϕ-factories seem to provide a natural context for testing complete positivity. Evidently, the formalism developed in this paper needs to be adapted in order to describe pairs of correlated kaons [10]. Preliminary ideas on the role of complete positivity in ϕ-factories setups are addressed in the next section.

In conclusion, concerning the single kaon dynamics, we have discussed how the condition of complete positivity on the non-standard quantum mechanical treatment can affect its time-evolution and as a consequence be experimentally detected. To our knowledge, it is the first time that such a condition can effectively be tested in an experimental setup that examines a sub-nuclear system.

6.2 Entangled Kaons

In this section we consider coupled kaons and argue that results from planned correlation experiments in ϕ-factories might be strongly influenced by the possible presence of a non-standard quantum mechanical evolution of positive, but not completely positive type as the one proposed by [4].

In sections 2.1 and 2.2 we have discussed the behavior of entangled vs uncorrelated states of compound systems when acted upon by simply positive, respectively completely positive maps. Let us then the linear map $T[\cdot]$

in (51) be of the form (60) with $\alpha \neq \gamma$, $\beta \neq 0$ and $\alpha\gamma \geq \beta^2$. Then, $T[\rho_S]$ is positive, but not completely positive. It follows that the evolution equation (50) generates a semi-group of dynamical maps L_t that are themselves simply positive, but not completely positive as outlined in section 2.1.

We start by considering the action on density matrices of the time-evolution

$$(69) \qquad L_t^0[\rho] := \exp(t\, T[\,\cdot\,])[\rho] \;,$$

generated by the non-standard positive part $T[\,\cdot\,]$ of (51) only. The solution can be explicitly worked out. Representing $L_t^0[\rho] = \rho_\mu(t)\, \sigma_\mu$ as a 4-vector by means of the Pauli matrices, then $\rho_\mu(t) = \tau_{\mu\nu}(t)\rho_\nu$ where

$$(70) \qquad [\tau_{\mu\nu}(t)] = \begin{pmatrix} 1 & 0 & 0 & 0 \\ 0 & 1 & 0 & 0 \\ 0 & 0 & A(t) & B(t) \\ 0 & 0 & B(t) & C(t) \end{pmatrix} \;,$$

with

$$(71) \qquad \begin{aligned} A(t) &= \frac{1}{\lambda_+ - \lambda_-} \left[(\lambda_+ + 2\alpha)\, e^{\lambda_- t} - (\lambda_- + 2\alpha)\, e^{\lambda_+ t} \right] \;, \\ B(t) &= \frac{2\beta}{\lambda_+ - \lambda_-} \left(e^{\lambda_- t} - e^{\lambda_+ t} \right) \;, \\ C(t) &= \frac{1}{\lambda_+ - \lambda_-} \left[(\lambda_+ + 2\alpha)\, e^{\lambda_+ t} - (\lambda_- + 2\alpha)\, e^{\lambda_- t} \right] \;. \end{aligned}$$

Notice that the quantities $\lambda_\pm = -(\alpha+\gamma)\pm\sqrt{(\alpha-\gamma)^2 + 4\beta^2}$ are both negative due to $\alpha\gamma \geq \beta^2$. If the linear maps L_t^0 were completely positive, they should act as in (14) and $[\tau_{\mu\nu}(t)]$ have the form (55). This is possible only when $\lambda_+ = \lambda_-$ or equivalently when $\alpha = \gamma$, $\beta = 0$.

The map L_t^0, as the *-anti-automorphism $\mathcal{V}$ in section 2.1, is a positive linear transformation which is not 2-positive. As discussed at the end of section 2.2, this has consequences on global states of the compound system $S + E_2$. Let then the 2-level system E_2 be itself another kaon system S and extend L_t^0 to $\tilde{L}_t^0 = L_t^0 \otimes \mathbf{1}_2$ acting on the four-dimensional system $S + S$ of the two coupled kaons. By using the Pauli matrices, the (entangled) singlet state (10) reads as follows

$$(72) \qquad \rho_{S+S} = \frac{1}{4}\left(\sigma_0 \otimes \sigma_0 - \sum_{i=1}^{3} \sigma_i \otimes \sigma_i \right) \;,$$

The action

$$(73) \qquad \begin{aligned} \tilde{L}_t^0[\rho_{S+S}] = \frac{1}{4}\Big(&\sigma_0 \otimes \sigma_0 - \sigma_1 \otimes \sigma_1 - A(t)\, \sigma_2 \otimes \sigma_2 \\ &-C(t)\, \sigma_3 \otimes \sigma_3 - B(t)\big(\sigma_2 \otimes \sigma_3 + \sigma_3 \otimes \sigma_2\big)\Big) \end{aligned}$$

does not preserve the positivity of the spectrum. This can be easily checked by inspection of the mean values

$$(74) \qquad \langle u|\tilde{\tau}_t[\rho_{S+S}]|u\rangle = -\langle v|\tilde{\tau}_t[\rho_{S+S}]|v\rangle = \frac{1}{2}\big(A(t) - C(t)\big) \,,$$

calculated with respect to the vectors

$$(75) \qquad
\begin{aligned}
|u\rangle &= \begin{pmatrix} 1 \\ 0 \end{pmatrix} \otimes \begin{pmatrix} 1 \\ 0 \end{pmatrix} + \begin{pmatrix} 0 \\ 1 \end{pmatrix} \otimes \begin{pmatrix} 0 \\ 1 \end{pmatrix} \\[2mm]
|v\rangle &= \begin{pmatrix} 1 \\ 0 \end{pmatrix} \otimes \begin{pmatrix} 0 \\ 1 \end{pmatrix} + \begin{pmatrix} 0 \\ 1 \end{pmatrix} \otimes \begin{pmatrix} 1 \\ 0 \end{pmatrix} \,.
\end{aligned}$$

The quantity in (74) never vanishes for $t > 0$ and has a definite sign depending on the relative magnitude of α and γ; therefore, if $A(t) - C(t)$ is positive in one case, it is negative in the other. This means that $\tilde{L}_t^0[\rho_{S+S}]$ develops negative eigenvalues.

At first sight, this conclusion might seem of little physical relevance. Indeed, the coupling of the subsystem of interest to an abstract n-level system is regarded as too artificial by those who consider the condition of complete positivity of the reduced dynamics as a mere technical request. The point is that in our case, the additional two-dimensional system has been taken to be another neutral kaon system. Precisely this physical situation is commonly encountered in the so-called ϕ-factories. In these experimental setups, a ϕ meson decays into an entangled antisymmetric state of two neutral kaons that in our formalism can be written as [10]

$$(76) \qquad |\Psi_S\rangle = \frac{1}{\sqrt{2}}\big(|K_1\rangle \otimes |K_2\rangle - |K_2\rangle \otimes |K_1\rangle\big) \,.$$

As a density matrix $|\Psi_S\rangle\langle\Psi_S|$ this state exactly corresponds to the "singlet" ρ_S in (72).

The global time-evolution of states like (76) is obtained by the tensor product $\Omega_t = L_t \otimes L_t$ of the time-evolutions of the single kaons. In the standard approach, with purely Weisskopf-Wigner dynamics, positive operators remain positive and negative eigenvalues are never generated. Indeed, this is guaranteed by any completely positive time-evolution L_t as it has been discussed in section 2.2 and comes out explicitly from (14):

$$(77) \qquad \Omega_t[\rho] = \sum_{ij} \big[V_i(t) \otimes V_j(t)\big] \rho \big[V_i^\dagger(t) \otimes V_j^\dagger(t)\big] \,.$$

On the contrary, the positivity of the evolving states need not be preserved by modified single kaon dynamics L_t that are only simply positive. Again, the single kaon time-evolution given by (71) provides us with an example of this fact. The action of $\mathcal{L}_t^0 := L_t^0 \otimes L_t^0$ on $|\Psi_S\rangle\langle\Psi_S| = \rho_{S+S}$ gives

$$\mathcal{L}_t^0[\rho_{S+S}] = \frac{1}{4}\Big[\sigma_0 \otimes \sigma_0 - \sigma_1 \otimes \sigma_1$$
$$-\big(A^2(t) + B^2(t)\big)\sigma_2 \otimes \sigma_2 - \big(B^2(t) + C^2(t)\big)\sigma_3 \otimes \sigma_3$$

(78)
$$-B(t)\big(A(t) + C(t)\big)\big(\sigma_2 \otimes \sigma_3 + \sigma_3 \otimes \sigma_2\big)\Big] ,$$

and, therefore, the presence of negative eigenvalues can be ascertained as before by computing the mean values

$$\langle u|\mathcal{L}_t^0[\rho_{S+S}]|u\rangle = -\langle v|\mathcal{T}_t[\rho_S]|v\rangle$$

(79)
$$= \frac{1}{2}\big(A^2(t) - C^2(t)\big) \neq 0 .$$

The unavoidable appearance of negative eigenvalues in the time evolution of some states of the coupled kaons cannot be cured by considering also the Weisskopf-Wigner contribution [22]. In the case of (60), the time-evolution generated by (61) makes some $\rho(t)$ have partly negative spectrum. Therefore, it is possible that non-trivial consequences might follow concerning the degree of violation of certain Bell-like inequalities (see for instance [32, 33]) that can be written for the entangled kaons in a ϕ-decay. As these inequalities may in principle be tested at ϕ-factories, it seems that simply positive non-standard quantum mechanical time evolutions might have visible manifestations. Plausibly, completely positive non-standard quantum mechanical evolutions should not affect too much the experimental evidence. Obviously, the whole non-standard approach developed for the case of single kaons must be adapted to the description of the dynamics of entangled ones.

References

1. T.D. Lee, *Particle Physics and an Introduction to Field Theory*, Harwood, New York 1981
2. T.D. Lee and C.S. Wu, Ann. Rev. Nucl. Sci. **16** (1966) 511
3. L. Maiani, *CP* and *CPT* violation in neutral kaon decays, in *The Second Daφne Physics Handbook*, L. Maiani, G. Pancheri and N. Paver, eds., (INFN, Frascati, 1995)
4. J. Ellis, J.S. Hagelin, D.V. Nanopoulos and M. Srednicki, Nucl Phys. **B241** (1984) 381
5. M. Srednicki, Nucl. Phys. **B410** (1993) 143
6. S. Hawking, Comm. Math. Phys. **87** (1983) 395
7. J. Ellis, N.E. Mavromatos and D.V. Nanopolous, Phys. Lett. **B293** (1992) 37
8. J. Ellis, J.L. Lopez, N.E. Mavromatos and D.V. Nanopolous, Phys. Rev. D **53** (1996) 3846
9. CPLEAR Collaboration, J. Ellis, N.E. Mavromatos and D.V. Nanopolous, Phys. Lett. **B364** (1995) 239
10. P. Huet and M.E. Peskin, Nucl. Phys. **B434** (1995) 3
11. V. Gorini, A. Kossakowski and E.C.G. Sudarshan, J. Math. Phys. **17** (1976) 821

12. E.B. Davies, *Quantum Theory of Open Systems*, (Academic Press, New York, 1976)
13. V. Gorini, A. Frigerio, M. Verri, A. Kossakowski and E.C.G. Surdarshan, Rep. Math. Phys. **13** (1978) 149
14. H. Spohn, Rev. Mod. Phys. **53** (1980) 569
15. R. Alicki and K. Lendi, *Quantum Dynamical Semigroups and Applications*, Lect. Notes Phys. **286**, (Springer-Verlag, Berlin, 1987)
16. M. Takesaki, *Theory of Operator Algebras I*, (Springr-Verlag, Berlin, 1979)
17. K. Kraus, *States, Effects, and Operations*, Lect. Notes Phys. **190** (Springer Verlag, Berlin, 1983)
18. F. Benatti and R. Floreanini, Complete positivity and the K-$\overline{K}$ system, Phys. Lett. B, to appear
19. F. Benatti and R. Floreanini, Completely positive dynamical maps and the neutral Kaon system, Nuc. Phys. B, to appear
20. A. Peres, Seaparability Criterion for Density Matrices, e-print archive quant-ph/9604005
21. M. Horodecki, P. Horodecki and R. Horodecki, Separability of mixed states: necessary and sufficient conditions, e-print archive quant-ph/9605038
22. F. Benatti and R. Floreanini, Testing complete positivity, Trieste-preprint, 1996
23. J. Ellis, N.E. Mavromatos and D.V. Nanopolous, Phys. Lett. **B293** (1992) 142
24. J. Ellis, N.E. Mavromatos and D.V. Nanopolous, Int. J. Mod. Phys. **A11** (1996) 1489
25. G. Lindblad, Comm. Math. Phys. **48** (1976) 119
26. H. Nakazato, M. Namiki and S. Pascazio, Int. J. Mod. Phys. **B10** (1996) 247
27. L. Fonda, G.C. Ghirardi and A. Rimini, Rep. Prog. Phys. **41** (1978) 587
28. G.C. Ghirardi, A. Rimini and T. Weber, Phys. Rev. D **34** (1986) 470
29. F. Benatti and H. Narnhofer, Lett. Math. Phys. **15** (1988) 325
30. T. Banks, L. Susskind and M.E. Peskin, Nucl. Phys. **B244** (1984) 125
31. W. Unruh and R.M. Wald, Phys. Rev. D **52** (1995) 2176
32. G.C. Ghirardi, R. Grassi and T. Weber, *in* Proc. Workshop on Physics and detectors for DAΦNE, Frascati April 1991, ed. G. Pancheri (INFN-LNF, Frascati)
33. A. Di Domenico, Nucl. Phys. **B450** (1995) 293

A. Appendix

We present here the first order approximations to the entries $\rho_i(t)$ of solutions of (61). Explicitly:

$$\rho_1(t) = e^{-\gamma_S t}\rho_1 + \frac{\gamma}{\Delta\Gamma}\left(e^{-\gamma_L t} - e^{-\gamma_S t}\right)\rho_2$$
$$+ \left(\epsilon_L^* - \frac{2C}{\Delta\Gamma_+}\right)\left(e^{-\Gamma_- t} - e^{-\gamma_S t}\right)\rho_3 + \left(\epsilon_L - \frac{2C^*}{\Delta\Gamma_-}\right)\left(e^{-\Gamma_+ t} - e^{-\gamma_S t}\right)\rho_4,$$

(A.1)

$$\rho_2(t) = e^{-\gamma_L t}\rho_2 + \frac{\gamma}{\Delta\Gamma}\left(e^{-\gamma_L t} - e^{-\gamma_S t}\right)\rho_1$$
$$+ \left(\epsilon_S - \frac{2C}{\Delta\Gamma_-}\right)\left(e^{-\Gamma_- t} - e^{-\gamma_L t}\right)\rho_3 + \left(\epsilon_S^* - \frac{2C^*}{\Delta\Gamma_+}\right)\left(e^{-\Gamma_+ t} - e^{-\gamma_L t}\right)\rho_4,$$

(A.2)

$$\rho_3(t) = e^{-(\Gamma_- + A - \gamma)t}\rho_3 - \frac{iB}{2\Delta m}\left(e^{-\Gamma_- t} - e^{-\Gamma_+ t}\right)\rho_4$$
$$- \left(\epsilon_S^* + \frac{2C^*}{\Delta\Gamma_+}\right)\left(e^{-\Gamma_- t} - e^{-\gamma_S t}\right)\rho_1 - \left(\epsilon_L + \frac{2C^*}{\Delta\Gamma_-}\right)\left(e^{-\Gamma_- t} - e^{-\gamma_L t}\right)\rho_2,$$

(A.3)

$$\rho_4(t) = e^{-(\Gamma_+ + A - \gamma)t}\rho_4 + \frac{iB^*}{2\Delta m}\left(e^{-\Gamma_+ t} - e^{-\Gamma_- t}\right)\rho_3$$
$$- \left(\epsilon_S + \frac{2C}{\Delta\Gamma_-}\right)\left(e^{-\Gamma_+ t} - e^{-\gamma_S t}\right)\rho_1 - \left(\epsilon_L^* + \frac{2C}{\Delta\Gamma_+}\right)\left(e^{-\Gamma_+ t} - e^{-\gamma_L t}\right)\rho_2,$$

(A.4)

where ρ_1, ρ_2, ρ_3 and $\rho_4 \equiv \rho_3^*$ are the initial values at $t = 0$. For the second order contributions we refer the interested reader to the Appendix of [19].

In the above expressions, the linear and quadratic time-dependence have been absorbed in the exponentials . Such an operation is always consistent, provided one suitably redefines the parameters in the time-exponential terms in such a way that γ_S, γ_L and Δm can be directly identified with the widths and mass difference of the K_S and K_L physical states as explained in the Appendix of [19].

B. Appendix

Using the results of the previous Appendix, we can write down the first order approximations to the matrix entries $\Delta_i(t)$, $\Sigma_j(t)$ in (43),(44). Explicitly:

$$\Delta_1(t) = \left[\left(\epsilon_L^* - \frac{2C}{\Delta\Gamma_+}\right) + \left(\epsilon_L - \frac{2C^*}{\Delta\Gamma_-}\right)\right] e^{-\gamma_S t}$$

(B.1)
$$- \left(\epsilon_L - \frac{2C^*}{\Delta\Gamma_-}\right) e^{-\Gamma_+ t} - \left(\epsilon_L^* - \frac{2C}{\Delta\Gamma_+}\right) e^{-\Gamma_- t} \; ,$$

$$\Delta_2(t) = \left[\left(\epsilon_S - \frac{2C}{\Delta\Gamma_-}\right) + \left(\epsilon_S^* - \frac{2C^*}{\Delta\Gamma_+}\right)\right] e^{-\gamma_L t}$$

(B.2)
$$- \left(\epsilon_S^* - \frac{2C^*}{\Delta\Gamma_+}\right) e^{-\Gamma_+ t} - \left(\epsilon_S - \frac{2C}{\Delta\Gamma_-}\right) e^{-\Gamma_- t} \; ,$$

(B.3)
$$\Delta_3(t) = -e^{-(\Gamma_- + A - \gamma)t} + \frac{iB}{2\Delta m} e^{-\Gamma_- t} - \frac{iB}{2\Delta m} e^{-\Gamma_+ t} \; ,$$

(B.4)
$$\Delta_4(t) = -e^{-(\Gamma_+ + A - \gamma)t} - \frac{iB^*}{2\Delta m} e^{-\Gamma_+ t} + \frac{iB^*}{2\Delta m} e^{-\Gamma_- t} \; ,$$

(B.5)
$$\Sigma_1(t) = \left(1 - \frac{\gamma}{\Delta\Gamma}\right) e^{-\gamma_S t} + \frac{\gamma}{\Delta\Gamma} e^{-\gamma_L t} \; ,$$

(B.6)
$$\Sigma_2(t) = \left(1 + \frac{\gamma}{\Delta\Gamma}\right) e^{-\gamma_L t} - \frac{\gamma}{\Delta\Gamma} e^{-\gamma_S t} \; ,$$

$$\Sigma_3(t) = -\left[\left(\epsilon_S^* + \frac{2C^*}{\Delta\Gamma_+}\right) + \left(\epsilon_L + \frac{2C^*}{\Delta\Gamma_-}\right)\right] e^{-\Gamma_- t}$$

(B.7)
$$+ \left(\epsilon_S^* + \frac{2C^*}{\Delta\Gamma_+}\right) e^{-\gamma_S t} + \left(\epsilon_L + \frac{2C^*}{\Delta\Gamma_-}\right) e^{-\gamma_L t} \; ,$$

$$\Sigma_4(t) = -\left[\left(\epsilon_S + \frac{2C}{\Delta\Gamma_-}\right) + \left(\epsilon_L^* + \frac{2C}{\Delta\Gamma_+}\right)\right] e^{-\Gamma_+ t}$$

(B.8)
$$+ \left(\epsilon_S + \frac{2C}{\Delta\Gamma_-}\right) e^{-\gamma_S t} + \left(\epsilon_L^* + \frac{2C}{\Delta\Gamma_+}\right) e^{-\gamma_L t} \; .$$

Chapter III

Resonances

Conventional and S-matrix Approaches to Hadronic Resonances

G. López Castro

[1] Departamento de Fisica, Cinvestav del IPN, México, D.F. Mexico
[2] Institut de Physique Théorique, Université Catholique de Louvain, Louvain-la-Neuve, Belgium

Summary. We present a comparison of the conventional and S-matrix formalisms for the description of hadronic resonances in particle physics. Using the ρ^0 meson resonance as an illustrative example, we argue that the pole of the S-matrix provides a definition for resonance properties with suitable physical properties, namely independent of the theoretical model and background subtraction from experimental data.

1. Introduction

Present experimental evidence indicates that only the electron, the photon, the proton, the neutrinos and possibly the graviton are absolutely stable states in particle physics [1]. All the other particles correspond to unstable states decaying through the effect of electroweak or strong interactions.

In addition to its mass (M) and other quantum numbers, unstable particles are characterized by a finite decay width (Γ) or lifetime (τ). Unstable particles can be classified into *quasi-stable* and *resonant* states. Quasi-stable particles, like the neutron or pions, are characterized by typical lifetimes larger than $10^{-16} \sim 10^{-17}$ seconds (or, equivalently, widhts smaller than a few eV), while resonances, like the massive gauge bosons ($W^\pm, Z$)[2], the Δ^{++} baryon or the ρ^0 meson, are characterized by a decay width larger than a few keV (or $\tau \leq 10^{-18}$ sec.). Notice that, for wide resonances, the relationship between the decay width and the lifetime (namely, $\tau = \hbar/\Gamma$) entering the exponential decay law is only an approximation valid for $\Gamma/M << 1$ and $\Gamma/|E_{th} - M| << 1$ (E_{th} is the threshold energy of decay products). Those last conditions are satisfied by quasi-stable particles.

The starting point of this talk is that the mass and decay width are *physical* intrinsic properties of resonances in particle physics. Thus, the determination of these properties should depend neither on the particular experiment nor on the specific theoretical model used to describe the production and decay of the resonance. In this paper we follow the argument that the pole position in the S-matrix amplitude provides a definition for mass and width of resonances that satisfies the above requirements [3]-[4] in contradistinction to the commonly used parameters entering a partial-wave amplitude with an energy-dependent width (see for example, p. 178 of [1]). The former (*pole*) definition of resonance parameters appears in the context of the relativistic S-matrix theory [3]-[4], while the second (*conventional*) definition is associated with the partial-wave analysis of scattering data.

In order to compare the definitions provided by the S-matrix and conventional approaches, we will consider the ρ^0 hadronic resonance observed in $\pi^+\pi^-$ pair production by electron-positron annihilation. Although a good fit to experimental data can be given in both approaches, the values extracted of resonance parameters can be largely different. Generally speaking, the numerical difference between parameters in both definitions increases as the resonance becomes wider (see also Refs. [2] and [5],[6]).

At a more theoretical level, the interest on this subject is connected with suggestions that in non-relativistic quantum mechanics, the pole of the S-matrix can be associated to states described by Gamow vectors [7].

2. Conventional and S-matrix Description of a Resonance

In the conventional approach, a model is required to describe the production and decay of a resonance. If, for example, the resonance is produced in a s-channel elastic scattering, the corresponding partial wave amplitude can be written as a Breit-Wigner formula (we assume only one decay channel for the resonance; see p. 178 of [1])

$$(1) \qquad a_l = \frac{-\sqrt{s}\Gamma(s)}{s - M_c^2 + i\sqrt{s}\Gamma(s)}.$$

Here, M_c and $\Gamma_c \equiv \Gamma(s = M_c^2)$ define the resonance parameters in the conventional scheme.

The energy-dependence of the total width $\Gamma(s)$ in Eq. (1), is fixed by the specific model used for the decay of the resonance. An energy-dependent width naturally arises in a field theoretical model for the resonance. Indeed, a resonance shape can be obtained in field theory by a Dyson summation of bubble graphs for the propagator [8]-[9], which transforms a bare propagator (with bare mass m_0) into a full one according to (for simplicity we consider a scalar resonance)

$$(2) \qquad \frac{1}{s - m_0^2} \longrightarrow \frac{1}{s - m_0^2 + \Pi(s)}.$$

By applying the following renormalization conditions

$$(3) \qquad \begin{aligned} m_R^2 &= m_0^2 - \mathrm{Re}\Pi(m_R^2) \\ Z^{-1} &= 1 + \mathrm{Re}\Pi'(m_R^2) \end{aligned}$$

in Eq. (2) and using the unitarity relation $Z\mathrm{Im}\Pi(s) = \sqrt{s}\Gamma(s)$, we are lead to the following renormalized propagator

(4)
$$\frac{1}{s - m_R^2 + i\sqrt{s}\Gamma(s)}$$

which is similar to Eq. (1). Note that Eq. (4) can always be written into a pole plus background term as follows [9, 2, 5, 6]:

(5)
$$\frac{1}{s - m_R^2 + i\sqrt{s}\Gamma(s)} = \frac{1}{s - m^2 + im\Gamma} + b(s).$$

As already mentioned, the relativistic S-matrix theory provides an alternative definition of resonance parameters. According to the axioms of the S-matrix theory [3], which are also fulfilled by quantum field theories, the relativistic scattering amplitudes must be analytic functions of Lorentz-invariant kinematical variables (extended to the complex plane) except for *poles* and branch cut singularities which are associated with the formation of resonances and appearance of new thresholds, respectively. In the specific case of a single resonance produced in a s-channel, the amplitude has the simple form [3]

(6)
$$\mathcal{A}(s) = \frac{R}{s - s_p} + B(s).$$

Since the background contribution $B(s)$ and the residue at the pole R are physical quantities [3], the complex pole position s_p is also a physical quantity. In the S-matrix approach, the mass and width of the resonance are defined from the pole position as

(7)
$$s_p \equiv M_p^2 - iM_p\Gamma_p.$$

Thus, when considering the formation of a resonance in a specific process we are faced with (at least) two definitions for resonance parameters. In the next sections we consider the formation of the ρ^0 resonance and discuss the advantages of the S-matrix formalism over the conventional approach.

3. Resonance in Conventional Approach

Our aim in this section is show that the ρ^0 resonance parameters in the conventional approach may depend on the specific process and also on the way to subtract background contributions from experimental data.

The ρ^0 meson has identical quantum numbers as the photon, but with a mass and decay width of around 770 MeV and 150 MeV (we use $\hbar = c = 1$ units), respectively. It can be produced, for example, in the following reactions:

$$\gamma C \rightarrow \rho^0 C$$
$$p\bar{p} \rightarrow \rho^0 \omega$$
$$e^+e^- \rightarrow \rho^0$$
$$\pi^+ n \rightarrow \rho^0 p$$

and decays, almost exclusively, to a pair of charged pions ($\rho^0 \to \pi^+\pi^-$). Experimental results summarized in Ref. [1], indicate that resonance conventional parameters of the ρ^0 spread over 764 to 776 MeV for the mass and over 130 to 155 MeV for the width depending of the process and the model used to extract those parameters.

In order to make evident the dependence of resonance parameters on background subtraction, consider for instance the reaction $e^+e^- \to \pi^+\pi^-$ in the ρ^0 resonance region. The total cross section for this process is:

$$\sigma(s) = \frac{\pi\alpha^2}{3s^{5/2}}(s - 4m_\pi^2)^{3/2}|F_\pi^{em}(s)|^2 \tag{8}$$

where $\sqrt{s}$ is the total center of mass energy and the electromagnetic form factor F_π^{em} describes the electric charge distribution of π^+ in momentum space.

To illustrate the dependence on background subtraction, we can consider for example a model based on vector meson dominance [10]:

$$F_\pi^{em}(s) = A(s) - \frac{g_{\rho\gamma}G_\rho}{D_\rho(s)} - \frac{g_{\omega\gamma}G_\omega}{D_\omega(s)} \tag{9}$$

where g's and G's denote some coupling constants. $A(s)$ is some explicit background contribution that is *not* fixed by the model. ω is a second vector resonance with its mass nearly equal to the ρ^0 mass, which distorts the ρ resonance shape near the peak and is present due to isospin symmetry breaking. The sum of $A(s)$ and ω terms can be considered as a total background to the ρ. The denominator $D_V(s)$ in Eq. (9) is given by

$$D_V(s) = s - m_V^2 + im_V\Gamma_V(s), \tag{10}$$

and the (energy-dependent) width $\Gamma_V(s)$ is fixed by the model [10].

The cross section experimental data can be fitted by specifying different choices of the background contribution and the results for the resonance parameters [10] are shown in the following table:

$A(s)$	m_ρ (MeV)	Γ_ρ (MeV)
0	780.8 ± 0.5	153 ± 2
$-c_0 - c_2 s$	768.7 ± 0.7	142.8 ± 2.4
$-(1 + s/m_\rho^2)^{-1}$	773.2 ± 0.6	148.3 ± 1.7

As observed in this table, the resonance parameters of the ρ^0 meson depends strongly on the choice of background contributions and therefore, can not provide a suitable definition of their physical properties.

4. ρ^0 Resonance in the S-Matrix Approach

Some of the difficulties encountered in the conventional scheme are not present in the S-matrix formalism. The pole position of the S-matrix amplitude is an intrinsic physical property that characterizes a resonance. Since the construction of the S-matrix amplitude is based on very general requirements of relativistic quantum theory [3]-[4], it is, *a priori*, model-independent. Thus, in this section we will use the S-matrix formalism to address the problem of background-dependence of resonance parameters.

According to the S-matrix theory, the pion form factor must be written as follows [5]

$$(11) \qquad F_\pi(s) = -\frac{am_\rho^2}{s - s_\rho} + B(s)$$

where $s_\rho \equiv m_\rho^2 - im_\rho\Gamma_\rho$, and $B(s)$ denotes the background term.

Since there is not a unique way to include the second pole corresponding to the ω meson, in Ref. [5] we have introduced this (small) term in several ways. We have performed 10 fits [5] to experimental data of $F_\pi^{em}(s)$ by using different choices of the background and the ω contribution; a few illustrative examples are the following:

1) *Constant background* ($b = constant$):

$$(12) \qquad F_\pi^{(1)} = \left(-\frac{am_\rho^2}{s - s_\rho} + b\right)\left(1 + \frac{ym_\omega^2}{s - s_\omega}\right)$$

2) *Polynomial background*:

$$(13) \qquad |F_\pi^{(2)}|^2 = \left|\frac{am_\rho^2}{s - s_\rho}\left(1 + \frac{ym_\omega^2}{s - s_\omega}\right)\right|^2 P\left(\frac{s - m_\rho^2}{m_\rho^2}\right)$$

where $P(x) = 1 + \sum_{j=1}^{N} c_j x^j$.

3) *Non-polynomial background*

$$(14) \qquad F_\pi^{(3)} = -\frac{am_\rho^2}{s - s_\rho}\left(1 + \frac{ym_\omega^2}{s - s_\omega}\right)\left[1 + b\left(\frac{s - m_\rho^2}{m_\rho^2}\right)\right]^{-1}$$

4) *Frozen $\rho - \omega$ mixing*

$$(15) \qquad F_\pi^{(4)} = -\frac{am_\rho^2}{s - s_\rho} + \epsilon e^{i\phi}\frac{m_\omega^2}{s - s_\omega} + b.$$

The values extracted for resonance parameters are shown in the following table (see Ref. [5] for further details on other free parameters and quality of the fits).

Background	m_ρ (MeV)	Γ_ρ (MeV)
Constant	756.76 ± 0.82	143.91 ± 1.15
Polynomial ($N = 3$)	758.60 ± 0.78	141.80 ± 1.93
Non-polynomial	757.03 ± 0.76	141.22 ± 1.19
Frozen $\rho - \omega$	757.00 ± 0.59	143.41 ± 1.27

The important point that we want to stress from these results is that *the pole position in the S-matrix approach is largely independent on the way to subtract the background contribution or to include the ω pole.*

To conclude, the pole position is a physical property of the S-matrix which provides a suitable definition for the properties of resonances in particle physics because these properties are model-independent, process-independent and independent of the precise choice for background subtraction.

Acknowledgments: I am indebted to J. Pestieau and A. Bernicha for a fruitful collaboration that originates this work. I would like to thank Prof. A. Bohm for the kind invitation to present our results at this meeting.

References

1. R. M. Barnett *et al.*, *Review of Particle Properties*, Phys. Rev. **D54**, Part I, (1996).
2. R. G. Stuart, Phys. Lett. **B262**, 113 (1991); Phys. Lett. **B272**, 353 (1991); Phys. Rev. Lett. **70**, 3193 (1993); A. Leike, T. Riemann and J. Rose, Phys. Lett. **B273**, 513 (1991).
3. R. J. Eden, P. V. Landshoff, P. J. Olive and J. C. Polkinghorne, *The Analytic S-Matrix* (Cambridge University Press, Cambridge, UK, 1966).
4. M. Veltman, Physica **29**, 186 (1963).
5. A. Bernicha, G. López Castro and J. Pestieau, Phys. Rev. **D50**, 4454 (1994).
6. A. Bernicha, G. López Castro and J. Pestieau, Nucl. Phys. **A597**, 623 (1996).
7. A. Bohm, *Quantum Mechanics*, (Springer Verlag, New York, 1979).
8. See for example: J. D. Bjorken and S. D. Drell, *Relativistics Quantum Fields*, (McGraw-Hill, 1964).
9. A. Sirlin, Phys. Rev. Lett. **67**, 2127 (1991); Phys. Lett. **B267**, 240 (1991).
10. M. Benayoun *et al.*, Zeit. Phys. **C58**, 31 (1993).

Z Boson Resonance Parameters

T. Riemann

DESY – Institut für Hochenergiephysik Platanenallee 6, D-15738 Zeuthen, Germany

Summary. The Z line shape is measured at LEP/CERN with an accuracy at the per mille level. Usually it is described in the Standard Model of electroweak interactions with account of quantum corrections. Alternatively, one may attempt an S-matrix based model-independent approach in order to extract quantities like mass and width of the Z boson. I describe the formalism and its application to data.

1. Introduction

Our present understanding of weak interactions is completely described by the Standard Model [1], a spontaneously broken locally gauge invariant, anomaly-free, renormalizable quantum field theory of pointlike leptons and quarks, the latter in three colors. The model contains fermions, vector bosons, and a scalar particle:

$$\text{fermions:} \quad \begin{pmatrix} \nu_l \\ l \end{pmatrix} \quad l = e, \mu, \tau$$

$$\begin{pmatrix} U \\ D \end{pmatrix} \qquad \left. \begin{array}{l} U = u, c, t \\ D = d, s, b \end{array} \right\} \text{in 3 colors}$$

vector gauge bosons: $W^{\pm}$, Z^0, γ

scalar Higgs boson: H

The particle's masses (and mixing angles) are free parameters. Their interactions are determined from the invariance of the Lagrangean under local gauge transformations with gauge group $SU(2)_L \times U(1)$ and associated gauge fields $W^{\pm}, W^0, B$. One may parameterize the model in terms of masses and mixing angles plus electromagnetic coupling constant α_{em}. Often, instead of the W boson mass, the Fermi constant is used:

- $\alpha_{em} = 1/137.036$
- $G_{\mu} = 1.16634 \times 10^{-5}$ GeV^{-2}
- m_f, including $m_t = 175$ GeV [2]
- $M_Z = 91.186$ GeV [2]
- $M_H \ldots$ unknown

Discovery and study of the Z resonance are part of the long history of weak interactions and of unification of forces. First observations of virtual Z exchange lead to the discovery of weak neutral current reactions in the scattering of neutrinos off electrons, $\nu_\mu + e^- \to \nu_\mu + e^-$, and off nucleons, $\nu_\mu + N \to \nu_\mu + N$ in 1973 (Gargamelle Collab.: F.J. Hasert et al., A. Benvenuti et al., B. Aubert et al.) at the proton accelerator PS (CERN). The cross-section measurements may be interpreted in terms of the weak mixing angle θ_w. This angle characterizes not only the mixing of photon and Z boson but also the strength of the weak neutral interactions and the relations of the gauge boson masses to the Fermi constant (and among themselves). In the Standard Model[1]:

$$\text{(1)} \qquad Z = \cos\theta_w\, W^0 - \sin\theta_w\, B$$

$$\text{(2)} \qquad \gamma = \sin\theta_w\, W^0 + \cos\theta_w\, B$$

$$\text{(3)} \qquad g\sin\theta_w = e = \sqrt{4\pi\alpha_{em}}$$

$$\text{(4)} \qquad a_{lept} = -\frac{1}{2}$$

$$\text{(5)} \qquad v_{lept} = -\frac{1}{2}\left(1 - 4\sin^2\theta_w\right)$$

$$\text{(6)} \qquad M_W = \sqrt{\frac{\pi\alpha}{G_\mu\sqrt{2}}}\,\frac{1}{\sin\theta_w} \geq 37.281 \text{ GeV}$$

$$\text{(7)} \qquad M_Z = \frac{M_W}{\cos\theta_w}$$

The theory predicts the gauge boson masses as soon as there is a numerical estimate for the weak mixing angle[2]. Thus, after a few weak neutral current events were observed, a lot of information could gained from this. From the cross-sections of 1973, one may derive $0.1 < \sin^2\theta_w < 0.6$. This corresponds to $M_W = 118\cdots48$ GeV and $M_Z = 125\cdots75$ GeV. Both particles were discovered at the specially designed $p\bar{p}$ collider SPS (CERN) in 1983 (UA1 Collab.: G. Arnison et al., UA2 Collab.: P. Bagnaia et al., M. Banner et al.).

After the discovery of the Z boson, its detailed study by a dedicated tool, an e^+e^- collider with a center of mass energy corresponding to the Z mass, became a dream of particle physicists. Since the advent of the e^+e^- colliders LEP (CERN) and SLC (SLAC) in 1989, about sixteen millions of Z bosons have been produced at LEP and hundreds of thousands at SLC. They are produced as a resonance peak in the cross-section of the reaction

$$\text{(8)} \qquad e^+ + e^- \to (\gamma, Z) \to \quad \text{anything}$$

LEP finished operation as a Z factory in 1995 and is now running at higher energies for the study of W pair production and searches for Higgs, susy, and other particles while SLC goes yet on for a while.

---

[1] The relations get modified by radiative corrections; see e.g. [3].
[2] An absolute lower limit is about 37 GeV, see (6).

Due to the impressive accuracy of the measurements it was possible to test the Standard Model at the level of quantum corrections. This raises the problem of the accurate description of unstable particles in a quantum field theory. In a quantitative sense, this has been done with great success. Practically all experimental results are described by the Standard Model consistently within the experimental errors. I should mention specially the recent discovery of the top quark with a mass of about 176 GeV at Fermilab (CDF Collab.: F. Abe et al. (1994), D0 Collab.: S. Abachi et al. (1995)). This value agrees nicely with that predicted from measurements of the *Z* resonance parameters when quantum corrections from virtual top quark exchange are taken into account in the Standard Model, $m_t \sim 147 - 167$ GeV [4]. The fits favor a light Higgs boson with $M_H = 121^{119}_{-68}$ GeV and the estimate $M_H < 430$ GeV at 95% C.L. However, one should note that there is no experimental hint for the existence of the Higgs boson, whose interactions are assumed to create all the particle masses.

In this contribution, the shape of the *Z* resonance excitation curve will be described[3]. It is analyzed to which extent the description is model-independent. Some emphasis will be given to an approach based on first principles as formulated in the S-matrix theory.

The *Z* resonance is part of our physical world. We are not faced with the problem of its existence but rather of its proper description. A rigid mathematical handling of unstable particles to which the efforts of many of the participants at this Symposium are devoted is certainly not developed within the framework of relativistic quantum field theory. I hope that my talk may serve as an introduction to the status of the study of the *Z* resonance. The presentation will reflect my working activities which are closely related to the interpretation of measured cross-sections and other observables in terms of theoretical quantities. This task deserves a close interaction of theoreticians and experimental physicists. For details about this cooperation interested collegues may consult e.g. [2, 3] and references therein.

With the advent of the high-precision data from LEP 1 on single *Z* boson production and the frequent *W* pair production at LEP 2, the problem of definition of their masses and widths in a renormalizable quantum field theory became an important issue. Experimentalists often use formulae in the on-mass-shell approach, while some theorists prefer the introduction of a complex particle pole prescription, proposed in [6, 7]. The first one is preferred by recent tradition and well-developed while the latter one looks more convincing from a conceptual point of view: the propagator may be constructed in an explicitely gauge invariant way. When used properly, both schemes will give gauge-invariant results in the relevant order of perturbation theory (see e.g. [6, 8, 9, 10]), but the numerical values for the *Z* mass differ significantly.

---

[3] For the description of hadron resonances see the contribution of G. López Castro [5] to this symposium.

This was observed first in [11, 12]. Quite recently, the relation of both schemes was discussed in detail [13]. Although I will not give an introduction to perturbative renormalization for unstable particles, few comments on it may be found in sections 5. and 7..

2. The Z Line Shape

Some of the predictions of the Standard Model have been mentioned in the Introduction. Particle masses are used as input parameters while their life times $\tau = 1/\Gamma$, Γ being the decay width, may be predicted. The Z boson decays nearly exclusively into pairs $f\bar{f}$ of leptons $(e, \nu_e, \mu, \nu_\mu, \tau, \nu_\tau)$ or colored quarks $(3 \times d, u, s, c, b)$. The inverse life time (total Z width) is the incoherent sum of all partial widths of the different decay channels. From the Lagrangean for the $Zf\bar{f}$ interactions

$$(9) \qquad \mathcal{L} = -\frac{ig}{2\cos\theta_w} Z_\mu \sum_f \bar{f}\gamma^\mu [v_f + a_f \gamma_5] f$$

one may derive [14, 3]

$$(10) \quad \Gamma_Z = \sum_f \Gamma_f =$$

$$\sum_f N_{color}^f \frac{G_\mu M_Z^3}{6\pi\sqrt{2}} \left[(v_f^{eff})^2 + (a_f^{eff})^2 \right] = 2.4946 \pm 0.0027 \text{ GeV [2]}$$

The notations indicate that the (effective) couplings are slightly modified by radiative corrections.

Unfortunately, the width of a particle is not directly measurable. The Z width may be derived from an analysis of the Z resonance measured at the accelerators LEP1 and SLC. There, the most frequent reaction is

$$(11) \qquad\qquad e^+ e^- \to (\gamma, Z) \to \bar{f}f(+n\gamma)$$

The cross-section is shown as a function of the beam energy for a wide energy range in Figure 1.

The mass M_Z and width Γ_Z may be determined from cross-sections obtained in a small region around the Z peak $(s = 4E_{beam}^2)$:

$$(12) \qquad\qquad |\sqrt{s} - M_Z| < 3 \text{ GeV}$$

For energies off the resonance the cross-section falls down rapidly.

Without radiative corrections, the isolated Z resonance shape may be fitted with the following ansatz:

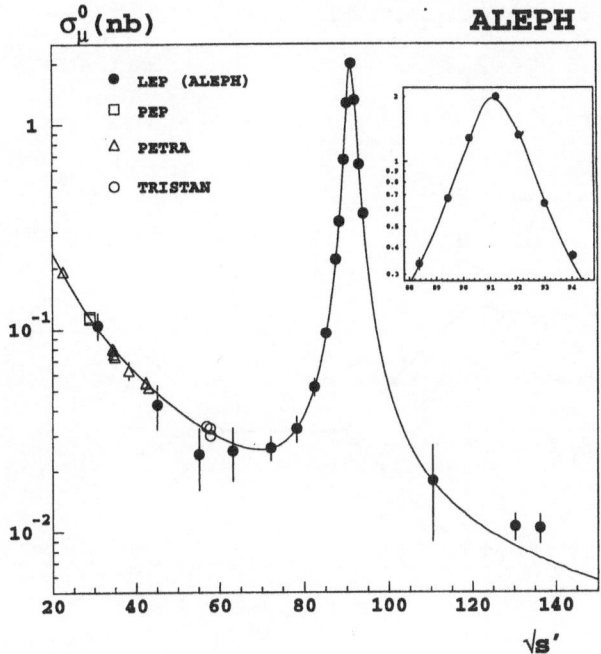

Fig. 1. The muon production cross-section over a wide energy range [14]

$$(13) \qquad \sigma_0^{(Z)}(s) \;=\; \frac{4\pi\alpha^2}{3s}|\chi(s)|^2\left(a_e^2+v_e^2\right)\left(a_f^2+v_f^2\right)N_{color}^f$$

with a Breit-Wigner shape function

$$(14) \qquad \chi(s) \;=\; \frac{G_\mu M_Z^2}{\sqrt{2}2\pi\alpha}\times\kappa(s), \qquad \kappa(s)=\frac{s}{s-M_Z^2+iM_Z\Gamma_Z(s)}$$

The natural appearance of an s dependence of the width function in a perturbative calculation was pointed out by Wetzel (1983) [16].

In Born approximation, the following expression in terms of partial Z widths is equivalent to (13):

$$(15) \qquad \sigma_0^{(Z)}(s) \;=\; \frac{12\pi}{s}|\kappa(s)|^2\frac{\Gamma_e\,\Gamma_f}{M_Z^2}$$

The cross-section values have to be related to the free parameters of the theory. In the Standard Model these are, e.g., $M_Z, M_H, m_t, \alpha_{strong}$ – and *not*, e.g., Γ_Z, or the partial widths Γ_f. When radiative corrections are taken into account – and they have to be – numerical differences may not be neglected.

Let me now mention some features of the Z line shape which make its analysis complicated. The first fact is that we want to study a $2 \to 2$ process with intermediate Z, but have also to take into account virtual photon exchange, see figure 2.

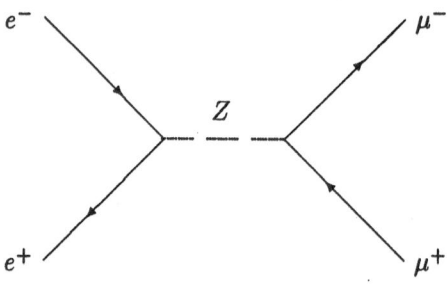

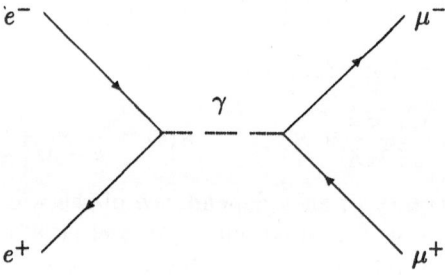

Fig. 2. Born contributions to the Z resonance shape

In addition, there are huge $2 \to 3, 4, \ldots$ contributions due to initial state radiation (ISR) and final state radiation (FSR), see figure 3.

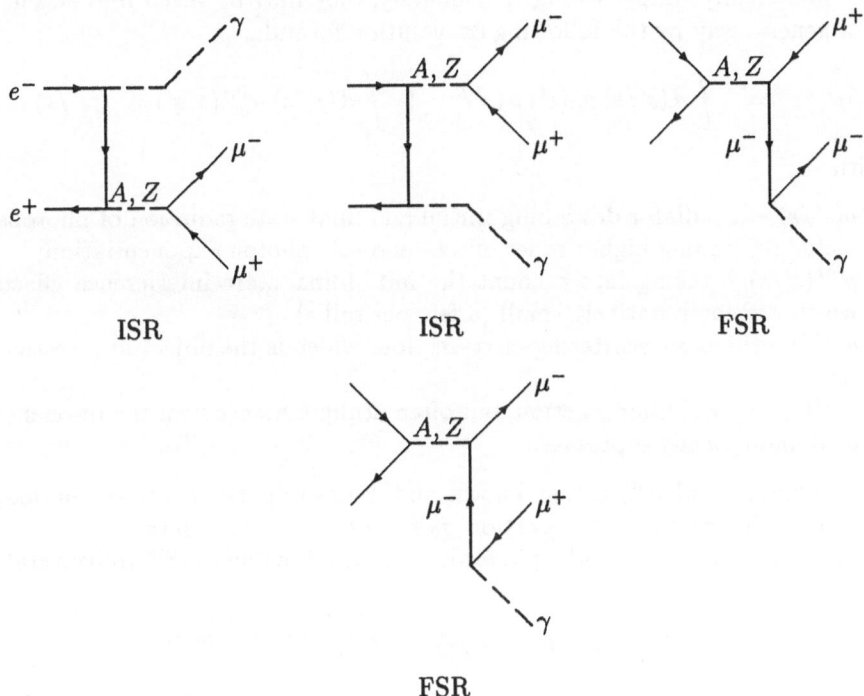

Fig. 3. QED corrections to the *Z* resonance shape

Further, many virtual corrections are not shown here but have to be included:

— vertex insertions
— self-energy insertions
— box diagrams
— for quarks: final state QCD corrections

In the next section, I will indicate the proper handling of the *Z* resonance as it is practiced by the LEP experimental groups ALEPH, DELPHI, L3, OPAL and the SLD group at SLC. Further details may be found in [2, 3] and, of course, in the articles of these collaborations.

3. Photonic Corrections to the Z Line Shape

Photonic corrections influence position and shape of the Z resonance heavily.
An analysis of data without proper treatment of them leads to results which
are numerically simply wrong. Fortunately, they may be taken into account
in a generic way by the following convolution formula:

$$(16) \quad \bar{\sigma}(s) \; = \; \int d\left(s'/s\right) \sigma_0(s')\, \rho\left(s'/s\right) \; + \; \int d\left(s'/s\right) \sigma_0^{int}(s, s')\, \rho^{int}\left(s'/s\right)$$

with

- $\rho(s'/s)$ – a radiator describing initial and final state radiation of photons,
  including leading higher order effects and soft photon exponentiation;
- $\rho^{int}(s'/s)$ – taking into account the initial-final state interference effects
  which are comparatively small (a few per mille);
- $\sigma_0(s')$ – the basic scattering cross-section, which is the object of investigation;
- $\sigma_0^{int}(s, s')$ – a similar function, but often negligible since near the resonance
  peak numerically suppressed.

If ρ is known and ρ^{int} is also known and, more important, small, one may
try to unfold the basic cross-section σ_0 from experimental data.

The dominant part of the photonic corrections is due to ISR (initial state
radiation):

$$(17) \quad \rho(s'/s) \; = \; \beta(1 - s'/s)^{\beta-1}\delta^{soft+virtual} + \delta^{hard}$$

where [3]

$$(18) \quad \beta \; = \; (2\alpha/\pi)\left[\ln\left(s/m_e^2\right) - 1\right]$$
$$(19)\; \delta^{soft+virtual} \; = \; 1 + (\alpha/(2\pi))\left[3\ln\left(s/m_e^2\right) + 2\pi^2/3 - 4\right] + \mathcal{O}\left(\alpha^2\right)$$
$$(20) \quad \delta^{hard} \; = \; -(\alpha/\pi)\left(1 + s'/s\right)\left[\ln\left(s/m_e^2\right) - 1\right] + \mathcal{O}\left(\alpha^2\right)$$

Near the resonance peak ISR leads to huge corrections of shape, position,
and height of the peak and cannot be neglected. This means that any serious
physical analysis is not only faced by higher order corrections but also by a
substantial admixture of $2 \rightarrow 3$ (and higher order) processes which may not
be experimentally extracted from the $2 \rightarrow 2$ process under investigation.

4. Approaching a Reasonable Formula for σ_0

In this section, I discuss the sensitivity of the determination of M_Z on the
theoretical ansatz[4]. At first glance, one may expect that the peak of the

---

[4] More details may be found in [17].

Z resonance is at $\sqrt{s_{max}} = M_Z$. This would reduce the determination of M_Z to a search of the peak location. This intuitive picture is modified by several effects. As already mentioned, photonic initial state radiation is rather influential. One may estimate the resulting shift of the Z resonance peak from (17):

$$(21) \qquad \sqrt{s_{max}} - M_Z \;=\; \delta_{QED} = \frac{\pi}{8}\beta\Gamma_Z + \text{small corr's.} \approx 106\,\text{MeV}$$

A simple and reasonable ansatz for $\sigma_0^{(Z)}$ is a pure Breit-Wigner function

$$(22) \qquad \sigma_0^{(Z)}(s) \sim \frac{M_Z^2 \cdot R}{|s - M_Z^2 + iM_Z\Gamma_Z|^2}$$

It may be shown (and will be made plausible in the next two sections) that the following ansatz is more realistic[5]:

$$(23) \qquad \sigma_0(s) = \frac{4}{3}\pi\alpha^2 \left[\frac{r^\gamma}{s} + \frac{s \cdot R + (s - M_Z^2) \cdot J}{|s - M_Z^2 + is\Gamma_Z/M_Z|^2}\right]$$

This line shape is characterized by five parameters:

- $r^\gamma \sim \alpha_{em}^2(M_Z^2)/\alpha_{em}^2$ – this γ exchange term may be assumed to be known
- M_Z, Γ_Z
- R – measure of the Z peak height
- J – measure of the γZ interference

Besides (21), from the replacements

$$(24) \qquad M_Z^2 \cdot R \to s \cdot R, \; iM_Z\Gamma_Z \to is\Gamma_Z/M_Z$$

additional shifts arise:

$$(25) \qquad \sqrt{s_{max}} - M_Z \;=\; \delta_{QED} \oplus \frac{1}{4}\frac{\Gamma_Z^2}{M_Z} \ominus \frac{1}{2}\frac{\Gamma_Z^2}{M_Z} \sim (90 + 17 - 34)\ \text{MeV}$$

Additionally, there is the effect of the γZ interference J:

$$(26) \qquad \sqrt{s_{max}} - M_Z \;=\; \delta_{QED} \oplus \frac{1}{4}\frac{\Gamma_Z^2}{M_Z}\left(1 + \frac{J}{R}\right) \ominus \frac{1}{2}\frac{\Gamma_Z^2}{M_Z}$$

$$\sim \left[90 + 17 \times \left(1 + \frac{J}{R}\right) - 34\right]\ \text{MeV}$$

If one wants to take into account the J, a model for its prediction is needed. Neglecting this interference (by setting $J{=}0$) leads to an erroneous systematic shift of the Z mass of 17 MeV$\otimes(J/R)$.

The value for hadron production in the Standard Model is, e.g., [21]:

$$(27) \qquad \frac{J}{R} \otimes 17\ \text{MeV} = \frac{0.22}{2.97} \otimes 17\ \text{MeV} = 0.07 \otimes 17\ \text{MeV} = 1.26\ \text{MeV}$$

---

[5] This or similar formulae have been proposed in [18, 19, 7, 20].

5. Z Boson Parameters (I):
The Standard Model Approach

A realistic scan of the Z line shape may be performed with the following ansatz derived from the Standard Model, including higher order radiative corrections [3]:

$$(28) \qquad \sigma_0(s) = \sigma_0^{(Z)}(s) + \sigma_0^{(\gamma Z)}(s) + \sigma_0^{(\gamma)}(s)$$

The dominating part is

$$(29) \quad \sigma_0^{(Z)}(s) = \frac{4\pi\alpha^2}{3s}|\chi(s)|^2|\rho_{ef}^{eff}|^2 \left(\frac{1}{16} + \frac{1}{4}|v_e^{eff}|^2 + \frac{1}{4}|v_f^{eff}|^2 + |v_{ef}^{eff}|^2 \right)$$

with an s-dependent width function in the Breit-Wigner shape (14). The width is obtained in perturbation theory by summing an infinite Dyson series of self-energy insertions to the Z boson propagator. In order to prevent gauge violation (in the given order of perturbation theory), one has to add up a minimal set of Feynman diagrams that is necessary for the compensation of gauge dependences. The decay width of the Z boson in lowest order is given by the imaginary parts of fermion loops in the one loop self-energy. For single Z boson production we may use[6]:

$$(30) \qquad \Gamma_Z(s) = \frac{s}{M_Z^2}\Gamma_Z$$

The simple s dependence is due to the smallness of the fermion masses allowing for the neglect of threshold effects. A complete two-loop calculation would also modify this. The virtual corrections are contained in the width Γ_Z and in four complex-valued form factors ρ_Z^{eff}, v_e^{eff}, v_f^{eff}, v_{ef}^{eff} which depend on beam energy and scattering angle.

To a good approximation, (28) agrees with (23). The cross-section values have to be related to the free parameters of the theory. A recent determination is [22]:

$$(31) \qquad M_Z = 91.188 \pm 0.002 \text{ GeV}$$
$$(32) \qquad \alpha_{strong}(M_Z^2) = 0.126 \pm 0.007 \pm 0.002 \text{ (Higgs)}$$
$$(33) \qquad m_t = 189 \pm 15 \text{ (exp.)} \pm 16 \text{ (Higgs) GeV.}$$

The Higgs mass is varied from 60 to 1000 GeV with central values for 300 GeV. Quite similar values have been quoted in [2]. For comparison I quote also the direct measurements of m_t by CDF, $m_t = 175.6 \pm 4.4 \pm 4.8$ GeV, and D0 collaborations, $m_t = 169 \pm 8 \pm 8$ GeV [23].

---

[6] For the by far more complex case of off-shell gauge boson pair production, the fermion-loop scheme [10, 13] solves the gauge problem satisfactorily.

A step towards a model-independent Z resonance analysis is the determination of some characteristic line shape parameters from table 6 of [2] (with indicated relative errors δ):

(34) $\qquad\qquad M_Z \;=\; 91.1863 \pm 0.0020$ GeV ($\delta = 0.0022$ %)

(35) $\qquad\qquad \Gamma_Z \;=\; 2.4946 \pm 0.0027$ GeV ($\delta = 0.11$ %)

(36) $\qquad\qquad \sigma_0^{had} \;=\; 41.508 \pm 0.056$ nb ($\delta = 0.13$ %)

(37) $\qquad R_l = \dfrac{\sigma_0^{had}}{\sigma_0^{lept}} \;=\; 20.778 \pm 0.029$ ($\delta = 0.14$ %)

(38) $\qquad\qquad A_{FB,0}^{lept} \;=\; 0.0174 \pm 0.0010$

Here, $M_Z, \Gamma_Z, \sigma_0^{had}$ are determined mainly from $\sigma^{had}(s)$, while R_l and A_{FB} from $\sigma^{lept}(s)$:

- $\sigma_0^{had(lep)}$ – hadronic (leptonic) peak cross-section
- $A_{FB,0}^{lept}$ – forward-backward asymmetry at the peak

These parameters are considered to be primary parameters in contrast to derived ones, e.g. the effective leptonic weak neutral current couplings of leptons (table 8 of [2]):

(39) $\qquad\qquad\qquad v_l \;=\; -0.03776 \pm 0.0062$

(40) $\qquad\qquad\qquad a_l \;=\; -0.50108 \pm 0.00034$

or the effective weak mixing angle (tables 1,4 of [2]):

(41) $\qquad \sin^2 \vartheta_W^{eff} \equiv \dfrac{1}{4}\left(1 - \dfrac{v_l}{a_l}\right) \;=\; 0.23165 \pm 0.00024$

The introduction of effective weak neutral couplings and the effective weak mixing angle comes back close to the language of the Standard Model.

6. Z Boson Parameters (II): The S-Matrix Approach

All the above results fit nicely with each other and strengthen the Standard Model's credit. Nevertheless, one may ask for an approach being independent of it. A tool with minimal assumptions is S-matrix theory [24, 25, 26]. The first application of S-matrix theory to the Z resonance is due to A. Martin (1985) [26] who studied the toponium-Z interference pattern assuming their masses to be of similar size. In 1991, R. Stuart proposed to consider the scattering matrix element for the process $e^+e^- \rightarrow Z \rightarrow f\bar{f}$ as a Laurent series with the Z boson as resonance [7]. This allowed him to collect gauge invariant pieces of the cross-section in perturbation theory [27] and to derive a simple cross-section formula similar to (22), but with small perturbations.

For an application to experimental data, a number of modifications have been added [20, 28] and the necessary software has been created [29]: consider the cross-section as an incoherent sum of four helicity scatterings; treat the photonic corrections properly, especially those due to initial state radiation; treat in the same manner as the total cross-section also asymmetries; try to include into the formula the fact that there is also photon exchange, i.e. that in reality one has the co-existence of two resonances. The first fit to LEP 1 data was performed in [20].

Consider four independent helicity amplitudes in the case of massless fermions f:

$$(42) \quad \mathcal{M}^{fi}(s) = \frac{R_\gamma^f}{s} + \frac{R_Z^{fi}}{s - s_Z} + \sum_{n=0}^{\infty} \frac{F_n^{fi}}{\overline{m}_Z^2} \left(\frac{s - s_Z}{\overline{m}_Z} \right)^n, \quad i = 0, \dots, 3$$

Without the first (photon) term, they are Laurent series. The position of the Z pole in the complex s plane is given by s_Z:

$$(43) \qquad\qquad s_Z = \overline{m}_Z^2 - i \overline{m}_Z \overline{\Gamma}_Z$$

The R_γ^f and R_Z^{fi} are complex constants characterizing the photon and the Z boson, respectively. For practical purposes one may truncate the series:

$$(44) \qquad \mathcal{M}^{fi}(s) \;=\; \frac{R_\gamma^f}{s} + \frac{R_Z^{fi}}{s - s_Z} + \frac{F_0^{fi}}{\overline{m}_Z^2} \approx \frac{R_\gamma^f}{s} + \frac{R_Z^{fi}}{s - s_Z}$$

There are four residua R_Z^{fi}:

$$(45) \qquad
\begin{aligned}
R_Z^{f0} &= R_Z(e_L^- e_R^+ \longrightarrow f_L^- f_R^+) \\
R_Z^{f1} &= R_Z(e_L^- e_R^+ \longrightarrow f_R^- f_L^+) \\
R_Z^{f2} &= R_Z(e_R^- e_L^+ \longrightarrow f_R^- f_L^+) \\
R_Z^{f3} &= R_Z(e_R^- e_L^+ \longrightarrow f_L^- f_R^+)
\end{aligned}$$

The amplitudes $\mathcal{M}^{fi}(s)$ give rise to four cross-sections σ_i:

$$(46) \qquad
\begin{aligned}
\sigma_T^0(s) &= & &= &+\sigma_0 + \sigma_1 + \sigma_2 + \sigma_3 \\
\sigma_{lr\text{-}pol}^0(s) &= &\sigma_{FB}^0(s) &= &+\sigma_0 - \sigma_1 + \sigma_2 - \sigma_3 \\
\sigma_{FB\text{-}lr}^0(s) &= &\sigma_{pol}^0(s) &= &-\sigma_0 + \sigma_1 + \sigma_2 - \sigma_3 \\
\sigma_{lr}^0(s) &= &\sigma_{FB\text{-}pol}^0(s) &= &-\sigma_0 - \sigma_1 + \sigma_2 + \sigma_3
\end{aligned}$$

Here, it is

σ_T^0 – the total cross-section,

σ_{FB}^0 – numerator of the forward-backward asymmetry,

σ_{pol}^0 – numerator of the final state polarization,

$\sigma_{FB\text{-}pol}^0$ – that of the forward-backward asymmetry of the final state polarization etc.

All these cross-sections may be parameterized by the following master formula:

$$(47) \quad \sigma_A^0(s) = \frac{4}{3}\pi\alpha^2 \left[\frac{r_A^{\gamma f}}{s} + \frac{s r_A^f + (s - \overline{m}_Z^2) j_A^f}{(s - \overline{m}_Z^2)^2 + \overline{m}_Z^2 \overline{\Gamma}_Z^2} + \frac{r_A^{f0}}{\overline{m}_Z^2} \right] + \dots$$

$$\approx \frac{4}{3}\pi\alpha^2 \left[\frac{r_A^{\gamma f}}{s} + \frac{s r_A^f + (s - \overline{m}_Z^2) j_A^f}{(s - \overline{m}_Z^2)^2 + \overline{m}_Z^2 \overline{\Gamma}_Z^2} \right], \quad A = T, FB, \dots$$

Thus we have re-derived (23) with one modification: the *Z* width function is treated as constant here. Now, the parameters r, j are related to the residua of the pole terms. The $r_A^{\gamma f}$ is the photon exchange term:

$$(48) \quad r_A^{\gamma f} = \frac{1}{4}c_f \sum_{i=0}^{3}\{\pm 1\}\left|R_\gamma^f\right|^2 R_{QCD}^A$$

It is known from QED for the total cross-section ($A = T$) and vanishes for all asymmetric cross-sections. Further, $c_f = 1, 3$ for leptons and quarks, respectively. QCD corrections for quarks are taken into account by the factor R_{QCD}. The Z exchange residuum r_A^f and the γZ interferences j_A^f are:

$$(49) \quad \begin{aligned} r_A^f &= c_f \left\{ \frac{1}{4} \sum_{i=0}^{3} \{\pm 1\} \left| R_Z^{fi} \right|^2 + 2 \frac{\overline{\Gamma}_Z}{\overline{m}_Z} \Im m C_A^f \right\} R_{QCD}^A \\ j_A^f &= c_f \left\{ 2 \Re e C_A^f - 2 \frac{\overline{\Gamma}_Z}{\overline{m}_Z} \Im m C_A^f \right\} R_{QCD}^A \\ C_A^f &= (R_\gamma^f)^* \left(\frac{1}{4} \sum_{i=0}^{3} \{\pm 1\} R_Z^{fi} \right) \end{aligned}$$

The factors $\{\pm 1\}$ in (48) and (49) indicate that the signs of $\left|R_\gamma^f\right|^2$, $\left|R_Z^{fi}\right|^2$, and of R_Z^{fi} correspond to the signs of σ_i in (46).

6.1 Asymmetries

Without QED corrections, asymmetries are defined by:

$$(50) \quad \mathcal{A}_A^0(s) = \frac{\sigma_A^0(s)}{\sigma_T^0(s)}, \quad A \neq T$$

They take a simple form around the Z resonance. For applications at LEP 1, they may be characterized by only two parameters [28]:

$$(51) \quad \mathcal{A}_A^0(s) = A_0^A + A_1^A \left(\frac{s}{m_Z^2} - 1 \right) + A_2^A \left(\frac{s}{m_Z^2} - 1 \right)^2 + \dots$$

$$\approx A_0^A + A_1^A \left(\frac{s}{m_Z^2} - 1 \right)$$

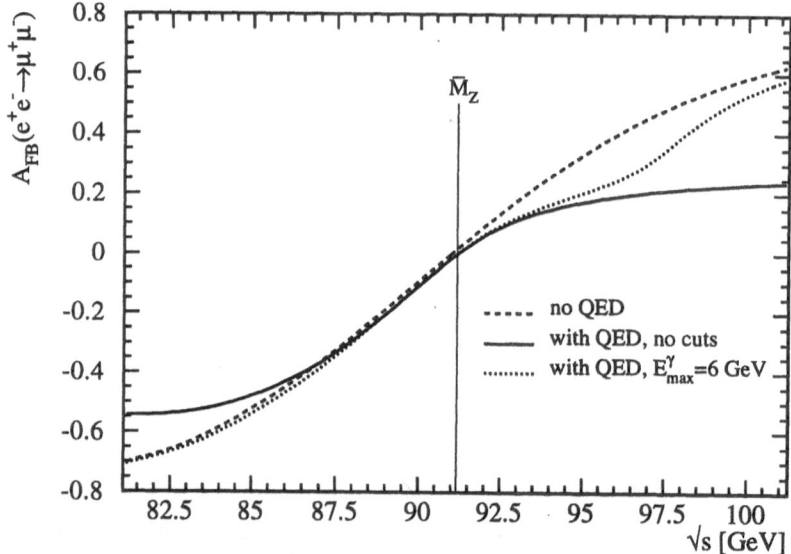

Fig. 4. The forward-backward asymmetry for the process $e^+e^- \to \mu^+\mu^-$ near the Z peak [26].

The higher order terms may be neglected since $(s/\overline{m}_Z^2 - 1)^2 = \sigma^2 < 2 \times 10^{-4}$. The first coefficients are:

$$(52) \qquad A_0^A = \frac{r_A^f}{r_T^f + \gamma^2 r_T^{\gamma f}} \approx \frac{r_A^f}{r_T^f}$$

$$(53) \qquad A_1^A = \left[\frac{j_A^f}{r_A^f} - \frac{j_T^f - 2\gamma^2 r_T^{\gamma f}}{r_T^f + \gamma^2 r_T^{\gamma f}} \right] A_0^A \approx \left[\frac{j_A^f}{r_A^f} - \frac{j_T^f}{r_T^f} \right] A_0^A$$

Here, the r_A^{0f} is neglected in both A_0 and A_1. Further, the definition $\gamma^2 = \overline{\Gamma}_Z^2 / \overline{m}_Z^2 \approx 0.75 \times 10^{-3}$ is used. The non-vanishing of the slope of the asymmetry shape is due to the γZ interference. From figure 4 one may see that the linear rise is damped right of the peak. This is due to amplified QED corrections to the pure Z exchange cross-sections versus non-amplified QED corrections to the γZ interferences [28].

6.2 Numerical Results in the S-matrix Approach

The first fit with the S-matrix approach to experimental data has been performed in 1991 [20]. The first experimental analysis of a LEP collaboration was due to L3 in 1993 [30]. Further systematic studies may be found in [31]. They helped to determine the appropriate number and location of energy points for a Z line shape scan. Recent experimental studies are e.g. [32, 33, 21, 34, 22]. Typically, results as in table 1 are obtained from the LEP1 and LEP1.5 Z line shape scans which were performed mainly in 1993 and 1995 (from table 6 of [21][7]). The biggest error correlations are shown in table 2 (from table 7 of [21]). We see here an essential difference to Standard Model fits which assume *fixed* relations among many of these parameters. They rely thus on stronger theoretical assumptions.

From the strong correlations in the S-matrix fit together with the excellent agreement of the central values of fitted parameters in both fit scenarios one may conclude that the two scenarios are highly compatible with each other.

Table 1. Results from a combined LEP1 line shape fit

Parameter	S-matrix fit	SM Prediction
M_Z [GeV]	91.1965±0.0048	–
Γ_Z [GeV]	2.4941±0.0033	2.4973
r_T^{had}	2.9644±0.0083	2.9681
j_T^{had}	0.22±0.25	0.22
r_T^{lept}	0.14245±0.00044	0.14268
j_T^{lept}	0.020±0.017	0.004
r_{FB}^{lept}	0.00315±0.00022	0.00271
j_{FB}^{lept}	0.793±0.016	0.799

Table 2. Biggest correlations in the S-matrix fit

Correlation	Value
$M_Z - j_T^{had}$	−0.89
$M_Z - j_T^{lept}$	−0.62
$\Gamma_Z - r_T^{had}$	0.77
$\Gamma_Z - r_T^{lept}$	0.69
$r_T^{had} - r_T^{lept}$	0.86
$j_T^{had} - j_T^{lept}$	0.62

Including into the analysis cross-sections measured at other energies may improve substantially e.g. the resolution of M_Z and j_T which are highly correlated (for a combination with data from the TOPAZ collaboration at KEK with $\sqrt{s} \sim 55$ GeV as shown in figure 1 see reference [35]; data from LEP 1.5 with $\sqrt{s} \sim 135$ GeV have been included already).

---

[7] Note that the table shows values of the on shell mass M_Z which were derived after the fit of the complex pole mass $\overline{m}_Z$ as explained in section 7..

7. Defining the Z Boson Mass

The complex Z pole definition in (42) with a constant width is natural in the S-matrix ansatz. It leads to different numerical values compared to the usual Standard Model, on mass shell approach as used in (28)–(30). The following discussion of this difference follows closely appendix D of [13] where more details may be found.

In the perturbative approach, the *complex pole* μ_Z of the propagator is defined as follows:

$$(54) \qquad \mu_Z - \mu_Z^0 + \Sigma_Z^0(\mu_Z, \mu_Z^0) \;=\; 0$$

$$(55) \qquad\qquad\qquad \mu_Z \;=\; \overline{m}_Z^2 - i\overline{m}_Z\overline{\Gamma}_Z$$

The bare Z boson mass is denoted by μ_Z^0 and Σ_Z^0 is the bare self-energy. The perturbative solution of the above equations is:

$$(56) \quad \overline{m}_Z^2 \;=\; \mu_Z^0 - \Re e \Sigma_Z^0(\overline{m}_Z^2) - \left[\Im m\Sigma_Z^0(\overline{m}_Z^2)\right]\left[\Im m\Sigma_Z^{0\prime}(\overline{m}_Z^2)\right] + \dots$$

$$(57) \; \overline{m}_Z\overline{\Gamma}_Z \;=\; \Im m\Sigma_Z^0(\overline{m}_Z^2)\Bigg\{1 - \Re e\Sigma_Z^{0\prime}(\overline{m}_Z^2) + \left[\Re e\Sigma_Z^{0\prime}(\overline{m}_Z^2)\right]^2$$

$$- \frac{1}{2}\left[\Im m\Sigma_Z^0(\overline{m}_Z^2)\right]\left[\Im m\Sigma_Z^{0\prime\prime}(\overline{m}_Z^2)\right] + \dots\Bigg\}$$

The *on shell mass and width* are defined as:

$$(58) \qquad\qquad M_Z^2 \;=\; \mu_Z^0 - \Re e\Sigma_Z^0(M_Z^2)$$

$$(59) \qquad\qquad M_Z\Gamma_Z \;=\; \frac{\Im m\Sigma_Z^0(M_Z^2)}{1 + \Re e\Sigma_Z^{0\prime}(M_Z^2)}$$

One may relate the two definitions and see that they differ by two-loop and higher order corrections. This has been discussed first in 1986 [6]. The expected experimental accuracy was about 10 MeV [36] at that time. Since the authors of [6] restricted themselves to the one loop order, they failed to observe the numerical significance of the difference in the definitions of about 35 MeV. There are also bosonic corrections in one-loop approximation. A systematic Dyson summation of bosonic self-energy corrections may be attempted without violating Ward identities in the context of the background field method [37]. For references to the application of so-called pinch techniques see also there. Since the on shell mass and the complex mass definitions are uniquely related order by order in perturbation theory, it is fair to say that either both or none of them has a gauge invariance problem, provided it is used properly. One may argue that a mass definition should be related to a structure like $\text{Const}/(s - s_0)$, with s_0 being a constant, but as long as relations are unique to another definition, there is some freedom of choice. Of course, the (perturbative) complex mass definition is conceptually closest to what one has in the S-matrix theory.

Around $s = M_Z^2$ the Z decays only into light fermions and it is

(60) $$\Im m \Sigma_Z^0(s) = s \, \Im m \Sigma_Z^{0\prime}(s) = s \, \frac{\Gamma_Z}{M_Z}$$

and thus

(61) $$\overline{m}_Z^2 \;=\; M_Z^2 - \Gamma_Z^2 + \dots$$

(62) $$\frac{\overline{\Gamma}_Z}{\overline{m}_Z} \;=\; \frac{\Gamma_Z}{M_Z}$$

The resulting numerical differences may well be approximated by the following relations:

(63)
$$\overline{m}_Z \;=\; M_Z - \frac{\Gamma_Z^2}{2M_Z} \;\approx\; M_Z \;-\; 34 \text{ MeV}$$
$$\overline{\Gamma}_Z \;=\; \Gamma_Z - \frac{\Gamma_Z^3}{2M_Z^2} \;\approx\; \Gamma_Z \;-\; 1 \text{ MeV}.$$

Equations (63) and the numerical values of the shifts were derived in 1988 in [12], where also the γZ mixing was taken into account, and repeatedly discussed later (see e.g. [38, 8, 7, 39, 40]). The Z resonance peak shift due to the difference of the two treatments of the Z boson self energy was numerically observed independently in [11] and in [12]. Both papers did not point out the significance of the complex pole mass definition, although it became obvious immediately after and was frequently discussed during the 1989 LEP 1 workshop organized by CERN. Similar derivations to the above may be found in [6, 8, 7, 9]. Gauge problems have been studied also in [43].

The observation of the sensitive dependence of numerical mass values on the definitions chosen in the Breit-Wigner shape function was made for hadron resonances by Gounaris and Sakurai (1968) [41]. See also the recent studies of hadron resonances reviewed in [5].

8. The Z Resonance and the Photon

Being strict, one may develop the S-matrix into a Laurent series as a function of s around *one* resonance only. Otherwise the coefficients are not uniquely determined. This has been stressed recently [42].

In order to be rigorous, one has to replace the ansatz (42) by one with $R_\gamma^f = 0$. The essential physical consequence is that the photonic cross-section becomes part of the background:

(64) $$\frac{r^\gamma}{s} \to \frac{r^\gamma}{M_Z^2}$$

What was understood so far as γZ interference J becomes a result of the background-Z interference.

I, personally, dislike this approach. The photon exists and we know how to describe it. So, I would prefer to see it treated as known input to an experimental analysis of the Z. But I agree that a detailed study of the resulting numerical differences between the rigorous S-matrix ansatz and that used by the LEP community at present could be of some interest. Perhaps it is worth to be mentioned that a consistent quantum mechanical description of two-resonance systems is possible; see e.g. chapter XX in [25].

9. Summary

Two different approaches to a numerical analysis of the Z boson line shape have been presented – the Standard Model of electroweak interactions and the S-matrix approach. The S-matrix approach allows to treat the γZ interference as an independent quantity, which enlarges the error for M_Z. Two different mass definitions may be used. Both agree in the numerical determination of the Z mass when the substantial difference in the mass definitions is taken into account. In the Standard Model the Z width is a derived quantity; the S-matrix approach allows a direct fit. Again, the two approaches agree numerically. The S-matrix approach shows that the Z line shape may be described by 4 independent parameters (per channel) – if QED is assumed to be a known phenomenon. Asymmetries may also be described by the S-matrix approach. They depend on two parameters (per channel). Their variation with s near the peak is due to the γZ interference. The S-matrix approach allows the combination of data from quite distinct kinematic regions. Finally, one should mention that the use of effective coupling constants leads to similarly reasonable numerical results.

Acknowledgments: I would like to thank the organizer of the Symposium on Semigroups and Resonances, Arno Böhm, for the kind invitation to the conference and to Wim Beenakker, Fred Jegerlehner, and Robin Stuart for careful reading the manuscript and suggestions.

References

1. S.L. Glashow, Nucl. Phys. 22 (1961) 579;
   S. Weinberg, Phys. Rev. Letters 19 (1967) 19;
   A. Salam, in: N. Svartholm (ed.), *Proc. of the Eighth Nobel Symposium* (Almqvist and Wiksell, Stockholm; Wiley, New York, 1968), p. 367;
   S.L. Glashow, J. Iliopoulos and L. Maiani, Phys. Rev. D6 (1970) 1285;
   C. Bouchiat, J. Iliopoulos and Ph. Meyer, Phys. Letters 38B (1972) 519.

2. A. Blondel, Rapporteur's Talk on *Experimental Status of Electroweak Interactions*, in: Z. Ajduk and A. Wroblewski (eds.), Proc. of the 28th Int. Conf. on High Energy Physics (ICHEP'96), Warsaw, Poland, 25-31 July 1996 (World Scientific, Singapore, 1997), Vol. I, p. 205, and references therein.

3. D. Bardin et al., ZFITTER – *An Analytical Program for Fermion Pair Production in e^+e^- Annihilation*, CERN preprint CERN-TH. 6443/92 (1992) [Electronic Archive: hep-ph/9412201]; D. Bardin, A. Olchevski, S. Riemann and T. Riemann, update in preparation;
   D. Bardin, W. Hollik and G. Passarino (eds.), *Reports of the Working Groups on Precision Calculations at the Z Resonance*, CERN Yellow Report CERN 96–01 (1996), and references therein.

4. The LEP Experiments: ALEPH, DELPHI, L3 and OPAL, the LEP Electroweak Working Group and the SLD Heavy Flavour Group, D. Abbaneo et al., *A Combination of Preliminary Electroweak Measurements and Constraints on the Standard Model*, preprint CERN-PPE/96-183 (Dec 1996), unpublished;
   J. Timmermans, *Precision Tests of the electroweak Interaction from e^+e^- Colliders*, talk held at Lepton-Photon Conference LP97, Hamburg, Germany, 28 July - 1 Aug 1997, see: http://www.desy.de/lp97/.

5. G. López Castro, talk at this Symposium, and references therein.

6. M. Consoli and A. Sirlin, *The role of the one-loop electroweak effects in $e^+e^- \rightarrow \mu^+\mu^-$*, in: J. Ellis and R. Peccei (eds.), Physics at LEP, CERN Yellow Report CERN 86–02 (1986), p. 63.

7. R. Stuart, Phys. Letters B262 (1991) 113.

8. F. Jegerlehner, *Renormalizing the Standard Model*, in: M. Cvetic and P. Langacker, Testing the Standard Model, Proc. of the 1990 Theoretical Advanced Study Institute in Elementary Particle Physics, Boulder, Colorado, USA, 3-27 June 1990 (World Scientific, Singapore, 1991).

9. A. Sirlin, Phys. Rev. Letters 67 (1991) 2127; Phys. Letters B267 (1991) 240.

10. E.N. Argyres, W. Beenakker, G.J. van Oldenborgh, A. Denner, S. Dittmaier, J. Hoogland, R. Kleiss, C.G. Papadopoulos and G. Passarino, Phys. Letters B358 (1995) 339.

11. F.A. Berends, G. Burgers, W. Hollik and W. van Neerven, Phys. Letters B203 (1988) 177.

12. D. Bardin, A. Leike, T. Riemann and M. Sachwitz, Phys. Letters B206 (1988) 539.

13. W. Beenakker, G.J. van Oldenborgh, A. Denner, S. Dittmaier, J. Hoogland, R. Kleiss, C.G. Papadopoulos and G. Passarino, *The fermion-loop scheme for finite-width effects in e^+e^- annihilation into four fermions*, preprint NIKHEF 96–031 (1996) [Electronic Archive: hep-ph/9612260].

14. A. Akhundov, D. Bardin and T. Riemann, Nucl. Phys. B276 (1986) 1;
    F. Jegerlehner, Z. Physik C32 (1986) 425, E: ibid. C38 (1988) 519;
    W. Beenakker and W. Hollik, Z. Physik C40 (1988) 141;
    J. Bernabéu, A. Pich and A. Santamaria, Phys. Letters B200 (1988) 569.

15. Figure by courtesy of Frederic Teubert, from: The ALEPH Collab. (R. Barate et al.), *Study of muon-pair production at centre-of-mass energies from 20 to 136 GeV with the ALEPH detector*, paper PA07-070, contributed to the ICHEP'96, Warsaw, July 1996. Updated version published: Phys. Letters B399 (1997) 329.

16. W. Wetzel, Nucl. Phys. B227 (1983) 1.

17. W. Beenakker, F.A. Berends and S.C. van der Marck, Z. Physik C46 (1990) 687.

18. A. Borrelli, M. Consoli, L. Maiani and R. Sisto, Nucl. Phys. B33 (1990) 357.

19. F. Jegerlehner, *Physics of precision experiments with Zs*, in: A. Faessler (ed.), Prog. Part. Nucl. Phys., vol. 27, p. 1 (Pergamon Press, Oxford, U.K., 1991).

20. A. Leike, T. Riemann and J. Rose, Phys. Letters B273 (1991) 513.
21. The S-Matrix Subgroup of the LEP Electroweak Working Group, A. Blondel et al., *An Investigation of the interference between Photon and Z-Boson Exchange*, preprint LEPEWWG/LS/96-01 (March 1996), unpublished.
22. The L3 Line Shape Group, G. Bobbink et al., *Preliminary L3 results on Electroweak Parameters using 1990–95 data*, L3 note # 1980 (Aug 1996), prepared for the ICHEP'96, Warsaw, unpublished.
23. The CDF Collab., F. Abe et al., Phys. Rev. Letters 74 (1995) 2626;
    The D0 Collab., S. Abachi et al., Phys. Rev. Letters 74 (1995) 2632;
    J. Lys (CDF) and S. Protopopescu (D0), contributions to: Z. Ajduk and A. Wroblewski (eds.), Proc. of the 28th Int. Conf. on High Energy Physics (ICHEP'96), Warsaw, Poland, 25-31 July 1996 (World Scientific, Singapore, 1997).
24. R.J. Eden, P.V. Landshoff, D.I. Olive and J.C. Polkinghorn, *The Analytic S-Matrix* (Cambridge University Press, Cambridge, 1966).
25. A. Böhm, *Quantum Mechanics – Foundations and Applications* (Springer, New York, 1994).
26. A. Martin, Phys. Letters B156 (1985) 411; *Unstable particles,* in: Maurice Levy, Jean-Louis Basdevant, Maurice Jacob, David Speiser, Jacques Weyers, Raymond Gastmans (eds.), Z0 Physics, Proc. of the NATO Advanced Study Inst., Series B: Physics, v. 261, Cargése Summer School, Cargése, France, Aug 13-25, 1990 (N.Y., Plenum Press, 1991), p. 483; *The rigorous analyticity-unitarity program: a historical account*, preprint CERN–TH.6894/93 (1993).
27. R. Stuart, Phys. Letters B272 (1991) 353.
28. T. Riemann, Phys. Letters B293 (1992) 451.
29. S. Kirsch and T. Riemann, Comput. Phys. Commun. 88 (1995) 89;
    M. Grünewald, S. Kirsch and T. Riemann, SMATASY – *a Fortran program for the model independent description of the Z resonance*, to be used together with ZFITTER [3].
30. The L3 Collab., O. Adriani et al., Phys. Letters B315 (1993) 494; Phys. Reports 236 (1993) 1.
31. S. Kirsch and S. Riemann, *A combined fit to the L3 data using the S-matrix approach (first results)*, L3 note # 1233 (Sep 1992), unpublished;
    G. Isidori, Phys. Letters B314 (1993) 139;
    M. Grünewald and S. Kirsch, *A Possible Modification of a LEP Energy Scan for an Improved Determination of Z-Boson Parameters*, preprint CERN-PPE/93-188 (1993), unpublished;
    M. Consoli and M. Piccolo, *Strategy for the 1995 LEP energy scan*, unpublished [Electronic Archive: hep-ph/9505261].
32. M. Martinez, *Electroweak results at LEP–I*, in: J.W.F. Valle and A. Ferrer (eds.), Proc. of Int. Workshop on Elementary Particle Physics: Present and Future, Valencia, Spain, 5-9 June 1995 (World Scientific, River Edge, N.J., 1996), p. 32.
33. The S-Matrix Subgroup of the LEP Electroweak Working Group, A. Blondel et al., *An Investigation of the interference between Photon and Z-Boson Exchange*, preprint LEPEWWG/LS/95-01 (Aug 1995), unpublished.
34. The OPAL Collab., G. Alexander et al., Phys. Letters B376 (1996) 232.
35. S. Dutta, S. Ganguli, M. Grünewald, A. Gurtu and C. Paus, *Preliminary Results on electroweak Parameters from L3*, L3 note # 1914 (Feb 1996), unpublished.
36. G. Altarelli, *Precision tests of the electroweak theory at the Z^0.*, in: J. Ellis and R. Peccei (eds.), Physics at LEP, CERN Yellow Report CERN 86–02 (1986), p. 3.

37. A. Denner, G. Weiglein and S. Dittmaier, Phys. Letters B333 (1994) 420;
    A. Denner and S. Dittmaier, Phys. Rev. D54 (1996) 4499.
38. G. Burgers, *The shape and size of the Z resonance*, in: G. Alexander et al.
    (eds.), Polarization at LEP, CERN Yellow Report CERN 88–06, Vol. 1, p. 121.
39. S. Willenbrock and G. Valencia, Phys. Letters B259 (1991) 373.
40. G. López Castro, J.L. Lucio M. and J. Pestieau, Mod. Phys. Letters A6 (1991)
    3679.
41. G.J. Gounaris and J.J. Sakurai, Phys. Rev. Letters 21 (1968) 244.
42. R.G. Stuart, Phys. Rev. D56 (1997) 1515.
43. H. Veltman, Z. Physik C62 (1994) 35.

Chapter IV

Causality, Semigroups and Resonance States

Quantum Theory in the Rigged Hilbert Space – Irreversibility from Causality

A. Bohm and N.L. Harshman

Physics Department, The University of Texas at Austin, Austin, Texas 78712

Summary. After a review of the arrows of time, we describe the possibilities of a time-asymmetry in quantum theory. Whereas Hilbert space quantum mechanics is time-symmetric, the rigged Hilbert space formulation, which arose from Dirac's bra-ket formalism, allows the choice of asymmetric boundary conditions analogous to the retarded solutions of the Maxwell equations for the radiation arrow of time. This led to irreversibility on the microphysical level as exemplified by decaying states or resonances. Resonances are mathematically represented by Gamow kets, functionals over a space of very well-behaved (Hardy class) vectors, which have been chosen by a boundary condition (outgoing for decaying states). Gamow states have all the properties that one heuristically needs for quasistable states. For them a Golden Rule can be derived from the fundamental probabilities $\mathcal{P}(t) = \text{Tr}(\Lambda(t)W^{\text{Gamow}}(t_0))$ that fulfills the time-asymmetry condition $t \geq t_0$ which could not be realized in the Hilbert space.

1. Preface – Time-Asymmetries

This preface was added to the paper in order to explain what we mean by the word irreversible in the title.

The term "irreversible" has two different uses and has been applied to several different phenomena [1, 3]. These different phenomena are also called different "arrows of time". These arrows are not unrelated to each other, but there is no consensus as to their exact relationships. Somehow, all these arrows seem to be connected to the vague, intuitive notion of causality.

Most of the time the word "irreversibility" is used to refer to these arrows of time, i.e. to the directedness of the time evolution of a physical system or of the state of a physical system, classical or quantum. An alternate use of the word "irreversible" is to describe non-invariance (of the observables) with respect to the (antilinear) time reversal transformation A_T. These two notions, though related are not the same and must be distinguished. Whereas the time reversal operator is defined by its relations to the other *observables* that generate the space-time symmetry transformations or by its relations with these transformations (Poincaré group, Galilean group), "irreversibility" above means the impossibility or improbability to create a *state* which evolves backward in time. Irreversible time evolution is not necessarily in contradiction with a T-invariant Hamiltonian if a mathematical theory is used that makes a distinction between states and observables. Time reversal invariance – or its violation – is not the key for understanding irreversibility.

We shall mention A_T transformations only briefly in Sect. 7.1. In this paper and in the title, irreversibility is the time-asymmetry due to a preferred direction of time evolution.

Of the different irreversible phenomena or the different "arrows of time", most prominent and oldest is the thermodynamic arrow of time (TA): the entropy S in an isolated (classical) physical system increases, $\frac{dS}{dt} > 0$, until it reaches equilibrium, $\frac{dS}{dt} = 0$ (2nd law of thermodynamics). Using (for the sake of simplicity) the time-symmetric equation of Newtonian mechanics for the scattering of molecules in a gas, it has been shown [3] that Boltzmann's Stosszahl-Ansatz implies assumptions (boundary conditions) from which the TA follows: "If one assumes that the gas was prepared in some manner in the past..., then it follows that correlations between molecules and scattering centers will arise only from past but not from future collisions," (R. Peierls in [3]). That initial conditions, not final conditions, are specified is part of the intuitive law of causality.

Another arrow of time of classical physics is the radiation arrow (RA). The Maxwell equations, like all local physical laws, are symmetrical in time, yet retarded radiation predominates. The phenomenological law according to which nature favors the retarded potential over the advanced solution of the Maxwell field equations is called the time arrow of radiation. An effective way to describe this time arrow is to formulate an additional axiom to the Maxwell equations – a particular, time-asymmetric, boundary condition. This boundary condition excludes the strictly incoming fields A_{in}^{μ} (Sommerfeld radiation condition). For a system of charged particle, the fields $A^{\mu}(x) = A_{adv}^{\mu} + A_{out}^{\mu} = A_{ret}^{\mu} + A_{in}^{\mu}$ acting on a particle at x, in *every* region of space-time, are only the retarded fields of the other particles in the region, $A_{in}^{\mu} = 0$. In other words, all fields possess advanced sources somewhere in the universe; *Radiation must first be emitted by a source before it can be detected by the receiver*. Other boundary conditions are imaginable, e.g. the time-symmetric boundary conditions for which all fields also possess sinks and will be absorbed somewhere. In the famous Einstein-Ritz arguments [4], Einstein thought that physics could be restricted to the time-symmetric case for which retarded and advanced fields are equivalent. Then the RA is based on probability (i.e., the TA). Ritz considered this restriction of the boundary conditions as not allowed, in which case phenomena demand the choice of the retarded fields as the only possibility and the TA has its origin in the RA.

The classic mathematical attempt to settle this argument and account for the predominance of retarded radiation was given by Wheeler and Feynman[5]. They start with time-symmetric boundary conditions for the field equations (time-symmetric electrodynamics) and introduce a cosmic absorber (of huge amounts of dust matter) which annihilates the advanced fields over the retarded. This gives the impression that " the physics of radiation can be regarded as, at bottom, time-symmetric with only the statistics of large numbers giving the appearance of asymmetry " (J.A. Wheeler in [1]).

However such an absorber has to contain some arrow of time. In order to provide the appropriate thermodynamic conditions for absorption to occur in the far future, a random initial state for the matter has to be postulated, and this initial state depends in turn on the retardation. Therewith one is back to the Einstein-Ritz controversy [4] as to whether irreversibility is exclusively based on probability considerations, TA $\Rightarrow$ RA, (which Einstein believed) or whether an initial condition and thus causality is the basis of irreversibility, in which case RA $\Rightarrow$ TA (Ritz's opinion). Up to today, there seems to be no agreement in the literature as to whether TA $\Rightarrow$ RA or RA $\Rightarrow$ TA, though the former is more prevalent [1]. Causality and probability may just be two aspects of one and the same principle.

For every situation with increasing entropy and retarded potentials one also has a completely time reversed situation with decreasing entropy and advanced potentials. This dichotomy is not only restricted to the electromagnetic field but can also be found in quantum physics. This is not surprising because the quantum theory of radiation cannot be expected to be free of an arrow of time if classical electrodynamics possesses one. However, the mathematical theory of the Hilbert space allows only time-symmetric boundary conditions. To accommodate time-asymmetric boundary conditions one has to extend the mathematical theory beyond the confines of the Hilbert space.

In addition to the two arrows of classical physics of relevance to time scales achievable in the laboratory, a third arrow exists at the cosmic time scale – the cosmological arrow of time (CA). This states that the universe expands (the contracting solutions of the equation of motion are excluded) or at least that we live in the expanding phase fairly close to the initial singularity of the big bang.

Usually the cosmological arrow is believed to be the master arrow from which the others follow [P.C.W. Davis in [1]]. Though there seems to be no consensus whether CA $\Rightarrow$ RA $\Rightarrow$ TA or whether CA $\Rightarrow$ TA $\Rightarrow$ RA, the most attractive scenario is the entropy gap: the expansion of the universe during the first three minutes was much faster than the relaxation time of the nucleosynthesis, leaving the majority of the cosmological material for aeons in a metastable state. The quasistable states of our sun are the metastable stuff that is driving the time-asymmetric processes. On the other hand, local arrows of time need not depend on asymmetric cosmologies. For example, in some simple models [6], the light cones in a closed Friedmann universe tend to imitate the expansion figure of space-time. As a consequence, outgoing electromagnetic waves occur near the big bang and incoming (in our local sense of time) waves occur close to the big crunch.

The CA and the big bang give us a means of defining the cosmic time and its origin $t = t_0 \equiv 0$. In order to define irreversible time evolution (time evolution that does not extend to $-\infty$ *and* to $+\infty$) one has to be able to choose such a reference time $t_0 \neq \pm\infty$ like the creation time of the universe. In the laboratory one fixes this time $t_0 = 0$ for each physical system individually as, for example, the time at minimum entropy or the time at which the radiation is emitted. In quantum physics (the concern of this article) this time t_0 is the time at which the state has been prepared (e.g. a resonance has been "created") and at which one can start the detection of an observable (e.g. of the decay products) in this state.

Before we turn to the quantum mechanical arrow of time we want to mention the psychological arrow. It is the arrow of time by which we remember the past and predict the future. It is physically not well-defined, but usually considered subsumed under the thermodynamic arrow which in turn is maintained by the expanding universe. Thus if in a (re)contracting universe the entropy decreases (which may or may not be the case) then the psychological arrow should also turn around to be again in the direction of entropy increase (Hawking in [1]). That is, in such a contracting universe – or contracting region of the universe – the direction of the psychological arrow should be the opposite of *our* subjective sense of time; initial conditions should turn into final conditions and vice versa. This then would be the realm of states with negative time (with $t = 0$ meaning the creation time of our universe), or the time beyond the "switch-over" point of maximal expansion, if we think of the cosmic time as cyclic. How this reversal of time could take place is hard to comprehend. It does not seem to be of much practical importance; probably it is just a feature of the time-symmetric differential equations, which are incomplete without boundary conditions. For the mathematical boundary condition we have a choice, but for the physical boundary condition in our world we have not. The physics of our universe is time-asymmetric. It is important to have a mathematical theory that is not too restrictive and allows one arrow of time. Then the opposite arrow of time is obtained by time reversal transformation of the boundary condition. Whether for every physical system or process there is also a time reversed one in *our* world, is a different question which is not answered by the T- or CPT-invariance of the Hamiltonian (differential operator). Solutions of time-symmetric dynamical equations with time-asymmetric boundary conditions come in pairs. With the choice of the boundary condition, one of the two time-asymmetric solutions is selected. This applies to the classical equation of general relativity (big bang–big crunch; black hole–white hole) and electromagnetism (retarded–advanced) and must as well apply to the mathematical theory of quantum physics. One should not restrict the mathematics of quantum mechanics to such an extent that time-asymmetric boundary conditions cannot be formulated with mathematical rigor. The Hilbert space theory of quantum mechanics does not allow such a time-asymmetric formulation, while the rigged Hilbert space theory allows time-asymmetry in either of the directions (which are related by T (or CPT) conjugation, cf. Sect. 7.1 and [48]). The physics in our world chooses one of these directions.

In analogy to the radiation arrow of time, the quantum mechanical arrow of time can be formulated without reference to the mathematical theory as:

a state first must be prepared by a preparation apparatus before an observable can be detected in it by the registration apparatus. We will call this the preparation→registration arrow of time [7]. The preparation→registration arrow of time is perhaps the closest we can come to a physical definition of the psychological arrow of time, because it is the experimentalist who first prepares the state and then activates the detector. Irreversibility that we mean in the title is the asymmetry of the time evolution based on this arrow of time. Its most prominent realization in quantum physics is the intrinsic time evolution of resonances and decaying states.

In contrast, conventional irreversibility in quantum physics has been considered to be extrinsic. Two forms of state changes are usually discussed in quantum mechanics:

1. A unitary time evolution generated by the Hamiltonian H of the quantum system S:

(1a) $$\rho \to \rho(t) = e^{-iHt}\rho(0)e^{iHt} = U^{\dagger}(t)\rho U(t), \quad -\infty < t < \infty$$

or

(1b) $$\phi(t) = e^{-iHt}\phi, \quad -\infty < t < \infty$$

where ρ is the statistical or density operator of a mixed state and ϕ the state vector if the state is pure, $\rho = |\phi\rangle\langle\phi|$.

This change of the state is reversible and could be equivalently described by the Schrödinger equation

(1c) $$i\frac{d\phi}{dt} = H\phi$$

with the "Hilbert space" boundary condition $\phi(t) \in \mathcal{H}$.

2. The "reduction of the state vector" on measurement, which in von Neumann's idealization is given by the following additional axiom ("collapse of the wave function" – ideal measurement of 1st kind): If P_{a_i} are the projection operators of the observable measured $A = \sum_i a_i P_{a_i}$, then

(2a) $$\rho \to \rho^{\text{after}}(t_0) = \sum_i P_{a_i}\rho(t_0)P_{a_i}$$

or, if the results of the measurement is a_j,

(2b) $$\rho \to \rho^{\text{after}}(t_0) = P_{a_j}\rho(t_0)P_{a_j}.$$

Here ρ^{after} is the state after the measurement. If one defines the (von Neumann) entropy by

(3) $$\mathcal{S}(\rho) = -\text{Tr}(\rho \ln \rho)$$

then for (2a) – but not for (2b) – one obtains: $\mathcal{S}[\rho^{\text{after}}(t_0)] > \mathcal{S}[\rho^{\text{before}}(t_0)]$, i.e. this entropy increases if the initial entropy is fixed and smaller than the maximal entropy. If the initial state in (2a) was a pure state $\rho = |\phi\rangle\langle\phi|$, then the state after measurement is a mixture:

(4)
$$|\phi\rangle\langle\phi| = \sum_i \sum_j |a_i)(a_i|\phi\rangle\langle\phi|a_j)(a_j| \rightarrow \rho^{\text{after}} = \sum_i |a_i)(a_i|\phi\rangle\langle\phi|a_i)(a_i|.$$

This decoherence (elimination of the interference terms $(a_i|\phi\rangle\langle\phi|a_j)$, $i \neq j$) is considered an irreversible process, though no time has actually elapsed.

The increase in von Neumann entropy is conventionally considered as the arrow of time in quantum mechanics. It is not due to time-asymmetric laws (equations and/or boundary conditions) but appears through the extraneous act of measurement. The intrinsic time evolution of a quantum physical system, i.e. the time evolution of a state of a "closed physical system", isolated from all external influences, is described in the Hilbert space by (1) and is reversible.

The change described by (2) leads to the following problem: if $\rho = \rho(t)$ is time dependent and given by (1a) (Schrödinger picture) then A and P_{a_i} are time independent and ρ^{after} in (2) is at the same time as $\rho(t)$. The collapse (2) is supposed to happen instantly and would not shed any light on the question of a time arrow. However, realistically, every measurement takes time. To avoid this inconsistency one considers the Heisenberg picture [8]: keep ρ time independent $\rho = \rho(t_0 = 0)$ and equal to the "initial" state of (1), and take for the projection operators the time evolved $P_{a_i}(t) = e^{iHt} P_{a_i} e^{-iHt}$. Then in place of (2b) on has

(5) $\rho^{\text{eff}}(t_a) = P_a(t_a)\rho P_a(t_a)$, with the condition that $t_0 < t_a$

(or at least $t_a \geq t_0$, but $t_a = t_0$ would be the unrealistic case of instant measurement). The change (5) combines the two conventional changes of states in quantum mechanics. Its generalization for a sequence of observables, $A(t_a)$, $B(t_b) = \sum_j b_j P_{b_j}(t_b)$, etc. to a "history" [8] $[\mathbf{P}] = (P_a(t_a)P_b(t_b)\cdots P_c(t_c))$ is:

(6a)
$$\rho^{\text{eff}}(t_c) = \sum_{i,j,\ldots,k} P_{c_k}(t_c)\cdots P_{b_j}(t_b)P_{a_i}(t_a)\rho(t_0)P_{a_i}(t_a)P_{b_j}(t_b)\cdots P_{c_k}(t_c).$$

with the condition

(6b) $t_0 < t_a < t_b < \cdots < t_c$

Since the times "after" $t_a, t_b, \ldots, t_c$ cannot precede the "initial" times $t_0 < t_a, t_a < t_b \ldots$ one obtains the time ordering (6b). In contrast to (2b), where the change is to happen instantly, for (5) and (6a) one can talk of a change in time. This time ordering follows from the idea that there is a knowable initial (prepared) state ρ from which the series of probabilities

(7a)
$$\mathcal{P}(a, t_a) = \mathrm{Tr}\,(\mathrm{P_a}(t_a)\rho) = \mathrm{Tr}\,(\mathrm{P_a}(t_a)\rho\mathrm{P_a}(t_a)) = \mathrm{Tr}\,(\rho^{\mathrm{eff}}(t_a))\,, \qquad t_a > t_0$$

(7b) $\mathcal{P}(b, t_b; a, t_a)$
$$= \mathrm{Tr}\,(\mathrm{P_b}(t_b)\mathrm{P_a}(t_a)\rho\mathrm{P_a}(t_a)\mathrm{P_b}(t_b)) = \mathrm{Tr}\,(\mathrm{P_b}(t_b)\rho^{\mathrm{eff}}(t_a))\,, \quad t_b > t_a > t_0,$$

and in general

(7c)
$$\mathcal{P}(c, t_c; \cdots; b, t_b; a, t_a) = \mathrm{Tr}\,(\mathrm{P_c}(t_c) \cdots \mathrm{P_b}(t_b)\mathrm{P_a}(t_a)\rho\mathrm{P_a}(t_a)\mathrm{P_b}(t_b) \cdots \mathrm{P_c}(t_c))$$

can be predicted (rather than retrodicted). Though it may be an oversimplification to think of $\rho^{\mathrm{eff}}(t)$ as the state of the system after the measurement of the observable P_a, the $\mathrm{Tr}(\rho^{\mathrm{eff}}(t_a))$ are meaningful physical quantities, namely the probabilities to observe with an apparatus the value a for the observable A (cf. Sect. 7.4 below).

The time ordering (6b) expresses an arrow of time. If one considers an experiment performed on a quantum system in the laboratory with the state ρ prepared by the preparation apparatus (accelerator) and $P_a(t)$ the observable registered by the registration apparatus (detector), then the condition $t_0 < t_a$ in (7a) expresses the preparation→registration arrow of time that we mention above and discuss in greater detail in Sect. 6. below, cf. eq. (40). This intuitively obvious time ordering and the arrow of time it implies is the basis of the quantum mechanical irreversibility discussed in our article. The initial time $t_0 = 0$ is the time at which the preparation of the state ρ is completed and the registration of the observable $P_a(t)$ can begin. The probability $\mathcal{P}(a, t_a)$ is the probability to register the observable $P_a(t_a)$ at the time t_a in a prepared state ρ. For example, if ρ describes the prepared state of a $K^0 - \bar{K}^0$ system and P_a the projection operator on the $\pi^+\pi^-$ subspace then $\mathcal{P}(a, t_a)$ is the probability for decay of the neutral K meson into $\pi^+\pi^-$, which can be detected only at a time $t_a > t_0$. Since the formation time scale of the neutral K is ten orders of magnitude less than that of the decay, the creation time t_0 is very precisely fixed.

The formula (7a) of the quantum mechanics of measured systems has been generalized in [8] to (7a), (7b),...,(7c) which have been understood to apply to the quantum theory of cosmology. The probabilities are the probabilities of histories $P_a(t_a)$, $P_a(t_a)P_b(t_b)$, $\cdots P_a(t_a)P_b(t_b) \cdots P_c(t_c)$ of such systems as our universe as a whole. The arrow of time expressed by the time ordering in (6a) and (6b) would in this case be the time-asymmetry of the universe, i.e. the cosmological arrow of time.

Nothing has been said about the probabilities (7) for $t_a < t_0$, ..., $t_c < \cdots < t_b < t_a < t_0 = 0$. In cosmology, where t_0 is the time of the big bang, we would not know. In the decay process of a laboratory experiment we could make the following conclusion: the detector does not click if it is switched on before the decaying state has been prepared at $t = 0$. It is therefore reasonable to assume that the probabilities (7) are zero for these negative values of t.

As will be discussed in Sect. 4. below, in the Hilbert space theory (with no other further assumptions than H is self-adjoint and semi-bounded) the probabilities of (7) are identically zero (for all t) if they are zero on a negative time interval (actually on any set of non-zero measure). Therefore one cannot implement the above program as a mathematical theory in the Hilbert space unless the probabilities $\mathcal{P}(a, t_a); \mathcal{P}(b, t_b; a, t_a) \cdots$ are defined to be different from zero for all $t_a < t_0, t_b < t_a$, etc. The Hilbert space does not allow for a theory in which both energy *as well as* time have a lower bound (creation time). If one wants to have a quantum mechanical arrow of time, i.e. a mathematical formulation of the obviously correct preparation→registration arrow, then one has to go beyond the Hilbert space formulation. This is the subject of the present paper indicated by the word rigged Hilbert space (also called Gelfand triplet) in the title.

Irreversibility in conventional quantum theory is always thought of as being due to external influences upon the non-isolated ("open") quantum system. The irreversible time evolution of open quantum systems is described by the master equation [9]

$$(8) \qquad \frac{\partial \rho(t)}{\partial t} = L\rho(t)$$

where $\rho(t)$ represents the state of the open system S, and the Liouville operator L is given by

$$(9) \qquad L\rho(t) = -\frac{i}{\hbar}[H, \rho(t)] + \delta H(\rho)$$

For $\delta H(\rho) = 0$, (8) with (9) is the von Neumann equation whose solution is the reversible time evolution of the isolated quantum system given by (1a). Equation (8) with (9) is the standard equation for extrinsic irreversibility under the influences of an external reservoir R (which could be, for example, a measuring apparatus) upon the system S. The term $\delta H(\rho)$ represents some complicated external effects of the reservoir R upon the quantum system S. Under particular assumptions about the term $\delta H(\rho)$ the irreversible time evolution of S can be shown to be described by a completely positive semigroup[1] generated by a Liouvillian L [10]):

$$(10) \qquad \rho(t) = \Lambda(t)\rho(0), \; \Lambda(t) = e^{Lt}, \text{ for } t \geq 0.$$

---

[1] According to A. Kossokowski (private communication) maps $\rho \to \rho(t) = \Lambda(t)\rho$ from the set density operators into itself could also describe "some kind of intrinsic irreversibility" if $\Lambda(t)$ is not completely positive, but only positive. But so far it is not clear whether these maps have any physical meaning.

This is the conventional semigroup evolution of open quantum systems (Sudarshan-Gorini-Kossakowski-Lindblad semigroup) [11].

This time evolution described by a Liouvillian L, where $L\rho$ is not just the commutator with the Hamiltonian of the system, $\frac{1}{i\hbar}[H,\rho]$, has also been applied to the decaying K^0-system. It can evolve a pure state $|\phi\rangle\langle\phi|$ into a mixture and has been called non-quantum mechanical [12].

The 'Irreversibility' that we mean in the title is not the non-quantum mechanical irreversibility described by (10) with (9) and it is not the increase of the von Neumann entropy (3) due to the collapse axiom (2) and (4)[2]. And again, the 'Semigroup' that we mean is not the semigroup (10) generated by the Liouvillian (9). Our semigroup is the semigroup generated by the Hamiltonian of the isolated quantum system:

$$(11) \qquad\qquad e^{-iH^\times t}, \; t \geq 0 (= t_0),$$

where $H^\times$ is the extension of a self-adjoint (semi-bounded) Hilbert space operator $\bar{H}$. The semigroup (11) is obtained from the same time-symmetric dynamical equations (the Schrödinger equation (1c) or the von Neumann equation) from which one obtained the unitary group in (1a) and (1b) by integration. However, whereas the unitary group evolution, (1a) and (1b), is obtained if one requires that the set of ϕ in (1c) are elements of the Hilbert space $\mathcal{H}$, the semigroup evolution (11) is obtained from (1c) if one requires that the set of allowed solution ϕ in (1c) are elements of a space $\Phi_+^\times$ that extends the Hilbert space $\mathcal{H}$ to a Gelfand triplet $\Phi_+^\times \supset \mathcal{H} \supset \Phi_+$. The arrow of time expressed by the time evolution semigroup (11) thus arises mathematically from time-symmetric dynamical equations solved with time-asymmetric boundary conditions. These time-asymmetric boundary conditions which follow from the preparation→registration arrow of time, cannot be mathematically formulated in the Hilbert space; in $\mathcal{H}$ the equations (1c) always integrate to the one-parameter group (1b).

The semigroup time evolution (11) describes irreversible time evolution on the microphysical level (if one interprets the solutions of the Schrödinger equation in the extension $\Phi^\times$ of $\mathcal{H}$ as microphysical states). Examples of isolated (closed, autonomous) microphysical systems with irreversible time evolution abound in the real world. Reversible time evolution is a feature of only a minority of closed microphysical systems, e.g. the ground states of molecules and nuclei and the few stable elementary particles. There is a much larger number of decaying states and resonances (excited states of molecules and nuclei, weakly decaying elementary particles, hadron resonances) which

---

[2] The measurement process which changes ρ to ρ' is a scattering process of a microsystem on a macrosystem ("measurement scattering" [7]) and does not actually fulfill the idealized measurement axiom (2). Every scattering experiment possesses an arrow of time; preparation must always precede registration. The arrow of time for the change of state due to measurement is thus a consequence of the preparation→registration arrow, G. Ludwig, private communication. "I know of no 'other' time arrow than the preparation→registration arrow."

are not less isolated than ground states; their time evolution is irreversible (which is sometimes mentioned in textbooks [13]).

It has been argued that a decaying state or resonance is something complicated. The difficulty is that in Hilbert space $\mathcal{H}$ one cannot find a simple initial condition for it. However, in its extension $\Phi^\times_+$ there is a simple (pure) initial state $\rho_i = |E_R - i\Gamma/2\rangle\langle E_R - i\Gamma/2|$ at the time t_0, the creation time of the resonance (e.g. the time when the K^0 is leaving the proton target in $\pi^- p \to \Lambda K^0$ of the many experiments on neutral Kaon decay and CP-violation). The decaying state vector $\psi^G = |E_R - i\Gamma/2\rangle \in \Phi^\times_+$ and is a generalized eigenvector of the self-adjoint Hamiltonian H with complex eigenvalues $z_R = E_R - i\Gamma/2$ associated to the second sheet pole of the S-matrix [14]. (A resonance is actually described by a pair of poles $z_R = E_R \mp i\Gamma/2$, a pair of vectors $\psi^G_\mp$ in a pair of spaces $\Phi^\times_\pm$ evolving by a pair of semigroups, cf. Sect. 7.1).

The semigroup time evolution of the Gamow vector, is derived as [14]:

$$(12) \qquad e^{-iH^\times t}\psi^G = e^{-iE_R t}e^{-\frac{\Gamma}{2}t}\psi^G, \text{ for } t \geq 0 \text{ only.}$$

This is a mathematical consequence of the (mathematical formulation of the) preparation→registration arrow of time and vice versa (derived using the Paley-Wiener theorem). The time-asymmetry of resonances expressed by the semigroup (12) is not the irreversibility of quantum mechanical measurements; resonances evolve in time and decay without being "looked at". But it is identical with the arrow of quantum cosmology if (5) and (6a) are applied to the initial state of the universe. The irreversibility that we mean is the asymmetric time evolution on the quantum level whose basis is the causality principle, which in turn can be inferred from a very special state of the universe at the beginning of time.

2. Introduction – Dirac Kets

"Physical causality can be traced directly to the existence of a simple initial condition of the universe. But how does that initial condition enter into the theory?"
—Murray Gell-Mann[3]

A mathematical theory of physics cannot be deduced from experiments alone; it will necessarily have to involve some idealizations. Mathematicians like to work with mathematical structures that are complete – algebraically complete and, in particular, topologically complete. Only with such structures can they prove existence theorems. Experiments can never be complete in this sense; physicists have only a finite amount (albeit very large sometimes) of experimental data with which to work. Mathematicians generalize

---

[3] p. 216 of M.Gell-Mann, The Quark and the Jaguar, (W.H. Freeman, New York, 1994).

to infinity through various means. The topology[4] of a linear space defines the meaning of convergence of infinite sequences and the topological completion of a space is the appendage to the linear space of the limit elements of all infinite convergent (to be precise, Cauchy) sequences. A finite number of experiments, even if the number is arbitrarily large, cannot tell how the convergence of infinite sequences of states should be defined. Therefore the choice of topology cannot be "deduced" from experiments and only the overall success of a mathematical theory can show that one topology is preferable to another.

This paper explores different choices of topological completions for the spaces of states and observables. One is given by von Neumann's Hilbert space completion. The others – actually a multitude of topologies, one for each particular kind of quantum physical system – are conjectured from Dirac's formalism.

Dirac [15] introduced such elements as bras $\langle E|$ and $\langle x|$, kets $|E\rangle$ and $|x\rangle$ and an algebra of observables generated by such fundamental operators as the Hamiltonian H, momentum P and position Q. Of these kets he demanded that they were eigenvectors

$$H|E\rangle = E|E\rangle$$
(13)
$$Q|x\rangle = x|x\rangle$$

and that they formed a complete basis system, i.e. that every vector ϕ could be written as

(14a)
$$\phi = \sum_{n=1}^{\infty} |E_n)(E_n|\phi) + \int dE\, |E\rangle\langle E|\phi\rangle,$$

(14b)
$$\phi = \int dx\, |x\rangle\langle x|\phi\rangle$$

or

(14c)
$$\langle x'|\phi\rangle = \int dx\, \langle x'|x\rangle\langle x|\phi\rangle,$$

where $(E_n|\phi)$, $\langle E|\phi\rangle \equiv \phi(E)$ and $\langle x|\phi\rangle \equiv \phi(x)$ (the wave function) were thought of as scalar products, $\phi(x) = (|x\rangle, \phi)$.

---

[4] Webster's dictionary gives three definitions of the word topology: (1) the study of those properties of geometric forms that remain invariant under certain transformations, as bending, stretching, etc.; (2a) the study of limits in sets considered as collections of points; (2b) a collection of open sets making a given topological space. Physicists usually associate definition 1. with the word topology, we here use only definition (2). Mathematical structures are combinations of three basic kinds of structures: algebraic, topological and ordering [16]. The rigged Hilbert Space is the completion of the same linear (algebraic) space equipped with three different topologies.

These intuitive constructions were the results of Dirac's unconfined vision, but not well-defined mathematical objects. A mathematical structure to envelope all Dirac's calculative tools was not available at that time. The delta function $\langle x'|x \rangle = \delta(x' - x)$ inspired the development of distribution theory by Schwartz [17] and the $|x\rangle$, $|E\rangle$ inspired the creation of the rigged Hilbert space by Gelfand and his school [18][18a].

In the following discussion, we leave aside the proper eigenvectors $H|E_n) = E_n|E_n)$ with discrete eigenvalues E_n, which are conventionally negative $E_n = -|E_n|$, and the energy continuum starts at zero so that the integral in (14a) extends from 0 to ∞ (there can also be discrete eigenvalues in the continuous spectrum and then these $|E_n)$ with $E_n > 0$ would also be included in the discrete sum of (14a)). In the S-matrix they correspond to poles on the negative real axis of the first sheet (bound state poles). Both the discrete and continuous values for the energy must be bounded from below (stability of matter) and by convention the lower bound for the energy continuum is chosen to be zero.

The discrete basis vectors describe stationary states: $W = \sum_n w_n|E_n)(E_n|$. The linear space Ψ^{disc} spanned by these eigenvectors of the discrete spectrum could also be completed into a Hilbert space $\mathcal{H}^{\mathrm{disc}}$ (or a rigged Hilbert space) which is orthogonal to the Hilbert space $\mathcal{H}$ of the continuous spectrum. In the following, we will only consider the linear space and their completions Ψ, Φ, $\mathcal{H}$ which belong to the absolutely continuous spectrum. The quantum mechanics of stationary states is not affected by our considerations here. Thus $\mathcal{H}$ in this paper denotes the Hilbert space of the absolutely continuous spectrum, often denoted as $\mathcal{H}^{\mathrm{a.c.}}$ in the literature.

3. Hilbert Space (HS) and Rigged Hilbert Space (RHS) Formulation of Quantum Mechanics

"I would like to make a confession which may seem immoral: I do not believe in Hilbert space any more."
—John von Neumann[5]

The first attempt at a rigorous mathematical theory for quantum physics was provided by Weyl [19] and von Neumann [20] using the mathematics that was available at that time, the Hilbert space (HS). The HS is the completion of a linear scalar product space (also called pre-Hilbert space) with respect to the topology[6] given by the norm $||\phi|| = \sqrt{(\phi, \phi)}$.

---

[5] In a letter to G. Birkhoff, quoted in G. Birkhoff, Proceedings of Symposia in Pure Mathematics, Vol. 2, ed. R.P. Dilworth, (American Mathematical Society, Providence, Rhode Island, 1961), p. 158. The letter is date Nov. 13, and Birkhoff believes the year was 1935.

[6] cf. footnote (4).

This norm topology is one of many possible choices and cannot be deduced from physical observations of quantum mechanical states and the observables represented by the operators. The HS properties are:

1. The wave functions $\psi(x) \equiv \langle x|\psi\rangle$ representing the probability $|\langle x|\psi\rangle|^2 \Delta x$ to detect the particle state ψ within the position interval Δx or the energy wave function $\phi(E) \equiv \langle E|\phi\rangle$ representing the the energy distribution in a particle beam are given in the HS by a *class* of Lebesgue square integrable functions $\{\psi(x)\}$ or $\{\phi(E)\}$ that differ on a set of Lebesgue measure zero.

$$\psi \in \mathcal{H} \quad \Leftrightarrow \quad \{\psi(x)\} \in \mathcal{L}^2(\mathbb{R})$$
(15)
$$\phi \in \mathcal{H} \quad \Leftrightarrow \quad \{\phi(E)\} \in \mathcal{L}^2(\mathbb{R})$$

In a scattering experiment, $|\psi(x)|^2$ represents the detector size, location and efficiency and the $|\phi(E)|^2$ represents the energy resolution of the accelerator. Unlike the classes of (15), the resolution of an experimental apparatus is always given by a single *smooth* function, e.g. $\psi(x) \in \mathcal{S}$ (Schwartz space), and not by a class $\{\psi_1(x), \psi_2(x), \psi_3(x), \dots\}$ of functions which can vary wildly on any set of Lebesgue measure zero, e.g. on the set of all rational numbers.

One can always choose the one smooth function $\psi(x) \in \mathcal{S}$ in this class $\{\psi(x)\} \in \mathcal{L}^2(\mathbb{R})$, but the set of smooth functions (or more precisely the set of classes of Lebesgue square integrable functions containing a smooth function $\psi(x) \in \mathcal{S}$) is not HS-complete. For the HS-completeness, one needs the Lebesgue integral [21].

The notion of Lebesgue integrability is physically counterintuitive. Physicists make a finite number of measurements and interpolate smoothly between the experimental data. They never compute Lebesgue integrals, but calculate Riemann integrals using the smooth functions associated to their experimental apparatus.

2. The most fundamental commutation relations, such as $[P, Q] = -i\mathbb{1}$ of momentum P and position Q, imply that these observables cannot be represented by continuous (and hence bounded) operators in HS. Therefore P and Q have only a limited domain of definition in $\mathcal{H}$. Physicists work with operators that one can add and multiply (Dirac's algebra of observables) and not with operators that cannot be defined on the whole space. Often such an algebra is the enveloping algebra of a symmetry group which represents physical transformations, for example, of the registration apparatus (detector) relative to the preparation apparatus (accelerator). One can always find a subspace $\Phi \subset \mathcal{H}$ on which these operators are defined (e.g. the space of differentiable vectors or the space of analytic vectors). Then one can define a topology[7] with respect to which these operators are continuous. The space Φ cannot be a Hilbert space unless it is a finite-dimensional representation space of a compact group.

---

[7] cf. footnote (4).

However, such a subspace Φ can be represented a space of the smooth functions, in many cases (e.g. for Q or P) by the space $\mathcal{S}$:

(16)
$$\phi \in \Phi \Leftrightarrow \phi(x) \in \mathcal{S}.$$

3. The HS does not contain eigenkets like $|E\rangle$ and $|x\rangle$ or the bras $\langle E|$ and $\langle x|$, with the properties (13) and (14). Physicists use them however, e.g. as scattering states, and (14) is the fundamental relation for computations in quantum theory. The kets provide an opportunity to describe states of single microphysical systems.

In von Neumann's formulation with $\phi, \psi \in \mathcal{H}$, ϕ represents the state of an ensemble and $|\psi\rangle\langle\psi|$ represents the observable on an ensemble. The probability to measure the observable $|\psi\rangle\langle\psi|$ in the ensemble ϕ is $|(\psi, \phi)|^2$. The vectors ψ and ϕ are defined by the experimental apparatuses. For example, if ϕ is the idealized representation of a beam of microphysical particles prepared by an accelerator (idealized, because real accelerators prepare mixtures $\sum |\phi_i\rangle\langle\phi_i| = W$), then

(17)
$$\frac{\int_{\Delta E} dE |\phi(E)|^2}{\int_{\text{all E}} dE |\phi(E)|^2}$$

is the fraction of this large number of particles that have energy in the interval ΔE. One never talks of a single microphysical particle but always of a large number, an ensemble. (Equivalently one can talk of an ensemble of experiments.) The notion of a single microsystem or a state of a single microsystem does not exist in the standard HS formulation; ϕ describes an ensemble and $\phi(E) \in \{\phi(E)\} \in \mathcal{L}^2(\mathbb{R}_+)$ represents the energy distribution of that ensemble for which the physicists always choose a smooth function $\phi^{smooth}(E)$. Physicists do not work with the *class* of Lebesgue square integrable functions $\{\phi_1(E), \phi_2(E), \ldots, \phi^{smooth}(E)\}$ which contains the smooth function. They cannot even isolate experimentally those classes of functions which do not contain a smooth function $\phi(E) = \phi^{smooth}(E) \in \mathcal{S}$, i.e. the elements of $\mathcal{L}^2$ which are not elements of $\mathcal{S}$. That means the class $\{\phi(E)\} \in \mathcal{L}^2(\mathbb{R}_+)$ which does not contain a $\phi^{smooth}(E)$ has no physical meaning.

The probability amplitude $\langle\psi|\phi\rangle$ of the physicist is therefore given by

(18a)
$$\langle\psi|\phi\rangle = \sum_{i=1}^{\infty} \langle\psi|i\rangle\langle i|\phi\rangle = \int_{\text{Riemann}} dE \langle\psi|E\rangle\langle E|\phi\rangle$$

$$= \int_{\text{Riemann}} dx \langle\psi|x\rangle\langle x|\phi\rangle = \int \int_{\text{Riemann}} dx \, dE \langle\psi|x\rangle\langle x|E\rangle\langle E|\phi\rangle$$

where $|i\rangle$, $i = 1, 2, \ldots, n, \ldots$, is some discrete basis system, where the wave functions $\psi(x) \equiv \langle x|\psi\rangle$ and $\phi(E) \equiv \langle E|\phi\rangle$ are smooth functions and where the integrals are Riemann. The probability amplitude in the HS formulation is calculated as:

$$(18b) \quad \langle \psi | \phi \rangle = \int_{\text{Lebesgue}} dx\, \psi^*(x) \phi(x) = \int_{\text{Lebesgue}} dE\, \psi^*(E) \phi(E).$$

The integrals are Lebesgue and the set of mathematical wave functions, i.e. the classes $\{\psi_1(x), \psi_2(x), \ldots\} \in \mathcal{L}^2$ in (18b), also includes elements that contain no $\psi^{smooth}(x) \in \mathcal{S}$. Since the physicist only deal with smooth energy distributions $|\phi(E)|^2$ or smooth position distributions $|\psi(x)|^2$ due to the capabilities attributed to their apparatuses, it is of no advantage to choose some arbitrary Lebesgue integrable $\phi(E)$ or $\psi(x)$ as the tool for the calculation of (18b). A much "more practical method of computation" and the method that a physicist always uses is to choose for the $\phi(E) \in \{\phi(E)\}$ the $\phi(E) = \phi^{\text{smooth}}(E) \in \mathcal{S}$. Then the Lebesgue integrals in (18b) become Riemann integrals and are identical with (18a)[8]. The other integrals in (18b), for which $\{\psi_1(x), \psi_2(x), \ldots\}$ does not contain a $\psi^{smooth}(x)$, have no observational counterpart anyway and are therefore physically useless. They have to be included in (18b) because only with them is the HS a topologically complete space.

Only for $\langle E | \phi \rangle = \phi^{smooth}(E) \in \mathcal{S}$, and only if one restricts oneself to the integrals of (18a), can one give to the symbol $\langle E|$ in $\langle E|\phi \rangle$ a separate mathematical meaning by interpreting $\langle E|\phi \rangle$ as the value of an anti-linear functional $\langle E|$ at the element $\phi \in \Phi$. This means one can define $\langle E|$ as an element of $\Phi^\times$, the space of continuous anti-linear functionals over Φ. Since this is only possible for the smooth $\phi(E)$, i.e. for the $\phi \in \Phi$, not for all elements of the HS, we obtain for the set $\Phi^\times$ a larger set than the set of anti-linear continuous functionals over $\mathcal{H}$, since from $\Phi \subset \mathcal{H}$ it follows $\mathcal{H}^\times \subset \Phi^\times$. Using the Frechet-Riesz theorem [21] we then identify $\mathcal{H} = \mathcal{H}^\times$ and obtain a triplet of spaces, the Gelfand triplet or rigged Hilbert space (RHS):

$$(19) \qquad\qquad \Phi \subset \mathcal{H} = \mathcal{H}^\times \subset \Phi^\times,$$

where $|E\rangle \in \Phi^\times$.

The desire for a "more practical method of computation" using Riemann integrals and observable quantities only has thus led to a choice of $\langle E|\phi \rangle$ which allowed us to give a mathematical meaning to the kets $|E\rangle$ and $|x\rangle$. These kets lie outside the HS, whose elements (and therewith also the $\phi \in \Phi \subset \mathcal{H}$) describe ensembles of microphysical systems. These new vectors, the kets $|E\rangle$, are available for a physical interpretation that goes beyond the ensemble interpretation.

Though one can only observe the probabilities (18a) measured with macroscopic apparatuses, it is intuitively attractive to imagine that the

---

[8] cf. footnote (26) of Sect. (7.) for the prerequisites on the operators H, Q to make this possible.

effect which the preparation apparatus causes on the registration apparatus is carried from one to the other by some physical entities. These objects, by which the preparation apparatus acts on the registration apparatus (i.e. the carriers of the action), are imagined to be the microphysical systems. Though one cannot see them directly, every physicist believes in them, e.g. believes that the track in a cloud chamber is caused by a single particle. In HS there is nothing that can describe them, but in RHS the kets of $\Phi^\times$ may.

Whereas the physical entities connected with an experimental apparatus, like the states ϕ defined by the preparation apparatus or the property ψ defined by the registration apparatus, are assumed to be elements of $\mathcal{H}$, or as described above even of Φ, the imagined entities connected with microphysical systems do not have to be in $\mathcal{H}$ or Φ because the energy distribution for a microphysical system does *not* have to be a measurable or even a continuous, infinitely differentiable, rapidly decreasing function of the physical values of E, like the $\langle E|\psi\rangle$ describing the energy resolution of the detector, or the $\langle E|\phi\rangle$ describing the energy distribution of the beam. For the hypothetical entities connected with single microphysical systems one can use Dirac's kets $|E_0\rangle$ with the energy distribution $\langle E|E_0\rangle \propto \delta(E - E_0)$. The energy eigenkets $|E, \Omega\rangle$ (or momentum kets $|\mathbf{p}\rangle$, $E = \mathbf{p}^2/2\mathbf{m}$, $\Omega = \mathbf{p}/|\mathbf{p}|$) represent then the microphysical states of momentum $\mathbf{p}$. These are not states one can prepare with a macroscopic apparatus, but something that the physicist imagines as single microsystems, something that can be associated with a cloud (or bubble) chamber track.

Summarizing, the HS is too big if one only admits quantities associated with macroscopic apparatuses because apparatuses have smooth energy distributions, $\phi(E) \in \mathcal{S}$ and not every element of $L^2(\mathbb{R})$ describes an energy distribution of an apparatus, i.e., a physically preparable state $\phi \in \Phi$. However, if one also wants to describe single microphysical systems, then the HS is too small, because microphysical states like Dirac's scattering states $|\mathbf{p}\rangle$ cannot be represented in $\mathcal{H}$. Neither can Gamow's decaying states $|E - i\Gamma/2\rangle$ representing microphysical systems with well defined resonance energy E and lifetime $\hbar/\Gamma$ be represented in $\mathcal{H}$. The Gamow kets are the principal objects of this paper.

Usually von Neumann's formulation of quantum mechanics entails some further idealizations in addition to the choice of the HS which one may or may not want to make, like:

a The one-to-one correspondence between the set of states (equivalence classes of preparation apparatuses) and the set of statistical operators W in $\mathcal{H}$, or the one-to-one correspondence between pure states and elements of $\mathcal{H}$. This is already an exaggeration on the pre-Hilbert space level, since not every finite superposition of physical state vectors which will represent a physical state as has recently been emphasized in the discussion of decoherence [22].

b The one-to-one correspondence between the original set of observables and the set of self-adjoint operators A in $\mathcal{H}$.

c The axioms of the idealized measurements (collapse of the wave function). Realistic experiments usually do not even attempt to fulfill the condition of an ideal measurement (of first or second kind) and according to present views this postulate is not needed, since the Schrödinger equation and a measurement scattering process fully describe the measuring act [7, 23].

Since these additional idealizations are not of direct relevance for our discussion of quantum mechanical irreversibility and have no bearing on the choice of the most suitable topology for the scalar product space, we shall not discuss them any further here.

After having discussed the basic features of the HS formulation we shall now discuss the RHS formulation of quantum mechanics [7, 25, 26], which has only become possible after the new mathematics of distributions and linear topological spaces had been introduced 30 years after von Neumann's HS.

The RHS formulation is also a mathematical idealization of the structure that one can deduce from physical observations. This idealization provides the fundamental properties (13) and (14) required by Dirac's formalism. It restricts the allowable vectors for the description of entities defined by the experimental apparatus to the subspace $\Phi \subset \mathcal{H}$; Φ is complete with respect to a different topology (meaning of convergence) than $\mathcal{H}$. But this is of no importance to physics because physics will not use all elements of Φ; important is that this topology is such that (13) can be mathematically defined (see Sect. 5. below) and (14) can be proved.

The space Φ is specific to the particular quantum physical system considered, and the topology of the space Φ is defined such that:

1. The algebra of observables of the quantum physical system is an algebra of continuous operators in Φ.
2. The Dirac basis vector expansion (14) holds as a theorem, the Nuclear Spectral Theorem.

The distinction between the RHS and the HS formulations of quantum mechanics is summarized in Table I.

For the quantum mechanics of stationary states and reversible processes (using no more than Dirac kets), the HS formulation and the RHS formulation lead to experimental predictions which are only imperceptibly different. The calculations can be written in terms of elements which can be defined with either the HS or the RHS. The standard HS quantum mechanics is just an approximate sub-theory of RHS quantum mechanics. This similarity reflects the similarity of von Neumann's HS formulation and Dirac's incomplete bra and ket formalism. The RHS formulation of quantum mechanics makes the Dirac formalism rigorous and provides a more "practical method

HS formulation of theo-retical quantities	Experimental quantities	RHS formulation of the-oretical quantities
Preparation Apparatus defines the		
Density operator W in $\mathcal{H}$	Physical states (mixtures)	Density operator W in Φ_-
ϕ element of $\mathcal{H}$ or pro-jection operator $\lvert\phi\rangle\langle\phi\rvert$	pure states	ϕ element of $\Phi_- \subset \mathcal{H}$
Registration Apparatus defines the		
Unbounded linear oper-ator A in $\mathcal{H}$	Physical observables	Element of an algebra of τ_Φ-continuous operators
Projection operator Λ or self-adjoint operators F with $0 \leq F \leq 1$ in $\mathcal{H}$		Projection operators or positive operators in Φ_+
$\lvert\psi\rangle\langle\psi\rvert$ one dim. projec-tor in $\mathcal{H}$	Yes-no observables (property or proposition)	$\lvert\psi\rangle\langle\psi\rvert$ with $\psi \in \Phi_+ \subset \Phi$

The spaces Φ_+ and Φ_-, with $\Phi = \Phi_- + \Phi_+$; $\Phi_- \cap \Phi_+ \neq 0$, will be defined in Sect. 5..

Table 1. Comparison of HS and RHS description of physical quantities.

of computation" in the domain where both theories overlap, e.g. by allowing the use of Riemann integrals in computations like (18) rather than Lebesgue integrals.

But with the new mathematical language that the RHS provides one can speak and think new physics. The microphysical scattering states (hyperbolic orbits of the corresponding classical system) described by the Dirac kets $\lvert E\rangle$ gave already an indication of this new domain. The really new physics of the RHS formulation is the microphysical irreversibility which is exemplified by the semigroup time evolution of the Gamow kets which represent decaying states or resonance states. This irreversible semigroup time evolution was un-thinkable in the old HS formulation, where decaying states had complicated, problematic features, as we shall discuss next.

4. Consequences of the HS Formulation and Some Views on Time-Asymmetry in Quantum Mechanics

Most computations in quantum theory do not use the completeness property of the HS; they work only with properties of a pre-Hilbert space. But there are some general results that one obtains from HS-completeness which unveil the problems of the HS formulation of quantum decay:

1. There is no vector that obeys the exact exponential decay law [27]. Math-ematically stated, there is no $\phi \in \mathcal{H}$ whose survival probability

$$\mathcal{P}_s \equiv \left|(\phi, e^{-iHt}\phi)\right|^2$$

has the property

(20) $$\mathcal{P}_s = e^{-\Gamma t}.$$

Although this "deviation from the exponential law" is unobserved, the theoretical prediction of it has led some to infer that exact exponential decay does not exist in nature, instead of suspecting a flaw in the mathematical idealization of the HS. Since magnitudes cannot be predicted from only mathematical properties, one can argue that this deviation from the exponential law is smaller than any experimental error. Thus these deviations could always be smaller than what can be experimentally ruled out and therefore cannot be tested[9]. In spite of the untestibility of these mathematical deviations, alternate explanations for the observed exponentiality have been proposed. For example, the observed exponential behavior has been ascribed to the influence of the environment, to the measurement process or to both [28].

Physicists usually demand even more of a decaying state than (20). As envisioned by Gamow [29], physicists would like to describe a decaying state by an eigenvector $\psi^G \equiv |E_R - i\Gamma/2\rangle$ of the self-adjoint Hamiltonian H with complex eigenvalue $(E_R - i\Gamma/2)$, i.e.

(21) $$H|E_R - i\Gamma/2\rangle = (E_R - i\Gamma/2)|E_R - i\Gamma/2\rangle,$$

and the exponential time evolution

(22) $$\psi^G(t) = e^{-iHt}\psi^G = e^{-iE_R t}e^{-\frac{\Gamma}{2}t}\psi^G$$

but such vectors do not exist in the HS.

Empirically, stable particles are not considered qualitatively different from quasi-stable particles, but are only quantitatively different by a zero or very small value of Γ. A particle decays if it can decay and it is stable if selection rules for some quantum numbers prevent it from decaying. Both stable and decaying states have been described in elementary textbooks [30], in successful phenomenological or effective theories [31] and in tables of experimental data [32], as autonomous entities characterized by E_R and Γ, where Γ can sometimes be equal to zero. A vector like ψ^G fulfilling (21) and (22) would have the suitable properties.

Though there are no ψ^G in the HS, the RHS contains these Gamow vectors ψ^G. They are in a conjugate space $\Phi_+^\times$ of a RHS, $\Phi_+ \subset \mathcal{H} \subset \Phi_+^\times$.

---

[9] Probabilities like $\mathcal{P}_s$ are always observed as ratios of (preferably large) integers $N(t)/N(0)$ and not as real numbers like $e^{-\Gamma t}$, therefore any discussion of deviations from the exponential law is futile as long as the deviations are not predicted with a magnitude that can be compared with $1/\Gamma$.

2. Decay probabilities in the HS theory are identically zero. The probability for the transition from a state $\psi(t) = U^\dagger(t)\psi = \exp(-iHt)\psi$ into the decay products described by the subspace $\Lambda\mathcal{H} \subset \mathcal{H}$, where Λ is the projection operator on the subspace of decay products (or Λ is a positive operator), is in quantum theory given by

(23) $$\mathcal{P}(t) = \mathrm{Tr}(\Lambda|\psi(t)\rangle\langle\psi(t)|) = \langle\psi|e^{iHt}\Lambda e^{-iHt}|\psi\rangle.$$

This is the probability to detect the observable Λ in the state $|\psi(t)\rangle\langle\psi(t)|$. The Hamiltonian H is always assumed to be self-adjoint and semibounded $H \geq 0$, (the condition for the stability of matter). The decay of a prepared quasi-stationary state is assumed to start at finite time, $t > t_2 > -\infty$, which is mathematically formulated as:

(24) $$\int_{t_1}^{t_2} \langle\psi(t)|\Lambda|\psi(t)\rangle dt = 0$$

for some t_1 such that $-\infty \leq t_1 < t_2$ (t_2 is usually chosen to be $t_2 = 0$). With these assumptions one can show [33]

(25) $\mathcal{P}(t) \equiv 0$ for all t (precisely, almost all t) and for *every* $\psi \in \mathcal{H}$.

This means that in HS the transition probability $\mathcal{P}(t) = \mathrm{Tr}(\Lambda(t)\rho)$ of any state $\rho = \sum_i w_i|\psi_i\rangle\langle\psi_i|$ is identically zero if the transition starts at a finite time. In particular there are no decaying states in $\mathcal{H}$. This mathematical result is not usually interpreted as a deficiency of the HS idealization but as being due to some problems with causality [34].

The way out of this dilemma is shown in Sect. 7.4. One replaces ψ of the HS in (23) by the Gamow vector ψ^G of the RHS, which is not in the HS. Then the transition probability $\mathcal{P}(t)$ can be shown to be nonzero and exponentially approaching unity, i.e., $\mathcal{P}(t) \propto [1 - \exp(-\Gamma t)]$, for $t > 0$.

3. In the HS formulation of quantum mechanics the symmetry transformations (e.g. Galilean transformations, Poincaré transformations) are described by a unitary, projective group representation in $\mathcal{H}$. This implies that the time evolution is unitary and reversible and given by $U^\dagger(t) = \exp(-iHt)$, $-\infty < t < \infty$.

Instead of recognizing that this may be a property of the mathematical idealization imposed by the HS, the widespread conclusion was that quantum mechanical irreversibility of isolated microphysical systems is impossible.

Lately, there have been several exceptions to the widespread conclusion that the time evolution described by the Hamiltonian must be time-reversible. Different people mentioned different reasons why a microphysical "arrow of time" should exist.

i Peierls [35] and his school emphasized the importance of the initial and boundary conditions. They chose purely outgoing boundary conditions for the solutions of the Schrödinger equation. According to Peierls' ideas, irreversibility is connected with the choice of the boundary or initial conditions (for the "wave functions").

ii T. D. Lee [36] explained that the time reverse of a decay process is impossible or highly improbable to attain due to the phase of the state vector of the quantum system. These arguments immediately extend to scattering experiments.

It is easy to prepare two uncorrelated incoming beams that scatter into strongly correlated outgoing spherical waves as done in a typical scattering experiment. It is experimentally hopeless to prepare a state consisting of two strongly correlated spherical waves (with fixed relative phase) in such a way that after the scattering of two uncorrelated plane waves emerge. The latter would be the time reverse of the setup for a typical scattering experiment.

iii Ludwig [7] had also noticed that in an experiment with quantum systems one could not time translate the trigger of a detector that registers an observable to a time before the preparation apparatus (e.g. accelerator) has been turned on. This preparation⟶registration arrow of time for the apparatuses ought to be transcribed into a semigroup evolution of the state (defined by the preparation apparatus) relative to the observables (defined by the detector)[10]. However, knowing that in HS the time evolution is represented by the unitary group, he extrapolated this semigroup to all times $-\infty < t < \infty$.

iv The semigroup generated by the Hamiltonian emerged unexpectedly in the mathematical derivation of the time evolution for the Gamow vectors (22) and therewith introduced microphysical irreversibility into the RHS formulation of quantum mechanics [37].

v Prigogine and his school had emphasized for a long time [38] that irreversibility is fundamental. It therefore should be intrinsic to the dynamics of the microsystems rather than being caused by external effects of a quantum reservoir or the environment (measurement apparatus). Irreversibility therefore should be connected with the Hamiltonian (or Liouvillian) and described by a non-unitary transformation.

vi I. Antoniou [39] showed that the formalism of subdynamics leads to the RHS and suggested for this non-unitary transformation the semigroup of the Gamow vectors generated by the extended Hamiltonian in the RHS.

vii Coming from an entirely different problem, Gell-Mann and Hartle [8] introduced an arrow of time in the quantum mechanics of cosmology. In order to avoid inconsistancies for the probabilities of histories, they required a time ordering for the projection operators in a history and the

---

[10] cf. Sect. 6., footnote (14).

initial states (cf. (5) and (6) of Preface). This arrow of time is identical to the preparation—→registration arrow of time if one applies the same formula (viz. (5) Preface) to the probability to register the observable (projection operator) in the prepared state ρ of a laboratory experiment.

All the examples above are different manifestations of a fundamental quantum mechanical distinction between the past and the future – different expression of causality in quantum physics. The paradigm of this microphysical irreversibility is the time evolution of resonances. Since resonances abound, this time-asymmetry is prevalent. HS quantum theory cannot describe it, however the same mathematical theory of the rigged Hilbert space that was introduced to justify the Dirac formalism also describes the irreversible decay of microsystems and allows for the mathematical transcription of the quantum mechanical arrow of time. It describes irreversibility on the microphysical level.

5. From a Pre-Hilbert Space to the Rigged Hilbert Space

"A role of rigorous mathematics in physical science is to make sense of heuristic ideas (i.e. find the 'correct setting') – not to assert they are nonsense."
—M. Fisher[11]

A pre-Hilbert space is a linear space Ψ with a scalar product. This scalar product is denoted by

$$(26) \qquad\qquad (\psi, F) \text{ or by } \langle \psi | F \rangle.$$

The pre-Hilbert space is without any topological structure[12]. Such mathematical concepts as neighborhoods, convergence of infinite sequences, topological completeness, continuous operators, continuous functionals, dense subspaces, etc. are not defined. This pre-Hilbert space Ψ is what one usually means in physics when one speaks of the Hilbert space, and this is what one mostly uses, together with the representations of ψ by square integrable functions and the calculation of scalar products as Riemann integrals. The full mathematical structure of the Hilbert space $\mathcal{H}$ is much more complicated, and the proof of statements like the deviation from the exponential decay law in Sect. 4.1 and the vanishing of the transition probability in Sect. 4.2 require the precise mathematical definition of $\mathcal{H}$.

The linear scalar product space Ψ can be endowed with various topologies, which means that various definitions for the convergence of infinite sequences

---

[11] From talk "What's Mathematical Physics to Physicists? Some examples from Past, Present and Future" by M. Fisher, VIIth International Congress on Mathematical Physics, Boulder, Colorado, August 1983.

[12] cf. footnote (4).

can be given. We denote these topologies by $\tau_\mathcal{H}$ for the HS, for example, and label the topological notions with it, such as $\tau_\mathcal{H}$-convergence. A space is completed by adjoining to it the limit elements of all Cauchy sequences not already contained in the space.

The Hilbert space $\mathcal{H}$ is the completion of the pre-Hilbert space Ψ with respect to the convergence defined by the norm $||\phi|| = \sqrt{(\phi, \phi)}$. If one represents the space Ψ by a space of functions and the scalar product by the usual integral, then one cannot complete the space Ψ into a Hilbert space $\mathcal{H}$ unless the integral is a Lebesgue integral (as opposed to the more frequently used Riemann integral). In this case, each element $\phi \in \mathcal{H}$ is represented by a *class* of Lebesgue square-integrable functions which differ on a set of Lebesgue measure zero (e.g. all rational numbers), not by one wave function. This is the "realization" of $\mathcal{H}$ by $L^2(\mathbb{R})$.

The RHS is the same linear space Ψ completed with respect to three different topologies: one stronger, one the same, and one weaker than the topology defined by the Hilbert space norm. The stronger topology τ_Φ, e.g. a topology defined by a countable number of norms, leads by completion to the space Φ, which, as a consequence of its stronger topology, has the property $\Phi \subset \mathcal{H}$. The topological dual to Φ, i.e. the space of *continuous* anti-linear functionals on Φ, is denoted by $\Phi^\times$ and since its topology is weaker than the Hilbert space topology we have $\mathcal{H} \subset \Phi^\times$. The antilinear τ_Φ-continuous functionals or functions $F(\phi)$ over the space of $\phi \in \Phi$ are denoted by $F(\phi) = \langle \phi | F \rangle$. This bra-ket is an extension of the scalar product (ϕ, f), $f \in \mathcal{H}$ to those elements $F \in \Phi^\times$ which are not elements of $\mathcal{H}$. By completing Ψ with respect to these three topologies, one obtains a triplet of spaces. This is the Gelfand triplet or rigged Hilbert space [18].

$$(27) \qquad \Phi \subset \mathcal{H} = \mathcal{H}^\times \subset \Phi^\times,$$

$$\text{with elements "bra" and "ket"} \quad \langle \phi | \in \Phi \quad |F \rangle \in \Phi^\times$$
$$(28) \qquad \text{or "ket" and "bra"} \quad |\phi \rangle \in \Phi \quad \langle F | \in \Phi^\times$$

The vectors $\phi \in \Phi$, either as a ket $|\phi\rangle$ or bra $\langle\phi|$, represent physical quantities connected with the experimental apparatuses. For example, in a scattering experiment the preparation apparatus (accelerator) defines the state ϕ and the observable $|\psi\rangle\langle\psi|$ is defined by the registration apparatus (detector), and both ϕ and ψ are elements of Φ. The vectors $|F\rangle$ or $\langle F| \in \Phi^\times$ represent quantities connected with the microphysical system, e.g. scattering states $|E\rangle$ or decaying states $|E_R - i\Gamma/2\rangle$.

The general observable is now represented by a continuous (also bounded) operator A in Φ (but in general by an unbounded operator $\overline{A}$ or $A^\dagger$ in $\mathcal{H}$) and one now has a triplet of operators corresponding the the triplet of spaces (27)

$$(29) \qquad A^\dagger|_\Phi \subset A^\dagger \subset A^\times.$$

The operator $A^\dagger$ is the Hilbert space adjoint of A (if A is essentially self-adjoint, then $A^\dagger = \overline{A}$, where $\overline{A}$ denotes the closure of the operator A in $\mathcal{H}$). The operator $A^\dagger|_\Phi$ is its restriction to the space Φ, and the operator $A^\times$ in $\Phi^\times$ is the conjugate operator of A defined by

$$(30) \qquad \langle A\phi|F\rangle = \langle\phi|A^\times F\rangle \text{ for all } \phi \in \Phi \text{ and all } |F\rangle \in \Phi^\times.$$

By this definition, $A^\times$ is the extension of the operator $A^\dagger$ to the space $\Phi^\times$ and not the extension of the operator A, which is more often used in mathematics. The operator $A^\times$ is *only defined* for an operator A which is *continuous* (and bounded) in Φ, so $A^\times$ is always a continuous operator in $\Phi^\times$. In quantum mechanics it is empirically impossible to restrict oneself to continuous (and therefore bounded) operators $\overline{A}$ in $\mathcal{H}$ (e.g. already the observables in $[P,Q] = -i1$ cannot be represented by $\tau_\mathcal{H}$-continuous operators in $\mathcal{H}$, but they can be represented by τ_Φ-continuous operators in a space Φ with stronger topology). But one can restrict oneself to algebras of observables $\{A, B, \ldots\}$ described by continuous operators in Φ if the topology of Φ is suitably chosen. Then $\{A^\times, B^\times, \ldots\}$ are defined and continuous in $\Phi^\times$. If A in (30) is not self-adjoint then $A^\dagger|_\Phi$ need not be a continuous operator in Φ even if A is, but one can still define the conjugate $A^\times$ using (30), which is a continuous operator in $\Phi^\times$.

A generalized eigenvector of an operator A is defined to be that $F \in \Phi^\times$ which fulfills

$$(31) \qquad \langle A\phi|F\rangle = \langle\phi|A^\times F\rangle = \omega\langle\phi|F\rangle \text{ for all } \phi \in \Phi,$$

where the complex number ω is called the generalized eigenvalue. This is also written as

$$(32) \qquad A^\times|F\rangle = \omega|F\rangle.$$

For an essentially self-adjoint operator $(A^\dagger = \overline{A} = \text{closure of } A)$ this is often also written as it is in Dirac's formalism, $A|F\rangle = \omega|F\rangle$), especially if one avoids mentioning the topological structure and works with just a linear scalar product space Ψ. The precise meaning of eigenvalue equations like (13),(21) and (22) is given by (31).

The generalized eigenvalue ω in (31) may belong to the continuous spectrum of the Hilbert space operator $A^\dagger$, as is usually assumed for Dirac kets, but it need not belong to the Hilbert space spectrum of $A^\dagger$, and for a self-adjoint operator $A^\dagger$ (i.e. if A is essentially self-adjoint) ω need not even be real. If $|F\rangle \in \mathcal{H}$, then (31) is identical to the definition of an ordinary eigenvector of $A^\dagger$ in $\mathcal{H}$ with discrete eigenvalue ω.

The possible set of generalized eigenvalues of an operator A is determined by the choice of the space $\Phi^\times$; or equivalently, by the choice of Φ (i.e. by the topology τ_Φ with which we choose to equip the linear space Ψ). One can choose the topology τ_Φ with respect to which one completes the pre-Hilbert space Ψ unwisely so that $\Phi^\times$ contains all kinds of things. Others have suggested restricting τ_Φ such that the set of generalized eigenvalues defined by (31) is identical with the Hilbert space spectrum of $A^\dagger$ [40]. This would reproduce the Dirac formalism to the extent that it has been used in the past, because the Dirac kets are always chosen such that they are connected with the (absolutely) continuous spectrum of an operator $A^\dagger = \bar{A}$. But this would not allow for Gamow kets like those in (21), and no new physics would be incorporated in such a "tight rigging".

We invoke the principle that the space Φ (and its dual $\Phi^\times$) should be chosen by physical arguments. For one "kind" of quantum physical system (where the term "kind" is left to the intuitive interpretation based on experimental experience) one takes as a mathematical image one particular space Φ. For instance, one could define the topology for Φ such that the observables of the physical system under consideration are continuous operators in Φ [25, 26]. We shall discuss a scattering experiment and additional physical arguments related to causality and initial conditions which will be used to specify Φ further. Causality and the choice of initial conditions are old principles of classical physics which have not been fully utilized in standard quantum mechanics.

6. Time-Symmetric Equations and Time-Asymmetric Boundary Conditions

"The miracle of the appropriateness of the language of mathematics for the formulation of the laws of physics is a wonderful gift which we neither understand nor deserve."
—E.P. Wigner[13]

Theoretical predictions are based on two prerequisites [41]: the *laws of nature* and the *initial (or boundary) conditions*. The laws of nature provide the observables, e.g. the algebra of operators derived from a space-time symmetry group or the Hamiltonian H and the dynamical equations, which are the same in both the HS and RHS formulations and given by the Schrödinger equation or the von Neumann equation:

$$(33) \qquad i\hbar\frac{d\phi(t)}{dt} = H\phi(t) \quad \text{or} \quad i\hbar\frac{dW(t)}{dt} = [H, W(t)].$$

The laws of nature are not subject to human influences; they are given once and forever. The initial and boundary conditions leave some freedom of choice but are limited by the achievability – in principle and in practice – of building experimental apparatuses and by causality [35].

In the standard HS formulation of quantum mechanics, one chooses the initial and boundary conditions for the $|\phi\rangle$ and the $|F\rangle$ such that both are

---

[13] E.P. Wigner, *Symmetries and Reflections*, (Ox Bow Press, Woodbridge, Connecticut, 1979), p. 237.

always elements in $\mathcal{H}$. In this way one does not take into consideration the variety of possible choices and limitations for the initial and boundary conditions due to causality. In the RHS formulation, the initial condition can be chosen more specific to the particular problem under consideration. We shall use these additional conditions to define the space Φ, or equivalently the RHS $\Phi \subset \mathcal{H} \subset \Phi^\times$, for a quantum scattering system.

The best known example of a RHS is the space in which Φ is realized by the Schwartz space of "well-behaved" functions, i.e. functions that have derivatives which are all continuous, smooth and rapidly decreasing, $\mathcal{H}$ is the space of Lebesgue square integrable functions, and $\Phi^\times$ is the space of tempered distributions, $\mathcal{S} \subset L^2(\mathbb{R}) \subset \mathcal{S}^\times$. This is the RHS suitable for the quantum oscillator whose algebra of observables is represented by τ_Φ-continuous operators.

In quantum theory, if one distinguishes between preparations and registrations[14], one can further specify the RHS, and one is led to a pair of RHS's: one representing the preparations and the other representing the registrations. As an example we will discuss a scattering experiment. The scattering experiment can be subdivided into a preparation stage and a registration stage. The in-state ϕ^+ that evolves from the prepared in-state ϕ^{in} outside the interaction region is determined by the preparation apparatus (the accelerator). The "out-state" ψ^-, detected as the "out-state" ψ^{out} outside the interaction region, is determined by the registration apparatus (the detector). According to the physical interpretation of the RHS formulation, real physical entities connected with the experimental apparatuses, e.g. the ensemble $|\phi\rangle\langle\phi|$ describing the preparation apparatus or the observable $|\psi\rangle\langle\psi|$ describing the registration apparatus are described by the *well-behaved* vectors $\phi, \psi \in \Phi$. However, states are different from observables and should be described by a different set of vectors. We denote the space of state vectors ϕ^+ by Φ_- and the space of observable vectors ψ^- by Φ_+, where $\Phi = \Phi_- + \Phi_+$ and $\Phi_- \cap \Phi_+ \neq 0$. We will call the elements of Φ_- and Φ_+ *very well-behaved* vectors from below and above, respectively, where below and above refer to the second sheet of the energy surface of the analytically continued S-matrix. In place of the single rigged Hilbert space we therefore have a pair of rigged Hilbert spaces:

(34) $\phi^+ \in \Phi_- \subset \mathcal{H} \subset \Phi_-^\times$ for ensembles or prepared in-states,

(35) $\psi^- \in \Phi_+ \subset \mathcal{H} \subset \Phi_+^\times$ for observables or registered "out-states".

---

[14]   The one feature on which most discussions of the foundations of quantum mechanics agree is the dichotomy of "state" and "observable". If one interprets quantum theory objectively from the outside in terms of classical physics, as in [7], then the state is defined by the preparation apparatus (e.g. accelerator) and the observable is defined by the registration apparatus (detector). Though distinct by their definition, the HS formulation blurs this differentiation between states and observables by not specifying which elements of the HS are allowed as "states" and which as "observables".

Here the Hilbert space $\mathcal{H}$ is the same for both triplets, and since we ignore the discrete eigenvalues here it is $\mathcal{H}^{\text{a.c.}}$ of the continuous spectrum.

Mathematically, these spaces can be defined by their realizations as function space as in the case of the Schwartz space triplet $\mathcal{S} \subset L^2(\mathbb{R}) \subset \mathcal{S}^\times$ above. The spaces Φ_+ and Φ_- are defined as the spaces that are realized by the well-behaved Hardy class functions [42] in the upper half plane, $\mathcal{S} \cap \mathcal{H}_+{}^2|_{\mathbb{R}+}$, and in the lower half plane, $\mathcal{S} \cap \mathcal{H}_-{}^2|_{\mathbb{R}+}$, respectively:

(36) $\qquad \phi^+ \in \Phi_-$ if and only if $\langle {}^+E|\phi^+\rangle \in \mathcal{S} \cap \mathcal{H}_-{}^2|_{\mathbb{R}+}$

(37) $\qquad \psi^- \in \Phi_+$ if and only if $\langle {}^-E|\psi^-\rangle \in \mathcal{S} \cap \mathcal{H}_+{}^2|_{\mathbb{R}+}$

The notation $|_{\mathbb{R}+}$ mean restriction to the positive real line, i.e. the physical values of energy. If $\mathcal{H}$ in (34) and (35) denotes the Hilbert space realized as the space of Lebesgue square integrable functions, $L^2[0,\infty) = L^2(\mathbb{R}^+)$, then one can show [43] that

(38) $\qquad \mathcal{S} \cap \mathcal{H}_\pm{}^2|_{\mathbb{R}+} \subset L^2(\mathbb{R}^+) \subset \left(\mathcal{S} \cap \mathcal{H}_\pm{}^2|_{\mathbb{R}+}\right)^\times$

are two rigged Hilbert spaces of functions. The two RHS's of the in-states $\{\phi^+\}$ (34) and of the out-states $\{\psi^-\}$ (35) are mathematically defined as those RHS's whose realizations are the two RHS's of $\mathcal{S} \cap \mathcal{H}_-{}^2|_{\mathbb{R}+}$ and $\mathcal{S} \cap \mathcal{H}_+{}^2|_{\mathbb{R}+}$, respectively.

The opposite sub- and superscripts for vectors and spaces comes from the opposite nomenclature in physics (scattering theory for ϕ^+, ψ^-) and mathematics (theory of Hardy class functions for $\Phi_- \doteq \mathcal{S} \cap \mathcal{H}_-{}^2|_{\mathbb{R}+}$ and $\Phi_+ \doteq \mathcal{S} \cap \mathcal{H}_+^2|_{\mathbb{R}+}$) which had been developed independently. Except for the nomenclature the spaces are the "same", i.e. $\{\phi^+\} = \Phi_-$ and $\{\psi^-\} = \Phi_+$. This coincidence is another example of "The Unreasonable Effectiveness of Mathematics in Natural Sciences"[15].

It is of course not obvious at all that the physical spaces (34) and (35), the mathematical images of the prepared states and the detected observables respectively, should have anything to do with the Hardy class spaces (38). Originally the Hardy class property was introduced in order to derive (21) as a generalized eigenvalue equation and obtain a Breit-Wigner energy distribution for the vector associated with the resonance pole of the S-matrix [37, 44]. But the Hardy class property of the in-states $\phi^+ \in \Phi_-$ and $\psi^- \in \Phi_+$ is much more generally valid and can be obtained as a consequence of the Paley-Wiener theorem [42] from a mathematical formulation of causality [45].

The condition of causality can be formulated without any reference to the mathematical theory of quantum mechanics. Therefore this formulation of causality does not depend on the choice of the HS or RHS formulation. It is phrased in terms of only the intuitive properties of the macroscopic

---

[15] cf. footnote (13).

preparation and registration apparatuses. We call this theory-independent statement of causality the preparation⟶registration arrow of time[16]:

Let $t = t_0 = 0$ denote the point in time before which the preparation is complete and after which the registration begins. Then:

1. Time translation of the registration apparatus relative to the preparation apparatus makes sense only by an amount $t \geq 0$.

A second version of this statement uses quantum theoretical notions; we call it the quantum mechanical arrow of time.

2-H. (Heisenberg picture) An observable $|\psi^-(t)\rangle\langle\psi^-(t)|$ can be measured in a state $\phi^+(= \phi^+(0))$ only after the state has been prepared, i.e., for $t \geq 0$.

or

2-S. (Schrödinger picture) A state $\phi^+(t)$ must be prepared before an observable $|\psi^-\rangle\langle\psi^-|(= |\psi^-(0)\rangle\langle\psi^-(0)|)$ can be measured in that state, i.e., ϕ^+ must be prepared during $t \leq 0$.

This has implications for the most fundamental quantities in quantum physics, the expectation values. These probabilities $\mathcal{P}(t)$ are observed experimentally as a ratio of (large) numbers (detector counts) $N(t)/N = \mathcal{P}(t)$ and calculated theoretically as

(39a) $\mathcal{P}(t) = \mathrm{Tr}(\Lambda(t)W) = \mathrm{Tr}(\Lambda W(t))$, where

(39b) $\Lambda(t) = e^{iHt}\Lambda e^{-iHt}$ and $W(t) = e^{-iHt}W e^{iHt}$.

where Λ denotes the observable and W the state.

In many cases, such as for stationary states and time-independent observables, the right hand side of (39a) can be calculated for any value of t, i.e. the equation is valid for $-\infty < t < \infty$. The operators act in the space $\mathcal{H}^{\mathrm{disc}}$ of the discrete spectrum and all calculations are the same in the RHS formulation and the HS formulation. The physics of stationary states is not affected by the arrow of time.

This does not hold generally. In a scattering experiment, for example, the state has to be prepared. If $t = t_0 = 0$ is the time at which the preparation of the state $W = |\phi^+\rangle\langle\phi^+|$ is complete and the registration of $\Lambda(t) \equiv |\psi^-(t)\rangle\langle\psi^-(t)|$ can begin, then

(40)
$$\frac{N(t)}{N} = \mathcal{P}(t) = \mathrm{Tr}(\Lambda W(t)) = \mathrm{Tr}(\Lambda(t)W) = |\langle\psi^-(t)|\phi^+\rangle|^2, \text{ for } t \geq 0 \text{ only}.$$

---

[16] It is curious that the theory-independent version (Version 1) of this arrow of time had been known for some time [7], but was then extrapolated "into the past" in order to obtain the unitary one-parameter group $U(t) = e^{iHt}$ for the time translation of the observables in the HS.

The left hand side, $N(t)/N$ is observed only for $t \geq 0$ and assumed to be zero for $t < 0$ (detector clicks before $t = 0$ are discounted as noise):

$$(41) \qquad \frac{N(t)}{N} = \mathcal{P}(t) = 0, \text{ for } t < 0.$$

In the HS, $W(t)$ of (39b) is in all cases calculated for positive as well as negative t. The RHS formulation allows for $W(t)$'s which can only be *calculated* for $t \geq 0$, e.g. the Gamow states $W(t) = |\psi^G(t)\rangle\langle\psi^G(t)|$ of Sect. 7.2. If the time scale of the formation of $W(t)$ around $t_0 = 0$ is orders of magnitude smaller that the time scale of decay it is always more practical to work with (41).

The validity of the preparation$\longrightarrow$registration arrow (Version 1) is obvious. With the association between the experimental and the theoretical quantities of Table I, Version 2 (Heisenberg picture) should be an obvious consequence of Version 1 if for all times t these associations are upheld.

In the HS formulation this is not possible, let $t = 0$ be the point in time before which the preparation is completed and after which the registration begins. Because in the HS formulation, the time translation operator $U^\dagger(t)$ for the state ϕ (and $U(t)$ for the observable ψ) is defined for $-\infty < t < +\infty$, one can calculate $|\psi(\tau)\rangle\langle\psi(\tau)| = U(\tau)|\psi\rangle\langle\psi|U^\dagger(\tau)$ for *any* negative value of τ, and its expectation value in every state $\phi \in \mathcal{H}$ is in general not zero for $\tau < 0$. Thus, in the HS formulation it is possible to calculate a non-zero expectation value for an observable at times $\tau < 0$, which one should not observe according to Version 1.

In the RHS formulation the quantum mechanical arrow of time can be implemented in the following way [45]: the prepared state vectors $\phi^+ \in \Phi_-$ representing the preparation apparatus should have a smooth energy wave function $\langle^+E|\phi^+\rangle \in \mathcal{S}|_{\mathbb{R}^+}$ whose Fourier transform $\mathcal{F}(\tau) \in \mathcal{S}$ should be zero for $\tau > 0$, because $\tau = 0$ is the time after which there is no preparation. The vector $\psi^- \in \Phi_+$ describing the observable defined by the registration apparatus should have a smooth energy wave function $\langle^-E|\psi^-\rangle \in \mathcal{S}|_{\mathbb{R}^+}$ whose Fourier transform $\mathcal{G}(\tau) \in \mathcal{S}$ is zero for $\tau < 0$ because $\tau = 0$ is the time until the registration apparatus remains turned off. From this property of the Fourier transforms it follows by the use of the Paley-Wiener Theorem that the wave functions must have the properties

$$\langle^+E|\phi^+\rangle \in \mathcal{H}_-^2 \quad \text{and} \quad \langle^-E|\psi^-\rangle \in \mathcal{H}_+^2$$

This means they have the properties (36) and (37) which leads to the two different spaces Φ_- and Φ_+ (34) and (35) for the states and observables, respectively. The use of different mathematical spaces for states (in-states) and observables (so-called "out-states") is one of the new features of the RHS formulation of quantum mechanics.

Neither Φ nor $\mathcal{H}$ can describe single microsystems as mentioned above in Sect. 3..3. However, it is intuitively attractive to imagine that the preparation apparatus acts on the registration apparatus by the action of single physical entities, the microphysical systems. The energy distribution for a microphysical system does not have to be a "well-behaved" function of the physical values of energy E. For the entities connected with microphysical systems the RHS formulation provides the elements of $\Phi^\times$, $\Phi_+{}^\times$ and $\Phi_-{}^\times$. In particular, Dirac's scattering states $|\mathbf{p}^\pm\rangle$ are elements of $\Phi_\mp^\times$, Gamow's decaying states $\psi^G = |E_R - i\Gamma/2^-\rangle$ are elements of $\Phi_+^\times$ and (since resonance poles come in pairs at the complex energies $(E_R \mp i\Gamma/2)$) the exponentially growing Gamow vectors $|E_R + i\Gamma/2^+\rangle$ are elements of $\Phi_-^\times$.

7. Consequences of the RHS formulation of Quantum Mechanics

7.1 Semigroup Time Evolution from Time Asymmetry and Some Remarks on Time Reversal Transformations

> "When in the 18th century Euler discovered those formulas which till to-day delight the mathematical phantasy, he seriously stated that his pencil was more clever than himself. This impression that mathematical structures can include a kind of self-determination concerns me at this time.... Mathematics and Philosophy attack the world's problems in different ways. Only by their complementary action do they give the right direction."
> —E. Kähler[17]

Once the spaces (34) and (35),

$$\Phi_- \subset \mathcal{H} \subset \Phi_-{}^\times \qquad \text{for ensembles or prepared in-states} - \{\phi^+\} \text{ and}$$
$$\Phi_+ \subset \mathcal{H} \subset \Phi_+{}^\times \qquad \text{for observables or registered "out-states"} - \{\psi^-\},$$

are chosen, which we did in Sect. 6. on the basis of some causality arguments, then it is easy to see mathematically that the extension of the time evolution operator $U^\dagger(t) = e^{-iHt}$ in $\mathcal{H}$ to an operator $U_+{}^\times(t)$ in $\Phi^\times{}_+$ – the conjugate of the operator $U(t)$ in Φ_+ as defined by (30) – is only a semigroup[18].

Historically, this was not the situation because ingrained in our thinking was the notion that the time evolution operator $U(t) = e^{iHt}$ in quantum mechanics – and the representation of any continuous symmetry transformation – are unitary (reversible) group operators. Philosophizing alone would not be enough to take a semigroup instead. To arrive at the semigroup, we start from the empirically desirable properties (21) and (22) of Gamow resonance states and let mathematics determine the path. Then the result will still be surprising, but now acceptable.

We use the definition of a resonance as a (pair of) first order poles in the analytically continued S-matrix, $S_{bb'}(E) = \langle b||S(E)||b'\rangle$. In the representation of the S-matrix

---

[17] Preprint (1996), translated by the authors.
[18] For a semigroup, the inverse $(U^\times(t))^{-1}$ does not have to exist.

(42) $(\psi^{out}, S\phi^{in}) = (\psi^-, \phi^+)$

$$= \sum_{bb'} \int_{\text{spectrum of } H} \langle\psi^-|b, E^-\rangle\langle b||S(E)||b'\rangle\langle^+b', E|\phi^+\rangle,$$

we deform the contour of integration from the cut $\{0 \leq E < \infty\}$ (the spectrum of H) into the second sheet of the lower complex plane. The Dirac kets $|E^-\rangle = |b, E^-\rangle$ (where b is the degeneracy quantum numbers) with $E \in \{\text{continuous (real) spectrum}\}$ are therewith continued to complex values. At the position of the resonance pole $E = z_R = E_R - i\Gamma/2$, they become – using Cauchy's formula – the Gamow kets $|z_R^-\rangle$. This is standard scattering theory except that we want to be careful about the $\langle\psi^-|E^-\rangle$ and the $\langle^+E|\phi^+\rangle$ and do the analytic continuation only for those ψ^- and ϕ^+ for which it is mathematically allowed. This means that $\langle\psi^-|E^-\rangle = \langle E^-|\psi^-\rangle^*$ and $\langle^+E|\phi^+\rangle$ must not only be elements of $\mathcal{S}$, but also must be the boundary value of an analytic function. We repeat the same process for $\langle H\psi^-|E^-\rangle = \langle\psi^-|H^\times|E^-\rangle = E\langle\psi^-|E^-\rangle$. In order that (21) holds as a general eigenvalue equation,

(43) $$\langle\psi^-|H^\times|z_R^-\rangle = z_R\langle\psi^-|z_R^-\rangle,$$

the $\langle\psi^-|E^-\rangle$ must further be required to be of Hardy class, $\langle\psi^-|E^-\rangle \in (\mathcal{S} \cap \mathcal{H}_-^2)$, in other words, ψ^- must be required to be a Hardy class vector, $\psi^- \in \Phi_+$. Similarly we must require $\phi^+ \in \Phi_-$.

Now, in order to calculate (22) in its mathematically precise form as a generalized eigenvector equation, one repeats the same process for ψ^- in (42) replaced by $\psi^-(t) = e^{iHt}\psi^-$ – the observable $|\psi^-\rangle\langle\psi^-|$ translated by time t. On the right hand side of (42) one then obtains $\langle\psi^-(t)|E^-\rangle = \langle\psi^-|e^{iH^\times t}|E^-\rangle = e^{-iEt}\langle\psi^-|E^-\rangle$ in place of $\langle\psi^-|E^-\rangle$. The contour deformations needed on the right hand side of (42) are now possible if and only if $t \geq 0$, because then and only then is $e^{-iEt}\langle\psi^-|E^-\rangle$ a Hardy class function. Then and only then, one obtains, in the same way as (43) was obtained, the following result[19]:

(44) $$\langle\psi^-|e^{-iH^\times t}|z_R^-\rangle = e^{-iz_Rt}\langle\psi^-|z_R^-\rangle, \text{ but only for } t \geq 0.$$

Thus the Gamow vectors are eigenvectors of the time evolution operator given by (44) and the time evolution operator on Gamow vectors, $U_+^\times(t) = e^{-iH^\times t}$, is only a semigroup. After one was led to this conclusion and has accepted the semigroup evolution for the Gamow vectors, one will have no problems to generalize to the whole space $\Phi^\times_+$.

The time evolution operator in the space of microphysical states $\Phi_+^\times$ (and also the time evolution operator in the space $\Phi_-^\times$) is described by a semigroup $U_+^\times(t)$ for $t \geq 0$ only (analogously in the space $\Phi^\times_-$ the time

---

[19] cf. Sect. XXI.4, 3rd Ed. of [7], and [37].

evolution is given by a semigroup $U_-^\times(t)$ for $t \leq 0$). In the RHS's (34) and (35) one has the two extensions of the Hilbert space operator $U^\dagger(t)^{20}$

(45) the conjugate of $U|_{\Phi_-}$: $U^\dagger(t) \subset U_-^\times = e^{-iH^\times t/\hbar}$; for $t \leq 0$

(46) the conjugate of $U|_{\Phi_+}$: $U^\dagger(t) \subset U_+^\times = e^{-iH^\times t/\hbar}$; for $t \geq 0$

where $U_\pm^\times$ denote the extensions of the unitary operator $U^\dagger(t)$ to the spaces $\Phi^\times{}_\pm$ [14]. Mathematically, $U_-^\times$ in $\Phi^\times{}_-$ can be defined by (30) only for values of parameters $t \leq 0$, since for $t > 0$ the operator $U(t)|_{\Phi_-}$ is not a continuous operator which maps Φ_- into Φ_-. By the same argument, $U_+^\times$ in $\Phi^\times{}_+$ can only be defined by (30) for values of parameters $t \geq 0$ because for $t < 0$, $U(t)|_{\Phi_+}$ is not a continuous operator which maps Φ_+ into Φ_+ as required by (30).

These are the mathematical arguments by which the semigroup time evolution is derived from the mathematical properties of the spaces Φ_- and Φ_+, which in Sect. 6. had been conjectured from the intuitive notion of causality. However without the straightforward mathematical derivation of (44) from (43) and just on the basis of "philosophical" causality arguments alone, one would probably not have been willing to come up with these mathematical properties of Φ_- and Φ_+ that have such drastic consequences like the semigroup evolution. As one can see from the details [14, 7] of the above derivation, the reverse conclusion is also correct: The choice of the Hardy class spaces and therewith the preparation→registration arrow follows if one requires the existence of the Gamow vectors with the above properties.

The semigroup time evolution of the Gamow vectors is the mathematical expression of irreversibility for the microphysical decaying states. As a

---

[20] Regarding the notation $e^{-iH^\times t}$: $U(t)$ is the unitary operator in $\mathcal{H}$ and $e^{iHt} \equiv \sum_{n=0}^\infty \frac{(it)^n}{n!} H^n$ where the generator of $U(t)$, H, is defined on a dense subspace of $\mathcal{H}$, the space of differentiable vectors $\mathcal{D}$, and the infinite series converges on the dense subspace $\mathcal{A} \subset \mathcal{D} \subset \mathcal{H}$ of "analytic vectors". The topological notions of dense, convergence, differentiable, etc., all refer to the topology in $\mathcal{H}$ given by the norm, i.e. $U(t)$ and $U^\dagger(t)$ are $\mathcal{H}$-continuous (implying $\mathcal{H}$-bounded) operators that are $\mathcal{H}$-dense, $\mathcal{H}$-convergent, etc., in $\mathcal{H}$. The restriction of $U(t)$ to the subspace Φ_+, $U_+(t) = U(t)|_{\Phi_+}$ for $t \geq 0$ only, is a Φ-continuous operator; $U_+^\times(t)$ is its conjugate as defined by (30). If one applies $U_+^\times(t)$ to a generalized eigenvector $F \in \Phi^\times{}_+$ of $H^\times$ with an eigenvalue ω as defined by (31), then one can show that $U_+^\times(t)F = e^{-i\omega t}F$. For this reason we use the notation $U_+^\times(t) = e^{-iH^\times t}$. The operator $H^\times$ is a $\Phi^\times$-continuous operator in $\Phi^\times{}_+$ and one can define $\sum_{n=0}^N \frac{(it)^n}{n!} H^n \equiv U_+^N$ for all $t \geq 0$. One would usually want to denote by the exponential $e^{-iH^\times t}$ the limit with respect to the topology in $\Phi^\times$ of the sequence U_+^N for $N \to \infty$. Whether $U_+^N \to U_+^\times(t)$ for $N \to \infty$ and/or whether there is a Φ-dense subspace of vectors $F \in \Phi^\times{}_+$ on which $U_+^N F$ (which one would then call $\Phi^\times{}_+$-analytic vectors) is not known to us.

consequence of the quantum mechanical irreversibility, one can no longer calculate for every state $W^G(t) = |\psi^G(t)\rangle\langle\psi^G(t)|$, with $\psi^G \in \Phi^\times_+$, another state $W^{\text{neg}}(t) \equiv |\psi^G(-t)\rangle\langle\psi^G(-t)|$. This is empirically correct but it also leads immediately to the question whether and how this irreversibility can be compatible with the definition of a time reversal operator A_T [21] , since the state obtained from a given state $W(t)$ by A_T transformation,

$$(47) \qquad\qquad W^T \equiv A_T^{-1} W(t) A_T,$$

is usually identified with the state $W^{\text{neg}} \equiv W(-t)$, i.e. is assumed to have the property

$$(48) \quad A_T^{-1} W(t) A_T = W(-t)$$
$$(\text{or } \phi^T(\mathbf{x}, t) = \phi(\mathbf{x}, -t) = \phi^*(\mathbf{x}, t) \text{ for wave functions}).$$

The answer to this question is that neither the time reversed state $W^T(t)$ nor the backward time translated state $W^{\text{neg}}(t)$ is in general physically defined. For example, in a typical scattering experiment the "out-states" represent highly correlated spherical waves whereas the prepared in-states are typically two uncorrelated plane waves (e.g., two colliding monochromatic beams). The time reversal of this experiment would require a preparation apparatus that prepares highly correlated (with fixed phase relationship) incoming spherical waves that would be scattered into two uncorrelated plane waves. An experimental setup that would accomplish this would have to be so complicated that it is impossible to build, at least in this world. Thus, not for every preparable W can one prepare a state which would be described by its time reversal transformed $W^T = A_T^{-1} W A_T$ (for another example see [36]).

The time reversal operator A_T is not defined by its action on states, but by its relation to the observables. Examples of these relations are

$$(49) \quad \begin{aligned} A_T P_i A_T^{-1} &= -P_i, & A_T J_i A_T^{-1} &= -J_i, \\ A_T U_P A_T^{-1} &= \varepsilon_T \varepsilon_I U_P, & A_T H A_T^{-1} &= H, \\ A_T H_0 A_T^{-1} &= H_0, & A_T S A_T^{-1} &= S^\dagger = S^{-1}, \end{aligned}$$

and they follow from the extended projective representations of the extended Poincaré group [46]. The generators P_i, H, J_i represent momentum, energy, angular momentum, respectively, the S-operator is a complicated function of the interaction Hamiltonian $V = H - H_0$ and U_P is the unitary and hermitian parity operator normalized to $U_P^2 = 1$. The quantities

$$(50) \qquad\qquad \varepsilon_T = A_T^2, \text{ and } \varepsilon_I = (U_P A_T)^2 \equiv A_I^2$$

are real phase factors which define the four different extensions of the restricted space-time symmetry transformations by space inversion $P = g$,

---

[21] In place of A_T one could as well use the CPT-operator here with minor modifications.

time inversion $T = -g$ and the space-time inversion $I = PT = -1$, which were derived by Wigner [47]. Of the four possible extensions characterized by the four pairs of phase factor $(\varepsilon_T, \varepsilon_I) = (\pm 1, \pm 1)$ the the only extensions used in relativistic field theory [46] are those characterized by

$$(51) \qquad \varepsilon_T = (-1)^{2j} \; \varepsilon_I = (-1)^{2j} \text{ where } j \text{ is the spin.}$$

With this choice the time reversal operator A_T has the following transformation property:

$$(52) \qquad A_T : \Phi_\mp \rightarrow \Phi_\pm; \quad \Phi_+ \ni \psi^- = A_T \phi^+, \; \phi^+ \in \Phi_-.$$

This is also the solution suggested by conventional scattering theory where the *in*-states ϕ^+ are the time reverse of the so-called *out*-states ψ^-. The "out-states" ψ^- are actually observables and not states because they are specified by the detector whereas in-states ϕ^+ are specified by the preparation apparatus (accelerator). In our interpretation, the transformation property (52) means that states are transformed into observables and vice-versa. This requires the identification of the set of states with the set of observables (i.e., no arrow of time) and to assign to every $W(t)$ a $W^T(t) \equiv A_T^{-1} W(t) A_T$ fulfilling (48). This means that in the case (51) one cannot have irreversibility in the sense described above, which as mentioned in Sect. 3 is in contradiction to at least some arguments concerning the improbability to prepare time reversed states (cf. remark above and Chap. 13.2 in [36]).

Fortunately, there are three other classes of representations derived by Wigner [47] and one can choose instead of (51) one of the non-standard extensions for A_T which lead to a doubling of the spaces (time reversal doubling [47]). We suggest the choice:

$$(53) \qquad \varepsilon_T = -(-1)^{2j}, \quad \varepsilon_I = -(-1)^{2j}.$$

The A_T which fulfills (53) can be shown to be compatible with the microphysical irreversibility of (45) and (46) [48].

7.2 Gamow Vectors

> "The data suggest that a particle decays if it can and that it is stable only if there is no state... to which it is allowed to decay. Stability does not appear to be a criterion for *elementarity*."
> —Frauenfelder and Henly[22]

The RHS formulation accounts for the Dirac kets; they are defined as generalized eigenvectors over the spaces Φ_+, Φ_-, and $\Phi = \Phi_+ + \Phi_-$ with generalized eigenvalues E of the Hamiltonian H:

$$(54) \qquad \langle H\phi | E^\pm \rangle = \langle \phi | H^\times | E^\pm \rangle = E \langle \phi | E^\pm \rangle \text{ for all } \phi \in \Phi,$$

---

[22] p. 91-92 of [30].

where the E represent the scattering energies (continuous spectrum of $\bar{H}$). This is no surprise because the RHS formulation was devised to provide a mathematical justification for the Dirac scattering states.

Unforeseen was that the RHS formulation also accounts for the Gamow kets. The decaying Gamow kets $\psi^G = |E_R - i\Gamma/2^-\rangle\sqrt{2\pi\Gamma} = |z_R^-\rangle\sqrt{2\pi\Gamma}$ are defined as generalized eigenvectors over the space Φ_+ with generalized eigenvalue $(E_R - i\Gamma/2)$:

(55a) $$\langle H\psi^- | E_R - i\Gamma/2^- \rangle \equiv \langle \psi^- | H^\times | E_R - i\Gamma/2^- \rangle$$
$$= (E_R - i\Gamma/2)\langle \psi^- | E_R - i\Gamma/2^- \rangle,$$

or its complex conjugate:

(55b) $$\langle \psi^G | H | \psi^- \rangle = \langle \psi^G | \psi^- \rangle (E_R + i\Gamma/2), \text{ for all } \psi^- \in \Phi_+.$$

These Gamow vectors $\psi^G = |E_R - i\Gamma/2^-\rangle\sqrt{2\pi\Gamma}$ have the following properties:

(1) They are derived as functionals of the resonance pole term at $z_R = E_R - i\Gamma/2$ in the lower half of the second sheet of the analytically continued S-matrix.

(2) They have Breit-Wigner energy distribution

(56) $$\langle {}^- E | \psi^G \rangle = i\sqrt{\Gamma/2\pi}\,\frac{1}{E - (E_R - i\Gamma/2)}, \quad -\infty_{II} < E < \infty,$$

where E is on the second sheet for negative values[23].

(3) The time evolution of the Gamow vectors is derived from the pole term of the S-matrix as:

(57a) $$\langle e^{iHt}\psi^- | z_R^- \rangle \equiv \langle \psi^- | e^{-iH^\times t} | z_R^- \rangle = e^{-iE_R t} e^{-(\Gamma/2)t} \langle \psi^- | z_R^- \rangle$$

or for the complex conjugate:

(57b) $$\langle {}^- z_R | e^{iHt} | \psi^- \rangle$$
$$= e^{iE_R t} e^{-(\Gamma/2)t} \langle {}^- z_R | \psi^- \rangle, \text{ for every } \psi^- \in \Phi_+ \text{ and for } t \geq 0.$$

---

[23] The "wave function" $\langle {}^- E | z_R^- \rangle$ of the generalized vector $|z_R^-\rangle$ in (56) is represented by a regular function (even of Hardy class $\mathcal{H}_+^2$) though one would expect it to be a non-regular distribution. The latter is indeed correct because (56) is not the actual wave function. The wave function is a function on the continuous spectrum $\{0 \leq E \leq \infty\}$ and over $\mathcal{S} \cap \mathcal{H}_+^2 |_{\mathbb{R}^+}$, i.e. as an element of $(\mathcal{S} \cap \mathcal{H}_+^2 |_{\mathbb{R}^+})^\times$, $\langle {}^- E | z_R^- \rangle$ is a distribution – not a function (regular distribution). However in (56) we need the whole real axis $\{-\infty < E < \infty\}$ to represent it as a regular function. Extending E in (56) to the negative real axis does however not mean that we have let the spectrum of H go to $-\infty$, because the negative values of E in (56) are the result of an analytic continuation into the second sheet. For more detail see p. 504-505 of [14], or see [49], where an explicit expression for the distribution $\langle {}^- E | z_R^- \rangle \in (\mathcal{S} \cap \mathcal{H}_+^2 |_{\mathbb{R}^+})^\times$ has been given.

The above properties – except for the semigroup property – are historically ascribed to the empirical notion of decaying states and resonances. Equations (55) and (57) give the precise mathematical form as a distribution. Therefore the Gamow kets justly deserve the name resonance state vectors with complex resonance energy $z_R = E_R - i\Gamma/2$. The property (2) identifies E_R as the resonance energy and Γ as the resonance width. The property (3) shows that $\tau_R = 1/\Gamma \, (= \hbar/\Gamma)$ is the lifetime of the decaying resonance state.

The semigroup evolution (57) expresses the time-asymmetry of the Gamow vector and the resonance state which it describes. It is not a condition that we demanded of the $|z_R^-\rangle$, like we demanded (40) or some other mathematical formulation of causality, but it can be derived from the Hardy class property (35) which is some mathematical statement of causality[24].

In the same way as stationary states are given by bound state poles of the S-matrix and described as eigenvectors of H with real (negative) eigenvalues, decaying states are given by resonance poles and described as generalized eigenvectors of H with complex eigenvalue. This puts stable and decaying particles on the same footing and shows that resonances can be thought of as autonomous quantum physical entities which are not less fundamental than stable particles.

Instead of defining the Gamow vectors from the poles of the analytically continued S-matrix, the Gamow vectors can also be defined from the poles of the extended resolvent of the Hamiltonian [50]. These definitions are equivalent in many cases in which the S-matrix is obtained from the Hamiltonian. The S-matrix definition can be used in more general settings, e.g. in the relativistic case when one can start from the unitary representations of the Poincaré group in $\mathcal{H}$ and then extend it to $\Phi^\times{}_\pm$.

One can generalize the Gamow vectors to Gamow-Jordan vectors associated to higher-order poles of the S-matrix [51] and obtain higher-order Gamow states [52]. These vectors are also functionals over Hardy class spaces and have semigroup time evolutions. They have some features which are heuristically associated with higher-order poles of the S-matrix [53] like polynomial time dependence. But non-reducible Gamow states that follow from the higher-order S-matrix pole term have purely exponential time evolution, which is a feature that was not expected. Since so far there is no empirical evidence for states associated with higher-order poles of the S-matrix, we shall not discuss them here any further.

---

[24] It is remarkable that the asymmetry of the time evolution of the resonance states (57) – or more generally of the semigroup (46) – is the same as the preparation→registration arrow of time (40). This is established mathematically by the use of the Hardy class RHS's (34) and (35). Since equation (40) applied to the quantum theory of cosmology is, according to (7), the time-asymmetry of the universe [8], the mathematical idealization given by the RHS formulation of quantum mechanics would establish a connection between the time-asymmetry of quantum cosmology and the irreversibility of quantum decay phenomena.

7.3 Generalized Basis Vector Expansions

The most important consequence of the RHS formulation are the generalized basis vector expansions. The expansion (14) that Dirac envisioned has been proven as the nuclear spectral theorem. This theorem requires the nuclearity of the space Φ (or a little less [18]), a mathematical subtlety unimportant for physics. What is important for physics is that (14) holds in the space Φ; therefore we will call a triplet of spaces (27) a RHS only if the necessary conditions for the nuclear spectral theorem are fulfilled.

All basis vector expansions are generalizations of the elementary basis vector expansion of a vector in $\mathbb{R}^3$,

$$(58) \qquad \mathbf{x} = \sum_{i=1}^{3} \hat{\mathbf{e}}_i (\hat{\mathbf{e}}_i \cdot \mathbf{x}) = \sum_{i=1}^{3} \hat{\mathbf{e}}_i \cdot \mathbf{x}_i,$$

where $\hat{\mathbf{e}}_i$ are chosen to be the physically distinguished basis vectors.

Earlier generalizations of this are the fundamental theorem of linear algebra which states that for every self-adjoint operator H in an n-dimensional Euclidean space $\mathcal{H}_n$ there exists a complete basis system $e_i \ldots e_n$ in $\mathcal{H}_n$ of eigenvectors:

$$H e_i \;=\; E_i e_i, \; e_i \in \mathcal{H}_n \; (i = 1, 2, \ldots, n), \text{ such that}$$

$$(59) \qquad f \;=\; \sum_{i=1}^{n} e_i (e_i | f) \; \text{ for every } f \in \mathcal{H}_n$$

This theorem generalizes to the infinite dimensional Hilbert space $\mathcal{H}$, but only for self-adjoint operators H which are completely continuous[25]. For an arbitrary self-adjoint operator H one cannot find a complete system of eigenvectors in $\mathcal{H}$ (complete in the sense that every $f \in \mathcal{H}$ can be expanded in the form (59)). Many physically important operators do not have even a single eigenvectors in $\mathcal{H}$. Because $|(e_i, f)|^2$ represents the probability to measure the value E_i of the observable H in the state f, one wants to use a basis system of eigenvectors for the distinguished observables not an arbitrary basis system, which one can always find by the Schmidt orthonormalization procedure. In order to obtain such a basis system of eigenvectors for an arbitrary observable, one has to go outside the HS; they are generalized eigenvectors as defined by (31).

The *first* step outside the HS is to use the nuclear spectral theorem to justify the Dirac basis vector expansion (14) in terms of well-defined mathematical quantities. The nuclear spectral theorem states that for every $\phi \in \Phi$ (not for every $f \in \mathcal{H}$) one can find a complete set of eigenvectors in $\Phi^\times$ (not in $\mathcal{H}$):

---

[25] These are also called compact operators and include Hilbert-Schmidt, nuclear and trace-class operators.

(60a) $\phi = \displaystyle\int_0^{+\infty} dE\, |E^+\rangle\langle{}^+E|\phi\rangle + \sum_n |E_n)(E_n|\phi)$ for every $\phi \in \Phi$.

Since $\phi^+ \in \Phi_- \subset \Phi$, of course the same expansion holds for ϕ^+:

(60b) $\phi^+ = \displaystyle\int_0^{+\infty} dE\, |E^+\rangle\langle{}^+E|\phi^+\rangle + \sum_n |E_n)(E_n|\phi^+).$

In the above expansion, $|E_n)$ are the discrete eigenvectors of the exact Hamiltonian $H = K + V$, $H|E_n) = E|E_n)$, which describe bound states. The generalized eigenvectors (Dirac kets) of H, $|E^+\rangle \in \Phi^\times$, fulfill $\langle H\chi|E^+\rangle = \langle\chi|H^\times|E^+\rangle = E\langle\chi|E^+\rangle$ for all $\chi \in \Phi$, cf. (31). The "coordinates" of the vector ϕ with respect to the continuous basis $|E^+\rangle$, i.e., the set of energy wave functions $\langle{}^+E|\phi\rangle$, form a realization of the space Φ by a space of functions (in the same way as the coordinates x_i form a "realization" of the vector $\mathbf{x}$). We call the vector $\phi \in \Phi$ "well-behaved" if $\langle{}^+E|\phi\rangle$ is a well-behaved function, i.e., $\langle{}^+E|\phi\rangle \in \mathcal{S}$. The $|E^+\rangle$ correspond to the continuous spectrum of H (the "scattering states") and the integration extends over the continuous spectrum [26] $:0 \leq E < \infty$ (the physical scattering energies).

---

[26] The nuclear spectral theorem actually asserts much less than (60) but it applies to a much larger set of operators A than the operators needed in quantum physics:

   Let A be *any* self-adjoint (or unitary or normal) operator in the Hilbert space $\mathcal{H}$ (the analogous statement holds for a strongly commuting family of operators $A_1, A_2, \ldots, A_N$) of the rigged Hilbert space $\Phi \subset \mathcal{H} \subset \Phi^\times$, and Λ its Hilbert space spectrum. Then there exists a unique measure μ on Λ such that for $\phi, \psi \in \Phi$

(60′) $(\psi, \phi) = \displaystyle\int_\Lambda d\mu(\lambda)\langle\psi|\lambda\rangle\lambda\phi$

   where the $|\lambda\rangle \in \Phi^\times$ are the generalized eigenvectors of A, i.e.

(60″) $\langle A\phi|\lambda\rangle \equiv \langle\phi|A^\times|\lambda\rangle = \lambda\langle\phi|\lambda\rangle$, for μ-almost every λ.

   The measure $d\mu(\lambda)$ depends upon the operator A, and for a general operator μ consists of three parts:

1. the discrete spectrum, $d\mu(\lambda) = \sum_n \delta(\lambda - \lambda_n)d\lambda$,
2. the absolutely continuous spectrum, $d\mu(\lambda) = \rho(\lambda)d\lambda$, and
3. the singularly continuous spectrum.

   The discrete spectrum corresponds to the sum in (60), the absolutely continuous spectrum corresponds to the integral in (60) and the singularly continuous spectrum has never occurred for operators that represent physical observables.

   The observables in quantum mechanics are usually derived from representations of space-time symmetry groups, spectrum generating groups, intrinsic symmetry groups or other Lie groups. Therefore in physics the measure $d\mu$ is the Plancherel measure of compact or non-compact Lie groups and their compact or non-compact subgroups. For the compact case, one has only case (1); for the non-compact Lie groups that have occurred in physics (e.g. all Abelian groups

If the *out*-wave functions are more readily available, one would have chosen the $|E^-\rangle$, defined as $|E^+\rangle = |E^-\rangle S(E + i0)$, in place of the $|E^+\rangle$ in (60), or one could have chosen any other vector which differs from $|E^+\rangle$ by an energy-dependent phase factor[27].

$$\phi = \int_0^{+\infty} dE\, |E^-\rangle S(E + i0)\langle^+ E|\phi\rangle + \sum_n |E_n\rangle(E_n|\phi)$$

$$(61)\quad \phi = \int_0^{+\infty} dE\, |E^-\rangle\langle^- E|\phi\rangle + \sum_n |E_n\rangle(E_n|\phi) \quad \text{for every } \phi \in \Phi$$

In a scattering experiment the following S-matrix elements are measured

$$(\psi^{\text{out}}, \phi^{\text{out}}) = (\psi^{\text{out}}, S\phi^{\text{in}}) = (\psi^-, \phi^+)$$

$$= \int_0^{+\infty}\int_0^{+\infty} dE\, dE'\, \langle\psi^{\text{out}}|E\rangle\langle E|S|E'\rangle\langle E'|\phi^{\text{in}}\rangle$$

$$(62)\qquad = \int_0^{+\infty} dE\, \langle^-\psi|E^-\rangle S(E + i0)\langle^+ E|\phi^+\rangle$$

In here ψ^- and ϕ^+ are very well-behaved vectors with:

$\phi^+ \in \Phi_-$ representing the state prepared by the accelerator, cf. (34)
and
$\psi^- \in \Phi_+$ representing the observable registered by the detector, cf. (35).

This means that the functions $\langle E|\psi^{\text{out}}\rangle = \langle^- E|\psi^-\rangle$ are "very well-behaved functions from above" and $\langle E|\phi^{\text{in}}\rangle = \langle^+ E|\phi^+\rangle$ are "very well-behaved functions from below", i.e., equations (36) and (37):

---

or all classical groups, semi-direct product and more) one has case (2) with $\rho(\lambda)$ being a continuous function. In these cases one has the "normalizations" of the generalized eigenvectors:

$$\langle\lambda_n|\lambda_{n'}\rangle = \delta_{nn'} \quad \langle\lambda|\lambda'\rangle = \rho^{-1}(\lambda)\delta(\lambda - \lambda').$$

After redefinition of the kets, $|\lambda\rangle \to |E^+\rangle = |\lambda\rangle\rho^{\frac{1}{2}}(\lambda)$ and $\lambda \to E$, this gives (60a) with $\langle E|E'\rangle = \delta(E - E')$. The integral in (61) is still a Lebesgue integral and (61) holds for $\phi, \psi \in \mathcal{H}$ if $\langle\lambda|\phi\rangle, \langle\lambda|\psi\rangle \in \mathcal{L}^2(\lambda)$. However now we can follow the procedure described in Sect. 3. and use only the smooth functions $\langle\lambda|\phi\rangle, \langle\lambda|\psi\rangle \in \mathcal{S}_\Lambda$ under the integral in (61). Then for the subspace Φ whose realization is given by the smooth functions $\mathcal{S}$ we can use in (61) Riemann integration with (61) holding for every λ (rather than μ-almost every λ). No example of a physical observable is known in which the integrals (60) can not be interpreted as Riemann integrals and the eigenvalue equation is valid only for μ-almost all λ. We will always presume that the observables have discrete and/or absolutely continuous spectra. Dirac's visionary tools of quantum mechanics, (13) and (14), have thus been made rigorous.

[27] If the additional quantum numbers (e.g., channel labels) η would be considered, then in place of the phase factor $S(E) = e^{2i\delta(E)}$ one would of course have a unitary matrix, $|E, \eta^+\rangle = |E, \eta'^-\rangle\langle\eta'|S(E)|\eta\rangle$.

(63a) $$\langle E|\phi^{\mathrm{in}}\rangle = \langle {}^{+}E|\phi^{+}\rangle \in \mathcal{S} \cap \mathcal{H}_{-}^{\;2}|_{\mathbb{R}+}$$

(63b) $$\langle E|\psi^{\mathrm{out}}\rangle = \langle {}^{-}E|\psi^{-}\rangle \in \mathcal{S} \cap \mathcal{H}_{+}^{\;2}|_{\mathbb{R}+}.$$

In the scattering experiment, the wave functions $\langle {}^{+}E|\phi^{+}\rangle$ (the components of the vector ϕ^{+} along $|E^{+}\rangle$) represent the probability $\int_{\Delta E} dE |\langle {}^{+}E|\phi^{+}\rangle|^{2}$ that the beam prepared by the accelerator has an energy in the interval ΔE, i.e., $|\langle {}^{+}E|\phi^{+}\rangle|^{2}$ is the energy distribution in the beam state ϕ^{+} and $|\langle {}^{-}E|\psi^{-}\rangle|^{2}$ is the energy resolution of the detector ψ^{-} (the detector efficiency). But the components of ϕ^{+} along $|E^{-}\rangle$, and of ψ^{-} along $|E^{+}\rangle$, do not just represent apparatus properties, but the properties of the scattering system.

If we consider for the vector ϕ in (61) a very well-behaved vector $\phi^{+} \in \Phi_{-} \subset \Phi$

(64) $$\phi^{+} = \int_{0}^{\infty} dE\,|E^{-}\rangle\langle {}^{-}E|\phi^{+}\rangle + \sum |E_{n}\rangle\langle E_{n}|\phi^{+}\rangle$$

then we obtain a Dirac basis vector expansion of ϕ^{+} with respect to the generalized eigenvectors $|E^{-}\rangle$ of H. Although both expansions (60b) and (64) use generalized eigenvectors of H, the basis system used in expansion (64) differs from that of (60b) by a phase factor[28]. Whereas the components $\langle {}^{+}E|\phi^{+}\rangle$ along $|E^{+}\rangle$ contain only information about the preparation apparatus, the components $\langle {}^{-}E|\phi^{+}\rangle$ are the interaction wave functions describing also the dynamics (analogous considerations hold for "out-state" (observable) vectors $\psi^{-} \in \Phi_{+} \subset \Phi$).

The *second* step on the way outside the HS is the *complex basis vector expansion*. It holds for "very well-behaved" vectors, i.e., for vectors of a subspace $\Phi_{-} \subset \Phi$ only. For every $\phi^{+} \in \Phi_{-}$ (and a similar expression holds also for every $\psi^{-} \in \Phi_{+}$), one obtains the following basis vector expansion for the case of an S-matrix with a finite number of resonance poles at the positions $z_{R_{i}} = E_{R_{i}} - i\Gamma_{i}/2$, $i = 1, 2, \ldots, N$:

$$\phi^{+} = \int_{0}^{-\infty_{II}} dE\,|E^{-}\rangle\langle {}^{-}E|\phi^{+}\rangle - \sum_{i=1}^{N} |z_{R_{i}}^{-}\rangle 2\pi\Gamma_{i}\langle {}^{+}z_{R_{i}}|\phi^{+}\rangle + \sum_{n} |E_{n}\rangle\langle E_{n}|\phi^{+}\rangle$$

(65)   for $\phi^{+} \in \Phi_{-}$.

Here $|z_{R_{i}}^{-}\rangle\sqrt{2\pi\Gamma_{i}} = \psi^{G_{i}} \in \Phi_{+}^{\times}$ are Gamow kets (55) representing decaying states (57a)[29]. The remarkable feature of the basis vector expansion (65) is

---

[28] cf. footnote (27).

[29] There is a corresponding basis vector expansion of $\psi^{-} \in \Phi_{+}$; in place of $|z_{R_{i}}^{-}\rangle$, it contains generalized eigenvectors $|z_{R_{i}}^{*}{}^{+}\rangle$ with eigenvalue $z_{R_{i}}^{*} = E_{R_{i}} + i\frac{\Gamma_{i}}{2}$. These are the Gamow vectors associated with the S-matrix pole at $z_{R_{i}}^{*}$. They have an exponentially growing semigroup evolution for $-\infty < t \leq 0$.

that the decaying states $|z_{R_i}^-\rangle$ appear on the same footing as the stationary states $|E_n\rangle$[30].

The last term in the expansions (60), (64) and (65) will be absent if there are no bound states; we shall omit this term in the following discussions. The first term in (65) is the background integral related with the background phase shifts. The integration, taken along the negative real axis in the second sheet (for which the values of $\langle^- E|\phi^+\rangle = S(E)\langle^+ E|\phi^+\rangle$ can be calculated from the experimental values $\langle E|\phi^{\text{in}}\rangle$ for positive energies using the van Winter theorem [54]), could be deformed into integration over many other equivalent contours in the lower half plane of the second sheet, if those would be more convenient to calculate.

For a $\phi^+ \in \Phi_- \subset \Phi$, both expansions (64) and (65), which use two different basis systems, hold. The expansion (65) separates the individual resonance poles, whereas (64) has the resonances contained together with the "background" in the energy wave function $\langle^- E|\phi^+\rangle$. The function $|\langle^- E|\phi^+\rangle|^2$ may have a bump at $E = E_R$, but $\langle^- E|\phi^+\rangle \in \mathcal{S}$ cannot be the Breit-Wigner

---

[30] The forms (60) and (65) of the generalized basis vector expansions assume that H is the only observable to be diagonalized (cyclic operator). If the complete system of commuting observables (c.s.c.o.) consists of $(N+1)$ operators $H, B_{(1)}, B_{(2)}, \ldots, B_{(N)} \equiv H, B$, then we have to make the following replacements for the projection operators

$$|E_n)(E_n| \to \sum_b |E_n, b)(E_n, b|,$$

where the sum extends over all values of the degeneracy quantum numbers $b = b_{(1)}, b_{(2)} \ldots b_{(N)}$ of the energy E_n. Similarly, in (60) to (65) we have to make the replacements:

$$(66)\, |E^+\rangle\langle^+ E| \;\to\; \sum_b |E, b^+\rangle\langle^+ E, b|$$

$$= \int d\mu(b_1, b_2, \ldots) \sum_{b_d, \ldots, b_N} |E, b_1 \ldots b_N{}^+\rangle\langle^+ E, b_1 \ldots b_N|$$

$$|z_R^-\rangle\langle^+ z_R| \;\to\; \sum_b |z_R, b^-\rangle\langle^+ z_R, b|$$

$$= \int d\mu(b_1, b_2, \ldots) \sum_{b_d, \ldots, b_N} |z_R, b_1 \ldots b_N{}^-\rangle\langle^+ z_R, b_1 \ldots b_N|$$

where $b_1, b_2, \ldots$ are the continuous and $b_d, \ldots, b_N$ are the discrete degeneracy quantum numbers. The operators B could be, e.g., the orbital angular momentum operators $J_3, \mathbf{J}^2$ if we have a spherically symmetric (spin-less) scattering system $[H, J_i] = 0$; then the quantum numbers $b = b_1, b_2$ are $b = j, j_3$. They could be the momentum operators P_i if $[H, P_1] = 0$; then $(E, b_1, b_2) = (p_1, p_2, p_3)$ or $(E, b_1, b_2) = (E, \theta_p, \varphi_p)$, where (θ_p, φ_p) are the spherical coordinates of $\mathbf{p}$ and $d\mu(b_1, b_2) = d\cos\theta_p d\varphi_p$. The labels b could also be some intrinsic quantum numbers, like charges or the channel label η.

distribution (56) characteristic of a resonance because the Breit-Wigner distribution is not a well-behaved function[31]. The complex energy basis vector expansion (65) is the much preferred representation for investigating resonances.

For the sake of definiteness we shall now assume that there are two decaying states $R_1 = S$ and $R_2 = L$ and no bound states. According to the expansion (65), the pure state (prepared by the experimental apparatus) has the following representation in terms of the Gamow vectors $\psi_L^G = -|z_L^-\rangle\sqrt{2\pi\Gamma_L}$, $\psi_S^G = -|z_S^-\rangle\sqrt{2\pi\Gamma_S}$ and the remaining part which we call $\phi_{\rm bg}^+$ (the background):

$$(67) \qquad \phi^+ = \psi_L^G b_L + \psi_S^G b_S + \int_0^{-\infty_{II}} dE\, |E^+\rangle\langle^+E|\phi^+\rangle.$$

In here b_L and b_S are some complex numbers that depend on the normalization of the Gamow vectors ψ_L^G, ψ_S^G (and of ϕ^+), and upon some phase convention. All the vectors in the generalized basis system expansion are (generalized) eigenvectors of the exact Hamiltonian, and, in particular, the Gamow vectors ψ_L^G, ψ_S^G are eigenvectors of the exact Hamiltonian with complex eigenvalue $(E_L - i\Gamma_L/2)$ and $(E_S - i\Gamma_S/2)$, respectively. If we ignore $\phi_{\rm bg}^+ = \int_0^{-\infty_{II}} dE\, |E^+\rangle\langle^+E|\phi^+\rangle$ then ϕ^+ in (67) is the superposition of two eigenvectors of H with complex eigenvalues z_L and z_S; the Hamiltonian matrix is complex and diagonalizable.

We now apply the time evolution operator to (67). Since the ψ_L^G, ψ_S^G are elements of $\Phi_+^\times$ we apply the operator $U_+^\times(t)$ of (45) and obtain:

$$(68) \qquad \phi^+(t) \equiv e^{-iH^\times t}\phi^+ = e^{-i(E_L - i\Gamma_L/2)t}\psi_L^G b_L$$
$$+ e^{-i(E_S - i\Gamma_S/2)t}\psi_S^G b_S + \phi_{\rm bg}^+(t);\ t \geq 0.$$

Since the time evolution semigroup (45) has the restriction $t \geq 0$, the same restriction must be used for (68). The time evolved background term is

$$(69) \qquad \phi_{\rm bg}^+(t) \equiv \int_0^{-\infty_{II}} dE\, e^{iEt}|E^-\rangle S(E)\langle^+E|\phi^+\rangle;\ t \geq 0.$$

These equations ((65), (67) through (69)) are understood as a functional equation over all $\psi^- \in \Phi_+$. This means that these expansions of $\phi^+(t) \in \Phi_- \subset \Phi_+^\times$ can be used to obtain $\langle\psi^-|\phi^+(t)\rangle$ (whose modulus square is the probability to find the time evolved state by a detector that detects the observable $|\psi^-\rangle\langle\psi^-|$) for any $\psi^- \in \Phi_+$, but *not* to calculate $\langle\psi|\phi\rangle$ for a

---

[31] Note that the Gamow ket has an "idealized" Breit-Wigner energy distribution $\langle^-E|\psi^G\rangle$ (56) which extends from $-\infty_{II}$ to $+\infty_{II} = \infty_I$. This is a regular function on $\mathbb{R}$, but ψ^G cannot be realized by a regular function on the physical energy values $\mathbb{R}_+ = \{E|\ 0 \leq E < \infty_I\}$ since ψ^G is a generalized vector and correspondingly $\langle^-E|\psi^G\rangle$ is a distribution over $\mathcal{S} \cap \mathcal{H}_+^2|_{\mathbb{R}^+}$, cf. also footnote 23.

$\psi \in \Phi_-$. This is not a problem because the expressions $|\langle \psi | \phi^+(t) \rangle|^2$, $\psi \in \Phi_-$ and $\phi^+ \in \Phi_-$, have no physical meaning, since they would represent the probability for finding an in-state ψ in an in-state $\phi^+(t)$ at $t \geq 0$ which is not measurable in a scattering experiment. The semigroup time evolution operator $U_+^\times(t)$ can also be applied to the basis vector expansion (64) with ϕ^+ understood as functional $\phi^+ \in \Phi_-^\times$ over the $\psi^- \in \Phi_+$ because (64) is a functional equation over Φ_+ since the $\langle {}^-E | \psi^- \rangle$ are according to (37) elements of $\mathcal{S} \cap \mathcal{H}_+^2|_{\mathbb{R}+}$. However the semigroup operator $U_+^\times(t)$ cannot be applied to the representation (60b) for $\phi^+ \in \Phi_- \subset \Phi_+^\times$, because it is *not* a functional equation over Φ_+ since the $\langle {}^+E | \psi^- \rangle$, with $\psi^- \in \Phi_+$, are not necessarily elements of $\mathcal{S} \cap \mathcal{H}_+^2|_{\mathbb{R}+}$.

Of the two exact but different basis vector expansions (60b) and (65) for the same $\phi^+ \in \Phi_-$, (60b) is the standard expansion and has a correspondence in the Hilbert space (spectral resolution of operators with a continuous spectrum). The expansion (65) is new and shows that the quasi-stationary states $|z_{R_i}^-\rangle$ can serve as basis vectors in very much the same manner as the stationary states $|E_n\rangle$ in the standard case. But in addition to the resonance states the new basis vector expansion (65) (for any $N \neq 0$, e.g. $N = 1$ or $N = 2$ as in (68)) also contains an integral over the negative real axis from $E = -\infty_{II}$ to 0 in the second sheet of the energy surface of the S-matrix. This integral depends on the preparation of the state and may be infinitesimally small, but cannot be zero. Its time dependence is non-exponential and could cause observable deviations from the exponential law for the transition rate of the prepared state ϕ^+. It may also have some other small but observable consequences.

The result (68) means that the semigroup time evolution of a superposition of two (or more) Gamow states does not regenerate one Gamow state from the background $\phi_{bg}^+(t)$ or from the other Gamow vector. In particular, if the state ϕ^+ can be prepared such that at some time $t_0 \geq 0$ the background term $\phi_{bg}^+(t)$ is practically zero, then it will remain practically zero for all $t > t_0$, and the two Gamow states will evolve separately according to separate exponential laws without regenerating each other:

$$(70) \qquad \phi^+(t) \approx e^{-i(E_L - i\Gamma_L/2)t}\psi_L^G b_L + e^{-i(E_S - i\Gamma_S/2)t}\psi_S^G b_S; \quad t \geq 0.$$

Approximations like (70) have been used for the time evolution of one- and two-resonance systems (like the K_L-K_S system with $\phi^+(t)$ representing the K^0 state) in theories with "effective Hamiltonians" given by a 2×2 complex diagonalizable matrix.

The full expansion (65) (again neglecting bound states) leads to a matrix representation of the self-adjoint semi-bounded Hamiltonian H in the following form:

$$
\begin{pmatrix} \langle H\psi^-|z^-_{R_1}\rangle \\ \langle H\psi^-|z^-_{R_2}\rangle \\ \vdots \\ \langle H\psi^-|z^-_{R_N}\rangle \\ \langle H\psi^-|E^-\rangle \end{pmatrix} = \begin{pmatrix} \langle \psi^-|H^\times|z^-_{R_1}\rangle \\ \langle \psi^-|H^\times|z^-_{R_2}\rangle \\ \vdots \\ \langle \psi^-|H^\times|z^-_{R_N}\rangle \\ \langle \psi^-|H^\times|E^-\rangle \end{pmatrix}
$$

$$
= \begin{pmatrix} z_{R_1} & & & & 0 \\ & z_{R_2} & & & \vdots \\ & & \ddots & & \vdots \\ & & & z_{R_N} & 0 \\ 0 & \cdots & \cdots & 0 & (E) \end{pmatrix} \begin{pmatrix} \langle \psi^-|z^-_{R_1}\rangle \\ \langle \psi^-|z^-_{R_2}\rangle \\ \vdots \\ \langle \psi^-|z^-_{R_N}\rangle \\ \langle \psi^-|E^-\rangle \end{pmatrix}
$$

(71) $\psi^- \in \Phi_+ \subset \Phi, \quad -\infty_{II} < E \le 0,$

where the lowest row represents the diagonal continuously infinite real energy matrix:

(72)
$$
(\langle H\psi^-|E^-\rangle) = (\langle \psi^-|H^\times|E^-\rangle) = (E)(\langle \psi^-|E^-\rangle); \quad \psi^- \in \Phi, \quad -\infty_{II} < E \le 0.
$$

Since the basis vector expansion (65) is an exact representation of $\phi^+ \in \Phi_-$, the matrix representation (71) also is an exact representation of the self-adjoint Hamiltonian. In the phenomenological descriptions by complex effective Hamiltonians, one uses a truncation of (71) which corresponds to omitting the background integral, i.e., omitting the whole continuously infinite diagonal matrix (E) (and sómetimes even some of the z_{R_i}).

The extra term ϕ^+_{bg} in (67) and (68) is not taken into consideration in any of the finite dimensional effective theories of complex Hamiltonians, and in particular not in the Lee-Oehme-Yang theory of the neutral Kaon system. This term, which comes from the integral along the negative real axis in the second sheet of the S-matrix, can be shown to be also decaying, i.e., $\left|\langle \psi^-|\phi^+_{bg}(t)\rangle\right|^2 \to 0$ for $t \to \infty$ for every $\psi^- \in \Phi_+$, but it decays more slowly than the exponential. The standard effective theories, like the enormously successful Lee-Oehme-Yang theory of the neutral K-system, emerge as subtheories of the exact complex basis vector expansion (65) or (67) in the N-dimensional space $\mathcal{M}_N = \left\{\phi|\phi = \sum_{i=1}^N |z^-_{R_i}\rangle c_i\right\}$ spanned by the Gamow vectors $|z^-_{R_i}\rangle$. However $\mathcal{M}_N$ is a subspace of $\Phi^\times_+$ and lies *outside* the standard HS. These effective theories are usually legitimized by the Wigner-Weisskopf approximation. In our irreversible quantum theory the expression (68) is exact and can be used to justify the effective theory (70) as a subtheory in $\mathcal{M}_N \subset \Phi^\times_+$ which remains invariant under time translations in the forward direction, (68). The expression (68) shows that the "deviation from the exponential decay law" does not arise for the resonance state but are the properties of the background terms (background phase shifts).

The emergence of such an enormously successful phenomenological description as the Lee-Oehme-Yang theory as a subtheory is an empirical validation of the complex basis vector expansion (65).

7.4 The Golden Rule from Fundamental Probabilities

"[T]he physical system leaves the [initial decaying] state $|\phi_i\rangle$ irreversibly."
—Cohen-Tannoudji, et al.[32]

The probabilities $\mathcal{P}(t) = \text{Tr}(\Lambda W)$ are the most fundamental quantities in quantum physics. They describe the probability to register or measure the observable Λ in the state W and represent the directly measured experimental quantities; $\mathcal{P}(t)$ is measured by the number of counts of a registration apparatus (detector) and its derivative, $\dot{\mathcal{P}} = \frac{d}{dt}\mathcal{P}(t)$ is observed as the normalized counting rate of the detector. We want to apply these probabilities to the case where Λ is a projection operator (or more generally a positive definite operator) on a subspace of non-interacting decay products registered by a detector and W is a quasistable state $W^D(t)$. In this case $\mathcal{P}(t)$ is the decay probability and $\dot{\mathcal{P}}(t)$ is the decay rate for the decaying state W^D into the decay products Λ registered by the detector. Thus Λ is experimentally defined by the detector and $W^D(0)$ by the preparation apparatus or preparation process of the quasistable state (e.g. a resonance produced in a scattering experiment or a metastable product of some ancient creation process which is present at some initial time $t = 0$).

Traditionally, W is an ensemble defined by a macroscopic preparation apparatus (single microsystems are not defined) and Λ and W are mathematically represented by operators in the Hilbert space. However, as we mentioned in Section 3, the decay probabilities within the HS formulation have severe problems, justifying a fresh approach. Our approach is to extend the basic formula $\mathcal{P}(t) = \text{Tr}(\Lambda W)$ to apply to the decaying state W^D mathematically represented by the Gamow vector of Section 7.2, $W^D = |\psi^G\rangle\langle\psi^G|$. We give the following interpretation to this basic formula applied to W^D: a single microphysical decaying system described by W^D has been produced by a macroscopic apparatus and a quantum scattering process, at a time $t = 0$. Each count of the detector in a decay experiment is the result of the decay of this single microsystem that has lived for a time t_a – the time that it took the decaying system to travel from the scattering center to the decay vertex. The integral

$$\int_{t_a - \frac{1}{2}\Delta t}^{t_a + \frac{1}{2}\Delta t} dt\, \dot{\mathcal{P}}(t) \approx \dot{\mathcal{P}}(t_a)\Delta t,$$

is proportional to the number of microsystems that have decayed in the time interval Δt, i.e. $\dot{\mathcal{P}}(t)$ is the normalized counting rate of the detector, $\dot{N}(t)/N(\infty)$. The lifetime τ_D (mean life) is the average of the t_a's:

---

[32] p. 1345 of [13].

$$\sum_a t_a \, \Delta t \, \dot{P}(t_a) \approx \int dt \, t \, \dot{P}(t) = \tau_D.$$

This is the way $\dot{P}(t)$, τ_D, etc. are obtained (defined) experimentally. The quantity $P(t)$ in the basic formula is thus experimentally the normalized number of counts $N(t)/N(\infty)$ (with N very large). Therefore the normalization condition of the W^D is such that

(73) $$P(\infty) = 1;$$

the probability of finding the state decayed is certainty. In addition $P(t)$ is subject to an initial boundary condition

(74) $$P(0) = 0;$$

the probability of finding one of the many (N) decay products already at $t = 0$ is zero. The reason for this is that $t = 0$ is the time at which the decaying state has been prepared and one starts counting the decay products (one assumes there are no other incoming decay products in the detector area).

In the mathematical theory of quantum mechanics, the quantity $P(t)$ is calculated from the operators Λ and W^D by the basic formula (39a) given by

(75)
$$P(t) = \mathrm{Tr}(\Lambda W^D(t)) = \mathrm{Tr}(\Lambda(t) W^D) = \langle \psi^D(t) | \Lambda | \psi^D(t) \rangle, \text{ for } t \geq 0 \text{ only.}$$

The formula for the decay rate $\dot{P}(t)$ should be obtained as the derivative of the right hand side of (75).

However, in the standard textbooks of quantum mechanics one does not calculate the transition probabilities (75) [33]. Instead one gives a Golden Rule [55] for the initial decay rate $\dot{P}(t = 0)$, and one justifies this Golden Rule with some ingenious heuristic arguments that were originated by Dirac [53]. In place of an initial (t=0) decaying state $W^D = |\psi^D\rangle\langle\psi^D|$ which evolves by the exact Hamiltonian $\psi^D(t) = e^{-iHt}\psi^D$, one usually chooses an eigenstate of the interaction-free Hamiltonian, $\psi^D \to f^D$, where

(76) $$H_0 f^D = E_D f^D \text{ with } H_0 = H - V.$$

In place of the projection operators for decay products, one uses the "improper" states, $\Lambda \to |E_b, b\rangle\langle E_b, b|$ which have dimension of (energy)$^{-1}$. The Golden Rule is then given as the decay rate energy-density

(77a) $$\dot{P}^{E_b}_{b\,D}(0) = \frac{2\pi}{\hbar} |\langle b, E_b | V | f^D \rangle|^2 \delta(E_D - E_b),$$

and the initial decay rate is obtained by integration and/or summation over all final quantum numbers E_b and b, as:

---

[33] One cannot calculate it in the HS formulation because of (25), [33].

(77b) $$\dot{\mathcal{P}}_D(0) = \frac{2\pi}{\hbar} \int dE_b \sum_b \left| \langle b, E_b | V | f^D \rangle \right|^2 \delta(E_b - E_D).$$

Alternatively, one chooses for the initial state also an improper "energy eigenstate" of the free Hamiltonian, $H_0|E_D, d\rangle = E_D|E_D, d\rangle$, or of the exact Hamiltonian, $H|E_D, d^+\rangle = E_D|E_D, d^+\rangle$. Then, for example, one can write the exact but highly singular Golden Rule as

(77c) $$\dot{\mathcal{P}}_b^{E_b E_D}_D(0) = \frac{2\pi}{\hbar}|T_{bD}|^2\delta(E_D - E_b), \text{ where}$$

$$T_{bD} = \langle b, E_b|V|E_D, d^+\rangle \approx \langle b, E_b|V|E_D, d\rangle.$$

The initial transition probability rate $\dot{\mathcal{P}}(0)$ is then obtained by integration/summation over the final quantum numbers E_b and b and averaging over the initial quantum numbers E_D and d.

The expression (77b) (and its variants) is one of the most important and most widely used formulas in quantum physics. It expresses the decay probability per unit time of the state $|f^D\rangle\langle f^D|$ at $t = 0$ (the time when the decaying state has been created and the registration of the decay products begins) into the non-interacting decay products described by the projection operator given by Λ:

(78) $$\Lambda = \sum_{\substack{\text{all } b \\ b \neq b^D}} \int_0^\infty dE\, |E, b\rangle\langle E, b|, \quad \text{where } H_0|E, b\rangle = E|E, b\rangle$$

The initial decay rate $\dot{\mathcal{P}}(t = 0) = \frac{d}{dt}\mathcal{P}(t)|_{t \to 0+}$ of (77b) should be the time derivative of the probability (75), at least approximately, or it should be the time derivative of a probability $\mathcal{P}(t) = \text{Tr}(\Lambda(t)|f^D\rangle\langle f^D|)$. This suggests the following program: Find a decaying state $W^D(t) = |\psi^D(t)\rangle\langle\psi^D(t)|$ that evolves in time according to the exact Hamiltonian $H = H_0 + V \geq 0$,

(79) $$W^D(t) = e^{-iHt}W^D(0)e^{iHt},$$

calculate the decay probability $\mathcal{P}(t)$ using the basic formula (75), and then take the time derivative of the probability, $\dot{\mathcal{P}}(t)$. This decay rate at $t = 0$, $\dot{\mathcal{P}}(0)$, should somehow resemble Dirac's Golden Rule (77b), at least in some approximation $\psi^D \to f^D$.

From the results discussed in Sect. 4., in particular from the result that $\mathcal{P}(t) \equiv 0$ for every $\psi^D \in \mathcal{H}$ [33], it is clear that such a program cannot be implemented by a mathematical theory in the Hilbert space. The futility of such attempts was the actual reason for the introduction of the Gamow vectors [7]. We shall now describe how the Gamow vectors and irreversible quantum mechanics in the RHS lead to a decay rate formula that reduces to Dirac's Golden Rule in the Born approximation.

For the detector Λ, one takes a positive operator

(80)
$$\Lambda = \sum_{\substack{\text{all } b \\ b \neq b^D}} \int_0^\infty dE\, \lambda_b(E) |E, b\rangle \langle E, b|,$$

where $H_0|E, b\rangle = E|E, b\rangle$ and where $\lambda_b(E)$ is a smooth and rapidly decreasing function of E (and of all the other (continuous) quantum numbers in b) that describes the detector efficiency. The summation in b extends over those quantum numbers (and momentum directions) of the decay products which are registered by the detector but the summation does not include the quantum numbers b^D of the decaying state.

For the decaying state $W^D(t) = |\psi^D(t)\rangle\langle\psi^D(t)|$ one takes the pure Gamow state ψ^G of Sect. 7.2:

(81)
$$\psi^D(t) = \frac{1}{f} \psi^G(t) \in \Phi^\times_+,$$

where the "normalization" factor f is chosen such that $\mathcal{P}(\infty) = 1$ when Λ describes all decay products This state has the time evolution (57a), which in operator form is:

(82a)
$$W^G(t) = |\psi^G(t)\rangle\langle\psi^G(t)|$$
$$= e^{-iH^\times t}|\psi^G\rangle\langle\psi^G|e^{iHt} = e^{-\Gamma t}|\psi^G\rangle\langle\psi^G|, \text{ for } t \geq 0 \text{ only.}$$

This is inserted for $W^D(t)$ in (75) where also $t \geq 0$. The equality (82a) is understood as a functional equation in the space Φ_+, as in (57a) and (57b), i.e. as

(82b) $\langle\psi_1^-|W^G(t)|\psi_2^-\rangle = e^{-\Gamma t}\langle\psi_1^-|\psi^G\rangle\langle\psi^G|\psi_2^-\rangle$ for $\psi_1^-, \psi_2^- \in \Phi_+.$

This means (82a) is only valid in matrix elements with the vectors $\psi^- \in \Phi_+$, which represent the out-states that are ultimately registered by the detectors. This presents no limitations, since only those matrix elements are experimentally accessible[34].

The decay probability (75), with (80) and (81), can now be derived in a lengthy calculation [7] using (82). This derivation makes use of the Lippmann-Schwinger equation in one of its standard (singular) forms:

(83)
$$|E, b^-\rangle = |E, b\rangle + \frac{1}{E - H - i\epsilon} V|E, b\rangle$$

where $H^\times|E, b^-\rangle = E|E, b^-\rangle$ and $H^\times_0|E, b\rangle = E|E, b\rangle$. The derivation also makes use of the relation

(84)
$$\langle {}^- E|f(H^\times)|\psi^G\rangle = f(z_R)\langle {}^- E|\psi^G\rangle,$$

---

[34] The matrix element $\langle\phi^+|W^G|\phi^+\rangle$ does not make any sense mathematically or physically.

which follows from (55) and can be proven as a functional equation over the $\langle^-\psi|E^-\rangle \in (\mathcal{S}\cap\mathcal{H}_-^2)_{\mathbb{R}^+}$. From (83) and (84) one obtains for sufficiently good interaction Hamiltonians V (such that $\langle E|V|\psi^G\rangle \in (\mathcal{S}\cap\mathcal{H}_-^2)^\times$):

$$(85) \qquad \langle\psi^G|E,b^-\rangle = \langle\psi^G|E,b\rangle + \langle\psi^G|V|E,b\rangle\frac{1}{E-(E_R+i\Gamma/2)-i\epsilon},$$

This is again understood as a functional equation over the $\langle^-\psi|E^-\rangle \in (\mathcal{S}\cap\mathcal{H}_-^2)_{\mathbb{R}^+}$. With these equations, and under the assumption that the mathematically singular expressions above can be rigorously justified[35], one calculates the following result:

$$(86) \qquad \mathcal{P}(t) = 1 - e^{-\Gamma t}\int_0^\infty dE \sum_{b\neq b^D} \lambda_b(E)|\langle E,b|V|\psi^G\rangle|^2$$

$$\times \frac{1}{(E-E_R)^2+(\Gamma/2)^2}; \; t\geq 0.$$

This is the probability for the transition of the decaying state W^D into all mixtures of decay products with the property Λ. Usually one omits the detector efficiency $\lambda_b(E)$ in formulas like (86) above and (87)...(89) and (98) below. This means one gives these formulas for an ideal detector with $\lambda_b(E) = 1$ for all b which are quantum numbers of decay products and with $\lambda_b(E) = 0$ otherwise. The detector efficiency $\lambda_b(E)$ is then used in the analysis of the experimental data to correct the observed events for detector efficiency. The reported counting rate has usually been corrected for this detection efficiency. The factor $\lambda_b(E)$ is then omitted in the theoretical formulas like (86), but this factor is always present and can be used in the mathematical calculations to make expressions like (85) less singular.

Taking the time derivative of (86) (setting $\lambda_b(E) = 1$), we obtain for the decay rate

$$(87)$$

$$\dot{\mathcal{P}}(t) = e^{-\Gamma t}2\pi\int_0^\infty dE \sum_{b\neq b^D} |\langle E,b|V|\psi^G\rangle|^2 \frac{\Gamma/2\pi}{(E-E_R)^2+(\Gamma/2)^2}; \; t\geq 0.$$

---

[35] In contrast to the statements 1. and 2. (eq.25) in Sect. 4.), whose proofs use only well defined mathematics of the HS, the statement (86) cannot be formulated as a mathematical theorem because its derivation requires such singular expressions as the Lippmann-Schwinger equation. While this equation is exact and well-accepted by the physics community, it has as yet not been given a mathematically rigorous foundation in either the HS or RHS. In this respect (86) and (87) are also different from the statements in the preceeding sections, like the semigroup time evolutions (45) and (46) in $\Phi_\mp \subset \mathcal{H} \subset \Phi^\times_\mp$, the exponential law (57) and (82), the Nuclear Spectral Theorem (60) and the "complex spectral theorem" (65) and (71), which are well-founded in the mathematics of the RHS. On the other hand, (87) reproduces in a reasonable approximation the well-proven Golden Rule, providing heuristic support.

Since $\lambda_b(E)$ has been omitted the decay rate in (87) is to be compared to the experimental counting rate corrected for detection efficiency. The formula (87), and also (86), we call the exact Golden Rule.

The initial decay rate is then obtained as

$$(88) \qquad \dot{\mathcal{P}}(0) = 2\pi \int_0^\infty dE \sum_{b \neq b^D} |\langle E, b|V|\psi^G\rangle|^2 \frac{\Gamma/2\pi}{(E - E_R)^2 + (\Gamma/2)^2}.$$

Since the probability to find the decay product at time $t \leq 0$ needs to be zero, $\mathcal{P}(0) = 0$, we obtain from (86) for $t = 0$:

$$(89) \qquad \frac{2\pi}{\Gamma} \int_0^\infty dE \sum_{b \neq b^D} |\langle E, b|V|\psi^G\rangle|^2 \frac{\Gamma/2\pi}{(E - E_R)^2 + (\Gamma/2)^2} = 1.$$

The sums in (86)...(89) extend over all quantum numbers and all decay products. Comparing (89) with (88), we obtain

$$(90) \qquad \dot{\mathcal{P}}(0) = \Gamma \quad (= \Gamma/\hbar),$$

and from the exponential time dependence in (86) or (87) we obtain the lifetime of the decaying state:

$$(91) \qquad \tau_G = \frac{1}{\Gamma} \quad (= \hbar/\Gamma),$$

The result (90) means that the imaginary part of the complex energy in (55), which is also the imaginary part of the S-matrix pole position and the width of the Breit-Wigner energy distribution (56) is equal to the initial rate $\dot{\mathcal{P}}(0)$ of the decay probability (75).

The formulas (86) and the formulas (87) and (88) give the *total* decay probability and the *total* decay rate, respectively, because we took for Λ the projection operator on *all* decay products and summed over *all* values of quantum numbers $b = \{b_1, b_2, \ldots, \eta\}$ of *all* the decay products. The partial decay rates are obtained if we sum in (87) and (88) only over part of the quantum numbers $b = \{b_1, b_2, \ldots, \eta\}$, e.g. over all $\hat{b} = \{b_1, b_2, \ldots\}$ but not over the quantum number η that characterizes the decay channels:

$$(92) \qquad \dot{\mathcal{P}}_\eta(t) = 2\pi e^{-\Gamma t} \int_0^\infty dE \sum_{b_1, b_2, \ldots} |\langle E, b_1, b_2, \ldots, \eta|V|\psi^G\rangle|^2$$

$$\times \frac{\Gamma/2\pi}{(E - E_R)^2 + (\Gamma/2)^2}.$$

The value $\dot{\mathcal{P}}_\eta$ is the partial decay rate into decay products with the quantum number η or into the decay channel characterized by η. The total decay rate ((87) or (88)) is then written as the sum over partial decay rates, with a corresponding relation for the "partial widths" $\Gamma_\eta \equiv \hbar \dot{\mathcal{P}}_\eta(0)$:

(93) $$\dot{\mathcal{P}}(t) = \sum_{\text{all } \eta} \dot{\mathcal{P}}_\eta(t), \quad \Gamma = \sum_{\text{all } \eta} \Gamma_\eta.$$

In analogy to (93), with (92) one can write for the decay probability (86) (setting $\lambda_b(E) = 1$)

(94) $$\mathcal{P}(t) = 1 - \sum_\eta \frac{2\pi}{\Gamma} e^{-\Gamma t} \int_0^\infty dE \sum_{b_1,\dots \neq b^D} |\langle E, b_1, \dots \eta | V | \psi^G \rangle|^2$$

$$\times \frac{\Gamma/2\pi}{(E - E_R)^2 + (\Gamma/2)^2}, \quad t \geq 0.$$

The derivation of the above formulas (86), (87), (88) and (92) did not make use of any approximations. We therefore call these formulas exact Golden Rules. Equation (86) is the exponential decay law with the directionality of time [56]

(95) $$\mathcal{P}(t) = 1 - e^{-\Gamma t}, \quad t \geq 0.$$

From the quantum mechanical arrow of time (Section 4), via the semigroup (46) and the evolution of the Gamow states (57a), we derived the time-asymmetry for the decay process: *If we start at $t = 0$ from a "decaying state" $|\psi^G\rangle\langle\psi^G|$, then the probability of finding the decay products Λ increases exponentially with time from $\mathcal{P}(0) = 0$ to $\mathcal{P}(t \to \infty) = 1$.*

In addition to these results concerning principles, we can also obtain a rule for the calculation of the initial decay rate or width in quantum theory. The exact Golden Rules are not of much practical use in calculations since these formula contain the unknown ψ^G and the natural line width under the integral. The state vector ψ^G is unknown and the values (E_R, Γ) are unknown until the eigenvalue equation

(96) $$H^\times \psi^G = \left(E_R - i\frac{\Gamma}{2}\right)\psi^G; \quad \psi^G \in \Phi^\times_+$$

has been solved. Therefore, to obtain a calculational tool from the above relations (87), (88) and (92), one uses the Born approximation on the right hand side: one expands the exact ψ^G in a perturbation series (for the interaction Hamiltonian V) in terms of the non-interacting "decaying state vector" f^D defined by (76), since the solutions of (76) are usually known.

The Born approximation is given by

(97a) $$\langle b, E | V | \psi^D \rangle \approx \langle b, E | V | f^D \rangle$$

(97b) $$\frac{\Gamma}{2E_R} \to 0$$

(97c) $$E_R \approx E_D$$

The Breit-Wigner energy distribution (natural line width) has the property:

(97d)
$$\lim_{\frac{\Gamma}{2E_R} \to 0} \frac{\Gamma/2\pi}{(E - E_R)^2 + (\Gamma/2)^2} = \delta(E - E_R).$$

Using (97), one obtains from (88) the initial decay rate in this Born approximation:

(98)
$$\dot{\mathcal{P}}(0) = \frac{2\pi}{\hbar} \int dE \sum_b |\langle b, E|V|f^D\rangle|^2 \delta(E - E_D),$$

and similar expressions for the other formulas. This is the standard Golden Rule (77b) again, only here it has been *derived* from the transition probability (75) and from the time evolution of the Gamow vectors (82) as the Born approximation of an exact Golden Rule.

The Gamow vector ψ^G provides the link that was missing from the HS formulation to connect the all-important empirical rules for the rates to the fundamental theoretical relations for the probabilities.

8. Summary

Dirac's bra and ket formalism can be given a mathematical justification by the rigged Hilbert space (RHS). The RHS does not only contain Dirac kets ("scattering states") but also Gamow vectors, which are eigenkets $|E_R - i\Gamma/2\rangle$ of self-adjoint Hamiltonians H with complex eigenvalue $(E_R - i\Gamma/2)$ representing decaying states or resonances with Breit-Wigner energy distribution of energy E_R and width Γ. The value Γ is also the initial rate of the decay probability $\mathcal{P}(t)$, $\Gamma = \hbar \frac{d\mathcal{P}}{dt}|_{t=0}$, for which an exact Golden Rule could be derived. In the Born approximation, this exact Golden Rule becomes Dirac's Golden Rule. These features are central requirements for a theory of resonances and decay, yet the standard quantum theory in Hilbert space (HS) can not produce them since the transition probabilities from a decaying state $W^D(t)$ into the observed decay products Λ, $\mathcal{P}(t) = \text{Tr}(\Lambda W^D(t))$, can be proven to be identically zero if they are zero at any time interval, e.g. before the decaying state has been prepared. Therefore, resonances in the HS formulation of quantum mechanics cannot be described by a state vector $\psi^D \in \mathcal{H}$ and had to be denied the status of autonomous microphysical entities in HS. This is contrary to the phenomenological observation that stability and the value of the lifetime are not criteria for elementarity and that quasistable states and stable states should be described on the same footing – as it is done in S-matrix theory.

The semigroup time evolution of the Gamow states sprung from the new mathematical language of the RHS and led to their exact exponential decay with lifetime $\tau = \hbar/\Gamma$

$$e^{-iH^\times t}|E_R - i\Gamma/2\rangle = e^{-iE_R t} e^{-(\frac{\Gamma}{2})t}|E_R - i\Gamma/2\rangle, \text{ for } t \geq 0 \text{ only.}$$

The semigroup $e^{-iH^{\times}t}$, $t \geq 0$, expresses intrinsic irreversibility on the microphysical level. This irreversibility is not restricted to the time evolution of Gamow states. It can be formulated in terms of time-asymmetric boundary conditions for states and observables and is related to a general arrow of time which can be expressed by the truism: A state needs to be prepared first before an observable can be registered in it. This can also be expressed as a condition of time ordering $t_a > t_0$ in the probabilities $\mathcal{P}(t) = \mathrm{Tr}(\Lambda(t_a)W(t_0))$, where $t = t_0$ is the time at which state W has been prepared and the registration of the observable Λ can begin (or the initial time of the universe $W(t_0) = \rho_i$ if the quantum mechanics of measured systems is extended to the quantum theory of cosmology). This intrinsic irreversibility seems to have a different origin than the standard extrinsic irreversibility of open quantum systems which results from the effect of an external reservoir or measurement apparatus. The latter has also a semigroup evolution which however is generated by a reservoir-dependent Liouvillian and not by the Hamiltonian of the quantum system. This intrinsic irreversibility also does not follow from the collapse axiom of a pure or less mixed state into a more mixed state. However, the change of state due to measurement scattering of a microsystem on a macrosystem possesses, like every scattering process, an arrow of time which has the same origin as the semigroup.

Since our time-asymmetry comes from the boundary conditions and the algebras of observables and the dynamical laws are time-symmetric, microphysical irreversibility is compatible with a time reversal transformation (or CPT) if one chooses for T a representation that doubles the space of states and observables. Such representations of T exist, but the interpretation of a time-reversal doubled world is not clear.

The Gamow kets describing Breit-Wigner resonances have been derived from first order poles of the analytically continued S-matrix. This can be generalized to quasi-stationary systems associated with higher order poles of the S-matrix. An S-matrix pole of r-th order at $z_R = E_R - i\Gamma/2$ leads to r generalized eigenvectors of order $k = 0, 1, \ldots, r - 1$. The Gamow vector of order k, $|z_R\rangle^{(k)}$, is the k-th derivative of the ordinary Gamow vector and, except for a normalization factor, also a Jordan vector of degree $(k+1)$ with generalized eigenvalue $z_R = E_R - i\Gamma/2$.

The Gamow vectors appear in the complex basis vector expansion, which is, for state vectors, an alternative to the Dirac basis vector expansion and contains Gamow kets instead of Dirac kets as basis vectors. The complex basis vector expansion is particularly useful for problems that involve resonances and decay. For a system with a finite number of resonance states (such as the neutral Kaon system), it leads to the "effective" theories with complex Hamiltonian matrices (e.g. Lee-Oehme-Yang theory of the K_L-K_S system) by truncating the complex basis vector expansion to the subspace spanned by the Gamow vectors, the resonance subspace. In addition to the Gamow kets the complex basis vector expansion of a prepared state vector also contains other terms called background. This background term varies with the preparation process of the state and its properties are not connected with the decaying Gamow state (or states, if more than one resonance is involved). In particular, the time evolution of this preparation-dependent background term is non-exponential.

All that has been achieved by the RHS is a mathematical theory that allows to separate out the exponentially decaying (and also the exponentially growing which have not been discussed here) states and to derive their Golden Rule from the fundamental probabilities of quantum mechanics. In the process of doing this, the time-asymmetry of quantum physics became apparent.

Acknowledgments: We would like to express our gratitude to the many friends, colleagues and teachers who helped us shape this paper into its present form. J.-P. Antoine, M. Gadella and O. Melsheimer read various stages of the manuscript, corrected mistakes and suggested changes. I. Antoniou, Y. Ne'eman and S. Wickramasekara read different parts and made valuable suggestions. C. Püntmann and M. Loewe contributed to problems addressed by this article. Discussions with J. B. Hartle, G. Ludwig and J. A. Wheeler about the arrow of time led to the writing (and rewriting) of the preface. H. Atmanspacher's historical expertise led to the inclusion of the literature quotations. Various topics addressed in this paper were discussed with the participants of the workshop on Recent Problems of Quantum Mechanics and Cosmology of the Fondation Peyresq Foyer d'Humanisme, Peyresq, St. André-les-Alpes, France. Their hospitality, and the financial support of NATO, is gratefully acknowledged. We are appreciative of the Welch Foundation for their financial support.

References

1. Physical Origins of Time Asymmetry, J.J. Halliwell, et al., eds., (Cambridge University Press, 1994).
2. P.C.W. Davis, The Physics of Time Asymmetry, (University of California Press, 1977); R. Penrose, Singularities and Time-asymmetry, in General Relativity: Einstein Centenary Survey, S.W. Hawking, et al., eds., (Cambridge University Press, 1979); H.D. Zeh, The Physical Basis of the Direction of Time, (Springer-Verlag, 1989).
3. R. Peierls, Surprises in Theoretical Physics, Sect. 3.8, (Princeton University Press, Princeton, 1979).
4. R. Ritz, Physikalische Zeitschrift 9 (1908) 903; A. Einstein, Physikalische Zeitschrift 10 (1909) 185; R. Ritz, Physikalische Zeitschrift 10 (1909) 224; R. Ritz and A. Einstein, Physikalische Zeitschrift 10 (1909) 323.
5. J.A. Wheeler and R.P. Feynman, Rev. Mod. Phys. 21 (1949) 425.
6. G. Süssmann, in Nonlinear, Deformed and Irreversible Quantum Systems, H.D. Doebner, et al., eds., (World Scientific Publ., 1995), p. 98. See also F. Hoyle and J.V. Narlikar, Proc. Roy. Soc. A 277 (1964) 1, and J.E. Hogarth, Proc. Roy.

Soc. A 267 (1962) 365, who discuss these questions within the framework of ref. [5].

7. G. Ludwig, Foundations of Quantum Mechanics, Volume I, (Springer-Verlag, Berlin, 1983) and Volume II, (1985); An Axiomatic Basis of Quantum Mechanics, Volume I, (Springer-Verlag, Berlin, 1983) and Volume II, (1987); K. Kraus, State, Effects and Operations, Springer Lecture Notes in Physics 190, (Springer-Verlag, Berlin, 1983).

8. M. Gell-Mann and J.B. Hartle, in Complexity, Entropy and the Physics of Information, SFI Studies in Science and Complexity Vol. VIII, W. Zurek, ed., (1990); M. Gell-Mann and J.B. Hartle in ref. [1], p. 311; R.B. Griffiths, J. Stat. Phys., 36 (1984) 219; R.B. Griffiths in Symposium on the Foundation of Modern Physics 1994, K.V. Laurikainen, et al., eds., (Editions Frontières, 1984), p. 85.

9. I. Prigogine, Non-Equilibrium Statistical Mechanics (Wiley, New York, 1962); E.B. Davis, Quantum Theory of Open Systems, (Academic Press, London, 1976) where detailed references to the original papers can be found.

10. K. Kraus, Ann. Phys. 64 (1971) 311; A. Kossakowski, Rep. Math. Phys. 3 (1972) 247; G. Lindblad, Commun. Math. Phys. 40 (1975) 147; 48 (1976) 119; V. Gorini, A. Kossakowski and E.C.G. Sudarshan, J. Math. Phys. 17 (1976) 821, A. Kossakowski, On Dynamical Semigroups and Open Systems, this volume and references thereof.

11. A special case of this irreversible time evolution is obtained if one chooses for the reservoir R the measuring apparatus. It has been shown that the collapse axiom (2) together with the Schrödinger equation (1c) leads to semigroup evolutions (10) generated by a Liouvillian L; G.C. Ghirardi, A. Rimini and T. Weber, Phys. Rev. D 34 (1986) 470; I. Antoniou and S. Tasaki, Int. J. Quant. Chem. 46 (1993) 425.

12. P. Huet and M.E. Peskin, Nucl. Phys. B 434 (1995) 3; J. Ellis, J.S. Hagelin, D.V. Nanopoulos and M. Srendicki, Nucl. Phys. B 241 (1984) 381; J. Ellis, N.E. Mavromatos and D.V. Nanopolous, Phys. Lett. B 293 (1992) 142; CERN-TH-6755/92 (1992); Fabio Benatti, Complete Positivity and Neutral Kaon Decay, this volume.

13. C. Cohen-Tannoudji, B. Diu and F. Laloë, Quantum Mechanics: Volume II (Wiley, New York, 1977), p. 1345.

14. A. Bohm, S. Maxson, Mark Loewe and M. Gadella, Physica A 236 (1997) 485.

15. P.A.M. Dirac, The Principles of Quantum Mechanics, (Clarendon Press, Oxford, 1930).

16. N. Bourbaki, Élements de Mathématique, (Hermann, Paris, 1953).

17. L. Schwartz, Théorie des Distributions, (Hermann, Paris, 1950).

18. I.M. Gel'fand and N. Ya. Vilenkin, Generalized Functions, Volume 4, (Moscow, 1961)(English trans. Academic Press, New York, 1964); K. Maurin, Generalized Eigenfunction Expansions and Unitary Representations of Topological Groups. (Polish Scientific Publishers, Warzawa, 1968).

19. H. Weyl, Gruppentheorie und Quantenmechanik, (S. Hirzel, Leipzig, 1928).

20. J. von Neumann, Mathematische Grundlagen der Quantentheorie, (Springer, Berlin, 1931) (English translation by R.T. Beyer) (Princeton University Press, Princeton, 1955).

21. K. Maurin, Analysis, Part II, (Polish Scientific, Warsaw, 1980), Chap. XIII; M. Reed and B. Simon, Methods of Modern Mathematical Physics, Vol. 1, (Academic Press, San Diego, 1980).

22. W.H. Zurek, Phys. Rev. D 26 (1982) 1862: W.H. Zurek and J.P. Paz, in : Proc. Symp. on the Foundations of Modern Physics, Cologne, June 1, 1993, Eds. P. Busch and P. Mittelstaedt, (World Scientific, Singapore, 1993), p. 458.

23. N. van Kampen, Physica A 153 (1988) 97; R. Omnès, Reviews of Modern Physics 64 (1992) 339.
24. A. Bohm, Quantum Mechanics, 1st Ed. (Springer, New York, 1979); 3rd Ed. (1993).
25. E. Roberts, J. Math. Phys. 7 (1966) 1097; A. Bohm, Boulder Lectures in Theoretical Physics 1966, Vol. 9A, (Gordon and Breach, New York, 1967); J.P. Antoine, J. Math. Phys. 10 (1969) 53; 10 (1969) 2276; see also O. Melsheimer, J. Math. Phys., 15 (1974) 902; 917.
26. A. Bohm and M. Gadella. Dirac Kets, Gamow Vectors and Gel'fand Triplets, Lecture Notes in Physics, Volume 348, (Springer-Verlag, Berlin, 1989).
27. L.A. Khalfin, JETP Lett. 15 (1972) 388; see also L. A. Khalfin, this volume.
28. L. Fonda, G.C. Ghirardi and A. Rimini, Repts. on Prog. in Phys., 41 (1978) 587, and references thereof.
29. G. Gamow, Z. Phys. 51 (1928) 204.
30. H. Frauenfelder and E. M. Henley, Subatomic Physics (Prentice Hall, Englewood Cliffs, N.J., 1991).
31. T.D. Lee, R. Oehme, and C.N. Yang, Phys. Rev. 106 (1957) 340.
32. Particle Data Group, Review of Partical Physics, Phys. Rev. D 54 (1996) 1.
33. G.C. Hegerfeldt, Phys. Rev. Lett. 72 (1994) 596.
34. D. Buchholz and J. Yugvason, Phys. Rev. Lett. 73 (1994) 613; P.W. Milonni, D.F.V. James and H. Fearn, Phys. Rev. A 52 (1995) 1525.
35. R. Peierls, More Surprises in Theoretical Physics, (Princeton University Press, 1992). Gamow's complex energy state have been defined as eigenstates of the Schrödinger equation with purely out-going boundary conditions in the following: P.L. Kapur and R. Peierls, Proc. Roy. Soc. A166 (1938) 277; R. Peierls, Proceedings of the 1954 Glasgow Conference on Nuclear and Meson Physics, E.M. Bellamy, et al., editors, (Pergamon Press, New York, 1955); G. Garcia-Calderon and R. Peierls, Nuclear Physics A 265 (1976) 443; E. Hernandez and A. Mondragon, Phys. Rev. C 29 (1984) 722; A. Mondragon and E. Hernandez, Annual der Physik 48 (1991) 503; G. Garcia-Calderon, Symmetries in Physics (Moshinsky Symposium), Eds. A. Frank and K.B. Wolf, (Springer-Verlag, Berlin, 1992).
36. T.D. Lee, Particle Physics and Introduction to Field Theory, Chapter 13, (Harwood Academic, New York, 1981). In this reference the quantum mechanical time reversed state is called complicated and improbable.
37. A. Bohm, J. Math. Phys. 22 (1981) 2813; Lett. Math. Phys. 3 (1978) 455 (1978).
38. I. Prigogine, From Being to Becoming, (Freeman, New York, 1980); G. Nicholis and I. Prigogine, Exploring Complexity, (Freeman, New York, 1988); I. Prigogine, Phys. Rep. 219 (1992) 93; I. Antoniou, Nature 338 (1989) 210; T. Pertosky, I. Prigogine and S. Tasaki, Physica A 173 (1991) 175; T. Pertosky and I. Prigogine, Physica A 147 (1988) 439; T. Pertosky and I. Prigogine, Physica A 175 (1991) 146; M. de Haan, C. George, and F. Mayné, Physica A 92 (1978) 584.
39. I. Antoniou, Proc. 2nd Internat. Wigner Symposium, Gosla 1991, Eds. H.D.Doebner, et al., (World Scientific, Singapore, 1992). I. Antoniou and I. Prigogine, Physica Ia, 192 (1993) 443. I antoniou and S. Tasaki, Intern. Journ. Quantum Chemistry 46 (1993) 425.
40. K. Napiorkowski, Bulletin of the Polish Academy of Sciences 22 (1974) 1215; 23 (1975) 251; Not much is gained with this change (except that Dirac's bra-ket formalism can be made rigorous).
41. E.P. Wigner, Symmetries and Reflections, (Indiana University Press, Bloomington, 1967), page 38.

42. P.L. Duren, $\mathcal{H}^p$ Spaces, (Academic Press, New York, 1970); K. Hoffman, Banach Space of Analytic Functions, (Prentice-Hall, Inc., Englewood Cliffs, N.J., 1962; Dover Publications, Mineola, NY, 1988).
43. M. Gadella, J. Math. Phys. 24 (1983) 1462.
44. A. Bohm, Proceedings, Group Theoretical Methods in Physics, Springer Lecture Notes in Physics, Vol. 94 (Springer, Berlin, 1978) 245; and also [7], 1st Ed., Chap. XXI. H. Baumgartel mentioned the Hardy class functions around 1977 in a private communication.
45. A. Bohm, I. Antoniou, P. Kielanowski, Phys. Lett. A189 (1994) 442; A. Bohm, I. Antoniou, P. Kielanowski, J. Math. Phys. 36 (1995) 2593.
46. S. Weinberg, The Quantum Theory of Fields, Vol. 1, Chapter 2, Appendix C, (Cambridge University Press, Cambridge, 1995); also in [7], chapter XIX.
47. E.P. Wigner, Group Theoretical Concepts and Methods in Elementary Particle Physics, Ed. F. Gürsey (Gordon and Breach, New York, 1994), p. 37.
48. A. Bohm, Phys. Rev. A 51 (1995) 1978; A. Bohm and Sujeewa Wickramasekara, Found. of Phys. 27 (1997) 969.
49. I. Antoniou, Z. Suchanecki, and S. Tasaki, to appear in Generalized Functions, Operator Theory and Dynamical Systems, eds. I. Antoniou and G. Lumer, (Addison Wesley Longman, London, 1998).
50. I. Antoniou, L. Dmetrieva, Yu. Kuperin and Yu. Melnikov, Comp. Math. Appl., 34 (1997) 399; see also I. Antoniou and Yu. Melnikov, Quantum Scattering Resonances: Poles of a continued S-matrix and Poles of an extended Resolvant, this volume.
51. I. Antoniou and M. Gadella, preprint (Intern. Solvay Institute, Brussels, 1995). Results of this preprint were published in: A. Bohm et al., Rep. Math. Phys. 36 (1995) 245.
52. A. Bohm, M. Loewe, S. Maxson, P. Patuleanu, C. Püntmann, and M. Gadella, on WWW at http://www.ph.utexas.edu/~bohmwww, JMP (1997) to be published.
53. M.L. Goldberger and K.M. Watson, Phys. Rev. B 136 (1964) 1472; M.L. Goldberger and K.M. Watson, Collision Theory, (Wiley, New York, 1964); R.G. Newton, Scattering Theory of Waves and Particles, 2nd ed. (Springer-Verlag, 1982).
54. C. van Winter, Trans. Am. Math. Soc. 162 (1972) 103; J. Math. Anal. and Appl. 47 (1974) 633.
55. E. Fermi, Nuclear Physics, (University of Chicago Press, 1950).
56. An exponential law for transition probabilities in atoms was first obtained by V.E. Weisskopf and E.P. Wigner, Zeitschr. f. Physik 63 (1930) 54; 65 (1930) 18, in the Weisskopf-Wigner approximation, cf. also Chap. 8 of the first reference of [53].

Causality, Particle Localization and Positivity of the Energy

G.C. Hegerfeldt

Institut für Theoretische Physik, Universität Göttingen Bunsenstr. 9, D-37073 Göttingen, Germany

Summary. Positivity of the Hamiltonian alone is used to show that particles, if initially localized in a finite region, immediately develop infinite tails.

1. Introduction

The concept of finite signal velocity or, more precisely, the speed of light as highest signal velocity is often called Einstein causality. If, in the special theory of relativity, there were superluminal signals then there could exist tachyons, i.e. superluminal particles, and the sequence of cause and effect could be reversed. On the other hand, if signals of arbitrarily high velocities existed, one could also argue that one could obtain absolute clock synchronization and absolute simultaneity, thus making a revision of special relativity necessary.

In usual quantum mechanics it is well known that wave functions, if initially localized in a finite region, immediately develop infinite tails. For a nonrelativistic theory this is of no concern. A similar phenomenon was then noted, however, for the Newton-Wigner position operator [1, 2, 3]. In fact, the question of localization in quantum field was recognized as a difficult problem quite early [4]. In particular, localization by means of a current-density four-vector were investigated in a class of models [5].

It was then recognized by the present author that the superluminal spreading had nothing to do with position operators, field equations, current densities or a particular notion of particle localization. In a model-independent way the theorem was proved that a free relativistic particle, if initially localized with probability 1 in a finite (bounded) region, immediately thereafter would have spread over all space [6]. An alternative proof of this theorem was given in [7] and a generalization to relativistic systems in [8]. The theorem was carried over to quite general interactions, not necessarily relativistic ones in [9]. It became apparent in that paper that the main ingredient was positivity of the energy, with translation invariance used as a technical tool.

In 1985 the present author [10] showed that the connection between localization and Einstein causality was more restrictive than previously thought. It was proved in [10] that relativistic systems with Gaussian-like bounded tails at $t = 0$ – provided they exist! – would lead to a superluminal probability flow.

The main purpose of this paper is to show that translation invariance is not needed for superluminal spreading of particles which are initially confined in a bounded region. For this purpose a recent result [11] of the present author for Fermi's two-atom system will be reformulated in Section 2 as an abstract mathematical theorem and applied to particle localization. It will then be shown that, as a consequence of positivity of the energy alone, a particle which is initially strictly localized in a finite region either stays there indefinitely or immediately develops infinite tails. In the last section the connection to Einstein causality is discussed. As in the author's result [11] on the Fermi problem several ways out are mentioned to avoid a conflict.

2. A Consequence of Positivity of the Energy

In this section we will prove a simple mathematical result on the temporal behavior of certain expectation values. If the time-development operator is positive one might expect analyticity properties, but for arbitrary expectation values this is not true. One has, however, the following result.

Theorem: Let H be a selfadjoint operator, positive or bounded from below, in a Hilbert space $\mathcal{H}$. For given $\psi_0 \in \mathcal{H}$ let $\psi_t, t \in \mathbb{R}$, be defined as

$$(1) \qquad \psi_t = e^{-iHt}\psi_0 \; .$$

Let A be a positive operator in $\mathcal{H}, A \geq 0$, and let $p_A(t)$ be defined as

$$(2) \qquad p_A(t) = \langle \psi_t, A\psi_t \rangle \; .$$

Then either

$$(3) \qquad p_A(t) \neq 0 \text{ for almost all } t$$

and the set of such t's is dense and open, or

$$(4) \qquad p_A(t) \equiv 0 \text{ for all } t \; .$$

The proof is based on an analyticity argument for which, however, a little care – and the positivity of A – is needed. Evidently, since $H \geq -c$, one can define $\exp\{-iH(t + iy)\}$ for $y \leq 0$, and $\exp\{-iHz\}$ is analytic in z for Im $z < 0$, and hence ψ_t can be analytically continued to the lower half-plane, with continuous boundary values on the real axis. However, the r.h.s. of Eq. (2) can in general not be analytically continued since it equals

$$\langle \psi, e^{iHt}Ae^{-iHt}\psi \rangle$$

and since $\exp\{i(H + iy)t\}$ is in general unbounded for $y < 0$. To by-pass this the positivity of A can be used. We write

$$(5) \qquad p_A(t) = \langle A^{1/2}\psi_t, A^{1/2}\psi_t \rangle$$

where $A^{1/2}$ is the positive square root of A, and denote by $\mathcal{N}_0$ the set of t's for which $p_A(t) = 0$. By continuity of $p_A(t)$, $\mathcal{N}_0$ is closed and its complement $\mathcal{N}_0^c$ is open. Eq. (5) now implies

$$(6) \qquad A^{1/2}\psi_t = 0 \text{ for } t \in \mathcal{N}_0 .$$

For fixed $\phi \in \mathcal{H}$ we define the function $F_\phi(z)$ for Im $z \leq 0$ by

$$(7) \qquad F_\phi(z) = \langle \phi, A^{1/2}e^{-iHz}\psi_0 \rangle .$$

By the above remark on $\exp\{-iHz\}$, $F_\phi(z)$ is a continuous function for Im $z \leq 0$ and is analytic for Im $z < 0$. By Eq. (6) one has

$$(8) \qquad F_\phi(t) = 0 \text{ for } t \in \mathcal{N}_0 .$$

Now let us assume that the complement $\mathcal{N}_0^c$ is not dense. Then $\mathcal{N}_0$ contains some interval I of nonzero length, and $F_\phi(z)$ vanishes on I. One can now directly employ the Schwarz reflexion principle [12] to conclude that $F_\phi(z) \equiv 0$ or proceed in a more pedestrian way as follows. One defines an extension of F_ϕ to the upper half plane by putting

$$(9) \qquad F_\phi(z) = F_\phi(z^*)^* \text{ for } \text{Im } z > 0 .$$

Since $F_\phi(t) = 0$ for $t \in I$ and thus, a fortiori, real for $t \in I$, the extension $F_\phi(z)$ continuous for $z \in I$. From this one easily shows [12] that $F_\phi(z)$ is analytic for $z \notin \mathbb{R}\backslash I$, and thus I is contained in the domain of analyticity. Since $F_\phi(z)$ vanishes on I it must therefore vanish on the analyticity domain, i.e. for $z \notin \mathbb{R}\backslash I$. By continuity $F_\phi(z)$ then vanishes everywhere. Since ϕ was arbitrary, we obtain $A^{1/2}\psi_z = 0$ for all t. Hence,

$$(10) \qquad A\psi_t = 0 \text{ for all } t$$

and thus $p_A(t) \equiv 0$ if $\mathcal{N}_0^c$ is not dense, i.e. alternative (ii) holds in this case.

Since a dense open set need not have full Lebesgue measure, it remains to show that $\mathcal{N}_0$ is a null set if alternative (ii) does not hold. To prove this we use the fact that, as a boundary value of a bounded analytic function, $F_\phi(t)$ satisfies the inequality [13]

$$(11) \qquad \int_{-\infty}^{\infty} dt \, \frac{\ln|F_\phi(t)|}{1+t^2} > -\infty$$

unless it vanishes identically. If $\mathcal{N}_0$ had positive measure the integral would be $-\infty$, and thus $F_\phi(t)$ would vanish for all t, for each ϕ. This would again imply alternative (ii). This proves the theorem.

The theorem is a more abstract version of a result in [11] on Fermi's two-atom problem. To check the speed of light in quantum electrodynamics, Fermi had considered two atoms, separated by a distance R and with no photons present initially. One of the atoms was assumed to be in its ground state, the other in an excited state. The latter could then decay with the emission of a photon. Fermi calculated the excitation probability of the atom which had initially been in its ground state. Using standard approximations he found the excitation probability to be zero for $t < R/c$.

Now, if one takes for ψ_0 in the theorem the initial state considered by Fermi and for A the operator describing the excitation probability, e.g. the projector onto the excited states, then $p_A(t)$ becomes the excitation probability, and the theorem states that this probability is immediately nonzero. Already in [11] it was discussed how to avoid a possible conflict with causality, and this was continued in more detail for example in [14, 15, 16, 17]. The upshot was that the immediate excitation could be understood in a field-theoretic context through vacuum fluctuations due to virtual photons. The part of the excitation due to the second atom behaves causally [16, 17]. Causality then holds for expectation values after the spontaneous part has been subtracted. This corresponds to the notion of weak causality, i.e. for expectation values, introduced in [2], which contrasts to the notion of strong causality, i.e. causality for individual events, as discussed in [14]. Fermi seems to have had strong causality in mind.

3. Application to Particle Localization and Spreading

We will now apply the above theorem to the question of particle localization. We note that the results hold independent of whether the theory is relativistic or not, or a field theory or not. Neither is the existence of a position operator assumed. The ingredient is just positivity of the energy. Translation invariance, which was used in previous treatments [6, 9, 10, 18], is not needed here.

Let us suppose that it makes sense to speak of particles inside some volume V, i.e. of the probability to find a particle at a given time in V. This is a highly nontrivial assumption. Indeed, it has been argued [19] that in algebraic quantum field theory the notion of local particle states may make sense only asymptotically for free particles.

In a quantum theory the probability to find a particle or system inside V should be given by the expectation of an operator, $N(V)$ say. Since probability lie between 0 and 1, one must have

$$(12) \qquad\qquad 0 \leq N(V) \leq 1 .$$

Now let us assume that the system, with state ψ_0 at $t = 0$, is strictly localized in a region V_0, i.e. with probability 1, so that

(13) $$\langle\psi_0, N(V_0)\psi_0\rangle = 1$$

or, equivalently,

(14) $$\langle\psi_0, (1 - N(V_0))\psi_0\rangle = 0 \ .$$

From Eq. (12) one has

(15) $$1 - N(V_0) \geq 0$$

and hence the theorem can be applied, with $A \equiv 1 - N(V_0)$. As a consequence one either has

(16) $$\langle\psi_t, N(V_0)\psi_t\rangle \equiv 1 \text{ for all } t$$

or

(17) $$\langle\psi_t, N(V_0)\psi_t\rangle < 1 \text{ for almost all } t \ .$$

The argument in Eqs. (14-17) is for pure states. It can easily be carried over to mixtures characterized by density matrices. Eq. (13) is then replaced by

(18) $$tr(\rho_0 N(V_0)) = 1 \ .$$

Writing

(19) $$\rho_0 = \sum \alpha_i |\psi_{i0}\rangle\langle\psi_{i0}|$$

with $\sum \alpha_i = 1$, Eq. (18) becomes

$$\sum_i \alpha_i \langle\psi_{i0}, N(V_0)\psi_{i0}\rangle = 1$$

which implies

$$\langle\psi_{i0}, N(V_0)\psi_{i0}\rangle = 1 \text{ for all } i \ .$$

Then one can proceed as before and obtains that either

(20) $$tr(\rho_t N(V_0)) \equiv 1 \text{ for all } t$$

(21) $$tr(\rho_t N(V_0)) < 1 \text{ for almost all } t \ .$$

Alternative (20) or (16) means that the particle or system stays in V_0 for all times, as might happen for a bound state in an external potential.

Now, if the particle or system is strictly localized in V_0 at $t = 0$ it is, *a fortiori*, also strictly localized in any larger region V containing V_0. If the boundaries of V and V_0 have a finite distance and *if finite propagation speed holds* then the probability to find the system in V must also be 1 for sufficiently small times, e.g. $0 \leq t < \epsilon$. But then the theorem, with

$A \equiv 1 - N(V)$, states that the system stays in V for *all* times. Now, we can make V smaller and let it approach V_0. Thus we conclude that if a particle or system is at time $t = 0$ strictly localized in a region V_0, then finite propagation speed implies that it stays in V_0 for all times and therefore prohibits motion to infinity. Or put conversely, if there exist particle states which are strictly localized in some finite region at $t = 0$ and later move towards infinity, then finite propagation speed cannot hold for localization of particles.

This can be formulated somewhat more strongly as follows. If at $t = 0$ a particle is strictly localized in a bounded region V_0 then, unless it remains in V_0 for all times, it cannot be strictly localized in a bounded region V, however large, for any finite time interval thereafter, and the particle localization immediately develops infinite "'tails"'. The spreading is over all space except possibly for "'holes"' which, if any, will persist for all times, by the same arguments as before. If the theory is translation invariant then there can be no holes, as shown in [9] under some mild spectrum conditions.

4. Discussion. Connection with Causality

As shown above, a particle or system, if initially strictly localized in a bounded region, will immediately develop infinite tails except in the exceptional case that it stays in the original region for all times. The latter seems to require external potentials, and this will be disregarded here. If a particle is part of a larger system, e.g. of a composite system, the results still apply, and the appearance of immediate infinite tails may be attributed to the center-of-mass motion.

In nonrelativistic quantum mechanics the immediate spreading of wave functions over all space is a well known phenomenon. In a relativistic theory this might lead to a conflict with Einstein causality if it were observable. Indeed, if a particle were initially strictly localized in some region on earth and if there were a nonzero probability, however small, to observe the particle a fraction of a second later on the moon, this could be used for the absolute synchronization of clocks. One simply would have repeat the experiment sufficiently often, preferably with distinguishable particles in order to be sure when a detected particle was originally released.

But isn't the Dirac equation for a free particle of spin 1/2 a counterexample to our results? This is, however, not so. Indeed, the Dirac equation is hyperbolic and thus satisfies finite propagation speed. If for the localization operator $N(V)$ one takes the characteristic function $\chi_V(\mathbf{x})$ then, for a wave function with initial support in a finite region, the localization does evolve causally. However, the Dirac equation contains positive and negative energy states, and we therefore can conclude from our results that positive-energy solutions of the Dirac equation always have infinite support (cf. also [20]).

This example suggests a simple solution to the causality problem seemingly connected with particle localization. If there would not exist any particle states localized with probability 1 in a bounded region or, more generally, if all systems were spread out over all space to begin with, then no problems would arise. This would imply in particular that $N(V)$ could not have the eigenvalue 1 and thus could not be a projector. As a consequence there would be no selfadjoint position satisfying causal requirements, not even if one allowed position operators with noncommuting components as suggested, e.g., in [21]. For if one had a selfadjoint component of a position operator then its spectral decomposition would yield localization operators $N(V)$ for V's being infinite slabs, and to these the results could be applied in the same way as to bounded regions.

If one adopts the standpoint that all particle states have infinite tails to begin with, one might also argue that these tails could be made to drop off as fast as one likes, thus approximating a strictly localized state to arbitrary accuracy. In [10] it was shown, however, that in a relativistic theory such tails cannot drop off arbitrarily fast.

In a field theoretic context the permanent infinite tails could be intuitively understood through clouds of virtual particles around real particles ("'dressed particle states"'). Sometimes it is simply argued that local commutation or anti-commutation relations of the fields must clearly ensure causal behavior of localized particles. This overlooks the possibility that the operator $N(V)$ – if it exists – might not be a local function of the fields. A more satisfactory discussion of causality aspects is given in algebraic quantum field theory in [2] and in an alternative algebraic framework in [22].

Instead of speaking about infinite tails one may also envisage that all particle detectors exhibit inherent noise due to vacuum fluctuations and that therefore localization with probability 1 or zero can never be recorded. This would essentially lead to the same conclusion as permanent infinite tails.

Finally, it may well be true that, as advocated in [19], the notion of localizable particles in field theory makes sense only for free particles.

References

1. G.N. Fleming, Phys. Rev. **139**, B 963 (1965)
2. S. Schlieder, in: *Quanten und Felder*, ed. by H.P. Dürr (Vieweg, Braunschweig, 1971), p. 145
3. S.N.M. Ruijsenaars, Ann. Phys. (N.Y.) **137**, 33 (1981), has shown for the Newton-Wigner operator that the amount of causality violation tends to zero asymptotically.
4. A.S. Wightman and S.S. Schweber, Phys. Rev. **98**, 812 (1955) and references therein.
5. B. Gerlach, D. Gromes, and J. Petzold, Z. Phys. **202**, 1 (1967); **204**, 1 (1967); **221**, 141 (1916)
6. G.C. Hegerfeldt, Phys. Rev. D**10**, 3320 (1974)
7. B. Skagerstam, Int. J. Theor. Phys. **15**, 213 (1976)
8. J.F. Perez and I.F. Wilde, Phys. Rev. D**16**, 315 (1977)

9. G.C. Hegerfeldt and S.N.M. Ruijsenaars, Phys. Rev. D22, 377 (1980)
10. G.C. Hegerfeldt, Phys. Rev. Lett. 54, 2395 (1985)
11. G.C. Hegerfeldt, Phys. Rev. Lett. 72, 596 (1994)
12. N. Levinson and R.M. Redheffer, *Complex Variables*, Holden-Day, San Francisco 1970
13. J.B. Garnett, *Bounded Analytic Functions*, Academic Press, New York 1981; p. 64
14. G.C. Hegerfeldt, in *Nonlinear, deformed, and irreversible quantum systems*, edited by H.-D. Doebner, V.K. Dobrev, and P. Nattermann, World Scientific, Singapore (1995), p. 253
15. D. Buchholz and J. Yngvason, Phys. Rev. Lett. 73, 613 (1994)
16. A. Labarbara and R. Passante, Phys. Lett. 206, 1 (1994)
17. P.W. Milonni, D.F.V. James, and H. Fearn, Phys. Rev. A52, 1525 (1995)
18. G.C. Hegerfeldt, Nuclear Phys. B 6, 231 (1989)
19. M. Redhead, Found. Phys. 25, 123 (1995)
20. B. Thaller and S. Thaller, Il Nuovo Cim. 82A, 222 (1984); B. Thaller, *The Dirac Equation*, Springer, Berlin 1992
21. H. Bacry, Ann. Inst. H. Poincaré 49, 245 (1988)
22. H. Neumannn and R. Werner, Int. J. Theor. Phys. 22, 781 (1983)

Quantum Scattering of Resonances: Poles of a Continued S-Matrix and Poles of an Extended Resolvent

I. Antoniou[1,2] and Yu. Melnikov[1,2,3]

[1] International Solvay Institutes for Physics and Chemistry, Campus Plaine ULB C.P.231, Boulevard du Triomphe, Brussels 1050, Belgium
[2] Theoretische Natuurkunde, Free University of Brussels, Brussels 1050, Belgium
[3] Laboratory of Complex Systems Theory, St. Petersburg State University, St. Petersburg 198904, Russia

1. Introduction

Since the development of quantum theory several theoretical discussions of resonances and resonance states have been proposed, like poles of a partial resolvent, points of spectral concentration, poles of the scattering matrix or poles of the extended power spectrum. For an overview of the literature see the volume edited by Brändas and Elander [1] and references therein.

In the paper [2] a unified operator theoretic formulation of resonances and resonance states in rigged Hilbert space was proposed. In the case of simple scattering resonances this formulation coincides with the Bohm-Gadella approach [3], [4] and Brussels/Austin approach [5], [6], [7], [8], [9]. The same holds true for multiple scattering resonances [10].

The approach developed in [2] is based on the extension of the resolvent of the generator of the evolution group to an appropriate rigged Hilbert space and on the analytic properties of the extended resolvent. Resonances are defined as poles of the extended resolvent outside the spectrum of the generator. The natural question is whether these poles coincide with the scattering resonances [11]. In accordance with the results presented in [2] this question is meaningful in the context of an appropriate class of rigged Hilbert spaces where the resolvent is extended. The objective of the present paper is to characterize the class of rigged Hilbert spaces for which the scattering resonances coincide with poles of the extended resolvent (Sect. 5). The result is illustrated by two examples, namely the Friedrichs model (Sect. 6) and a sperical-symmetric two-body scattering (Sect. 7). In order to have a self-contained paper we include some facts from the scattering theory (Sect. 2), scattering resonances (Sect. 3) and a discussion of resonances as poles of the extended resolvent (Sect. 4).

2. Necessary Facts from Scattering Theory

Conventional scattering theory [12], [13] actually compares the asymptotic evolution U_t generated by the selfadjoint operator H, with the free evolution U_t^0 generated by the selfadjoint operator H_0 on a separable Hilbert space $\mathcal{H}$. The wave operators W_{in}, W_{out} intertwine the selwadjoint generator H_0 with $H = H_0 + V$,

$$H = W_{in}^\dagger H_0 W_{in} = W_{out}^\dagger H_0 W_{out} \ .$$

These wave operators transform the absolutely-continuous subspace of the operator H_0 onto the absolutely continuous subspace of the operator H,

$$W_{in}, W_{out} : \mathcal{H}^{ac}(H_0) \to \mathcal{H}^{ac}(H) \ .$$

The wave operators are defined as follows

$$W_{in} = (s) \lim_{t \to -\infty} e^{iHt} e^{-iH_0 t} \mathcal{P}_0^{ac} = \lim_{\varepsilon \downarrow 0} \varepsilon \int_{-\infty}^0 e^{\varepsilon t} e^{-iH_0 t} \mathcal{P}_0^{ac} dt \ ,$$

$$W_{out} = (s) \lim_{t \to \infty} e^{iHt} e^{-iH_0 t} \mathcal{P}_0^{ac} = \lim_{\varepsilon \downarrow 0} (-\varepsilon) \int_{\infty}^0 e^{-\varepsilon t} e^{-iH_0 t} \mathcal{P}_0^{ac} dt \ ,$$

where $\mathcal{P}_0^{ac}$ is the orthoprojection onto the absolutely continuous subspace $\mathcal{H}_{ac}$ of H_0.

The spectral decomposition of the free Hamiltonian is usually known and it is given by the Gelfand-Maurin formula [14], [M] which gives meaning to Dirac formula [D] in terms of rigged Hilbert spaces or Gelfand triplets. The absolutely continuous part of H_0 is:

$$\mathcal{P}_0^{ac} H_0 = \int_{\sigma_{ac}(H_0)} dE E \int_{\Omega_E} ds |\phi_{Es}\rangle\langle\phi_{Es}| \ ,$$

where Ω_E are the compact manifolds labeling the degeneracy subspaces. The generalized eigenvectors $|\phi_{Es}\rangle$ form a generalized complete orthonormal system:

$$\int_{\sigma_{ac}(H_0)} dE \int_{\Omega_E} ds |\phi_{Es}\rangle\langle\phi_{Es}| = \mathcal{P}_0^{ac} \ ;$$

$$\langle\phi_{Es}|\phi_{E's'}\rangle = \delta(E - e')\delta(s - s') \ .$$

The generalized eigenvectors $|\phi_{Es}\rangle$ give a concrete expression to the spectral representation of the free operator H_0. We denote by $\mathcal{U}$ the unitary transformation which diagonalizes the free operator H_0:

$$\mathcal{U} : \mathcal{H} \to L^2_{\sigma(H_0), d\mu} \ ;$$

$$\mathcal{U} : F \mapsto F(E, s) = \langle F|\phi_{Es}\rangle \ .$$

The spectral representation transforms any operator A to the integral operator $\mathcal{A} = \mathcal{U} A \mathcal{U}^{-1}$ with kernel:

$$\mathcal{A}(E, s|E', s') = \langle\phi_{Es}|A|\phi_{E's'}\rangle .$$

The kernels of the wave operators in the spectral representation of H_0 are:

$$W_{in}(E, s|E', s') = \lim_{\varepsilon\downarrow0}\varepsilon\int_{-\infty}^{0} e^{\varepsilon t}\langle\phi_{Es}|e^{iHt}e^{-iH_0t}\mathcal{P}_0^{ac}|\phi_{E's'}\rangle dt =$$

$$= \lim_{\varepsilon\downarrow0}\varepsilon\int_{-\infty}^{0} e^{\varepsilon t}e^{-iEt}\langle\phi_{Es}|e^{iHt}\mathcal{P}_0^{ac}|\phi_{E's'}\rangle dt =$$

$$= \lim_{\varepsilon\downarrow0}(-i\varepsilon)\langle\phi_{Es}|R(E - i\varepsilon)P_0^{ac}|\phi_{E's'}\rangle ,$$

where $R(z) = (H - z)^{-1}$ stands for the resolvent of the operator H. Similary

$$W_{out}(e, s|E', s') = \lim_{\varepsilon\downarrow0}i\varepsilon\langle\phi_{Es}|R(E + i\varepsilon)P_0^{ac}|\phi_{E's'}\rangle .$$

We shall need the well known [12], [13] relation of the resolvent with the T-matrix

$$R(z) = R_0(z) - R_0(z)T(z)R_0(z) ; \quad R_0(z) = (H_0 - z)^{-1}$$

which gives

(1)
$$W_{in}(E, s|E', s') = \left[\delta(E - E')\delta(s - s') - \frac{T(E, s|E', s'; E' + i0)}{E' - E - i0}\right]\mathbf{1}_{ac}(E', s') ,$$

$$W_{out}(E, s|E', s') = \left[\delta(E - E')\delta(s - s') - \frac{T(E, s|E', s'; E' - i0)}{E' - E + i0}\right]\mathbf{1}_{ac}(E', s') ,$$

where $\mathbf{1}_{ac}(E, s) \stackrel{\text{def}}{=} \|(\mathcal{P}_0^{ac}|\phi_{Es}\rangle\langle\phi_{Es}|)\|$ is an indicator of the absolutely-continuous subspace of the operator H_0 and $T(E, s|E', s'; z) = \langle\phi_{Es}|T(z)|\phi_{E's'}\rangle$.

For the scattering operator $S = (W_{in})^\dagger W_{out}$ therefore we have

$$S(E, s|E', s') = \int_{\sigma(H_0)} d\nu \int_{\Omega_E} dq(W_{in}^*(E, s|\nu, q)W_{out}(\nu, q|E', s')$$

$$= \left[\delta(E - E')\delta(s - s') - \frac{T(E, s|E', s'; E + i0)}{E' - E - i0} - \frac{T(E, s|E', s'; E' + i0)}{E - E' - i0}\right.$$

$$+ \int_{\sigma(H_0)} d\nu \int_{\Omega_E} dq$$

(2)
$$\left.\frac{T(E, s|\nu, q; E + i0)T(\nu, q|E', s'; E' + i0)}{(\nu - E' - i0)(\nu - E - i0)}\right]\mathbf{1}_{ac}(E, s)\mathbf{1}_{ac}(E', s').$$

On the other hand, using the Hilbert identity for the resolvent

$$\frac{R(z_1) - R(z_2)}{z_1 - z_2} = R(z_1)R(z_2)$$

one gets [13] a similar identity for the T-matrix,

$$\frac{T(z_1) - T(z_2)}{z_1 - z_2} = -T(z_1)R_0(z_1)R_0(z_2)T(z_2) \ .$$

Applying this formula for $z_1 = E + i0$, $z_2 = E' + i0$, we have

(3)
$$\frac{T(E, s|E', s'; E + i0) - T(E, s|E', s'; E' + i0)}{E' - E - i0}$$

$$= \int_{\sigma(H_0)} d\nu \int_{\Omega_E} dq \frac{T(E, s|\nu, q; E + i0)T(\nu, q|E', s'; E' + i0)}{(\nu - E - i0)(\nu - E' - i0)} \ .$$

Combining Eqs.(2) and (3) we get a well-known expression [12], [13]:

(4)
$$\mathcal{S}(E, s|E', s') = \delta(E - E')\delta(s - s')\mathbf{1}_{ac}(E, s)\mathbf{1}_{ac}(E', s) -$$

$$- T(E, s|E', s'; E' + i0) \left(\frac{1}{E' - E - i0} - \frac{1}{E' - E + i0} \right) \mathbf{1}_{ac}(E, s)\mathbf{1}_{ac}(E', s')$$

$$= \delta(E - E') \left[\delta(s - s') - 2\pi i T(E, s|E', s'; E' + i0) \right] \mathbf{1}_{ac}(E, s)\mathbf{1}_{ac}(E', s') \ .$$

The scattering operator is represented therefore as the direct integral

(5)
$$\mathcal{S} = \Gamma S \Gamma^{-1} = \int_{\sigma(H_0)} \oplus S_E \, dE \ .$$

Eq. (4) shows that the operator $\mathcal{S} = \mathcal{U}S\mathcal{U}^{-1}$ acts on any function $F_{Es} \in \mathcal{U}\mathcal{H}\mathcal{U}^{-1}$ as

$$(\mathcal{S}F_{Es})$$

$$= F_{Es}\mathbf{1}_{ac}(E, s) - 2\pi i \mathbf{1}_{ac}(E, s) \int_{\Omega_E} F_{Es'}T(E, s|E, s'; E + i0)\mathbf{1}_{ac}(E, s')ds' \ .$$

In the latter formula E can be treated as a parameter. Hence the representation (5) is valid with the operators $S_E : L_2(\Omega_E) \to L_2(\Omega_E)$ determined by their kernels

(6) $S_E(s, s') = \left[\delta(s - s') - 2\pi i T(E, s|E, s'; E + i0) \right] \mathbf{1}_{ac}(E, s)\mathbf{1}_{ac}(E, s') \ .$

Operators S_E are defined for any E in the absolutely continuous spectrum.

3. Scattering Resonances

Scattering resonances are singularities of the scattering operator extended to the complex plane [1], [11]. The extension of the scattering operator is easely understood in terms of formula (6) as the extension of the operator-valued function S_E to the complex plane: $S_E \to \tilde{S}_z$, $z \in \mathbf{C}$.

The perturbation operator V in the spectral representation of H_0 has a kernel

$$V(E, s | E', s') = \langle \phi_{Es} | V | \phi_{E's'} \rangle \, ,$$

therefore expressing the T-matrix in terms of the resolvent [12], [13]

$$T(z) = V - V R(z) V$$

we get from Eq. (6)

$$S_E(s, s') = [\delta(s - s') - 2\pi i \langle \phi_{Es} | V | \phi_{Es'} \rangle +$$

(7) $$+ 2\pi i \langle V \phi_{Es} | R(E + i0) | V \phi_{Es'} \rangle] \mathbf{1}_{ac}(E, s) \mathbf{1}_{ac}(E, s') \, .$$

This relation is valid for any $E \in \sigma_{ac}(H_0)$. Let us see when it can be extended to the complex plane.

4. Resonances as Intrinsic Properties of Dynamics

We have recently proposed [2] an operator approach to resonances as singularities of the extended resolvent of the evolution operator. We briefly review the main results which we shall need in order to compare with scattering resonances.

For any operator H the spectral properties are expressed in terms of the analytic structure of the resolvent

$$R(z) = (H - zI)^{-1} \, .$$

The resolvent is an analytic function on the whole complex plane except on the spectrum of H. Resonances will be defined in terms of the resolvent extended in a suitable rigged Hilbert space [14], [3]. The key idea is a generalization of a well known fact: the residue of the resolvent of any normal operator H in the discrete spectral point λ is the projection on the corresponding eigenspace $\mathcal{H}_\lambda$,

$$\mathrm{res}\, R(z)|_{z=\lambda} = -\frac{1}{2\pi i} \oint_{C_\lambda} R(\zeta) d\zeta = P_\lambda \, ,$$

where the contour C_λ encircles the point λ in the positive orientation. In the case of a simple eigenvalue λ we have the one-dimensional projection:

$$\operatorname{res} R(z)|_{z=\lambda} = |\phi_\lambda\rangle\langle\phi_\lambda| ,$$

$$H|\phi_\lambda\rangle = \lambda|\phi_\lambda\rangle \quad |\phi_\lambda\rangle \in \mathcal{H} .$$

The extension $\tilde{H}$ of the dynamical operator H to some suitable rigged Hilbert space $\Phi \subset \mathcal{H} \subset \Phi^\times$ is defined as follows [14], [3]:

Definition 1. The space Φ is a test space for the extension of the operator H to the dual $\Phi^\times$ if:

1. Φ is a locally convex topological vector space with topology stronger than the Hilbert space topology;
2. Φ is continuously and densely embedded in $\mathcal{H}$;
3. The adjoint operator $H^\dagger$ does not lead out of Φ, i.e. $H^\dagger\Phi \subset \Phi$.

Definition 2. The operator $\tilde{H}$ is called an extension of the operator H to the dual space $\Phi^\times$ if $\tilde{H}$ acts on linear functionals $\langle f| \in \Phi^\times$ as:

$$\langle\tilde{H}f|\phi\rangle = \langle f|H^\dagger\phi\rangle$$

for all test functions $\phi \in \Phi$.

We use the following definition of the extended resolvent to a rigged Hilbert space [2]:

Definition 3. We call the operator-valued function $\tilde{R}(z)$ in a suitable rigged Hilbert space $\Phi \subset \mathcal{H} \subset \Phi^\times$ an extended resolvent of the operator H if:

1. $\tilde{R}(z)$ satisfies the resolvent equality in the weak sence outside the singularities

$$\langle u|\tilde{R}(z)(H-z)|v\rangle = \langle u|v\rangle ,$$

   for all $u, v \in \Phi$;
2. $\tilde{R}(z)$ satisfies a completeness condition in a weak sence, i.e.

$$-\frac{1}{2\pi i}\oint_\Gamma \langle u|\tilde{R}(z)|v\rangle dz = \langle u|v\rangle ,$$

   for all $u, v \in \Phi$, where the contour Γ encircles all the singularities of the integrand with positive orientation.

Following [2], we define resonances and resonance states as the singularities of the extended resolvent $\tilde{R}(z)$ to a rigged Hilbert space $\Phi \subset \mathcal{H} \subset \Phi^\dagger$. Resonances associated with n-order poles will be called pole resonances of n-th order. Resonances associated with cuts will be called continuous resonances.

The choice of the test space Φ depends upon the operator H and the physical observations on the system. Therefore the concept of resonance depends on the observables on physical processes involved. This is in conformity with the fact that the resonances appear in agiven in experimental context.

5. The Relation Between the Scattering Resonances and the Poles of the Extended Resolvent

The relation of the scattering resonances with the poles of the extended resolvent is summarized by the following theorem.

Theorem. For any selfadjoint operators H_0, $H = H_0 + V$ satisfying the conditions of scattering theory discussed in Sect. 2 we have:

1. If the extended resolvent [2] $\tilde{R}(z)$ in the rigged Hilbert space $\Phi \subset \mathcal{H} \subset \Phi^\times$ has a pole at $z = z_0$ in the lower halfplane such that

   i) function $\gamma_V(E) \overset{\text{def}}{=} \int_{\Omega_E} \int_{\Omega_E} |\langle \phi_{Es}|V|\phi_{Es'}\rangle|^2 ds ds'$ admits a bounded continuation to a vicinity of the point z_0;

   ii) functions ϕ_{Es} admit a continuation to a vicinity of the point z_0;

   iii) $V\phi_{z_0s}$ belong to the test space Φ for all $s \in \bigsqcup_{E \in \sigma(H_0)} \Omega_E$;

   iv) integral $\oint_{C_{z_0}} \langle V\phi_{z_0s}|R(z)|V\phi_{z_0s'}\rangle dz$ does not vanish at a set of s, s' of a finite measure;

   then the extended S-matrix has also a pole at the same point $z = z_0$.

2. If the extended S-matrix has a pole at $z = z_0$ and the conditions i)–iii) for a test space Φ, then the extension $\tilde{R}(z)$ to the rigged Hilbert space $\Phi \subset \mathcal{H} \subset \Phi^\times$ has also a pole at $z = z_0$.

Proof

In a vicinity of the pole $z = z_0$ the extended resolvent $\tilde{R}(z)$ can be represented as the sum of the regular part $R^{reg}(z)$ and the singular term (for the sake of simplicity we suppose that there is a simple pole):

$$\tilde{R}(z) = R^{reg}(z) + \frac{\Pi_{z_0}}{z - z_0} \,,$$

where

$$\Pi_{z_0} = -\frac{1}{2i\pi} \oint_{C_{z_0}} \tilde{R}(z) dz$$

is the residue of the extended resolvent $\tilde{R}(z)$ in the pole $z = z_0$. Using Eq. (7) for the continuation of the scattering matrix and in the vicinity of the point z_0 we have

$$S_z(s,s') = \delta(s - s') - 2i\pi\langle \phi_{z_0s}|V|\phi_{z_0s'}\rangle + \langle V\phi_{z_0s}|R^{sm}(z)|V\phi_{z_0s'}\rangle +$$

$$+ 2i\pi\langle V\phi_{z_0s}|\Pi_z+|V\phi_{z_0s'}\rangle + O(|z - z_0|) \,.$$

Therefore the pole z_0 of the extended resolvent $\tilde{R}(z)$ generates the same pole of the continued scattering matrix under conditions i) and ii) if the factor $\langle V\phi_{z_0s}|\Pi_{z_0}|V\phi_{z_0s'}\rangle$ is not equal to zero at least for a non-zero measure set of s, s'.

Conversely now, let us suppose that a continued S-matrix has a pole in some point $z = z_0$ and the conditions i) – iii) are satisfied for given test space Φ and extended resolvent $\tilde{R}(z)$. Due to the condition i) the term $\langle \phi_{z_0 s} | V | \phi_{z_0 s'} \rangle$ is the kernel of a bounded operator. Hence the term $\langle V \phi_{z_0 s} | \tilde{R}(z) | V \phi_{z_0 s'} \rangle$ has to be singular in the point $z = z_0$. Due to the condition ii) that is possible only if the extended resolvent $\tilde{R}(z)$ is singular in that point.

Now we are going to illustrate the general discussion in two examples, namely the Friedrichs model and the two-body scattering by a spherically-symmetric potential.

6. The Friedrichs Model

We consider the simplest Friedrichs model [2], [6], [5], [15], [16]. The Hilbert space of pure states is taken to be an orthogonal sum of two Hilbert spaces (channels), $\mathcal{H} = \mathcal{H}_0 \oplus \mathbf{C}$ and the unperturbed Hamiltonian is

$$(8) \qquad H_0 = h \oplus \alpha \mathbf{1} = \begin{pmatrix} h & 0 \\ 0 & \alpha \end{pmatrix}.$$

Here α is some positive constant and $h = \int_0^\infty E |E\rangle\langle E| dE$ is a selfadjoint operator in the space $\mathcal{H}_0$ with homogeneous spectrum of multiplicity one filling the positive semiaxis. The perturbation V is

$$(9) \qquad V = \lambda \begin{pmatrix} 0 & \int_0^\infty V_E |E\rangle dE \\ \int_0^\infty V_E \langle E| dE & 0 \end{pmatrix},$$

where V_E is an L_2-function on $\mathbf{R}_+$ and the coupling parameter λ is a real constant. Therefore we have

$$\sigma(H_0) = \mathbf{R}_+ \cap \{\alpha\} \; ; \; \mathcal{P}_0^{ac} = \begin{pmatrix} I & 0 \\ 0 & 0 \end{pmatrix}.$$

The manifolds Ω_E for $E \neq \alpha$ are trivial (consist of one point) and the generalized eigenfunctions are

$$|f_E^c\rangle = \begin{pmatrix} |E\rangle \\ 0 \end{pmatrix},$$

whereas for $E = \alpha$ it cosists of two points corresponding to the eigenfunctions

$$|f_\alpha^c\rangle = \begin{pmatrix} |\alpha\rangle \\ 0 \end{pmatrix}, \; |f^d\rangle = \begin{pmatrix} 0 \\ 1 \end{pmatrix}.$$

The operator $\mathcal{U}$ maps the space $\mathcal{H}$ onto $L_2(\mathbf{R}_+) \oplus \mathbf{C}$. The standard representation [5], [6] of the perturbed Friedrichs model Hamiltonian $H = H_0 + V$ in our notations is

(10)

$$H = \int_0^\infty E|f_E^c\rangle\langle f_E^c|dE + \alpha|f^d\rangle\langle f^d| + E\int_0^\infty V_E\left(|f_E^c\rangle\langle f^d| + |f^d\rangle\langle f_E^c|\right)dE .$$

One can check that the only nontrivial block of the scattering operator acts from the channel $\mathcal{H}_0$ onto $\mathcal{H}_0$ and is an operator of multiplication fy a function in the unperturbed representation. By straightforward calculation we get $\langle f_E^c|V|f_E^c\rangle \equiv 0$ and applying Eq. (7) we obtain

(11) $$S_E = 1 + 2\pi i E^2|V_E|^2\langle f^d|R(E+i0)|f^d\rangle .$$

In the paper [2] we have shown that

(12) $$\langle 1|R(z)|1\rangle = Q^{-1}(z) = \left(\alpha - z - \lambda^2\int_0^\infty \frac{|V_\omega|^2}{\omega - z}d\omega\right)^{-1} ,$$

hence

(13) $$S_E = 1 + \frac{2\pi i\lambda^2|V_E|^2}{Q(E+i0)} = \frac{\alpha - E - \lambda^2 \text{V.P.}\int_0^\infty \frac{|V_\omega|^2}{\omega - E}d\omega + i\pi\lambda^2|V_E|^2}{\alpha - E - \lambda^2 \text{V.P.}\int_0^\infty \frac{|V_\omega|^2}{\omega - E}d\omega - i\pi\lambda^2|V_E|^2} .$$

From the latter equality one can see that continuation of the scattering matrix is related to the continuation of the function $Q(z)$. This function has a cut on the positive semiaxis and if one continues it through the cut from the upper halfplane to the lower, the new function

$$\tilde{Q}(z)^+ = \alpha - z - \lambda^2\text{V.P.}\int_0^\infty \frac{|V_\omega|^2}{\omega - z}d\omega - 2i\pi\lambda^2|V_z|^2$$

has a root z_0 there in the lower halfplane. Here V_z stands for a continuation of the function V_E into the complex plain which is supposed to exist. In case $\lambda \ll 1$ in the first order approximation the root is

$$z_0 = \alpha + \pi\lambda^2|V_\alpha|^2 - i\lambda^2\text{V.P.}\int_0^\infty \frac{|V_\omega|^2}{\omega - 0\alpha}d\omega + O(\lambda^4) .$$

This root obviously generates a pole of the scattering matrix continuation. On the other hand, it generates also a pole of the extended resolvent $\tilde{R}^+(z)$, as in [2] it is shown that

(14) $$R(z) = \begin{pmatrix} \frac{1}{E-z} & 0 \\ 0 & 0 \end{pmatrix} + \frac{1}{Q(z)}\begin{pmatrix} \lambda^2|\psi_z\rangle\langle\psi_{z*}| & -\lambda|\psi_z\rangle \\ -\lambda\langle\psi_{z*}| & 1 \end{pmatrix} ,$$

where $\psi_z = \frac{V_E}{E-z}$.

As mentioned above, $\langle f_E^c|V|f_E^c\rangle \equiv 0$, so $\gamma_V(E) \equiv 0$ and condition i) is trivial. One can calculate that

$$V|f_E^c\rangle = \begin{pmatrix} 0 \\ V_E \end{pmatrix} ,$$

therefore condition ii) is satisfied for any test space Φ if the function V_E admits a bounded continuation to a vicinity of the point z_0.

7. Two-Body Scattering
by a Spherical - Symmetric Potential

We consider two-body potential scattering in the c.m. frame in $\mathbf{R}^3$ under condition of eistence and completeness of wave operators. The free Hamiltonian is the Laplace operator

$$(15) \qquad H_0 = -\Delta \ .$$

The perturbed Hamiltonian is

$$(16) \qquad H = H_0 + V(x) \ , \quad x \in \mathbf{R}^3 \ .$$

As well-known [12], [13], the spectrum of the operator H_0 is $\sigma(H_0) = \mathbf{R}_+$, $P_0^{ac} = I$, and the manifolds Ω_E, $E \in \mathbf{R}_+$, are actually two-dimensional sphere Ω_E. The generalized eigenfunctions of H_0 are

$$(17) \qquad \phi_{Es}(x) = \exp\{i\sqrt{E}|x|\cos\widehat{(x,s)}\} \ ,$$

so the spectral representation transform $\mathcal{U}$ is the three-dimensional Fourier transform from the coordinate representation to the momentum representation.

Suppose that for a given test space Φ the continued resolvent $\tilde{R}(z)$ has a pole in some point $z = z^+$ in the lower halfplane. Then, if

$$(18) \qquad \exp\{|\Im\sqrt{z^+}||x|\}\, V(x) \in \Phi$$

condition ii) is satisfied. On the other hand, as this function belong to the test space Φ, it can be tested by all the functionals φ from the dual space $\Phi^\times$ [2] so that

$$(19) \qquad \langle\varphi|\exp\{|\Im\sqrt{z_0}||x|\}V(x)\rangle < \infty \ .$$

If Φ is an appropriate test space for the extended resolvent $\tilde{R}(z)$ and this extended resolvent can be continued into a vicinity of the point $z = z_0$, formal solutions of the Schrödinger equation with the complex spectral parameter $z = z_0$ should be kernels of the functionals from the dual space $\Phi^\times$ [2]. These formal solutions have an asymptotic behaviour $\phi_{zs}(x) \sim \exp\{i\sqrt{z}|x|\cos\widehat{(x,s)}\}$, as $|x| \to \infty$, therefore from the inequality (17) we have

$$\int_{\mathbf{R}^3} \exp\{|\Im\sqrt{z_0}||x|\}\,\exp\{i\sqrt{z^+}|x|\cos\widehat{(x,s)}\}d^3x < \infty \ ,$$

which immediately implies the condition i). Therefore the objects in the right hand side of Eq. (7) are well-defined under continuation $E \to z_0 \in \mathbf{C}$, so one can use Eq. (7) for the continuation of the scattering matrix.

8. Conclusions

The above discussion demonstrates that the question of the equivalence of resonances defined through the poles of a continuation of the scattering matrix and through the poles of an extended resolvent is meaningful only if one specifies an appropriate rigged Hilbert space for the extended resolvent. This means that the test space Φ with an appropriate topology should be specified. If it is done our theorem (see Sect. 5) gives the sufficient conditions for the identity of the scattering resonances with the poles of the extended resolvent in this rigged Hilbert space.

Acknowledgments: We thank Prof. I. Prigogine for several inspiring discussions and to Profs. A. Bohm, M. Gadella, B. Pavlov, G. Pronko, who shared their insight on resonances with us.

This work was supported by the Comission of the European Communities in the frame of the ESPRIT project 21042 CTIAC and by the Belgian Interuniversity Attraction Poles.

References

1. E. Brändas and N. Elander (eds.), *Resonances*, Springer Lect. Notes in Physics **325**, Springer-Verlag, Berlin 1989.
2. I. Antoniou, L. Dmitrieva, Yu. Kuperin and Yu. Melnikov, *Resonances and the Extension of Dynamics to Rigged Hilbert Space*, Int. J. Comput. Math. Appl. (1997) (to appear).
3. A. Bohm, *Quantum Mechanics, Foundations and Applications*, 3^ded., Springer, Berlin 1993.
4. A. Bohm and M. Gadella, *Dirac Kets, Gamow Vectors and Gelfand Triplets*, Springer Lect. Notes on Physics **348**, Berlin 1989.
5. I. Antoniou and I. Prigogine, Physica **A 192** 443-464 (1993).
6. I. Antoniou and S. Tasaki, Int. J. Quantum Chemistry **46** 425-474 (1993).
7. I. Prigogine, *From Being to Becoming*, Freeman, New York 1980.
8. I. Prigogine, Phys. Reports **219** 93-120 (1992).
9. T. Petrosky, I. Prigogine and S. Tasaki, Physica **A 173** 172-242 (1991).
10. I. Antoniou, M. Gadella and G. Pronko, *Gamow Vectors for Degenarate Scattering Resonances*, Ann. Phys. (submitted).
11. V.I. Kukulin, V.M. Krasnopol'sky and J. Horacek, *Theory of Resonances. Principles and Applications*, Kluwer Academic Publishers, Dordrecht/Boston/London 1989.
12. J.M. Jauch, *Foundations of Quantum Mechanics*, Addison-Wesley Pub. Co. 1973.
13. S.P. Merkuriev and L.D. Faddeev, *Quantum Scattering Theory for Several Particle Systems*, Kluwer Academic Publishers, London, 1993.
14. I. Gelfand and N. Vilenkin, *Generalized Functions* Vol. 4, Academic Press, New York 1964.
15. K. Friedrichs, Commun. Pure Appl. Math. **1** 361-406 (1948).
16. L. Horwitz and J.P. Marchand, Rocky Mnt. J. Math. **1** 225-253 (1971).

Accidental Degeneracy and Berry Phase of Resonant States*

A. Mondragón and E. Hernández

Instituto de Física, UNAM, Apartado Postal 20-364, 01000 México D.F., Mexico

Summary. We study the complex geometric phase acquired by the resonant states of an open quantum system which evolves irreversibly in a slowly time dependent environment. In analogy with the case of bound states, the Berry phase factors of resonant states are holonomy group elements of a complex line bundle with structure group $\mathbf{C}^*$. In sharp contrast with bound states, accidental degeneracies of resonances produce a continuous closed line of singularities formally equivalent to a continuous distribution of "magnetic" charge on a "diabolical" circle, in consequence, we find different classes of topologically inequivalent non-trivial closed paths in parameter space.

1. Introduction

For many years now, it has been appreciated that there are distinct advantages in describing quantum resonances and the quantum phenomena associated to the production, evolution and decay of resonances in terms of resonant or Gamow states, since many physical effects are then readily expressed and evaluated[1]. In this work, we will give closed analytical expressions for the complex Berry phase of an open quantum system in a resonant state of a Hermitian Hamiltonian with non-self-adjoint boundary conditions, and we will discuss some of its properties.

During the last fourteen years the geometric phase factors arising in the adiabatic evolution of quantum systems[2] have been the subject of many investigations[3, 4]. The early literature was mostly concerned with the geometric phase factors of closed systems driven by Hermitian Hamiltonians[4]. More recently there has been a substantial interest in the complex geometric phase acquired by the eigenstates of open quantum systems. This problem arises naturally in connection with various experiments which, by their very essence, require the observation of the geometric phase in metastable states. The Berry phase in the optical supermode propagation in a free laser, which is a classical system described by a Schrödinger-like equation with a non-Hermitian Hamiltonian, was studied by Dattoli et al.[5]. The measurement of the geometric phase in atomic systems with two energy levels, one of which at least is metastable, was also described in terms of a non-Hermitian Hamiltonian by Miniatura et al.[6]. The validity of the adiabatic approximation for dissipative, two level systems driven by non-Hermitian Hamiltonians was examined by Nenciu and Rasche[7], and by Kvitsinsky and Putterman[8], who

---

* This work was partially supported by CONACYT (México) under contract No. 4964-E9406

also established that the Berry phase is complex in this case. Sun[9] proposed a higher-order adiabatic approximation for two-level non-Hermitian Hamiltonians, and showed that the holonomy structure associated to the Berry phase factor in this non-Hermitian case is the non-unitary generalization of the holonomy structure of the Hermitian case.

In two previous papers[10, 11], we gave explicit expressions for the geometric phase of true resonant states, defined as complex energy eigenstates of a Hermitian Hamiltonian which satisfy purely outgoing wave boundary conditions at infinity[12], and pointed out some of the mathematically interesting and physically relevant properties resulting from the extended nature of the singularities in parameter space associated with the occurrence of accidental degeneracies of two resonances. In this conection, in another paper we showed that the codimension of the accidental degeneracy of n resonances differ significantly from those of bound states, we also showed that, close to a crossing of two resonant states, the topological structure of the energy hypercomplex surfaces differ significantly from the double conical point singularity typical of bound states[13]. Later, by means of a numerical analysis of the experimental data on the 2^+ doublet of resonances with $T = 0, 1$ in the energy spectrum of ^{8}Be, we showed, in a realistic example, that a true crossing of resonances mixed by a Hermitian interaction may be brought about by the variation of only two real linearly independent parameters[14]. In this paper we show that the geometric interpretation of the Berry phase factor, first given by B. Simon[15] for the adiabatic evolution of closed quantum systems, may be generalized, in a very natural way, to the case of resonant states of open quantum systems. That is, the adiabatic evolution of resonant states may be interpreted as parallel transport in a complex line bundle defined over the space of parameters with structure group $\mathbf{C}^*$ (the multiplicative group of the non-zero complex numbers). Then, the Berry phase factors of resonant states arise as the holonomy group elements due to a connection in the bundle such that during the adiabatic evolution the resonant state is parallel transported along the fibre. The horizontal spaces are perpendicular to the fibre with respect to a generalized inner product of resonant states defined in a rigged Hilbert space.

2. Resonant States

2.1 Resonant States in a Slowly Time Evolving Environment

Let us consider the time evolution of a quantum system in a state which is a superposition of unstable eigenstates moving in some strong external field of force which changes slowly with time. In order to have some concrete example in mind, although a very hypothetical one, we may think of an ^8Be nucleus which has only unstable energy eigenstates moving in the field of forces of a double magic nucleus, like ^{208}Pb, in a peripheral collision in which the distance between the two nuclei is never smaller than the sum of the nuclear radii. In a semiclassical treatment of the collision, when the centers of the nuclei move along classical trajectories, the parameters in the nucleus-nucleus interaction change with time[16].

The evolution of the system under the influence of the external perturbation is governed by the time dependent Schrödinger equation

$$(1) \qquad i\hbar\frac{\partial \Psi}{\partial t} = H\Psi.$$

The Hamiltonian H is the sum of the time-independent Hamiltonian H_0 describing the evolution of the unperturbed system plus a perturbation term H_1 which is a function of a number N of "external" parameters $\{X_1, X_2...X_N\}$ which may change with time,

$$(2) \qquad H(t) = H_0 + H_1(X_i(t)).$$

The energy eigenfunctions of the unperturbed Hamiltonian are the solutions of the equation,

$$(3) \qquad H_0\varphi_m(\xi_i, \eta_j, r) = \mathcal{E}_m\varphi_m(\xi_i, \eta_j, r),$$

the wave functions $\varphi_m(\xi_i, \eta_j, r)$ satisfy the boundary conditions appropriate to a decaying state.

Assuming that, in the absence of perturbation, the unstable system decays spontaneously in two stable fragments, the unperturbed energy eigenfunctions may be written as cluster model wave functions[17]

$$(4) \qquad \varphi_m(\xi_i, \eta_j, r) = \mathcal{A}\left\{\phi_A(\xi_i)\phi_B(\eta_j)\frac{u_{ml}(r)}{r}\mathcal{Y}_{Jls}^M(\hat{r})\right\}.$$

where $\phi_A(\xi_i)$ and $\phi_B(\eta_j)$ are the wave functions of the clusters A and B, $u_{ml}(r)$ is the radial part of the wave function of the relative motion of the two clusters, $\mathcal{Y}_{Jls}^M(\hat{r})$ is a spherical harmonic and $\mathcal{A}$ is the antisymmetrizer. In our example, φ_m would be the eigenfunction of a state of ^{8}Be which decays spontaneously in two ^{4}He clusters. In this case $u_{ml}(r)$ is a Gamow function.

2.2 A Few Facts About Gamow Functions

Gamow functions are the eigenfunctions of the time independent Schrödinger equation which vanish at the origen, and behave as purely outgoing waves for large values of the relative distance r,

(5) $$u_{ml}(0) = 0,$$

and

(6) $$\lim_{r \to \infty} [u_{ml}(r) - O_l(k_m, r)] = 0,$$

where the function $O_l(k_m, r)$ is an outgoing spherical wave of complex wave number k_m and angular momentum l[12].

The boundary condition (6) is not self-adjoint, in consequence, the energy eigenvalues are complex, with $Re\mathcal{E}_m > 0$ and $Im\mathcal{E}_m < 0$. Hence, Gamow functions are a generalization of bound state eigenfunctions in that they belong to complex wave numbers, $k_m = \kappa_m - i\gamma_m$, with $\kappa_m > \gamma_m > 0$, instead of purely imaginary ones. This generalization leads out of the Hilbert space based quantum mechanics. Therefore, the quantum mechanical rules for normalization, orthogonality and completeness in their usual form, do not apply. Nevertheless, bound and resonant states form a bi-orthonormal set with their adjoints, which may be amended by a continuum to a complete basis system of generalized eigenvectors of H in terms of which any well behaved wave function of a state vector may be expanded [12].

The symmetry properties of the Schrödinger equation and the boundary conditions under the operations of complex conjugation and reversal of time suggest a definition for the adjoint $\tilde{u}_{ml}(k_m, r)$ of the Gamow function $u_{ml}(k_m, r)$[12],

(7) $$\tilde{u}_{ml}(k_m, r) = u_{ml}^*(-k_m^*, r).$$

Now, if the adjoint $\tilde{u}_{ml}(k_m, r)$ is identified with the bra-eigenfunction, the quantum mechanical inner product, or bra-c-ket rule, may be generalized to include Gamow eigenfunctions. With this prescription, matrix elements of operators and bra-c-kets are computed in configuration space representation as integrals over the radial variable r. Since the Gamow function $u_{ml}(k_m, r)$ and its adjoint $\tilde{u}_{ml}(k_m, r)$ oscillate between envelopes that grow exponentially with r, the integrals over r must be properly defined. This may be done, either by analytic continuation in the complex k-plane from above [12, 18] or by means of an Abel regulator[19] with a suitable convergence factor and a limiting procedure[20, 21]. Both procedures give the same result[22, 23]. In this paper, we will adopt the second definition.

Then, it may be shown that

(8) $$\lim_{\mu \to 0} \int_0^\infty e^{-\mu r^2} u_{ml}^*(-k_m^*, r) u_{nl'}(k_n, r) dr = \delta_{mn} \delta_{ll'}$$

i.e, the Gamow eigenfunctions are orthonormalized in a generalized sense. It may also be shown[12, 24] that, for any two square integrable functions, $f(r)$ and $g(r)$, the following relation holds

$$\int_0^\infty f^*(r)g(r)dr$$

$$= \sum_n \left\{ \left[\int_0^\infty f^*(r)u_{nl}(k_n,r)dr \right] \times \left[\int_0^\infty u_{nl}^*(-k_n^*,r')g(r')dr' \right] \right\}$$

$$(9) \qquad + \int_c dk \left[\int_0^\infty f^*(r)\phi_l(k,r)dr \right] \left[\int_0^\infty \phi_l^*(-k^*,r')g(r')dr' \right].$$

In this expression, the functions $u_{nl}(r)$ are bound states or Gamow state eigenfuctions belonging to real negative or complex eigenvalues $\mathcal{E}_n$, the functions $\phi_l(k,r)$ are scattering partial wave functions of angular momentum l and complex wave number k. The integration contour C, in the wave number plane k, starts from the origin as a straight line with slope -1, it goes down to $Im\,k = \alpha$ and then, it continues as a straight line parallel to the real axis[12]. Finally, the square brackets around the integrals mean that, when necessary, the integrals are defined by means of a gaussian regulator and a limit as in (8) or by analytical continuation from above as in Mondragón and Hernández[12, 24].

Since $f(r)$ and $g(r)$ are arbitrary very well behaved functions, we are justified in writing the expansion

$$(10) \qquad g(r) = \sum_m u_{ml}(r) <u_{ml}|g> + \int_c \phi_l(k,r) <\phi_l(k)|g> dk,$$

the index m runs over bound and resonant states.

The expansion coefficients are given by

$$(11) \qquad <u_{ml}|g> = \lim_{\mu\to 0} \int_0^\infty e^{-\mu r^2} u^*(-k_m^*,r)g(r)dr,$$

$$(12) \qquad <\phi_l(k)|g> = \lim_{\mu\to 0} \int_0^\infty e^{-\mu r^2} \phi_l^*(-k^*,r)g(r)dr.$$

Once the validity of the quantum mechanical inner product has been generalized to apply to bound and resonant eigenfunctions of the relative motion of the two clusters, the generalization of (9) and (10) to expansions of many body wave functions in terms of cluster model bound and resonant eigenfunctions is fairly straightforward[12, 24].

2.3 The Mixing Matrix

We may, now, go back to our problem, namely the time evolution of the resonant states of a many body system moving in a slowly time-dependent external field of force. Since we are interested in the time evolution of a state

Ψ which is a superposition of unstable states, we make an expansion of the wave function Ψ in terms of bound and resonant states of H_0,

$$(13) \qquad \Psi = \sum_m a_m(t)\varphi_m(\xi_i, \eta_j, r) + \int_c b(k;t)\varphi^{(+)}(k;\xi_i, \eta_j, r)dk.$$

In general, the index m runs over bound and resonant states. In our example, φ_m would be the complex energy eigenfunctions of the ^{8}Be nucleus which decays spontaneously in two ^{4}He clusters. The scattering states $\varphi^{(+)}(k, \xi_i, \eta_j, r)$ of complex wave number k and the integration contour C are defined in the previous subsection.

Substitution of (13) in (1) gives the set of coupled equations

$$\frac{da_m(t)}{dt} = -\frac{i}{\hbar}\mathcal{E}_m a_m(t) - \frac{i}{\hbar}\sum_n < \varphi_m|H_1(t)|\varphi_n > a_n(t)$$

$$(14) \qquad -\frac{i}{\hbar}\int_c < \varphi_m|H_1(t)|\varphi^{(+)}(k) > b(k;t)dk$$

and a similar expression for $db(k,t)/dt$.

We will use the notation $|\varphi_m(\xi_i, \eta_j, r) >$ for the Gamow function, and $< \varphi_m(\xi_i, \eta_j, r)|$ for its adjoint. Hence, the matrix element of the perturbation term $H_1(t)$ taken between bound or resonant states of the unperturbed system is given by

$$\langle\varphi_m|H_1(t)|\varphi_n\rangle = \int \cdots \int \langle\varphi_m(\xi_i, \eta_j, r)|H_1(t)|\varphi_n(\xi_i, \eta_j, r)\rangle$$

$$(15) \qquad \times d^3\xi_1 \cdots d^3\xi_{A-1}d^3\eta_1 \cdots d^3\eta_{B-1}d^3r,$$

the integral over the radial variable r is defined as in (8).

When the interactions are time reversal invariant, the dual of the complex Gamow function $u_{ml}(r)$ is the same function[12, 21, 24]. But, when the interactions are not time reversal invariant, the Gamow function and its dual are not the same function.

Therefore, when the forces acting on the system are time reversal invariant, the complex matrix $\mathbf{H}$, with matrix elements

$$(16) \qquad \mathbf{H}_{mn}(t) = \mathcal{E}_m\delta_{mn} + \langle\varphi_m|H_1(t)|\varphi_n\rangle$$

is symmetric but non-Hermitian. When the forces acting on the system are not time reversal invariant, $\mathbf{H}$ is, in general, complex, non-symmetric and non-Hermitian.

The contribution of the non-resonant background integral over the continuum of scattering wave functions in (13) and (14) will not be relevant to the following discussion. Therefore, to ease the notation, we will disregard the background integrals in (13) and (14). With this truncation,

$$(17) \qquad H\Psi = \sum_m |\varphi_m(\xi_i, \eta_j, r)> \left[\sum_n \mathbf{H}_{mn} a_n(t) \right].$$

Then, the set of coupled equations(14) reduces to

$$(18) \qquad \dot{a}_m(t) = -\frac{i}{\hbar} \sum_n \mathbf{H}_{mn} a_n(t),$$

where $\dot{a}_m(t)$ is the time derivative of $a_m(t)$.

3. Geometric Phase of a Resonant State

The time evolution of the quantum system is governed by the Hamiltonian H, or the matrix $\mathbf{H}$, which are functions of N real, linearly independent parameters. Therefore, we may consider the matrix $\mathbf{H}$ embedded in a population of $\mathbf{H}$ matrices smoothly parametrized by N external parameters which take values in some domain D of a manifold or parameter space. Each point in D represents an $\mathbf{H}$ matrix. When the external parameters change with time, the system traces a path C in parameter space. In the following, we will study the behaviour of the system in the time interval [0,T], assuming that at the initial time, $t = 0$, it is in an eigenstate of $\mathbf{H}$, at $t = T$ the parameters $(X_1, X_2, ...X_N)$ have returned to their initial values, and the adiabatic theorem holds. Then, the unstable system traces a closed path C in parameter space while it remains in an eigenstate of $\mathbf{H}(t)$ at all times.

In the absence of symmetry, $\mathbf{H}$ has no repeated eigenvalues at almost all points in the domain D. The set of points in D where $\mathbf{H}$ has one twofold repeated eigenvalue is a subdomain $D' \subset D$. If D has N dimensions, and D' has N' dimensions, then $N = N' + \kappa$, κ is the codimension of the twofold degeneracy. Similar relations hold in the case of an m-fold degeneracy.

It will be assumed that the complex non-Hermitian matrix $\mathbf{H}$ has no repeated eigenvalues at all points on the path C. In consequence, at all points on C, it may be brought to diagonal form by means of a similarity transformation

$$(19) \qquad \mathbf{K}^{-1}\mathbf{H}\mathbf{K} = \mathbf{E},$$

where $\mathbf{E}$ is the diagonal matrix of the complex energy eigenvalues. The columns in the matrix $\mathbf{K}$ are the instantaneous right eigenvectors of $\mathbf{H}$. In an obvious notation

(20) $$\mathbf{K} = \left(|\phi^{(1)}\rangle, |\phi^{(2)}\rangle, \ldots |\phi^{(s)}\rangle \ldots |\phi^{(n)}\rangle \right).$$

which satisfy the eigenvalue equation

(21) $$\mathbf{H}\left(X_l(t)\right) |\phi^{(s)}(t)\rangle = \hat{\mathcal{E}}_s(t) |\phi^{(s)}(t)\rangle.$$

The rows in $\mathbf{K}^{-1}$ are the corresponding left eigenvectors of $\mathbf{H}$, properly normalized,

(22) $$\left\langle \phi^{(i)} | \phi^{(j)} \right\rangle = \delta_{ij}.$$

The adiabatic basis, $\{|\hat{\varphi}_s\left(\xi_i, \eta_j, r; X_l(t)\right)\rangle\}$, of instantaneous bound and resonant energy eigenstates of the complete Hamiltonian H, is obtained from the set of unperturbed bound and resonant states with the help of the matrix $\mathbf{K}$

(23) $$|\hat{\varphi}_s(\xi_i, \eta_j, r; X_l(t))\rangle = \sum_m |\varphi_m(\xi_i, \eta_j, r)\rangle \mathbf{K}_{ms}(t)$$

and their adjoints (duals) are given by

(24) $$\langle \hat{\varphi}_s(\xi_i, \eta_j, r; X_l(t))| = \sum_n (\mathbf{K}^{-1}(t))_{sn} \langle \varphi_n(\xi_i, \eta_j, r)|.$$

It follows that the instantaneous bound and resonant energy eigenstates of $H(t)$ satisfy an orthogonality relation similar to (8), which they inherit from the unperturbed complex energy eigenfunctions

$$\langle \hat{\varphi}_s(\xi_i, \eta_j, r; X_l(t)) | \hat{\varphi}_{s'}(\xi_i, \eta_j, r; X_l(t)) \rangle$$
$$= \sum_{m,n} (\mathbf{K}^{-1})_{sn} \langle \varphi_n(\xi_i, \eta_j, r) | \varphi_m(\xi_i, \eta_j, r) \rangle \mathbf{K}_{ms'}$$

(25) $$= \sum_{m,n} (\mathbf{K}^{-1})_{sn} \delta_{nm} \mathbf{K}_{ms'} = \delta_{ss'}.$$

If we multiply both sides of eq.(23) by $\mathbf{K}^{-1}$, we obtain

(26) $$|\varphi_m(\xi_i, \eta_j, r)\rangle = \sum_s |\hat{\varphi}_s(\xi_i, \eta_j, r; X_l(t))\rangle (\mathbf{K}^{-1})_{sm}.$$

Substitution of this expresion in (13), gives the expansion of Ψ in instantaneous energy eigenfunctions of $\mathbf{H}$,

(27)
$$\Psi = \sum_s |\hat{\varphi}_s\, (\xi_i, \eta_j, r; X_l(t))\rangle \hat{a}_s(t),$$

where

(28)
$$\hat{a}_s(t) = \sum_n \left(\mathbf{K}^{-1}(t)\right)_{sn} a_n(t).$$

In (27) we have kept only the summation over bound and resonant states and, as in the previous section, we have disregarded the contribution from the integral over the continuum of scattering functions of complex wave number. Similarly, the expansion of $H\Psi$ becomes

(29)
$$H\Psi = \sum_s |\hat{\varphi}_s\, (\xi_i, \eta_j, r; X_l(t))\rangle \hat{\mathcal{E}}_s(t)\hat{a}_s(t).$$

Substitution of (27) and (29) in the time dependent Schrödinger equation gives the set of coupled equations

(30)
$$\frac{d\hat{a}_s(t)}{dt} + \sum_{m=1} \langle \hat{\varphi}_s | \nabla_{\Re} \hat{\varphi}_m \rangle \cdot \frac{d\vec{\Re}}{dt} \hat{a}_m(t) = -i\hat{\mathcal{E}}_s(t)\hat{a}_s(t)$$

where, $|\nabla_{\Re}\hat{\varphi}_m >$ is the gradient of $|\hat{\varphi}_m >$ in parameter space,

(31)
$$\frac{d|\hat{\varphi}_m >}{dt} = \sum_{i=1}^{N} \left(\frac{\partial}{\partial X_i}|\hat{\varphi}_m >\right) \frac{dX_i}{dt} = |\nabla_{\vec{\Re}}\varphi_m > \cdot \frac{d\vec{\Re}}{dt}$$

It will be assumed that the non-adiabatic transition amplitudes are very small

(32)
$$\frac{1}{|\hat{a}_s|} \left| \langle \hat{\varphi}_s | \nabla_{\Re} \hat{\varphi}_m \rangle \cdot \frac{d\vec{\Re}}{dt} \right| << 1, \qquad m \neq s.$$

Then, we can make the approximation

(33)
$$\frac{1}{\hat{a}_s} \frac{d\hat{a}_s}{dt} \simeq -i\hat{\mathcal{E}}_s(t) - \langle \hat{\varphi}_s | \nabla_{\Re} \hat{\varphi}_s \rangle \cdot \frac{d\vec{\Re}}{dt}.$$

Integrating both sides, we get

(34)
$$\hat{a}_s(t) = exp[-\frac{i}{\hbar} \int_{t_0}^{t} \hat{\mathcal{E}}_s(t')dt']exp[i\gamma_s]a_s(0),$$

the first factor is the complex dynamical phase, whereas the second one is the complex Berry phase given by

$$(35) \qquad \gamma_s = i \int_c \langle \hat{\varphi}_s | \nabla_{\Re} \hat{\varphi}_s \rangle \cdot d\vec{\Re}.$$

Direct evaluation of $\nabla_{\Re} | \hat{\varphi}_s \rangle$ requires a locally single valued basis for $| \hat{\varphi}_s \rangle$ and can be awkward. Such difficulties are avoided by transforming the path integral into a surface integral with the help of Stokes theorem

$$(36) \qquad \gamma_s = i \sum_{m \neq s} \int_\Sigma \int_{\partial \Sigma = c} \langle \nabla_{\Re} \hat{\varphi}_s | \hat{\varphi}_m \rangle \wedge \langle \hat{\varphi}_m | \nabla_{\Re} \hat{\varphi}_s \rangle \cdot d\vec{\Sigma},$$

where Σ is any surface in parameter space whose boundary is the curve $\mathbf{C}$, and $\wedge$ means wedge product.

This expression may be written in a more convenient form by means of the identity

$$(37) \qquad \langle \hat{\varphi}_s | \nabla_{\Re} \hat{\varphi}_m \rangle = \frac{1}{\hat{\mathcal{E}}_m - \hat{\mathcal{E}}_s} \langle \hat{\varphi}_s | \nabla_{\Re} H_1 | \varphi_m \rangle,$$

then,

$$(38) \qquad \gamma_s = i \sum_{m \neq s} \int_\Sigma \int_{\partial \Sigma = c} \frac{\langle \hat{\varphi}_s | \nabla_{\Re} H_1 | \hat{\varphi}_m \rangle \wedge \langle \hat{\varphi}_m | \nabla_{\Re} H_1 | \hat{\varphi}_s \rangle \cdot d\vec{\Sigma}}{\left(\hat{\mathcal{E}}_s - \hat{\mathcal{E}}_m \right)^2},$$

provided the surface $\sum$ does not cross any point in D where the denominator vanishes. Since the dependence on $| \nabla_{\Re} \varphi_s \rangle$ has been eliminated in (38), phase relations between eigenstates with different parameters are no longer important, any complete set of instantaneous eigenfunctions of the time dependent Hamiltonian H may be used to evaluate the integral in (38).

Explicit expressions for the Berry phase in terms of our parametrization of the interaction Hamiltonian may easily be obtained. Since the unperturbed Hamiltonian and its bound and resonant states are independent of time, the time dependence of the bound and resonant instantaneous energy eigenstates of the perturbed system is entirely contained in the matrix $\mathbf{K}$, which is a function of t through the time dependence of the external parameters. Therefore, from eqs.(23), (24) and (25) we get

$$(39) \qquad \gamma_s = i \int_c \left[\mathbf{K}^{-1}(\nabla_{\Re} \mathbf{K}) \right]_{ss} \cdot d\vec{\Re}.$$

The columns in $\nabla_{\Re} \mathbf{K}$ are the gradients of the instantaneous eigenvectors of $\mathbf{H}(t)$. Therefore, γ_s may also be written as

$$(40) \qquad \gamma_s = i \int_c \langle \phi_s | \nabla_{\Re} \phi_s \rangle \cdot d\vec{\Re}.$$

Furthermore, from (36) we may derive a sum rule for the geometric phases of the interfering resonant states. Writing the integrand in the right hand side of (36) in terms of $\mathbf{K}$ and $\nabla_{\Re}\mathbf{K}$, and taking the sum over s, we get

$$(41) \qquad \sum_s \gamma_s = i \int_\Sigma \int_{\partial\Sigma=c} tr\left[(\nabla_{\Re}\mathbf{K}^{-1}) \wedge (\nabla_{\Re}\mathbf{K})\right] \cdot d\Sigma$$

but

$$(42) \qquad -\frac{1}{2\pi} tr\left[(\nabla_{\Re}\mathbf{K}^{-1}) \wedge (\nabla_{\Re}\mathbf{K})\right] = C_1(N),$$

is the first Chern class of a complex vector bundle defined by the matrix $\mathbf{H}$ over the parameter space M. Its integral is the first Chern number[25]

$$(43) \qquad -\frac{1}{2\pi} \int_\Sigma \int_{\partial\Sigma=c} tr\left[(\nabla_{\Re}\mathbf{K}^{-1}) \wedge (\nabla_{\Re}\mathbf{K})\right] \cdot d\vec{\Sigma} = c_1.$$

Hence,

$$(44) \qquad \sum_s \gamma_s = i \oint tr\left[\mathbf{K}^{-1}(\nabla_{\Re}\mathbf{K})\right] \cdot d\vec{\Re}. = -2\pi c_1,$$

is a topological invariant.

4. Berry Phase Factors of Resonant States and Holonomy in a Complex Line Bundle

4.1 Resonant States as Elements of a Rigged Hilbert Space

Since 1983, B. Simon[15] pointed out that for Hermitian closed quantum systems, the adiabatic evolution can be interpreted as a parallel translation in a Hermitian line bundle and the Berry phase factor is the holonomy in such a bundle. In this section , it will be shown that this geometric interpretation may be generalized in a very natural way to the Berry phase factor of resonant states of open quantum systems.

The resonant or Gamow functions $\{|\phi_m(\xi_i, \eta_j, r >\}$ of the unperturbed system are eigenfunctions of the self-adjoint unperturbed Hamiltonian H_0, which satisfy purely outgoing wave boundary conditions for large values of the separation distance r of the decay fragments. Since the boundary condition is not self-adjoint, the corresponding energy eigenvalues $\mathcal{E}_m$, are complex, with $Re\mathcal{E}_m > 0$ and $Im\mathcal{E}_m < 0$. In configuration space representation, the Gamow functions, as functions of r, behave as outgoing waves which oscillate between envelopes that increase exponentially. In consequence, Gamow functions are not square integrable and cannot be characterized as elements of a Hilbert space.

Gamow functions are usually associated with the resonance poles of the scattering matrix and the resolvent operator of the time independent Schrödinger equation which lie in the lower half-plane of the unphysical sheet of the Riemann energy surface[1, 12]. In order to give a proper mathematical characterization of Gamow functions as elements of a space, one has to specify a rigged Hilbert space in which the Gamow states $\{|\varphi_m(\xi_i, \eta_j, r) >\}$ are defined as generalized eigenvectors of the Hamiltonian H_0 with generalized complex eigenvalues $\mathcal{E}_m$.

Following Bohm and Gadella[26], we will associate to the resonant poles of the resolvent operator of the Schrödinger equation of the unperturbed system, a rigged Hilbert space

$$(45) \qquad \Phi_+ \subset \mathcal{H} \subset \Phi_+^*,$$

in this expression, Φ_+ is the space of well-behaved functions of the position coordinates (ξ_i, η_j, r) which are Hardy class 2 functions of the complex energy E from above, $\mathcal{H}$ is the Hilbert space of square integrable functions and Φ_+^* is the space of continuous antilinear functionals on Φ_+. In this way, the Gamow states of the unperturbed system are defined as continuous antilinear functionals on Φ_+, that is,

$$(46) \qquad |\varphi_m(\xi_i, \eta_j, r) > \epsilon\Phi_+^*.$$

The adiabatically evolving Gamow functions $|\hat{\varphi}_s(\xi_i, \eta_j, r; \vec{\Re}(t)) >$ of the complete Hamiltonian H, introduced in section 3, eq.(23), are linear combinations of the Gamow functions $|\varphi_m(\xi_i, \eta_j, r) >$ of H_0,

$$(47) \qquad |\hat{\varphi}_s(\xi_i, \eta_j, r; [\vec{\Re}(t)]) >= \sum_m |\varphi_m(\xi_i, \eta_j, r) > \mathbf{K}_{ms}[\vec{\Re}(t)],$$

where $\mathbf{K}[\vec{\Re}(t)]$ is the matrix which diagonalizes the complex non-Hermitian matrix $\mathbf{H}[\vec{\Re}(t)]$.

Therefore, the adiabatically evolving Gamow functions are also elements of Φ_+^*,

$$(48) \qquad |\hat{\varphi}_s(\xi_i, \eta_j, r; [\vec{\Re}(t)]) > \epsilon\Phi_+^*.$$

The dual of the Gamow eigenvector $|\hat{\varphi}_s[\vec{\Re}]>$ is the eigenvector $<\hat{\varphi}_s[\vec{\Re}]| = \sum_m (\mathbf{K}^{-1})_{sm} <\varphi_m(\xi_i,\eta_j)|$ defined in eq.(24), corresponding to the same complex eigenvalue $\mathcal{E}_s[\vec{\Re}]$. Since, by assumption $\mathbf{H}[\vec{\Re}]$ has no repeated eigenvalues for $\vec{\Re}\epsilon C \subset M$, the adiabatically evolving Gamow eigenvectors and their duals satisfy the orthogonality relation

$$(49) \qquad <\hat{\varphi}_s[\vec{\Re}]|\hat{\varphi}_{s'}[\vec{\Re}]>= 0, \qquad\qquad \hat{\mathcal{E}}_s \neq \hat{\mathcal{E}}_s'$$

inherited from the Gamow eigenfuctions $|\varphi_m(\xi_i,\eta_j,r)>$ of H_0, and they may be normalized to one

$$(50) \qquad\qquad <\hat{\varphi}_s[\vec{\Re}]|\hat{\varphi}_s[\vec{\Re}]>= 1$$

Having characterized the adiabatically evolving Gamow functions as elements of a rigged Hilbert space we may turn to the question of the geometric interpretation of the Berry phase factors of resonant states.

4.2 Adiabatic Evolution and Parallel Translation

Let us suppose that, as time varies, the self-adjoint Hamiltonian $H[\vec{\Re}(t)]$ and its instantaneous energy eigenstates $|\hat{\varphi}_s(\xi_i,\eta_j,r;\vec{\Re}(t))>$ make a cyclic excursion in a closed circuit C in parameter space. We assumed that the time dependence of the Hamiltonian justifies the adiabatic approximation. We will further assume that, if $\vec{\Re}(t)$ is any point on C, the complex matrix $\mathbf{H}[\vec{\Re}(t)]$ has no repeated eigenvalues in an open neighbourhood of $\vec{\Re}\epsilon$ C, that is, the Hamiltonian $H(t)$ has no repeated instantaneous complex eigenvalues when $\vec{\Re}\epsilon$ C, and that both the generalized eigenvectors $|\hat{\varphi}_s(\xi_i,\eta_j,r;\vec{\Re}(t))>$ and the eigenvalues $\mathcal{E}_s[\vec{\Re}(t)]$ are smooth functions of $\vec{\Re}\epsilon C$. Moreover, we will also assume that the evolution of each state is such that there are no level crossings along C.

To avoid a clumsy notation in the rest of this section we will not write the position coordinates (ξ_i,η_j,r).

Let us call F_s the complex line bundle defined by the Gamow eigenfunction $|\hat{\varphi}_s[\vec{\Re}(t)]>$ over the parameter space M of the adiabatically evolving system

$$(51) \qquad F_s = \left\{ (\vec{\Re},|\hat{\varphi}_s[\vec{\Re}]>)|H[\vec{\Re}]|\hat{\varphi}_s[\vec{\Re}]>= \mathcal{E}_s[\vec{\Re}]|\hat{\varphi}_s[\vec{\Re}]>, \vec{\Re}\epsilon M \right\}$$

Its fibre is a complex, one-dimensional linear space

$$(52)$$
$$L^s_{\Re} := \left\{ |\hat{\psi}_s[\vec{\Re}]> \, ||\hat{\psi}_s[\vec{\Re}]>= e^{i\alpha}|\hat{\varphi}_s[\vec{\Re}]>, |\hat{\varphi}_s[\vec{\Re}]> \epsilon \Phi^*_+, \alpha[\vec{\Re}]\epsilon \mathbf{C}^*, \vec{\Re}\epsilon M \right\}$$

where Φ_+^* is the space of antilinear functionals defined over the space Φ_+ of well-behaved functions of Hardy class 2 from above and $\alpha[\vec{\Re}]$ is a complex function of $\vec{\Re}\epsilon M$.

Under the assumptions made above, the quantum number s, labelling the eigenvalue $\hat{\mathcal{E}}_s[\vec{\Re}]$ and the eigenstate $|\hat{\varphi}_s[\vec{\Re}] >$ of the instantaneous Hamiltonian, is an adiabatic invariant. Hence, a system prepared in a state $|\hat{\psi}_s[\vec{\Re}(t)] >$ such that

$$(53) \qquad |\hat{\psi}_s[\vec{\Re}(t_o)] >= C_s[\vec{\Re}(t_o)]|\hat{\varphi}_s[\vec{\Re}(t_o)] >$$

will evolve with $H[\vec{\Re}(t)]$ and be in a state

$$(54) \qquad |\hat{\psi}_s[\vec{\Re}(t)] >= C_s[\vec{\Re}(t)]|\hat{\varphi}_s[\vec{\Re}(t)] >$$

at t.

Now, consider the decomposition of a tangent vector in vertical and horizontal parts

$$(55) \qquad \frac{d}{dt}|\hat{\psi}_s(t) >= \left(\frac{d}{dt}|\hat{\psi}_s(t)\right)_{||} + \left(\frac{d}{dt}\hat{\psi}_s(t) >\right)_{\perp}$$

The horizontal part, orthogonal to the fibre is

$$(56) \qquad \left(\frac{d}{dt}|\hat{\psi}_s(t) >\right)_{\perp} = \sum_{m \neq s} < \hat{\varphi}_m[\vec{\Re}]|\frac{d}{dt}|\hat{\psi}_s(t) > |\hat{\varphi}_m[\vec{\Re}] >$$

and the vertical part, along the fibre is

$$(57) \qquad \left(\frac{d}{dt}|\hat{\psi}_s(t) >\right)_{||} =< \hat{\varphi}_s[\vec{\Re}]|\frac{d}{dt}|\hat{\psi}_s(t) > |\hat{\varphi}_s[\vec{\Re}] >$$

If $|\hat{\psi}_s(t) >$ is an evolution state in an adiabatic change,

$$(58)$$
$$\left(\frac{d}{dt}|\hat{\psi}_s(t) >\right)_{||} = \left(\frac{dC_s[\vec{\Re}(t)]}{dt}+ < \hat{\varphi}_s[\vec{\Re}]|\frac{d}{dt}\hat{\varphi}_s[\vec{\Re}(t)] > C_s[\vec{\Re}]\right) |\hat{\varphi}_s[\vec{\Re}] >$$

The condition for parallel translation along the curve C is

$$(59) \qquad < \hat{\varphi}_s[\vec{\Re}]|\frac{d}{dt}\hat{\psi}_s[\vec{\Re}] >= 0, \qquad \vec{\Re}\epsilon C$$

which gives a one-form equation

(60)
$$dC_s[\vec{\mathfrak{R}}] + < \hat{\varphi}_s[\vec{\mathfrak{R}}] | d\hat{\varphi}_s[\vec{\mathfrak{R}}] > C_s[\vec{\mathfrak{R}}] = 0$$

Its solution gives the Berry phase factor of the resonant state

(61)
$$exp[i \int_c i < \hat{\varphi}_s | d\hat{\varphi}_s >] = exp[i\gamma_s[\vec{\mathfrak{R}}]].$$

In the case of a cyclic evolution, such that $\vec{\mathfrak{R}}(0) = \vec{\mathfrak{R}}(T)$ and C is a closed curve, the complex phase γ_s may be expressed as

(62)
$$\gamma_s(C) = i \oint_c < \hat{\varphi}_s | d\hat{\varphi}_s > .$$

Now, it will be shown that the condition for the adiabatic evolution of a resonant state is equivalent to the condition for parallel translation along C.

The condition for the adiabatic time evolution of a resonant state, equation (33), may be written as

(63)
$$e^{\frac{-i}{\hbar} \int_0^t \hat{\mathcal{E}}_s(t')dt'} \left\{ \frac{d}{dt} \left(\hat{a}_s(t) e^{\frac{i}{\hbar} \int_0^t \hat{\mathcal{E}}_s(t')dt'} \right) \right.$$
$$\left. + < \hat{\varphi}_s | \frac{d}{dt} \hat{\varphi}_s > \hat{a}_s(t) e^{\frac{i}{\hbar} \int_0^t \hat{\mathcal{E}}_s(t')dt'} \right\} = 0.$$

Now, making use of the normalization condition, eq.(25), this expressions may be written as

(64)
$$e^{-\frac{i}{\hbar} \int_0^t \hat{\mathcal{E}}_s(t')dt'} < \hat{\varphi}_s[\vec{\mathfrak{R}}(t)] | \left\{ |\hat{\varphi}_s[\vec{\mathfrak{R}}(t)] > \frac{dC_s(t)}{dt} + \frac{|d\varphi_s[\vec{\mathfrak{R}}(t)] >}{dt} C_s(t) \right\} = 0,$$

where $C_s(t)$ is given by

(65)
$$C_s(t) = e^{\frac{i}{\hbar} \int_0^t \hat{\mathcal{E}}_s(t')dt'} \hat{a}_s(t),$$

multiplying both sides of (64) by the dynamic phase factor, the condition for adiabatic evolution of a resonant state takes the form

(66)
$$< \hat{\varphi}_s | \frac{d}{dt} \hat{\psi}_s >= 0, \qquad \vec{\mathfrak{R}} \epsilon C$$

which is the condition for parallel translation of the resonant state $|\hat{\psi}_s[\vec{\Re}(t)] >$ $= \hat{C}_s[\vec{\Re}(t)]|\hat{\varphi}_s[\vec{\Re}(t)] >$ along the curve C.

It follows that, except for the effect of the dynamical phase factor, the condition of adiabatic evolution of a resonant state is equivalent to the condition of parallel translation on the complex line bundle F_s.

The geometric phase factors $exp[i\gamma_s(c)]$, occurring in the adiabatic evolution, are holonomy group elements of the complex line bundle F_s [25].

5. Accidental Degeneracy of Resonances

As is apparent from (38), non-trivial phase factors of the energy eigenvectors or eigenfunctions are related to the occurrence of accidental degeneracies of the corresponding eigenvalues. In the absence of symmetry, degeneracies are called accidental for lack of an obvious reason to explain why two energy eigenvalues $\mathcal{E}_s$ and $\mathcal{E}_m$ of **H** should coincide. However, if the matrix **H** is embedded in a population of complex non-Hermitian matrices $[\mathbf{H}(X_1, X_2, ...X_N)]$ smoothly parametrized by N external parameters $(X_1, X_2, ...X_N)$, degeneracy in the absence of symmetry is a geometric property of the hypersurfaces representing the real or complex eigenvalues of **H** in a $(N + 2)$-dimensional Euclidean space with Cartesian coordinates $(X_1, X_2, ...X_N, Re\mathcal{E}, Im\mathcal{E})$. The energy denominators in (38) show that when the circuit C lies close to a subdomain $D' \subset D$, in parameter space, where the matrix **H** has an m-fold repeated eigenvalue, and the state $|\hat{\varphi}_s\rangle$ is involved in this degeneracy, the Berry phase $\gamma_s(c)$ is determined by the geometry of the energy hypersurfaces close to the crossing of eigenvalues and the other (m-1) states involved in the degeneracy.

It is in this connection of the Berry phase with the accidental degeneracy of energy eigenvalues that the non-Hermiticity of the matrix **H** plays an important role. In contrast with the case of Hermitian matrices, square, complex, non-Hermitian matrices with repeated eigenvalues cannot always be brought to diagonal form by a similarity transformation. However, any n-dimensional, square complex matrix **H** may always be brought to a Jordan canonical form **E** by means of a similarity transformation.

If **H** has ν ($\nu \leq n$) different eigenvalues, $\mathcal{E}_1, \mathcal{E}_2, ...\mathcal{E}_\nu$, with multiplicities $\mu_i(E_i)$, the Jordan canonical form **E** is the direct sum of ν square Jordan blocks $\mathbf{E}_i$. Each Jordan block $\mathbf{E}_i$ is the sum of a diagonal matrix $\mathcal{E}_i\mathbf{I}_{\mu_i \times \mu_i}$, and a nilpotent matrix N_{μ_i}. Corresponding to each Jordan block, there is a cycle of lenght μ_i of generalized eigenvectors. When the lenght μ_i of the cycle is ≥ 2, the codimension of the accidental degeneracy, the geometry of the energy hypersurfaces at the crossing and the properties of the generalized eigenvectors involved in the degeneracy of resonant states differ substantially from those of bound states([13]). Rather than trying to develop a theory of the most general case, in the following we will examine the simplest possible case, namely the accidental degeneracy of two resonances and the topology of the energy surfaces close to a crossing of two resonances in parameter space.

6. Degeneracy of Two Resonances

Let us consider a system with two resonant states strongly mixed by a Hermitian interaction, all other bound or resonant eigenstates being non-degenerate. We may suppose that we already know the correct eigenvectors of $\mathbf{H}$ for all the real and complex eigenenergies $\mathcal{E}_s$, except for the two the crossing of which we want to investigate. Using for this two states two vectors which are not eigenvectors but which are orthogonal to each other and to all other eigenvectors, we obtain a complete basis to represent $\mathbf{H}$. In this basis, $\mathbf{H}$ will be diagonal except for the elements $\mathbf{H}_{12}$ and $\mathbf{H}_{21}$. The diagonal elements $\mathbf{H}_{11}$ and $\mathbf{H}_{22}$ will, in general, be non-vanishing and different from each other. There is no loss of generality in this supposition, since any complex matrix $\mathbf{H}$ may be brought to a Jordan canonical form by means of a similarity transformation. When the eigenvalues are equal, $\mathbf{H}_{2\times 2}$ is either diagonal or equivalent to a Jordan block of rank two. Hence, we need consider only the conditions for degeneracy of the submatrix $\mathbf{H}_{2\times 2}$.

The matrix $\mathbf{H}_{2\times 2}$ may be written in terms of the Pauli matrix valued vector $\vec{\sigma} = (\sigma_1, \sigma_2, \sigma_3)$ and the 2×2 unit matrix as

$$(67) \qquad \mathbf{H}_{2\times 2} = \mathcal{E}\mathbf{1} + \left(\vec{R} - i\frac{1}{2}\vec{\Gamma}\right) \cdot \vec{\sigma},$$

where $\vec{R}$ and $\vec{\Gamma}$ are real vectors with cartesian components (X_1, X_2, X_3) and $(\Gamma_1, \Gamma_2, \Gamma_3)$. When the forces acting on the system are time reversal invariant, X_2 and Γ_2 vanish.

In the absence of more specific information about the external parameters X, we will parametrize $\mathbf{H}_{2\times 2}$ in terms of $\vec{R}$ and $\vec{\Gamma}$. Then,

$$(68) \qquad \vec{\mathfrak{R}} = \vec{R} - i\frac{1}{2}\vec{\Gamma}.$$

From (67), the eigenvalues of $\mathbf{H}_{2\times 2}$ are given by

$$(69) \qquad \mathbf{E}_{1,2} = \mathcal{E} \mp \epsilon.$$

where

$$(70) \qquad \epsilon = \mp\sqrt{\left(\vec{R} - i\frac{1}{2}\vec{\Gamma}\right)^2}.$$

Then, $\mathbf{E}_1$ and $\mathbf{E}_2$ coincide when ϵ vanishes. Since, real and imaginary parts of ϵ should vanish, the condition for accidental degeneracy of the two interfering resonances may be written as

$$(71) \qquad R_d^2 - \frac{1}{4}\Gamma_d^2 = 0,$$

and

$$(72) \qquad \vec{R}_d \cdot \vec{\Gamma}_d = 0.$$

These equations admit two kinds of solutions corresponding to $\mathbf{H}_{2\times 2}$ being or not being diagonal at the degeneracy:

i) When both $\vec{R}_d$ and $\vec{\Gamma}_d$ vanish, eqs. (71) and (72) define a point in parameter space, $\mathbf{H}$ is diagonal at the degeneracy and the submatrix $\mathbf{H}_{2\times 2}$ has two cycles of eigenvectors of lenght one each. The vanishing of Γ_d^2 implies that $Im\mathcal{E}$ also vanish[13], therefore the two complex eigenvalues of $\mathbf{H}$ which become degenerate fuse into one real positive energy eigenvalue embedded in the continuum. Since in this case all the Cartesian components of $\vec{R}_d$ and $\vec{\Gamma}_d$ should vanish, the minimum number of linearly independent, real, external parameters that should be varied to produce a degeneracy (codimension) of two resonances to form a bound state embedded in the continuum is four or six depending on the quantum system being or not being time reversal invariant.

ii) In the second case, when the degeneracy conditions (71) and (72) are satisfied for non-vanishing $\vec{R}_d$ and $\vec{\Gamma}_d$, these equations define a circle in parameter space. In this case $\mathbf{H}_{2\times 2}$ is not diagonal at the degeneracy, it is equivalent to a Jordan block of rank two and has one cycle of generalized eigenvectors of lenght two. In this case, the two complex eigenvalues become one two-fold repeated eigenvalue $\mathcal{E}$. Since, the two linearly independent conditions (71) and (72) should be satisfied for non-vanishing values of $\vec{R}_d$ and $\vec{\Gamma}_d$, at least two real, linearly independent parameters should be varied to produce a rank two degeneracy of resonances. Hence, the codimension of a degeneracy of resonances of second rank is two, independently of the time reversal invariance character of the interactions.

6.1 Computation of the Geometric Phase

Let us consider now the computation of the Berry phase in the simplest situation when the degeneracy involves only two resonant states. The matrix $\mathcal{H}$ that mixes the two interfering resonances off degeneracy is

$$(73) \qquad \mathcal{H} = \begin{pmatrix} Z - i\frac{1}{2}\Gamma & X - iY \\ X + iY & -Z + i\frac{1}{2}\Gamma \end{pmatrix}.$$

This matrix has two right and two left eigenvectors and may be diagonalized by a similarity transformation

$$(74) \qquad K^{-1} \mathcal{H} K = \begin{pmatrix} -\epsilon & 0 \\ 0 & \epsilon \end{pmatrix},$$

where

$$(75) \qquad K = \frac{1}{\sqrt{2\epsilon}} \begin{pmatrix} \sqrt{\epsilon + \eta} & \sqrt{\epsilon - \eta} \\ \sqrt{\epsilon - \eta} \frac{\xi^*}{|\xi|} & \sqrt{\epsilon + \eta} \frac{\xi^*}{|\xi|} \end{pmatrix},$$

ξ and η are short hand for $X - iY$ and $Z - i\frac{1}{2}\Gamma$ respectively. The matrices K^{-1} and $\nabla_R K$ are readily obtained from (70) and (75). If we assume that the interfering resonances are mixed by a Hermitian interaction, $\vec{\Gamma}$ will be kept fixed. Then a straightforward calculation gives

$$(76) \qquad \gamma_1 = -\frac{1}{2} \int_c \frac{1}{\Gamma \epsilon (\epsilon + \eta)} \left(\vec{\Gamma} \times \vec{R} \right) \cdot d\vec{R},$$

and

$$(77) \qquad \gamma_2 = -\frac{1}{2} \int_c \frac{1}{\Gamma \epsilon (\epsilon - \eta)} \left(\vec{\Gamma} \times \vec{R} \right) \cdot d\vec{R}.$$

These expressions are very similar to the well known results obtained for the geometric phase of bound states[2]. An obvious difference is that the geometric phase of resonant states is complex since ϵ and η are complex functions of the parameters $\vec{R}$ and $\vec{\Gamma}$. There is another important but less apparent difference: In the case of an accidental degeneracy of resonances ($\Gamma \neq 0$), the denominator in the right hand side of (76) and (77) vanishes on the continuous line of singularities defined by eqs.(71) and (72), which will be called the diabolical circle, and not at one isolated point as is the case for bound states. It follows that two kinds of non-trivial, topologically inequivalent closed paths are possible. First, those paths which surround the diabolical circle but are not linked to it. Second, the closed paths which are linked to the diabolical circle. Paths of the first kind are clearly analogous to the non-trivial paths that go around the diabolical point of bound state degeneracies while paths of the second kind have no analogue in accidental degeneracies of bound states.

For paths of the first kind it is always possible to find a surface Σ which spans the closed path C and does not cross the diabolical circle. Then, Stokes theorem applied to (76) and (77) gives

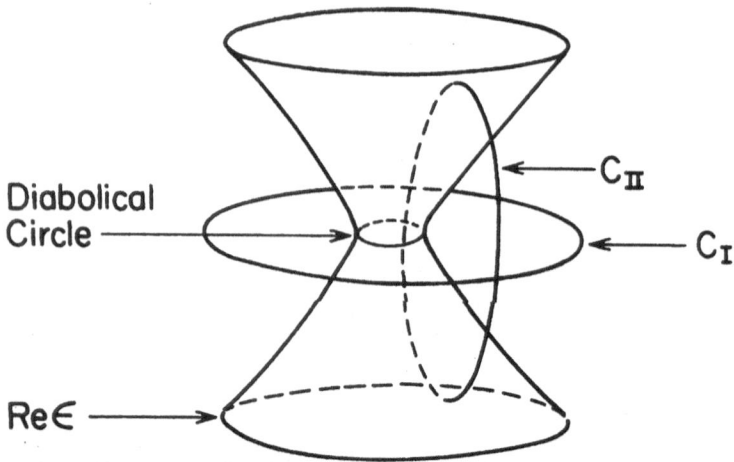

Fig. 1. In the evaluation of the Berry phase of two interfering resonant states there are two kinds of non-trivial, topologically inequivalent closed paths in parameter space. First, those, like $C^{(I)}$, which go around the diabolical circle but are not linked to it. Second, those, like $C^{(II)}$, which turn around the diabolical circle and are linked to it.

$$(78) \qquad \gamma_s = \frac{(-1)^s}{2} \int_{\Sigma_s} \int_{\partial \Sigma_s = c} \frac{\left(\vec{R} - i\frac{1}{2}\vec{\Gamma}\right) \cdot d\vec{\Sigma}}{\left[\left(\vec{R} - i\frac{1}{2}\vec{\Gamma}\right)^2\right]^{\frac{3}{2}}},$$

where $s = 1, 2$. Since γ_2 changes into γ_1 when Σ_1 and Σ_2 are exchanged and the sign of $d\vec{\Sigma}$ is changed, the normals for Σ_2 and Σ_1 should be oppositely oriented. If we say that $\vec{\Gamma}$ points upwards, then Σ_1 is above and Σ_2 is below C.

This is the same result as would have been obtained from the general expression (38) and our parametrizacion of the perturbation term in the Hamiltonian; no summation over intermediate states occurs in (78) since in the simple case of only two interfering resonant states, the summation in (38) has only one term.

Adding γ_1 and γ_2, we get

$$(79) \qquad \gamma_1 + \gamma_2 = -\frac{1}{2} \int_{\Sigma} \frac{\left(\vec{R} - i\frac{1}{2}\vec{\Gamma}\right) \cdot d\vec{\Sigma}}{\left[\left(\vec{R} - i\frac{1}{2}\vec{\Gamma}\right)^2\right]^{\frac{3}{2}}},$$

where Σ is a sphere with the diabolical circle completely contained in its interior. The integral is easily computed when $R > \frac{1}{2}\Gamma$, the result is

(80)
$$-\frac{1}{2\pi}(\gamma_1 + \gamma_2) = 1,$$

in agreement with our identification of the integral in (44) as the first Chern number of the complex line bundle defined by the eigenvector of the non-Hermitian matrix $\mathbf{H}_{2\times 2}$.

It is now easy to show that the resonance degeneracy produces a continuous distribution of singularities on the diabolical circle. The surface integral in (79) may be written as a volume integral by means of Gauss theorem. Then,

(81)
$$\gamma_1 + \gamma_2 = -\frac{1}{2} \int \int \int_V \left(\nabla_R \cdot \frac{\left(\vec{R} - i\frac{1}{2}\vec{\Gamma}\right)}{\left[\left(\vec{R} - i\frac{1}{2}\vec{\Gamma}\right)^2\right]^{\frac{3}{2}}} \right) dV,$$

where V is the volume inside Σ and bounded by it. The term in round brackets under the integration sign vanishes when $\left(\vec{R} - i\frac{1}{2}\vec{\Gamma}\right)^2 \neq 0$. Therefore, the non-vanishing value of $\gamma_1 + \gamma_2$ implies the occurrence of δ-function singularities of the integrand on those points where $\left(\vec{R} - i\frac{1}{2}\vec{\Gamma}\right)^2$ vanishes.

Hence,

(82)
$$\nabla_R \cdot \left[\frac{\vec{R} - i\frac{1}{2}\vec{\Gamma}}{\left[\left(\vec{R} - i\frac{1}{2}\vec{\Gamma}\right)^2\right]^{\frac{3}{2}}} \right] = -\frac{\delta\left(R - \frac{1}{2}\Gamma\right)}{R^2}\delta\left(\cos\theta\right),$$

the factor R^{-2} multiplying the delta function is needed to reproduce the value 2π of $\gamma_1 + \gamma_2$.

Let us turn to the case of closed paths of the second kind, that is, those paths which are linked to the diabolical circle. In this case there is no surface Σ which spans the closed path c^{II} without crossing the diabolical circle. Therefore, it is not possible to use Stokes theorem to convert the path integral into a surface integral. However, we may still compute the geometric phase from the path integral. To this end, we change from Cartesian coordinates (X, Y, Z) with OZ parallel to $\vec{\Gamma}$, to spherical coordinates in parameter space. Then, eqs. (76) and (77) take the form

(83)
$$\gamma_{1,2} = -\frac{1}{2}\int_c d\varphi \mp \frac{1}{2}\int_c \frac{\left(R\cos\theta - i\frac{1}{2}\Gamma\right) d\varphi}{\sqrt{R^2 - \frac{1}{4}\Gamma^2 - i\Gamma R\cos\theta}},$$

the path c is specified when R and θ are given as functions of φ.

From (83), it follows that, for closed paths which are linked to the diabolical circle,

$$(84) \qquad \gamma_1 + \gamma_2 = -\int_{c^{(II)}} d\varphi = 0, \qquad c^{(II)} \text{ of second kind}$$

since, in this case, the angle φ starts out at some value φ_0 and, as the system traces the path c^{II}, it oscillates between a minimum and a maximum values and finally ends at the same initial value φ_0. There is no analogue to this case in the geometric phase of bound states.

7. Results and Conclusions

The purpose of the foregoing has been to discuss the geometric phase acquired by the resonant states when they are adiabatically transported in parameter space around a degeneracy of resonances .

As in the case of bound states, the condition of adiabatic evolution of resonant states may be given a geometric interpretation as parallel translation in a complex line bundle F_s. The geometric phase factors acquired by the resonant states when adiabatically transported in a closed circuit in parameters space are holonomy group elements of the complex line bundle F_s.

In the case of two resonant states mixed by a Hermitian interaction we find two kinds of accidental degeneracies which may be characterized by the number and length of the cycles of instantaneous energy eigenfunctions at the degeneracy. In the first case there are two linearly independent eigenfunctions belonging to the same real positive repeated energy eigenvalue, that is, two cycles of lenght one. In the second case there is only one resonant eigenstate and one generalized resonant eigenstate belonging to the same degenerate (repeated) complex energy eigenvalue, i.e. one cycle of lenght two. At degeneracy, the Hamiltonian matrix has one Jordan block of second rank.

In the generic case of a non-time reversal invariant system, when the degeneracy is of the second rank, the topology of the energy surfaces is different from that at a crossing of bound states. The energy surfaces of the two resonant states that become degenerate touch each other at all points in a circle. Close to the crossing, the energy hypersurface has two pieces lying in orthogonal subspaces in parameter space. The surface representing the real part of the energy has the shape of a hyperbolic cone of circular cross section, or an open sandglass, with its waist at the diabolical circle. The surface of the imaginary part of the energy is a sphere with the equator at the diabolical circle. The two surfaces touch each other at all points on the diabolical circle[13].

In the case of two interfering resonant states, the geometric phase acquired by the resonant states when transported around the diabolical circle in a

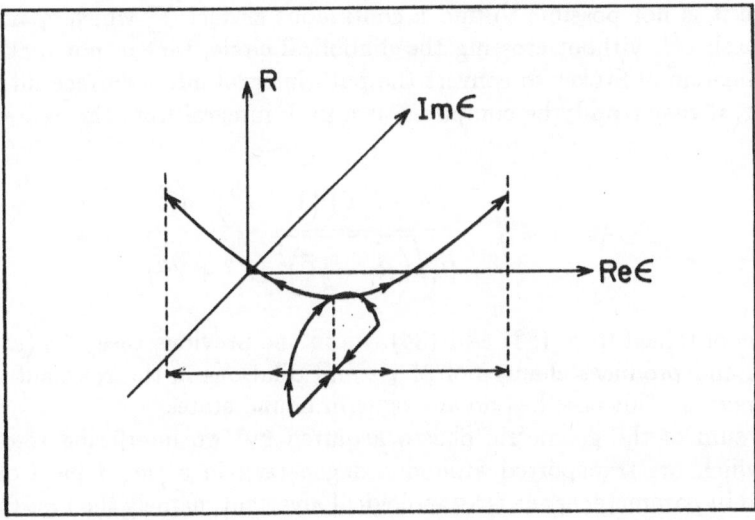

Fig. 2. Two interfering resonances which initially have equal level energies but different half widths are mixed by the Hermitian interaction $\vec{R} \cdot \vec{\sigma}$. As R increases from 0 to $\frac{1}{2}\Gamma$, the points representing $\mathcal{E}_1$ and $\mathcal{E}_2$ approach each other along a meridian circle on the sphere representing $Im\epsilon$ in parameter space, until they meet at a point on the equator, which is the diabolical circle. At this point the two resonances become degenerate and the mixing matrix $\mathbf{H}_d$ is equivalent to a Jordan block of rank two. When R becomes larger than $\frac{1}{2}\Gamma$ the points representing $\mathcal{E}_1$ and $\mathcal{E}_2$ move away from each other on the hyperbolic cone representing $Re\epsilon$.

closed path which is not linked to it, may be written as the sum of two terms.

$$(85) \qquad \gamma_{1,2}^{res}\left(\mathbf{c}^I\right) = \gamma_{1,2}^{bound}\left(\mathbf{c}^I\right) \pm \Delta\gamma\left(\mathbf{c}^I\right).$$

The first term, $\gamma_{1,2}^{bound}\left(\mathbf{c}^I\right)$, is the real geometric phase which a negative energy eigenstate would have acquired when transported around a diabolical point in a closed path in the same parameter space. The second term is complex, it gives rise to a change of the phase and a dilation of the resonant state eigenfunction. Its imaginary part may be positive or negative, in consequence, it may produce an amplification or a damping of the wave function which may compensate or reinforce the attenuation due to the imaginary part of the dynamical phase factor.

When the resonant states are transported in a closed path $\mathbf{c}^{II}$ which does not go around the diabolical circle but is linked to it, the geometric phase they acquire is

$$(86) \qquad \gamma_{1,2}^{res}\left(\mathbf{c}^{II}\right) = \pm\Delta\gamma\left(\mathbf{c}^{II}\right).$$

Since it is not possible to find a continuous surface $\sum$ which spans the closed path $\mathbf{c}^{II}$ without crossing the diabolical circle, we can not make use of the theorem of Stokes to convert the path integral into a surface integral. However, it may readily be computed as a path integral from the expression

$$(87) \qquad \Delta\gamma\left(\mathbf{c}^{II}\right) = \int_{\mathbf{c}^{II}} \frac{\left(Z - i\frac{1}{2}\Gamma\right)\left(\vec{\Gamma} \times \vec{R}'\right) \cdot d\vec{R}'}{\Gamma\sqrt{\left(\vec{R} - i\frac{1}{2}\vec{\Gamma}\right)^2}\left(X^2 + Y^2\right)},$$

which is obtained from (83) and (84). As in the previous case, $\Delta\gamma\left(\mathbf{c}^{II}\right)$ is complex and produces changes of phase and dilations in the resonant state wave function. This case has no analogue in bound states.

The sum of the geometric phases acquired by two interfering resonant states which are transported around a degeneracy in a closed path of the first kind in parameter space is a topological invariant, namely the first Chern number[25]. Its value is the "magnetic charge" on the diabolical circle. For paths of the second kind the sum of the geometric phases vanishes.

Acknowledgments: This work was partially supported by CONACYT (México) under contract No. 4964-E9406

References

1. Bohm A, *Quantum Mechanics: Foundations and Applications*, 3rd Edn. New York, Springer-Verlag 1993, Ch. XXI
2. Berry M. V. *Proc. Roy. Soc.* **A392**,45, (1984)
3. For reviews on Berry's phase see: Jackiw, R. *Commun. At. Mol. Phys.* **27**, 71, (1988); Vinitskii S.J., Derbov V.L., Dubovik V.N., Markovski B.L. and Stepanovskii Yu. P. *Sov.Phys. Usp.* **33**, 403, (1990):
   Moore J.D. *Phys. Rep.* **210** , 1, (1990).
   Zwanziger J.W., Rucker S.P. and Chingas, G.C. *Phys. Rev.* **A43**, 3232, (1991)
   Alden Mead, G. *Rev. Mod. Phys.*, **64**, 51, (1992)
4. For collections of basic papers on geometric phases see the monographs: Wilczek F. and Shapere A. (Eds) *Geometric Phases in Physics* Singapore: World Scientific, 1989.
   Markovsky B. and Vinitsky V.I. (Eds) *Topological Phases in Quantum Theory.* Singapore: World Scientific 1989.
5. Dattoli G., Mignani R. and Torre A. *J. Phys. A: Math Gen.* **23**, 5795, (1990)
6. Miniatura Ch., Sire C., Baudon J. and Bellissard J. *Europhys. Lett.* **13**, 199, (1990).
7. Nenciu, G. and Rasche G. *J. Phys. A: Math. Gen.* **25**, 5741, (1992).
8. Kvitsinsky, A. and Putterman, S. *J. Math. Phys.* **32**, 1403, (1991).
9. Sun, C.P. *Phys. Scr.* **48**, 393, (1993).
10. Hernández E., Jáuregui A. and Mondragón A. *Rev. Mex. Fis.* **38** (S2), 128, (1992).
11. Mondragón A. and Hernández E. *J. Phys. A: Math. Gen.* **29**, 2567, (1996).

12. Mondragón A., Hernández E. and Velázquez-Arcos J.M. *Ann. Phys. (Leipzig)* **48**, 503, (1991).
13. Mondragón A. and Hernández E. *J. Phys. A: Math. Gen.* **26**, 5595, (1993).
14. Hernández E. and Mondragón A. *Phys. Lett.* **326B**, 1, (1994).
15. Simon B. *Phys. Rev. Lett.* **51**, 2167 (1983).
16. Satchler, G.R. *Direct Nuclear Reactions.* London. Oxford University Press 1983 Ch. 3 and references therein.
17. Tang Y.C. *Microscopic description of nuclear cluster theory in: Topics in Nuclear Physics II* Lectures Notes in Physics 145 Eds. T.T.S. Kuo and S.S.M. Wong New York Springer 1981.
18. Romo W.J. *Nucl. Phys.* **A116**, 618, (1968).
19. Hardy G.H, *Divergent Series.* Oxford, Clarendon Press 1949.
20. Zel'dovich Ya. B. *JETP (Sov. Phys.)* **12**, 542, (1961).
21. Berggren T. *Nucl. Phys.* **A109**, 265 (1968).
22. Gyarmati B. and Vertse T. *Nucl. Phys.* **A160**, 523, (1971).
23. Gyarmati, B., Kruppa A.T. and Papp Z. *Phys. Rev.* **C31**, 2317, (1985).
24. Hernández E. and Mondragón A. *Phys. Rev.* **C29**, 722, (1984).
25. Nakahara N. *Geometry, Topology and Physics* Bristol: Adam Hilger, 1990.
26. Bohm, A. and Gadella M. *Dirac Kets, Gamow Vectors and Gel'fand Triplets* in: Lecture Notes in Physics **348**. Berlin Springer-Verlag 1989.

The Global Nature of the Arrow of Time and the Bohm-Reichenbach Diagram

M.A. Castagnino

Instituto de Astronomía y Física del Espacio, Casilla de Correos 67, Sucursal 28, 1428 Buenos Aires, Argentina.

Summary. The importance of the global nature of the arrow of time is shown. Classical Reichenbach diagram and quantum Bohm-Reichenbach diagram, for the universe are introduced. They are used to show the increase of entropy in closed systems, the global nature of the quantum measurement, and the relation among the different arrows of time.

1. Introduction

This is a conceptual essay about time asymmetry with practically no equations (these equations can be found in other contributions to this volume or in the literature quoted in the bibliography). The main thesis is that, even if the problem of the definition of the local arrow of time would be completely solved (which is not yet the case since there is not an unanimous agrement on the subject), it would not be enough to understand time-asymmetry. In fact, the arrow of time necessarily has a global nature and we sustain that the best structure to explain this arrow of time is a Reichenbach global system. Let us explain these two statements:

2. The Arrow of Time Is Global

Let us suppose that the arrow of time would be local. Then, it would be possible to consider two laboratories and to define, in each one, an arrow of time independently. Moreover let us suppose that the two laboratories are perfectly isolated. Then we can ask ourselves: Are the two arrows of time pointing to the same direction? Of course, this question has no answer since, as the two laboratories are isolated, it is impossible to compare one arrow with the other. If we would like to compare the two arrows of time an interaction must be introduced between the two laboratories and we are forced to consider the global system of the two laboratories and the interaction [1]. We can repeat the same story if we add a third isolated laboratory and so forth.

---

[1] From observational evidence we know that all the laboratories in the universe have the same arrow of time. E.g.: if not a radiastronomer would not see the condensation of gas clouds but the opposite, an astronomers would have found stars evolving in a direction opposite to the usual one (Hertzsprung-Russel), or any other sign of the inequality of the local arrows would be detected.

Then *the* arrow of time will only be well defined if we consider all the possible laboratories and the interactions among each other, namely the whole universe. (See the coincident opinion of Feynman in [1]).

3. Reichenbach Global System

The global arrow of time is best represented by a Reichenbach global system ([2],[3] page 127): the system of all branching irreversible processes within the universe, such that any process of the system begins in an unstable state that was produced using energy coming from another process of the global system. E.g.: the famous Gibbs ink drop in the glass of water (initial unstable state) evolves towards a final equilibrium state, the homogeneous mix of ink and water (final stable state) showing that we are dealing with an irreversible process. But the ink drop was not produced by an extremely improbable fluctuation that concentrates the ink in the glass. It was obtained from an ink factory where, to get the necessary energy for the factory coal (initial unstable state) was burnt in an oven until it became ashes (final equilibrium state). *Furthermore the system "ink-water in the glass" only exists as such after the instant when we put the ink drop into the water.* Before this instant a much more complex system exists, that eventually contains the ink factory, the oven, the coal burning, etc. In turn coal was not produced by a fluctuation, quite on the contrary, it was produced using the energy coming from the sun in geological ages. The necessary energy was provided by the light of the sun, where H (initial unstable state) was burnt until it became He and finally Fe (final equilibrium state). Finally H was produced using the energy coming from the unique initial global state of the whole global system: a cosmological initial instability. This initial unstable state can be explained, after decoupling time, by the effect of the gravitational field that takes the gas and radiation, in equilibrium before that time, into a state of hot condensed clouds of matter surrounded by cold radiation, in an expanding geometry, [4]. If we want to go beyond decoupling time we can consider the nucleosynthesis period [3] or, going closer to the beginning, we can consider Big-Bang quantum cosmological models, which also have an unstable unique initial state [5], [6]. Then through this hierarchical chain, that begins in the cosmological instability and contains all the irreversible processes, where each process begins where the corresponding creation device has finished its task, the irreversible nature of the universe and the origin of any irreversible process in it can be explained. Therefore Gibbs ink drop only exists because there was a primordial cosmological instability and Irreversible Statistical Mechanics can not be explained without Irreversible Cosmology. The global system can be symbolized as in fig. 1, which has a clear time symmetry: the branch arrow of time (BAT), which points in opposite direction to the unique initial cosmological instability and follows the evolution of the hierarchical chain towards equilibrium. Reichenbach global system is clearly a realistic model

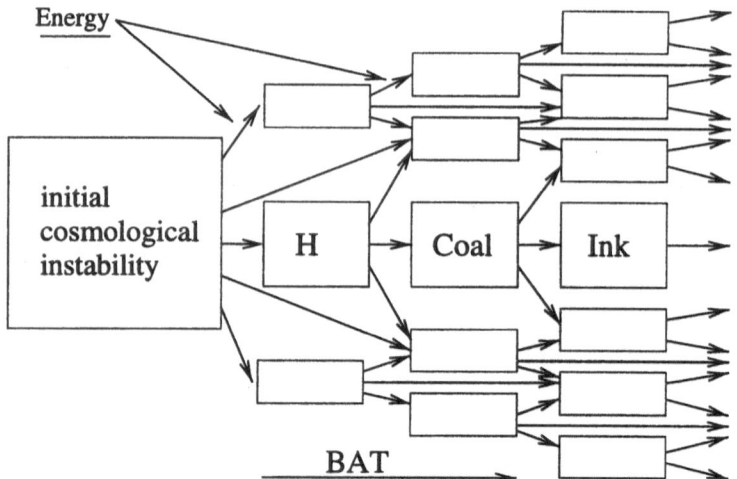

Fig. 1. The classical Reichenbach diagram.

of the set of irreversible processes within the universe. Let us now see the quantum implication of this idea.

4. Bohm Diagrams

Let us consider a usual scattering system (with its continuous energy spectrum as those of ref. [7]) and its diagram, with ingoing stable states a_1, a_2, ...and outgoing stable states b_1, b_2, ... and a central black box symbolizing any interaction (fig. 2). As it is a reversible process there is no modification

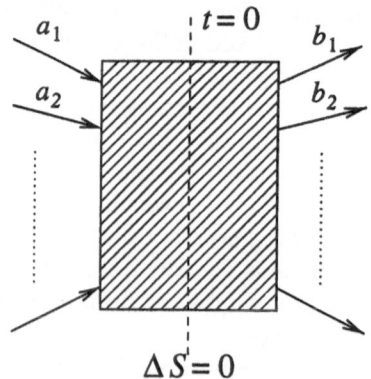

Fig. 2. A usual scattering diagram.

of the entropy from the initial to the final states and the eventual evolutions belong to a group. But A. Bohm [7] cuts the box, at a time $t = 0$, in two pieces by a vertical line. The l. h. s. of the this cut figure is a diagram representing the creation of unstable states u_1, u_2, ... from stable states a_1, a_2, .. (fig. 3), with a decreasing entropy [8], [9], [10] and a evolution that corresponds to a creation semigroup, for $t < 0$ only [7], [11], [12], [13]. The r.h.s. of fig. 2 is a diagram representing the decaying of unstable states u_1, u_2, ...[2] into

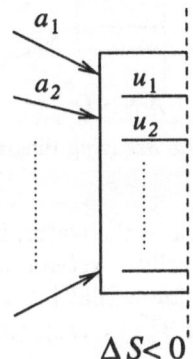

Fig. 3. The creation diagram.

stable states b_1, b_2, ...with a growing entropy that corresponds to a decaying semigroup for $t > 0$ (fig. 4).

The mathematical structures that correspond to the growing and decaying processes can be essentially obtained from the quantum version of Reichenbach idea. Let us first consider the spontaneous decaying states of fig. 4 [3]. It is obvious that these states only exist after the creation time at $t = 0$, because before that time the system was producing growing states. Then the probability to observe a decaying state $|\varphi(t)_- >$ before $t = 0$ in, e.g., any energy out-going eigenstate $|\overset{\circ}{\omega}_- >$, namely the out-going Lippmann-Schwinger state, is zero. The classical analogy would be to ask what the probability is to find a particular configuration of the distribution of the ink drop in the glass of water before the ink would be put in the water. This probability is

---

[2] The decaying states are obtained through the interaction of the system, with the continuous spectrum, that plays the role of the "environment" [7]. E.g. the system would be a H atom and the environment would be the electromagnetic radiation that makes all energy levels unstable but the fundamental one [13]. In this sense, even if the whole system is closed (since it contains the atom and the electromagnetic radiation), it evolves as the "open" system of other formalisms. As a branch system it is therefore temporarily isolated, even if we know that a completely isolated system does not exist. But, as explained in [3] (page 125) it is not the interaction with the rest of the universe (e.g. distant galaxies) the one that produces the decaying, but the interaction with the environment just defined.

[3] In this period the system is quasi-isolated, and theoretically it will be considered completely isolated, as explained in the previous footnote.

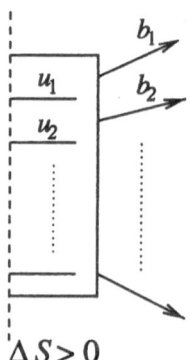

$$\Delta S > 0$$

Fig. 4. The decaying diagram.

obviously zero since there is no ink in the water, in complete agreement with Reichenbach idea that the irreversible systems *only exist as such after the creation instant.* Then if $t < 0$ we know that $| < \omega_- |\varphi(t)_- > |^2 = 0$, therefore also $\int_{-\infty}^{\infty} < \omega_- |\varphi(t)_- > d\omega = 0$ or $\int_{-\infty}^{\infty} < \omega_- |\varphi(0)_- > e^{-i\omega t} d\omega = 0$. So, from the Paley-Wiener [7] theorem we know that $< \omega_- |\varphi(0)_- > \in H_+^2$ the Hardy class from above [7], [14], [15] (see also [16] about the relation of outgoing states and Hardy classes in Lax-Phillips scattering theory). Then if we call Φ_+ the space of states endowed with this last property any decaying state is $|\varphi_- > \in \Phi_+$ and decaying states can be studied considering the Gel'fand triplet $\Phi_+ \subset \mathcal{H}_+ \subset \Phi_+^\times$ where $\mathcal{H}_+$ is the Hilbert space outgoing states [16] and $\Phi_+^\times$ is the space of antilinear functional over Φ_+. On the other hand, ideal growing states [4] of fig. 3 cannot exist after $t = 0$. So repeating the above reasoning we can define the space of growing states Φ_- as the states $|\varphi_+ >$ such that $< \omega_+ |\varphi_+ > \in H_-^2$, the Hardy class from below. Growing states can be studied using the Gel'fand triplet $\Phi_- \subset \mathcal{H}_- \subset \Phi_-^\times$, where $\mathcal{H}_-$ is the Hilbert space incoming [16] states and $\Phi_-^\times$ is the space of antilinear functional over Φ_-. Using these mathematical structures the statements about semigroups and entropy can easily be proved [10]. But other mathematical structures can be used instead of the Gel'fand triplet [17], [18], nevertheless the semigroups and the statements about entropy always remain the same. There are other solutions to the problem of the local arrow of time [19], unfortunately the relation and validity of all these solutions is not yet completely understood [9].

---

[4] These states are just ideal since in the growing period the system is never isolated but it is necessarily receiving energy from a source, i.e. from another branch system within the global Reichenbach system. So these states can only be considered, in the simple model of fig. 2, if we neglect the energy source.

5. Bohm-Reichenbach Diagram

From the classical Reichenbach diagram of fig. 1 and the Bohm diagrams of the previous section we can obtain a quantum diagram for the universe, first introduced in paper [8], that we will call the Bohm-Reichenbach diagram for the universe, precisely fig. 5. It begins with the cosmological unique primor-

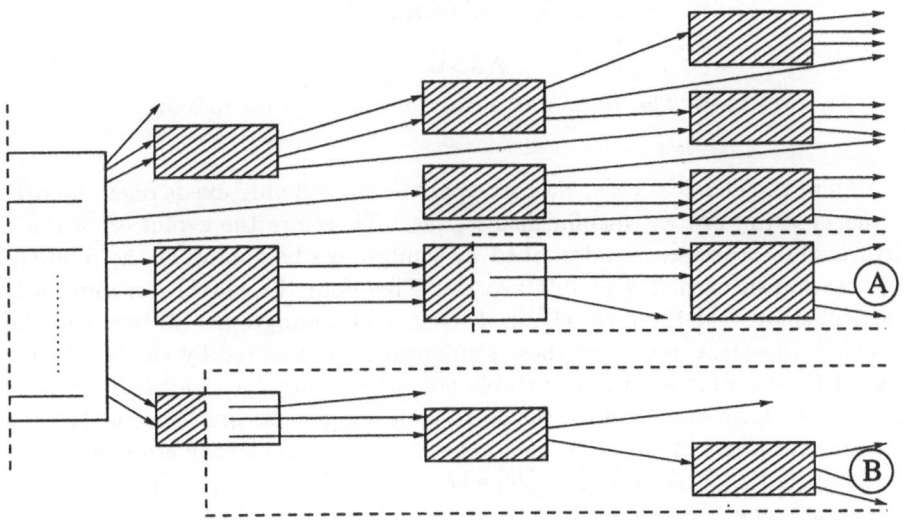

Fig. 5. The quantum Reichenbach diagram or Bohm-Reichenbach diagram.

dial unstable decaying state: the r. h. s. cut box in the far left of the figure, followed by all the scattering processes within the universe, all connected among themselves. This global process yields a final thermic equilibrium and therefore a growing of entropy which, defines the thermodynamical arrow of time (TAT), that goes from the unique initial unstable state towards equilibrium. So, from the simple inspection of the figure, we can conclude that BAT≡TAT.

Entropy also grows in temporally closed and isolated branch systems of the universe [5] as the one of the first dotted box (A) of fig. 5, that we reproduce in fig. 6. This would be the simplest closed branch system. It is a scattering process, such that the outgoing states go towards equilibrium, and it also includes the source of energy that it is used to prepare the ingoing state: the r. h. s. cut box in the far left of the figure. The process has an overall growing of entropy due to the decaying process of the initial cut box. Then in any realistic (i.e. containing an irreversible process) closed subsystem of the universe entropy always grows and the Second Law of Thermodynamics is contained, for all closed systems, in Bohm-Reichenbach diagram.

---

[5] These systems are similar to those of the previous section, so they have all the features described in the second footnote.

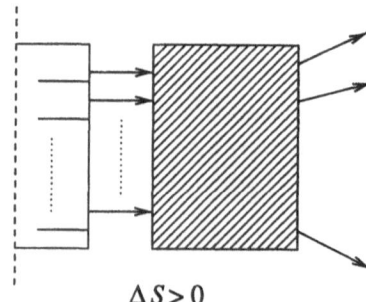

$$\Delta S > 0$$

Fig. 6. The diagram of a close subsystem of the universe.

As in the classical case, each irreversible process only exists once the previous creating process has finished its task. Therefore the evolution of these irreversible processes are described by semigroups beginning at the moment of creation of each irreversible process. Therefore the universe evolution is described by a hierarchical chain of irreversible semigroups all based in the same Hardy class, (since all these semigroups are oriented by the BAT, produced by the unique initial unstable states) and not by a reversible group. Even if we have not, by now, the corresponding general mathematical model that would prove of this statement, this scenario was already studied in several papers [20], [21], [22], [23], [24] where a global space Φ_+ is defined for the whole universe. Moreover, it is quite logical that the global Hardy class of the universe would endow with the same properties of analyticity all the local spaces Φ_+ of all the branch systems within the Reichenbach global system.

6. The Quantum Measurement Process

Let us consider a simple example of a measurement process: Stern-Gerlach experiment of fig. 7. If we would like to consider the complete preparation-

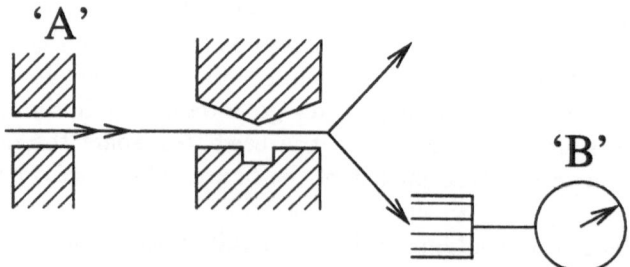

Fig. 7. Stern-Gerlach experiment.

measurement process we must consider not only the scattering process itself, but also the accelerator 'A' that prepares the beam, with its source of energy, and the measurement apparatus, namely the detector and counter 'B'. The accelerator obtains its energy from a source, where a decaying process takes place, and in the detector a creation and a decaying process occurs, e.g.: the particles of the deflected lower beam interact with some atmosphere where some states are excited (creation process) and then they decay [6]. So the complete process of preparation-measurement corresponds to the dotted box (B) of fig. 5, that we reproduce in fig. 8. Therefore every preparation-measurement

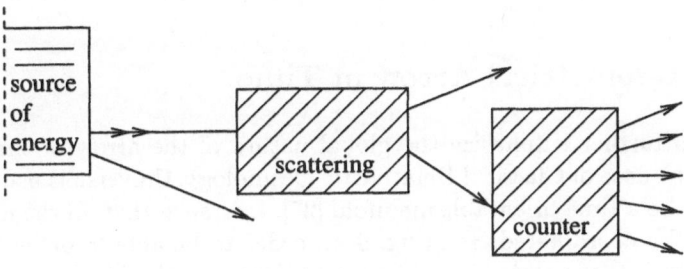

Fig. 8. The Bohm-Reichenbach diagram of the Stern-Gerlach experiment.

process takes place within the Reichenbach global system, since the energy comes from a source that can only be found in this system. Then the preparation procedure turns out to be essentially different from the measurement one: the preparation needs the energy that comes from the hierarchical chain, the measurement is a decaying process, where, even if some part of the energy is used activating the counter, the rest is degraded. Since the quantum arrow of time (QAT) ([15], [25], [26]), goes from preparation to measurement it necessarily coincides with BAT, so QAT≡BAT≡TAT.

On the other hand, if Reichenbach picture is not used someone would probably say that the difference between preparation state and measurement state is just a technological one i. e.: the outcome of a scattering process is a state very difficult to prepare, but anyhow, it is such that it can be prepared with a highly refined technology. Then QAT would not be an essential asymmetry of nature but just a technological one. This objection disappears if we consider the preparation-measurement process within the Reichenbach system: the difference between preparation and measurement becomes essential:

---

[6] This decaying process takes place in the detector being the environment the atmosphere within the detector. This environment is the one that transform the closed Stern-Gerlach apparatus in an "open" one, in the usual parlance. Anyhow we can consider every thing that is inside box (B) of fig. 5 as a closed system. Again the interaction with the rest of the universe (e.g. distant galaxies) in unimportant.

preparation needs energy coming from the primordial instability, independently of the level of technology we use; in the measurement process we do not need energy, which will be degraded in the direction of the final equilibrium state of the universe. Then no confusion is possible between quantum preparation and quantum measurement.

We also see that only in highly idealized scattering processes, with neither preparation nor measurement (fig. 2), the evolution is described by a group. In complete scattering processes, with the preparation and the measurement included, the semigroup structure appears naturally.

7. The Geometrical Arrow of Time

Finally, to further emphasize the global nature of the arrow of time let us consider the case of Classical Relativistic Cosmology. Universe is usually considered to be a time-orientable manifold [27], i. e.: such that all the null semicones can be coordinated, as in fig. 9, in order to be able to define a global arrow of time, that points from the past semicones to the future ones, that we will call the geometrical arrow of time (GAT). If this would not be the case,

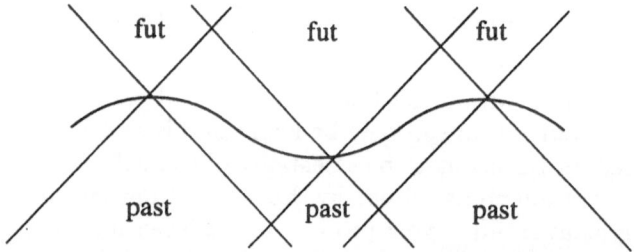

Fig. 9. Schematic representation of an orientable manifold.

namely if the manifold could not be oriented, we would have non-causal loops, as O→A→O'→B→O of fig. 10 (where the arrows always point towards the future, in some null cone, and nevertheless the cycle begins and ends at O). Then if we like to have a causal universe like ours (where, e.g., measurements always follow preparations) this universe must be a time-orientable manifold and the Reichenbach system, in its relativistic version, must be described in this manifold in such a way that GAT≡BAT. As a manifold can only be either orientable or non-orientable and, as our universe is certainly a time-orientable manifold, it necessarily has a global time-orientation, not a local one, because orientation problems are always global (think in Möbius strip!). So GAT is global and therefore BAT ≡TAT≡QAT must also be global.

The uniqueness of the initial unstable state and the global time orientation of the universe are, therefore, the two bases of the BAT.

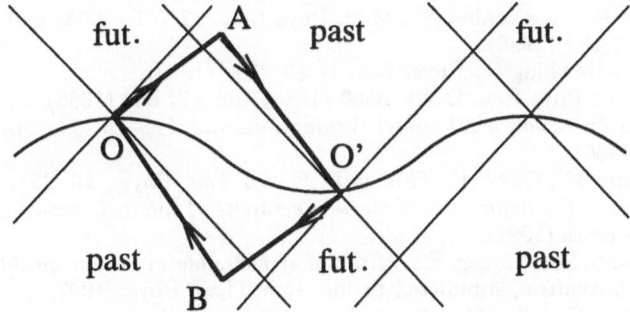

Fig. 10. Schematic representation of a non-orientable manifold with a non-causal loop.

We have not yet developed a mathematical structure for this heuristic model, but a first step to introduce this structure is done in papers [28], [29] where the two semigroups are found for the time-like directions of the null cones (one for the future and one for the past) in Relativistic Quantum Mechanics.

8. Conclusion

The morale of this essay is therefore that the arrow of time must be considered and defined as a global object in the whole universe, most likely using the Reichenbach global system, for the classical case, or the Bohm-Reichenbach diagram, for the quantum case. Therefore all attempts to define an intrinsic arrow of time based just in local reasonings are necessarily incomplete and probably damned to failure.

Acknowledgments: I am very grateful to the organizers of G21 for their kind invitation to participate in this symposium. This work was partially supported by grants: CI1*-CT94-0004 of the European Community, PID-0150 of CONICET (National Research Council of Argentina), EX-198 of the Buenos Aires University, and 12217/1 of Fundación Antorchas and the British Council.

References

1. Feynman L. P., Leighton R. B., Sand M., *The Feynman lectures on physics,* volume 1, Addison-Wesley, New York (1964).
2. Reichenbach H., *The direction of time,* Univ. of California Press, Berkeley, (1956).
3. Davies P., *Stirring up trouble,* in Physical Origin of Time Asymmetry, Halliwell J. J. et al. ed. Cambridge Univ. Press, Cambridge (1994).

4. Aquilano R., Castagnino M., Mod. Phys Lett. **11** 755 (1966) and Astroph. Sp. Sci., **238**, 159 (1996).
5. Hartle J., Hawking S., Phys. Rev. D **28** 2960 (1983).
6. Vilenkin A. Phys Rev. D **33**, 3560 (1986) and **37**, 888 (1986).
7. Bohm A. *Quantum Mechanics: Foundations and Applications,* Springer Verlag, Berlin (1986).
8. Castagnino M., Gaioli F., Gunzig E. Found. Cos. Phys., **16**, 221, (1996)
9. Ordoñez A., *Equivalence between two versions of intrinsic irreversibility,* Physica A, in press (1997).
10. Castagnino M., Gunzig E., *Minimal irreversible quantum mechanics: an axiomatic formalism,* submitted to Int. Jour Theo. Phys., 1997.
11. Bohm A., Gadella M., *Dirac kets, Gamov vectors, and Gel'fand triplets,* Springer-Verlag, Berlin (1989)
12. Gadella M., these proceedings, volume II, p.296, 1997.
13. Castagnino M., Laura R., Phys. Rev. A., **56**, 108, 1997.
14. Antoniou I., Bohm A., Kielanovsky P., Jour. Math. Phys., **36,** 1 (1995).
15. Bohm A. Phys. Rev. A., **51,** 1758 (1996).
16. Reed M., Simon B., *Methods of modern mathematical physics, volume. III,* Accad. Press, New York (1979).
    Lax P. D., Phillips R. S., *Scattering theory,* Acad. Press, New York (1967).
17. Courbage M. Physica A, **122**, 459 (1983) and
    Dynamics and the arrow of time, in The Paradigm of Self-Organization II, 94, Gordon & Breach, London (1994).
18. Castagnino M. Domenech G. Levinas M. L., Umerez N., Jour. Math. Phys.**37,** 2107 (1996).
19. Halliwell J. J. et al. ed., *Physical Origin of Time Asymmetry,* Cambridge Univ. Press, Cambridge (1994).
20. Castagnino M. Gunzig E., Nardone P., Prigogine I., Tasaki S., *Quantum cosmology and large Poincaré systems,* in Quantum, Chaos, and Cosmology, p. 3, Namiki M. et al. eds., AIP press, Woodbury, New York, (1996)
21. Castagnino M., Gunzig E., Lombardo F., Gen. Rel. Grav., **27**, 257 (1995).
22. Castagnino M., Lombardo F., Gen. Rel. Grav., **28**, 263 (1996).
23. Castagnino M., Gaioli F., Sforza D.,Gen. Rel. Grav. **28**, (1966).
24. Castagnino M., *The mathematical structure of quantum super space as a consequence of time asymmetry,* Phys. Rev. D, in press, 1997.
25. Bohm A., Loewe M., Maxson S., *Microphysical irreversibility in quantum mechanics,* Rep. Theo. Phys., in press (1995).
26. Bohm A., this volume.
27. Lichnerowicz A., Ann.. Inst. H. Poincaré (A), **I**, 233 (1964)
28. Pronko J., these proceedings, volume I, (1997).
29. Castagnino M., Gaioli F., García Alvarez E., these proceedings, volume II, p. 517, (1997).

Chapter V

Exact Quantum Theory of the Neutral K-System – Hilbert Space Versus Rigged Hilbert Space Formulation

CP-Violation Problem Beyond the Standard Lee-Oehme-Yang Theory

L.A. Khalfin

St.Petersburg Department Steklov Mathematical Institute. Russian Academy of Sciences, Fontanka 27, St.Petersburg 191011, Russia

Summary. The CP-violation problem is investigated in the exact Quantum Theory (QT) beyond the Lee–Oehme–Yang (LOY) theory, which is usually used and which is based on the famous Weisskopf–Wigner (W.W.) approximation. New unconditional CP-violation effects were derived, independent of the ones known before, as well as new unconditional tests of the CPT- and T-invariances and new results for the K^0, $\bar{K}^0$ correlations were found. On the base of these new theoretical results, propositions for experiments at CP LEAR and the ϕ-B factories are given. The new results are interesting not only for the CP-violation problem itself, but also provide possibilities for testing the violation of the standard quantum dynamics connected with ideas of quantum gravity, and for testing the Rigged Hilbert Space (RHS) formulation of QT. In this paper the standard Hilbert Space (HS) formulation of QT is used.

1. The CP-violation problem is one of the oldest [1] unsolved problems of modern physics. The very specific property of the K^0–$\bar{K}^0$, B^0–$\bar{B}^0$, and D^0–$\bar{D}^0$ mesons is an overlapping of their mass distributions. From this fact it follows that such mesons are a very sensitive "interferometer" in particle physics, which gives possibility for: a) detailed investigation of the CP-violation problem, including the ϵ'/ϵ problem, and so for the testing of a possible "new physics" beyond the Standard Model (see, for example, some articles in [2]), b) tests of the fundamental CPT- and T-invariance problems [2], c) tests of the Superposition Principle [3]–[6], d) tests of new (non-Hamiltonian) quantum dynamics, connected with ideas of quantum gravity (see, for example, [7]–[9] and recent reports [10, 11]), e) general tests of QT – classical Bell's and quantum Tsirelson's inequalities [12]. Up to the present, the standard LOY theory [13] is usually used for theoretical investigations of these fundamental problems, which is essentially based on the famous W.W. approximation (see, for example, [2, 9]). From the fundamental character of these problems it is evident that it will be very essential to investigate these points in the exact QT which goes beyond the LOY theory. This has been done in works by the author between 1968 and the present (see, for example, [14]–[18] and another references therein, and also some works [19]–[21], which were induced by the author's works). It is very essential that the exact theory beyond the LOY theory gives not only some small quantitative perturbations of the effects known before from the LOY theory, but rather some new qualitative effects.

In this paper, exact QT means the usually used Hilbert Space (HS) formulation of Quantum Theory, and all results of this paper were obtained within this usual HS QT. Some time ago, however, another formulation of QT, the Rigged Hilbert Space (RHS) formulation, was proposed (see [22] other references therein), which, in contrast to the usual HS QT, provides a theory for unstable states (particles) and, more generally, a theory of nonstationary solutions of the QT. Because the investigations of the CP-violation problem use the deep properties of the nonstationary solutions of the QT, this opens the possibilities to test experimentally which formulation of the QT - the usual HS or the RHS - is more adequate to reality. I shall point out especially such new predictions of the HS QT, which are different from the predictions of the RHS QT.

In this paper I cannot give all new results with corresponding mathematical proofs (see for some proofs the original works). I mainly discuss here the standing points of the problems, main theoretical results with the proofs and experimental consequences of these results. I especially point out the differences between these new results and the ones previously obtained from the LOY theory and from the RHS QT.

2. For the usual (one-level) unstable particles a new result beyond the W.W. approximation is the nonexponentiality of a decay law (first derived in [23]). This nonexponentiality induced some delicate experimental works and was verified for some specially prepared unstable states (see corresponding references in [24]). For the usual (one-level) unstable states (particles) with simple Breit-Wigner energy (mass) distributions such nonexponentiality is of the order of the dimensionless parameter Γ/m, which is very small. However, it does not necessarily follow that such small nonexponentiality cannot be amplified for some special problems, but in this paper I shall not investigate such possibilities. At the same time, for such unusual (multilevel) unstable particles like K^0–$\bar{K}^0$, B^0–$\bar{B}^0$, D^0–$\bar{D}^0$, we have many dimensionless parameters, some of which are not very small like $\Gamma_L/\Gamma_S \simeq 10^{-3}$, $\Gamma_L/\Delta m \simeq 10^{-3}$ for K^0–$\bar{K}^0$ and even bigger $\Gamma_L/\Gamma_S \simeq 1$, $\Gamma/\Delta m \simeq 10^{-1}$ for B^0–$\bar{B}^0$. So, from the magnitudes of dimensionless parameters, it is possible that there exist some new effects in the exact theory beyond the LOY (W–W approximation) which are not too small. It was proved (see, for example, [14]–[18]) that such possibilities are true in some natural model assumptions. All results in this paper are given for K^0–$\bar{K}^0$, but corresponding results are also true for B^0–$\bar{B}^0$, D^0–$\bar{D}^0$.

3. As it was proved [14]–[18] the main new effects exist for K^0, $\bar{K}^0$ correlations data. This led the author in 1968 [25] to propose correlations factory experiments like $e^-e^+ \rightarrow \phi \rightarrow K^0\bar{K}^0(K_S K_L)$ (independently from Lipkin proposal [26]). Analogical new possibilities are open also by CP LEAR [27]. In the near future it will be possible to do correlations experiments in Frascati ϕ factory [2] and later in SLAC and KEK B factories.

4. The main statements of the LOY theory for the CP-violation, in the CPT-invariant case (corresponding results are also true for the CPT-noninvariant case, but in the main part of this paper I shall investigate the

CPT-invariant case) are the following: there exist superpositions K_S and K_L of K^0, $\bar{K}^0$

(1)
$$\begin{cases} |K_S\rangle = p|K^0\rangle + q|\bar{K}^0\rangle, \; |K_L\rangle = p|K^0\rangle - q|\bar{K}^0\rangle, \\ \langle K_S|K_S\rangle = \langle K_L|K_L\rangle = |p|^2 + |q|^2 = 1, \; \langle K_S|K_L\rangle = |p|^2 - |q|^2 \neq 0, \\ \langle K_S|K_L\rangle = \langle K_L|K_S\rangle \quad (CPT), \end{cases}$$

which are usual (one-level) unstable particles, such that:

a) K_S, K_L (contrary to K^0, $\bar{K}^0$) do not regenerate to each other in the vacuum, but only decay

(2)
$$\begin{cases} |K_S(t)\rangle = p_{SS}(t)|K_S\rangle + \sum_F \sigma_{SF}(t)|F\rangle, \\ |K_L(t)\rangle = p_{LL}(t)|K_L\rangle + \sum_F \sigma_{LF}(t)|F\rangle, \end{cases}$$

where $|F\rangle$ are the state vectors of the decay products, for which it is true that $\langle F|K_S\rangle = 0 = \langle F|K_L\rangle$.

b) the decay laws of K_S, K_L are simple exponential:

(2a) $\quad p_{SS}(t) = \exp(-im_S t - \Gamma_S t), \qquad p_{LL}(t) = \exp(-im_L t - \Gamma_L t).$

In (1) p, q are some complex numbers, which are connected with m_S, m_L, Γ_S, Γ_L through a non-Hermitian effective Hamiltonian H_{eff} (W.W. approximation).

Let us now use the spectral language for the description of K_S, K_L $(K^0, \bar{K}^0)$

(3) $\quad H|\varphi_\alpha(m)\rangle = m|\varphi_\alpha(m)\rangle, \qquad \langle \varphi_\beta(m')|\varphi_\alpha(m)\rangle = \delta_\alpha^\beta \delta(m - m'),$

where H, contrary to H_{eff}, is a Hermitian Hamiltonian of all interactions, with absolute continuous spectrum. We have

(4)
$$\begin{cases} |K_S\rangle = \int_{\text{Spec } H} \sum_\alpha c_{S,\alpha}(m)|\varphi_\alpha(m)\rangle \, dm, \\ |K_L\rangle = \int_{\text{Spec } H} \sum_\beta c_{L,\beta}(m)|\varphi_\beta(m)\rangle \, dm, \\ |K^0\rangle = \int_{\text{Spec } H} \sum_\alpha c_{K^0,\alpha}(m)|\varphi_\alpha(m)\rangle \, dm, \\ |\bar{K}^0\rangle = \int_{\text{Spec } H} \sum_\beta c_{\bar{K}^0,\beta}(m)|\varphi_\beta(m)\rangle \, dm. \end{cases}$$

The LOY theory is equivalent to the assumption that the mass distributions ω_S, ω_L of K_S, K_L are simple one-pole (Breit–Wigner)

(5)

$$\begin{cases} \omega_S(m) = \sum_\alpha |c_{S,\alpha}(m)|^2 = \dfrac{\Gamma_S}{\pi}\left[(m-m_S)^2 + \Gamma_S^2\right]^{-1}, & \Gamma_S \ll m_S, \\[2mm] \omega_L(m) = \sum_\beta |c_{L,\beta}(m)|^2 = \dfrac{\Gamma_L}{\pi}\left[(m-m_L)^2 + \Gamma_L^2\right]^{-1}, & \Gamma_L \ll m_L. \end{cases}$$

The exponential decay laws (2a) follow from (5) if the very small nonexponential terms [23] are neglected. In the LOY theory, the Bell–Steinberger (B.St.) unitarity relation [28]

$$(6)\ \langle K_S|K_L\rangle\left[i\,(m_S - m_L) - (\Gamma_S + \Gamma_L)\right] = \sum_F \langle F|T|K_S\rangle^*\langle F|T|K_L\rangle,$$

which is usually used as CPT and T invariance tests [2], is derived. Let us point out that for CP-invariant theory the B. St. relation is trivial $0 = 0$.

From the main statements (1), (2), (2a) of the LOY theory, one obtains the usual formulas

$$(6a)\ |K^0(t)\rangle = \exp(-iHt)|K^0\rangle$$
$$= p_{K^0 K^0}(t)|K^0\rangle + p_{\bar{K}^0 K^0}(t)|\bar{K}^0\rangle + \sum_F \sigma_{K^0 F}(t)|F\rangle$$
$$= \frac{1}{2}\left[\exp(-im_S t - \Gamma_S t) + \exp(-im_L t - \Gamma_L t)\right]|K^0\rangle$$
$$+ \frac{q}{2p}\left[\exp(-im_S t - \Gamma_S t) - \exp(-m_L t - \Gamma_L t)\right]|\bar{K}^0\rangle$$
$$+ \sum_F \sigma_{K^0 F}(t)|F\rangle;$$

$$(6b)\ |\bar{K}^0(t)\rangle = \exp(-iHt)|\bar{K}^0\rangle$$
$$= p_{\bar{K}^0 \bar{K}^0}(t)|\bar{K}^0\rangle + p_{K^0 \bar{K}^0}|K^0\rangle + \sum_F \sigma_{\bar{K}^0 F}(t)|F\rangle$$
$$= \frac{1}{2}\left[\exp(-im_S - \Gamma_S t) + \exp(-im_L t - \Gamma_L t)\right]|\bar{K}^0\rangle$$
$$+ \frac{p}{2q}\left[\exp(-im_S t - \Gamma_S t) - \exp(-im_L t - \Gamma_L t)\right]|K^0\rangle$$
$$+ \sum_F \sigma_{\bar{K}^0 F}(t)|F\rangle.$$

Such expressions are usually used for the interpretation of the three classical effects of the CP-violation [1]: 1) 2π-meson decay mode for long times ($t \simeq \tau_L \gg \tau_S$), 2) interference in the 2π decay mode, 3) charge asymmetry in the semileptonic decay modes (mainly for long time $t \simeq \tau_L$). From expressions (6a, 6b) results of the LOY theory follow.

Comment 1: The following identity follows from (6a, 6b):

$$(7)\qquad\qquad p_{K^0 K^0}(t) = p_{\bar{K}^0 \bar{K}^0}(t).$$

Comment 2: The vacuum regeneration amplitudes $p_{\bar{K}^0 K^0}(t)$, $p_{K^0 \bar{K}^0}(t)$ have the same time dependence. So in the LOY theory the ratio

(8) $$r(t) = p_{\bar{K}^0 K^0}/p_{K^0 \bar{K}^0} = \text{const} = q^2/p^2 \neq 1.$$

5. Now I formulate the main new results of the exact QT beyond the LOY theory. In 1972 [29] a new unconditional test of the CPT-invariance was proved :

(9a) $$\omega_{K^0}(m) = \omega_{\bar{K}^0}(m),$$

(9b) $$p_{K^0 K^0}(t) = p_{\bar{K}^0 \bar{K}^0}(t).$$

It generalizes the famous Lee–Yang theorem on the identity of masses and life-times of particles and antiparticles to the identity of the mass distributions ω_{K^0}, $\omega_{\bar{K}^0}$ of "unusual" unstable particles K^0, $\bar{K}^0$. It is essential that this test is independent (contrary to B. St. test (6)) of wether CP is violated or not. The unconditional test obtained in (9a) and (9b) is true for *any* mass distributions of $K^0, \bar{K}^0$ (K_S, K_L). Test (9b) was also derived by Kabir [3], but within the LOY theory (see (7)), and so for one-pole Breit-Wigner mass distributions (5) of K_S, K_L.

Main Theorem (M.TH.) *Within the CPT-invariant, but CP-noninvariant QT there do not exist any superpositions K_S, K_L of K^0, $\bar{K}^0$ (see (1)) such that*

A. they do not regenerate (contrary to K^0, $\bar{K}^0$) to each other in the vacuum,

B. they have the single one-pole (Breit–Wigner) mass distributions and the corresponding exponential decay laws.

The analog of this M.TH. with statement A. for the CPT-noninvariant case [30] is also true.

Proof [14]–[18]. I shall give here only the proof of statement A.

We have

(10) $$\begin{cases} |K^0(t)\rangle = p_{K^0 K^0}(t)|K^0\rangle + p_{\bar{K}^0 K^0}(t)|\bar{K}^0\rangle + \sum_F \sigma_{K^0 F}(t)|F\rangle \\ |\bar{K}^0(t)\rangle = p_{\bar{K}^0 \bar{K}^0}(t)|\bar{K}^0\rangle + p_{K^0 \bar{K}^0}(t)|K^0\rangle + \sum_F \sigma_{\bar{K}^0 F}(t)|F\rangle \end{cases}$$

On the base of the fundamental Superposition Principle we obtain

(11) $$\begin{cases} |K_S(t)\rangle = \exp(-iHt)|K_S\rangle = p|K^0(t)\rangle + q|\bar{K}^0(t)\rangle \\ |K_L(t)\rangle = \exp(-iHt)|K_L\rangle = p|K^0(t)\rangle - q|\bar{K}^0(t)\rangle \end{cases}$$

Now we use (10), reformulate $|K^0\rangle$, $|\bar{K}^0\rangle$ through $|K_S\rangle$, $|K_L\rangle$ and after long but simple algebraic transformations obtain such expressions for $|K_S(t)\rangle$, $|K_L(t)\rangle$

(12a)

$$
\begin{cases}
|K_S(t)\rangle = \frac{1}{2}[p_{K^0K^0}(t) + \frac{q}{p}p_{K^0\bar{K}^0}(t) + p_{\bar{K}^0\bar{K}^0}(t) + \frac{p}{q}p_{\bar{K}^0K^0}(t)]\,|K_S\rangle \\
+\frac{1}{2}[p_{K^0K^0}(t) + \frac{q}{p}p_{K^0\bar{K}^0}(t) - p_{\bar{K}^0\bar{K}^0}(t) - \frac{p}{q}p_{\bar{K}^0K^0}(t)]\,|K_L\rangle \\
+\sum_F \sigma_{SF}(t)\,|F\rangle \\
= p_{SS}(t)\,|K_S\rangle + p_{LS}(t)\,|K_L\rangle + \sum_F \sigma_{SF}(t)\,|F\rangle
\end{cases}
$$

(12b)

$$
\begin{cases}
|K_L(t)\rangle = \frac{1}{2}[p_{\bar{K}^0\bar{K}^0}(t) - \frac{q}{p}p_{K^0\bar{K}^0}(t) + p_{K^0K^0}(t) - \frac{p}{q}p_{\bar{K}^0K^0}(t)]\,|K_L\rangle \\
+\frac{1}{2}[p_{K^0K^0}(t) - \frac{q}{p}p_{K^0\bar{K}^0}(t) - p_{\bar{K}^0\bar{K}^0}(t) + \frac{p}{q}p_{\bar{K}^0K^0}(t)]\,|K_S\rangle \\
+\sum_F \sigma_{LF}(t)\,|F\rangle \\
= p_{LL}(t)\,|K_L\rangle + p_{SL}(t)\,|K_S\rangle + \sum_F \sigma_{LF}(t)\,|F\rangle
\end{cases}
$$

So, in general, we see the possibility of the vacuum regeneration of $|K_S\rangle$ and $|K_L\rangle$ if the amplitudes of the vacuum regenerations $p_{SL}(t)$, $p_{LS}(t)$

(13)
$$
\begin{cases}
p_{SL}(t) = \frac{1}{2}[p_{K^0K^0}(t) - \frac{q}{p}p_{K^0\bar{K}^0}(t) - p_{\bar{K}^0\bar{K}^0}(t) + \frac{p}{q}p_{\bar{K}^0K^0}(t)] \\
p_{LS}(t) = \frac{1}{2}[p_{K^0K^0}(t) + \frac{q}{p}p_{K^0\bar{K}^0}(t) - p_{\bar{K}^0\bar{K}^0}(t) - \frac{p}{q}p_{\bar{K}^0K^0}(t)]
\end{cases}
$$

are nonzero. Let us now investigate such possibility. From the CPT-invariance (see (9b)) we obtain on the base of (12a,12b) for $t \in [0, \infty)$

(14)
$$
\begin{cases}
p_{SL}(t) = \frac{1}{2}[\frac{p}{q}p_{\bar{K}^0K^0}(t) - \frac{q}{p}p_{K^0\bar{K}^0}(t)] \\
p_{LS}(t) = \frac{1}{2}[\frac{q}{p}p_{K^0\bar{K}^0}(t) - \frac{p}{q}p_{\bar{K}^0K^0}(t)]
\end{cases}
$$

Recalling the main formulas (6a,6b) of the LOY theory for $p_{\bar{K}^0K^0}(t)$, $p_{K^0\bar{K}^0}(t)$, we see from (14) that in the LOY theory (W–W approximation) the vacuum regeneration of $|K_S\rangle$ and $|K_L\rangle$ is, as well known, absent. Let we now assume that this is also true in the exact QT beyond the LOY theory

(15) $p_{SL}(t) = 0, p_{LS}(t) = 0;$ for $t \in [0, \infty)$

From this assumption on the base of (14) we obtain

(16) $\frac{p}{2q}p_{\bar{K}^0K^0}(t) = \frac{q}{2p}p_{K^0\bar{K}^0}(t);$ for $t \in [0, \infty)$

but, with the use of the Fourier integrals, we obtain for $p_{\bar{K}^0K^0}(t)$, $p_{K^0\bar{K}^0}(t)$

(17)
$$
\begin{cases}
p_{\bar{K}^0 K^0}(t) = \langle \bar{K}^0 | K^0(t) \rangle = \langle \bar{K}^0 | \exp(-iHt) | K^0 \rangle \\
\quad = \int\limits_{\operatorname{Spec} H} \sum_\beta c^*_{\bar{K}^0,\beta}(m) c_{K^0,\beta}(m) \exp(-imt) dm \\
p_{K^0 \bar{K}^0}(t) = \langle K^0 | \bar{K}^0(t) \rangle = \langle K^0 | \exp(-iHt) | \bar{K}^0 \rangle \\
\quad = \int\limits_{\operatorname{Spec} H} \sum_\alpha c^*_{K^0,\alpha}(m) c_{\bar{K}^0,\alpha}(m) \exp(-imt) dm
\end{cases}
$$

Now we use the fundamental Spectral Principle of QT

(18)
$$\operatorname{Spec} H \geqslant 0$$

Let us point out here that in the LOY theory this fundamental Spectral Principle of QT was not taken into account, but (18) is used in the exact HS QT. From (16) and (18), on the base of Titchmarsh theorem, it follows that the function

(19)
$$f_{LS}(t) = \frac{q}{2p} p_{K^0 \bar{K}^0}(t) - \frac{p}{2q} p_{\bar{K}^0 K^0}(t)$$

is the boundary value of an analytic function in the $Im\ t < 0$ complex t half-plane. From the assumption that K_S and K_L do not regenerate to each other in the vacuum (see (15)) it follows that

(20)
$$f_{LS}(t) = 0, \qquad \text{for} \quad t \in [0, \infty)$$

But from (18) and analyticity (the Paley-Wiener theorem) it follows directly that $f_{LS}(t)$ must be identically zero for all $t \in (-\infty, +\infty)$

(21)
$$f_{LS}(t) = 0 \qquad \text{for} \quad t \in (-\infty, +\infty)$$

Then from the definition of $f_{LS}(t)$ (19) and the expressions (17) by the inverse Fourier transform we obtain

(22)
$$\frac{q}{2p} \sum_\alpha c^*_{K^0,\alpha}(m) c_{\bar{K}^0,\alpha}(m) = \frac{p}{2q} \sum_\beta c^*_{\bar{K}^0,\beta}(m) c_{K^0,\beta}(m), \quad \text{for} \quad m \in [0, \infty)$$

or

(23)
$$\frac{p^2}{q^2} = \frac{\sum_\alpha c^*_{K^0,\alpha}(m) c_{\bar{K}^0,\alpha}(m)}{\sum_\beta c^*_{\bar{K}^0,\beta}(m) c_{K^0,\beta}(m)} \qquad \text{for} \quad m \in [0, \infty)$$

Then immediately from (23) it follows that

(24)
$$|p|^2 = |q|^2 \Rightarrow \langle K_S | K_L \rangle = |p|^2 - |q|^2 = 0$$

in contradiction to the initial assumption of the M.TH. on the CP-violation. By this contradiction we proved part A. of the M.TH.

Comment 1. The M.TH. is true for any mass distributions of K^0, $\bar{K}^0$ (K_S, K_L) and so for any Hamiltonian H or for any model of interactions (superweak, Standard Model, ...), which induced the CP-violation.

Comment 2. In the CPT and CP invariant QT the M.TH. is trivially not true (because in this case $K_S = K_0^1$, $K_L = K_0^2$).

Comment 3. The B.St. unitarity relation (6) is not true in the exact theory. This Comment is essential, because the B.St. unitarity relation was proposed as the test of the fundamental CPT and T invariances.

From the M.TH. it follows directly (see, for example, [14]–[18]) that, contrary to the LOY theory prediction (8), the ratio $r(t)$ must be time-dependent

$$(25) \qquad\qquad r(t) \neq \text{const}$$

This is the first new CP-violation effect independent of the three classical ones [1]. It follows from Comment 1 that this new CP-violation effect is theoretically unconditional. Moreover, it follows from the M.TH. that the ratio $r(t)$ cannot be time independent for *any* interval of time t. So it is possible to test the CP-violation also for small times interval, where statistics is much bigger than for the long times ($t \simeq \tau_L \gg \tau_S$) , where the first CP-violation effect for the 2π decay mode was observed. Recent results from the CP LEAR experiment [31] do not contradict this new prediction, but the accuracy of the CP LEAR experiment up to now is not sufficient for more exact conclusion.

6. From the method of the proof of the M.TH. follows also very

General Theorem (G.TH.) *Let* $|a\rangle$ *and* $|b\rangle$ *be unstable states. If the amplitude* $p_{ab}(t) = \langle a| \exp(-iHt)|b\rangle$ *and the amplitude of the inverse (time-reversal) regeneration* $p_{ba}(t) = \langle b| \exp(-iHt)|a\rangle$ *have the same time-dependence*

$$(26) \qquad\qquad r(t) = p_{ab}/p_{ba} = const$$

then $|r| = 1$.

This G.TH. is interesting independently of the CP-violation problem and as it follows from the formulation is independent of the CPT and CP invariances assumptions. The G.TH. is essential for the problem of T-invariance and could be used from this point of view also for experiments on the neutron-antineutron and neutrinos oscillations. If $|a\rangle = |K^0\rangle$ and $|b\rangle = |\bar{K}^0\rangle$ then this G.TH. transforms to (26). In [3] the condition (8), which was obtained within the approximate LOY theory, was proposed as direct test of the T-noninvariance. As it follows from the M.TH. and G.TH. the direct test of the T-noninvariance within the exact QT predicts that $p_{\bar{K}^0 K^0}$ and $p_{K^0 \bar{K}^0}$ must be different not only in magnitudes (as predicted in [3]), but also in time-dependence. This is essential to point out, because Kabir test (8) will be tested in future factory experiment [2] and was tested by CP LEAR. As pointed out above the current accuracy of the CP LEAR experiment [31] is not sufficient to distinguish between the approximate LOY (8) and exact theory (25) predictions.

In [3], apart from the CPT and CP-invariance problems, a direct test of the fundamental Superposition Principle of the QT, namely

(27) $p_{\bar{K}^0 K^0}(t)/p_{K^0 \bar{K}^0}(t) = \text{const},$

was proposed. An analogical test (for two times case) was proposed in [4]. As it follows from the M.TH. such tests cannot be exactly tests of such fundamental Superposition Principle. Contrary to this, as it follows from the proof of the M.TH., the Superposition Principle was essentially used and so the new effect of the CP-violation (25) can be used also as the direct test of the Superposition Principle of QT.

7. We see that the M.TH. directly *contradicts* the main statements of the approximate LOY theory and also the predictions of the RHS formulation of the QT [22]. Because the LOY theory is an approximation within the standard HS QT we must solve two principal problems: 1) to explain why, in spite of this contradiction, the LOY theory gave true interpretation of three classical effects of the CP-violation [1], 2) to propose new experiments, which could distinguish between predictions of the LOY and the exact HS QT theories. On the other hand if the predictions of the exact HS QT, different from the predictions of the approximate LOY theory and from the predictions of the RHS QT, are experimentally verified, then this will support the commonly used standard HS QT.

First of all, it follows from the M.TH. that the superpositions K_S, K_L (1) regenerate to each other in the vacuum, so that

(28) $\begin{cases} |K_S(t)\rangle = p_{SS}(t)|K_S\rangle + p_{LS}(t)|K_L\rangle + \sum_F \sigma_{LF}(t)|F\rangle, \\ |K_L(t)\rangle = p_{LL}(t)|K_L\rangle + p_{SL}(t)|K_S\rangle + \sum_F \sigma_{LF}(t)|F\rangle, \end{cases}$

where the regeneration amplitudes $p_{SL}(t)$, $p_{LS}(t)$ cannot be zero in any interval of time t. Within CPT-invariant, but CP-noninvariant QT it is possible to prove on the base of the G.TH. that the same is true for any superpositions of K^0, $\bar{K}^0$. Now we explain why from the phenomenological point of view, in which p, q, m_S, m_L, Γ_S, Γ_L are phenomenological parameters, the predictions of the LOY theory and the exact theory are the same for the three classical CP-violation effects for usual experiments with pure K^0 (or $\bar{K}^0$) meson. In the LOY theory we have

(29)

$$|K^0(t)\rangle = \frac{1}{2p}\left[\exp(-im_S(t) - \Gamma_S t)|K_S\rangle + \exp(-im_L t - \Gamma_L t)\right]|K_L\rangle$$

$$+ \sum_F \sigma_{K^0 F}(t)|F\rangle.$$

In the exact theory we obtain

(30)

$$|K^0(t)\rangle = \frac{1}{2p}\left[p_{SS}(t)|K_S\rangle + p_{LS}(t)|K_L\rangle + p_{LL}(t)|L\rangle + p_{SL}(t)|K_S\rangle\right]$$

$$+ \sum_F \sigma_{K^0 F}(t)|F\rangle.$$

But if $p_{SS}(t)$, $p_{LL}(t)$, $p_{LS}(t)$, $p_{SL}(t)$ are some superpositions of two functions $\exp(-im_S t - \Gamma_S t)$, $\exp(-im_L t - \Gamma_L t)$ (this is true for some models with pole approximation [14]–[18], [21], see also below 8.), then it is possible on the base of the linearity of (30) to introduce the new superpositions $\widetilde{K}_S$, $\widetilde{K}_L$ such that (30) transforms to

$$(30a) \quad |K^0(t)\rangle = \frac{1}{2p} \Big[\exp(-im_S t - \Gamma_S t)|\widetilde{K}_S\rangle$$

$$+ \exp(-im_L t - \Gamma_L t)|\widetilde{K}_L\rangle \Big] + \sum_F \sigma_{K^0 F}(t)|F\rangle.$$

But it is impossible that the *same* $\widetilde{K}_S$, $\widetilde{K}_L$ transforms

$$(31) \quad |\bar{K}^0(t)\rangle = \frac{1}{2q} \Big[p_{SS}(t)|K_S\rangle + p_{LS}(t)|K_L\rangle - p_{LL}(t)|K_L\rangle$$

$$- p_{SL}(t)|K_S\rangle \Big] + \sum_F \sigma_{\bar{K}^0 F}(t)|F\rangle$$

to

$$(31a) \quad |\bar{K}^0(t)\rangle = \frac{1}{2q} \Big[\exp(-im_S t - \Gamma_S t)|\widetilde{K}_S\rangle$$

$$- \exp(-im_L t - \Gamma_L t)|\widetilde{K}_L\rangle \Big] + \sum_F \sigma_{\bar{K}^0 F}(t)|F\rangle.$$

This consideration explains why the LOY theory and the exact theory give the same phenomenological predictions for three classical CP-violation effects for usual experiments with pure K^0 (or $\bar{K}^0$) meson. At the same time, this consideration implies that the predictions of the LOY and the exact theory are different for $K^0 - \bar{K}^0$ correlations. The typical example of such different predictions is the prediction for the ratio $r(t)$ (25). Another example of different predictions is that the LOY theory predicts for the CPT-invariant case that the interference term in the 2π decay modes for $K^0(t)$ and $\bar{K}^0(t)$ are only different in sign. This is not true in the exact theory. The essential difference is for factory experiments like $e^- e^+ \to \phi \to K_S, K_L(K^0 \bar{K}^0)$. From the Pauli principle follows (see, for example, [2])

$$|\phi\rangle = \frac{1}{\sqrt{2}} \frac{1}{2pq} \Big[K_L(\vec{p}) K_S(-\vec{p}) - K_S(\vec{p}) K_L(-\vec{p}) \Big].$$

Let us investigate the decay mode f_1 in the direction $(-\vec{p})$ at the time t_1 correlated with the decay mode f_2 in the direction $(\vec{p})$ at time t_2. Let $A(f_1, (-\vec{p}), t_1; f_2, (\vec{p}), t_2) = A(f_1, t_1; f_2, t_2)$ be the corresponding correlation amplitude. In the LOY theory we have [2, 32]

$$(32) \quad A(f_1, t_1; f_2, t_2) = \frac{1}{\sqrt{2}} \frac{1}{2pq} \Big[\langle f_1|K_S\rangle\langle f_2|K_L\rangle p_{SS}(t_1) p_{LL}(t_2)$$

$$- \langle f_1|K_L\rangle\langle f_2|K_S\rangle p_{LL}(t_1) p_{SS}(t_2) \Big].$$

But in the exact theory, taking into account the vacuum regeneration of K_S, K_L, we obtain

$$(33) \quad A(f_1, t_1; f_2, t_2) = \frac{1}{\sqrt{2}} \frac{1}{2pq} \Big\{ \Big[\langle f_1 | K_S \rangle \langle f_2 | K_L \rangle p_{SS}(t_1) \, p_{LL}(t_2)$$

$$- \langle f_1 | K_L \rangle \langle f_2 | K_S \rangle p_{LL}(t_1) \, p_{SS}(t_2) \Big] + \langle f_1 | K_L \rangle \langle f_2 | K_S \rangle \Big[p_{LS}(t_1) \, p_{SL}(t_2) \Big]$$

$$- \langle f_1 | K_S \rangle \langle f_2 | K_L \rangle \Big[p_{SL}(t_1) \, p_{LS}(t_2) \Big] + \langle f_1 | K_L \rangle \langle f_2 | K_L \rangle \Big[p_{LS}(t_1) \, p_{LL}(t_2)$$

$$- p_{LL}(t_1) \, p_{LS}(t_2) \Big] + \langle f_1 | K_S \rangle \langle f_2 | K_S \rangle \Big[p_{SL}(t_2) \, p_{SS}(t_1) - p_{SS}(t_2) \, p_{SL}(t_1) \Big] \Big\}.$$

The last two terms for $t_1 \neq t_2$ are not zero and look as "violation" of the Pauli principle.

8. The new phenomena in the exact theory of the CP-violation problem and the new effects of the CP-violation, as it was pointed out, are model independent. But the magnitudes of these new effects are model dependent and can be very different for different model assumptions. I give here only the estimate of the new CP-violation effect (25) [17]. We have such expressions for the vacuum regeneration amplitudes

(34a)

$$p_{K^0 \bar{K}^0}(t) = \frac{1}{4p^* q} \int\limits_{\text{Spec } H} \sum_\alpha \Big\{ |c_{S,\alpha}(m)|^2 - |c_{L,\alpha}(m)|^2 - c^*_{S,\alpha}(m) \, c_{L,\alpha}(m) +$$

$$c_{S,\alpha}(m) \, c^*_{L,\alpha}(m) \Big\} \exp(-imt) \, dm;$$

(34b)

$$p_{\bar{K}^0 K^0}(t) = \frac{1}{4pq^*} \int\limits_{\text{Spec} H} \sum_\beta \Big\{ |c_{S,\beta}(m)|^2 - |c_{L,\beta}(m)|^2 - c_{S,\beta}(m) c^*_{L,\beta}(m) +$$

$$c^*_{S,\beta}(m) \, c_{L\beta}(m) \Big\} \exp(-imt) \, dm.$$

The terms $c^*_{S,\alpha}(m) \, c_{L,\alpha}(m)$, $c_{S,\alpha}(m) \, c^*_{L,\alpha}(m)$ depend on the overlapping of K_S, K_L mass distributions and give after integration in (34a,34b) the new effects of the order $\Gamma_L / \Delta m \simeq 10^{-3}$.

Let us assume the one-pole mass distributions for K_S and K_L (this contradicts statement B. of the M.TH., but taking into account the small value of the CP-violation this model is sufficiently good), including, in general, some nonsingular "preparation" functions $\xi_{S,\alpha}(m)$, $\xi_{L,\beta}(m)$ in $c_{S,\alpha}(m)$, $c_{L,\beta}(m)$

$$(35) \quad \begin{cases} c_{S,\alpha}(m) = \sqrt{\dfrac{\Gamma_S}{\pi}} \, \dfrac{\xi_{S,\alpha}(m) \, A_{S,\alpha}}{|\xi_{S,\alpha}(m_S - i\Gamma_S)| \, (m - m_S + i\Gamma_S)} \\[2ex] c_{L,\beta}(m) = \sqrt{\dfrac{\Gamma_L}{\pi}} \, \dfrac{\xi_{L,\beta}(m) \, A_{L,\beta}}{|\xi_{L,\beta}(m_L - i\Gamma_L)| \, (m - m_L + i\Gamma_L)} \\[2ex] \sum_\alpha |A_{S,\alpha}|^2 = 1, \qquad \sum_\beta |A_{L,\beta}|^2 = 1 \end{cases}$$

It is essential to point out that such preparation functions do not change the exponential terms in the K_S, K_L decay laws (2a). In this model assumption (35) we obtain from (34a,,34b), by taking into account only the exponential terms, the expressions

$$(36a) \begin{cases} p_{K^0 \bar{K}^0}(t) \simeq \dfrac{1}{4p^*q}[\exp(-im_S t - \Gamma_S t) \\[2mm] [1 + \dfrac{2i\sqrt{\Gamma_S \Gamma_L}}{(m_L - m_S) + i(\Gamma_S + \Gamma_L)} \sum_\alpha \xi_{L,\alpha}(m_S - i\Gamma_S) A_{S,\alpha} A_{L,\alpha}^*] \\[2mm] - \exp(-im_L t - \Gamma_L t) \\[2mm] [1 - \dfrac{2i\sqrt{\Gamma_S \Gamma_L}}{(m_L - m_S) - i(\Gamma_S + \Gamma_L)} \sum_\beta \xi_{S,\beta}(m_L - i\Gamma_L) A_{S,\beta}^* A_{L,\beta}]] \end{cases}$$

and

$$(36b) \begin{cases} p_{\bar{K}^0 K^0}(t) \simeq \dfrac{1}{4pq^*}[\exp(-im_S t - \Gamma_S t) \\[2mm] [1 - \dfrac{2i\sqrt{\Gamma_S \Gamma_L}}{(m_L - m_S) + i(\Gamma_S + \Gamma_L)} \sum_\alpha \xi_{L,\alpha}(m_S - i\Gamma_S) A_{S,\alpha} A_{L,\alpha}^*] \\[2mm] - \exp(-im_L t - \Gamma_L t) \\[2mm] [1 + \dfrac{2i\sqrt{\Gamma_S \Gamma_L}}{(m_L - m_S) - i(\Gamma_S + \Gamma_L)} \sum_\beta \xi_{S,\beta}(m_L - i\Gamma_L) A_{S,\beta}^* A_{L,\beta}]] \end{cases}$$

From (36a,36b) we see that the main terms in $p_{K^0 \bar{K}^0}(t)$, $p_{\bar{K}^0 K^0}(t)$, corresponding to $|c_{S,\alpha}(m)|^2$, $|c_{L,\alpha}(m)|^2$ do not change, but the terms corresponding to the new CP-violation effects, connected with $c_{S,\alpha}^*(m) c_{L,\alpha}(m)$, $c_{S,\alpha}(m) c_{L,\alpha}^*(m)$ are multiplied by the factors $\xi_{S,\alpha}(m_L - i\Gamma_L)$, $\xi_{L,\beta}(m_S - i\Gamma_S)$. From this fact it follows that the new CP-violation effects depend on the mass distribution of K_S at $m \simeq m_L$ and from the mass distribution of K_L at $m \simeq m_S$. In other words we can investigate by the new CP-violation effects the tails of mass distributions. Such intriguing possibility was never realized before. Today HS QT cannot predicts in general the tails of the mass distributions of the unstable particles and usually uses only pole approximations near the pole. On the another hand, the RHS QT assumed from the beginning that the mass distributions of the unstable states are pure Breit- Wigner simple pole distributions (5) not only near the poles but for any mass m.

If we assume for simplicity the trivial preparation functions $\xi_{S,\alpha}(m) = 1$, $\xi_{L,\beta}(m) = 1$, then after some long but trivial calculations of $\left|p_{\bar{K}^0 K^0}(t)\right|^2$, $\left|p_{K^0 \bar{K}^0}(t)\right|^2$, using (36a,36b) and taking into account only experimental data of the main 2π decay modes, we obtain [17]:

(37)

$$
\begin{cases}
\left| p_{K^0\bar{K}^0}(t) \right|^2 \simeq \dfrac{1}{4}\Big\{ \exp(-2\Gamma_S t) + \exp(-2\Gamma_L t) - 2\exp(-(\Gamma_S + \Gamma_L)t) \\
\times \Big[\cos(\Delta m\,t) + 0.4 \cdot 10^{-3}\,\sin(\Delta m\,t) \Big] \Big\}, \\[2mm]
\left| p_{\bar{K}^0 K^0}(t) \right|^2 \simeq \dfrac{1}{4}\Big\{ \exp(-2\Gamma_S t) + \exp(-2\Gamma_L t) - 2\exp(-(\Gamma_S + \Gamma_L)t) \\
\times \Big[\cos(\Delta m\,t) - 0.4 \cdot 10^{-3}\,\sin(\Delta m\,t) \Big] \Big\}.
\end{cases}
$$

We see that the new effect is of the order of $\Gamma_L/\Delta m \simeq 10^{-3}$ and that in this model the ratio $r(t)$ is time-dependent, as predicted by M.TH.

As it was pointed out above the magnitudes of the new CP-violation effects depend on the model assumptions. For some different from (35) model assumptions [14, 19], in which the derivative of the form factor has the order of $m_{S,L}^{-1}$, the new CP-violation effects are very small as the usual nonexponential terms in the decay law [23].

9. The exact theory beyond the LOY theory is very essential for investigating the problem of the CPT-invariance and the ϵ'/ϵ problem. As it follows directly from the M.TH. the usual [2] experimental tests of the CPT-invariance, formulated in terms of the LOY theory (for example, the B.St. unitarity relation (6) and the test [2] based on the properties of $\pi^+\pi^-$ and $\pi^0\pi^0$ decay modes of K^0 meson), must indicate the "violation" of the CPT-invariance. But this is not *real* CPT-violation but an "induced" CPT-violation , connected with the fact that the LOY theory main statements cannot be true in the exact CPT-invariant, but CP-noninvariant HS QT (see [14] and recent [20], [33]–[35]). At the same time the unconditional test (9a,9b) of the CPT-invariance is the exact direct test. It is essential that the unconditional test (9a,9b), which was derived within the exact HS QT, as it was pointed out above (see in 5.), is the same as the direct Kabir's test [3], which was derived in the approximate LOY theory. Recent experimental results from CP LEAR [31] on the base of this direct test support the CPT-invariance. Recently, the very attractive idea for the new tests of the CPT-invariance by investigating the charge asymmetry in the semileptonic decay of K_S in factory correlations, was proposed [32]. This idea was based on the possibility to create pure K_S in factory correlation. This is true within the LOY theory prediction (see (32)), but, as follows from the M.TH., it cannot be true in the exact theory beyond the W.W. approximation, because K_S regenerate to K_L for any $t \neq 0$ (see (28) and (33)).

We see that some effects within exact CPT-invariant but CP-noninvariant HS QT are the same as some effects of the CPT and CP-noninvariant approximate LOY theory. The essential question connected with this is : is it true that the exact CPT-invariant QT of the CP-violation is equivalent to the CPT-noninvariant LOY theory? Negative answer to this problem gives the following result :

Theorem [34]–[36]. *Within CPT invariant, but CP-noninvariant HS QT do not exist any superpositions $\widetilde{K}_S$, $\widetilde{K}_L$ of K^0, $\bar{K}^0$*

$$(38) \qquad |\widetilde{K}_S\rangle = p|K^0\rangle + q|\bar{K}^0\rangle, \qquad |\widetilde{K}_L\rangle = s|K^0\rangle - r|\bar{K}^0\rangle$$

such that they do not regenerate to each other in the vacuum for any complex parameters p, q, s, r, which satisfy to natural conditions

$$(39) \qquad \begin{cases} \langle \widetilde{K}_S | \widetilde{K}_S \rangle = |p|^2 + |q|^2 = 1 = |s|^2 + |r|^2 = \langle \widetilde{K}_L | \widetilde{K}_L \rangle \\ \langle \widetilde{K}_S | \widetilde{K}_L \rangle = p^*s + q^*r \neq 0 \end{cases}$$

Two different proofs [34]–[36] of this Theorem are based on the M.TH. and G.TH..

The analogical situation is the investigation of the delicate ϵ'/ϵ problem. Usual methods of the experimental determination of ϵ'/ϵ are based [2, 37, 38] on the main statements of the LOY theory for 2π decay modes for K^0 experiments and also on corresponding factory correlation experiments. It was derived [35], in particular, that the exact HS QT predicts an additional term to $Re(\epsilon'/\epsilon)$ of the order $2 \cdot 10^{-4}$. This fact is essential, because the accuracy of recent experimental works [37, 38] is close to this order.

10. The results of the exact theory beyond the LOY theory are very essential also for the possibility, by investigation of K^0, $\bar{K}^0$ dynamics, to test possible violation of the standard quantum dynamics, connected with effects of quantum gravity [7]–[11]. First of all, in [7]–[9], [11] the W.W. approximation was used, as well as the corresponding theoretical statement of the existence of superpositions K_S, K_L of K^0, $\bar{K}^0$, which do not regenerate to each other in the vacuum and have simple exponential decay laws. Of course, the exact dynamics of the nonstandard quantum dynamics for $K^0(t)$, $\bar{K}^0(t)$ and the exact dynamics of the standard QT are in general different. However, some predictions of the exact QT dynamics for K^0, $\bar{K}^0$ look as some (the main parts of the specific effects from [7]–[9], [11]) predictions of the nonstandard quantum dynamics in the W.W. approximation. I shall not go here into a detailed investigation of this problem [34, 39], but only indicate one of such effect. In [9] it was pointed on the base of results from [32], that the full time dependence of the decay into identical states, for example, to (π^+, π^-), (π^+, π^-) in corresponding correlation factory experiment is independent of the CP-violation parameters. This is easy to see from the LOY prediction (32). At the same time in [9] it was pointed out that when we allow for a departure from standard quantum mechanics, this is no longer true. But, as follows from the exact theory, the standard HS quantum dynamics also predicts (see (33)) such new effect. On the other hand, it is interesting to point out that the exact HS QT dynamics predicts for identical decay channels $f_1 = f_2$ at equal times $t_1 = t_2$ zero correlation amplitude (see (33)). So the experimental indication of a nonzero effect for this correlation will [10] indicate the violation of the standard quantum mechanics.

11. Some proposals for CP LEAR and factory experiments follow from the new theoretical results of the exact HS QT. Such future experiments could give possibility to test what formulation (standard Hilbert Space or Rigged Hilbert Space) is more adequate to the physical reality.

By CP LEAR and factory experiments, it is possible to perform a direct unconditional test (9a,9b) of the CPT-invariance. It is also very interesting to test the new CP-violation effect (25). The experimental investigation of this effect will give also direct tests of the T-noninvariance and the Superposition Principle. It is also interesting to see by CP LEAR that the interference terms in the 2π decay modes for $K^0(t)$ and $\bar{K}^0(t)$ are not different only in sign. It is also interesting to test by CP LEAR and by factory correlations new effects [34, 35, 40] in the time dependence of the charge asymmetry in the semileptonic decay channels for $K^0(t)$ and $\bar{K}^0(t)$. Factory experiments provide also the possibility to test quantum Tsirelson's inequalities [12], which could give indication for very general violation of the first principles of standard QT. The detail investigations of new proposals for CP LEAR and factory experiments will be published in a separate publication.

References

1. J. H.Cristensen, J. W. Cronin, V. L. Fitch and R. Turlay, Phys. Rev. Lett., v.13, 138 (1964).
2. The DAϕNE PHYSICS HANDBOOK, L. Maiani, G. Pancheri and N. Paver, eds. (INFN, Frascati, 1992).
3. P. K. Kabir, Phys. Rev., D2, 540 (1970).
4. G. V. Dass, Phys. Rev., D45, 980 (1992).
5. Leonid A. Khalfin, "On the Test of Superposition", preprint Research Institute for Theoretical Physics, The University of Helsinki, HU-TFT-92, Helsinki, March 1992.
6. Leonid A. Khalfin, - in Abstracts of reports on the XIX International Colloquium on Group Theoretical Methods in Physics, v. 1, 25 (1992), Salamanka (Spain).
7. J. Ellis, J. S. Hagelin, D.N.Nanopouls and M. Srednicki, Nucl. Phys., B241, 381 (1984).
8. J. Ellis, N. E. Mavromatos and D. N. Nanopoulos, Phys. Lett., B293, 142 (1992).
9. Patrick Huet and Michael E. Peskin, "Violaton of CPT and Quantum Mechanics in the K^0 - $\bar{K}^0$ System", preprint SLAC-PUB-6454, March 1994. Nucl. Phys.,B434, 3 (1995).
10. P. Eberhard , "Tests of Quantum Mechanics at a ϕ- factory", report at EURODAϕNE Collaboration Meeting, Frascati, April 19-22, 1994.
11. N. Mavromatos, "Violations of Quantum Mechanics and CPT from Plankian Scale Physics at DAϕNE", report at EURODAϕNE Collaboration Meeting, Frascati, April 19-22, 1994.
12. Leonid A. Khalfin and Boris S. Tsirelson, Found. Phys., v.22, 879 (1992).
13. L. A. Khalfin, "Theory of K Yang, Phys. Rev. D beyond Weisskopf-Wigner approximation and the CP-violation problem", preprint LOMI, P-4-80, Leningrad, 1980.

15. L. A. Khalfin, - in Proceedings of Seminar "Group Theoretical Methods in Physics", M. Markov ed., v.2, 608, "Nauka", M., 1986.
16. L. A. Khalfin, "New Results on the CP-violation Problem", preprint CPT, The University of Texas at Austin, DOE- ER 40200-211, Austin, February 1990.
17. L. A. Khalfin, "A New CP-violation Effect and a New Possibility for Investigation of K_S, K_L (K^0, $\bar{K}^0$) Decay Modes", preprint CPT, The University of Texas at Austin, DOE- ER 40200-246, Austin, March 1991.
18. Leonid A.Khalfin, "CP-violation Problem Beyond Standard Lee- Oehme- Yang Theory", June 1994, submitted to "The DAFNE Physics Handbook"
19. C. B. Chiu and E. C. G. Sudarshan, Phys. Rev., D42, 3712 (1990).
20. E. C. G. Sudarshan and C. B. Chiu, Phys. Rev., D47, 2602 (1993).
21. Ya. Azimov, JETPh Lett., v. 58, 159 (1993).
22. A. Bohm, "Quantum Mechanics", 3rd Edition, Springer, NY (1994); A. Bohm, I. Antoniou, P. Kielanowski, J. Math. Phys., v.36,2593 (1995).
23. L. A. Khalfin, DAN USSR, v. 115, 277 (1957): JETPh, v. 32, 1371 (1958).
24. L. A. Khalfin, Uspekhi Physicheskikh Nauk, v.166, N6, 688 (1996).
25. L. A. Khalfin, report on scientific session the Nuclear Physics Department Academy of Sciences USSR, 1968; "Investigations of the quantum theory of unstable elementary particles", Doctor of Sciences dissertation, FIAN (1968)- LTF JINR (1972).
26. H. Lipkin, unpublished, 1964.
27. CP LEAR Collaboration, R. Adler et al., Phys. Lett., B286, 180 (1992).
28. J. S. Bell, J. Steinberger, Proceedings of the Oxford International Conference on Elementary Particles, 1965.
29. L. A. Khalfin, JETPh Lett., v. 25, 349 (1972).
30. Leonid A. Khalfin, "An Exact Theory of the CP-violation Problem within CPT-noninvariant Quantum Theory", unpublished, 1994.
31. P. Pavlopoulos, "Measurements of CP and T VIolation Parameters in the Neutral Kaon System at CP LEAR"-in "Workshop on Physics and Detectors for DAFNE'95", Frascati Physics Series, v.IV, eds. R. Baldini, F. Bossi, G. Capon, G. Pancheri,55 (1995); R. LeGac, "Study of CP, T and CPT in the Neutral K System"-report on Workshop on K Physics, Orsay, May 1996.
32. C. D. Buchanan, R. Cousin, C. O. Dib, R. D. Peccei and J. Quackenbush, Phys. Rev., D45, 4088 (1992).
33. L. Khalfin, "CP-violation and CPT-invariant Tests beyond Standard Lee–Ohme–Yang Theory", report at EURODAφNE Collaboration Meeting, Frascati, April 19-22, 1994- in LNF-94/033 (IR), 249, 7 giugno 1994.
34. Leonid A.Khalfin "CP-violation Problem Beyond Approximate Lee- Oehme- Yang Theory", report on Workshop on Physics and Detectors for DAFNE'95, April 1995.
35. Leonid A.Khalfin, "Unconditional Tests of Discrete Symmetries by K Mesons Beyond the Approximate Lee- Oehme- Yang Theory", report on Workshop on K Physics, Orsay, May 1996 (will be published).
36. Leonid A.Khalfin, "On the CPT-invariant Theory of CP-violation", PDMI PREPRINT-3/1995. 30. Leonid A. Khalfin, "The Exact Theory of the CP-violation Problem and Violation of Quantum Mechanics", unpublished, 1994.
37. E371; L. K. Gibbons et al., Phys. Rev. Lett., v. 70, 1199 and 1203 (1993).
38. NA31; G. D. Barr et al., Phys. Lett., B317, 233 (1993).
39. Leonid A.Khalfin, "The Exact Theory of the CP-violation Problem and Violation of Quantum Mechanics", 1994 (unpublished).
40. Leonid A.Khalfin, "On Charge Asymmetry in Semileptonic Decay Channels in the Exact Theory of the CP-violation Problem", 1994 (unpublished).

$K^0 - \overline{K^0}$ System in the Rigged Hilbert Space Quantum Mechanics

P. Kielanowski[1,2]

[1] Departamento de Física, Centro de Investigación y de Estudios Avanzados del IPN, Ap. Postal 14-740, 07000 México D.F., Mexico D.F., Mexico
[2] Institute of Theoretical Physics, Warsaw University, ul. Hoża 69, 00-681 Warszawa, Poland

Summary. The neutral K meson system is analyzed in the framework of the Rigged Hilbert Space quantum mechanics. The K mesons that are unstable are described as the Gamow states which are the generalized eigenstates of the hamiltonian with complex energy. It is shown that the generalized eigenfunction expansion of a coherent superposition of K^0 and $\overline{K^0}$ contains the usual K_1 and K_2 states *and* the background term which has non vanishing coupling to the 2π meson state. This can produce effect similar to the CP violating hamiltonian.

1. Introduction

1.1 CP violation

The discrete symmetries in quantum mechanics were studied by Wigner [1] who has shown that the Schrödinger equation is invariant under the P and T transformations. It was widely believed that P and T are exact symmetries of all interactions until Lee and Yang [2] pointed out that these symmetries had not been tested in weak interactions. This led to the discovery of P [3] violation and later to CP violation [4].

CP violation has been observed until now only for the neutral K system. The peculiar properties of the neutral K meson system have been predicted by Gell-Mann and Pais [5]. They have shown after the discovery of neutral K mesons that K^0 and $\overline{K^0}$ states which are produced in strong interactions and have definite strangeness combine upon weak interactions into two different states K_1 and K_2 with different decay modes and different lifetimes. The origin of this behavior was the assumed CP conservation in weak interactions. K_1 and K_2 states are the eigenstates of the CP operator with the eigenvalues ± 1. They have shown that K_1 decays into 2 π mesons while the decay of K_2 into 2 π's was forbidden. The experiment [4] has shown that the state K_2 decays into 2 π's and it is interpreted as the evidence of the CP violation. The magnitude of CP violation is small and is measured by the following observables

$$
(1a) \qquad \eta_{+-} = |\eta_{+-}| e^{i\phi_{+-}} = \frac{A(K_L \to \pi^+ \pi^-)}{A(K_S \to \pi^+ \pi^-)},
$$

(1b) $$\eta_{00} = |\eta_{00}|e^{i\phi_{+-}} = \frac{A(K_L \to \pi^0 \pi^0)}{A(K_S \to \pi^0 \pi^0)}$$

which are of the order 10^{-3} and have the following values

$$|\eta_{+-}| = (2.269 \pm 0.023) \times 10^{-3}, \quad \phi_{+-} = (44.3 \pm 0.8)°$$
$$|\eta_{00}| = (2.259 \pm 0.023) \times 10^{-3}, \quad \phi_{00} = (43.3 \pm 1.3)°$$

The dynamical origin of the CP violation is an open problem. This is because of the scarcity of the experimental information which is well described by various plausible models. These models predict transition matrix elements which are then compared with the experimental data. The process of the comparison is the quantum mechanical procedure which for the K meson system has been formulated by Lee, Oeheme and Yang [6]. This theory is approximate and uses the Hilbert space language that is not suitable for the description of the unstable states and resonances. Thus there is a possibility that the effects attributed to the CP violation may have its origin in the limitations of the method and a more appropriate method can explain them without introducing CP violation in the hamiltonian. In such a way the observation of the 2π decay mode of K_L would only be a quantum mechanical phenomenon.

In this paper we discuss the Rigged Hilbert Space (RHS) [7] description of the K-meson system. In the RHS quantum mechanics there are states corresponding to the resonances and we identify K mesons with such states.

The organization of the paper is the following. In Section 2 we recapitulate the Lee-Oeheme-Yang theory, Section 3 introduces the basic definitions of the RHS. Section 4 contains the main result of this paper which is the generalized eigenfunction expansion of the coherent superposition of the K^0 and $\overline{K^0}$ mesons. This expansion contains the usual K_1 and K_2 states and the background integral. The background integral has a non vanishing matrix element for the 2π final state and can be the source of the final 2π states for large times in the K_1 meson decay. This would produce the same effect as CP violation but would be a quantum mechanical effect that does not require CP violation. Section 5 is devoted to the interpretation of the background integral and Section 6 contains conclusions.

1.2 Unstable Particles

The unstable particles are characterized by its mass m and the width $\Gamma = \frac{1}{\tau}$, where τ is its lifetime. m and Γ are combined into one complex number $z_R = m - i\frac{\Gamma}{2}$. The value of z_R is identified with the position of the S-matrix pole on the second sheet of the complex energy. This is the generalization of the bound state poles of the S-matrix. The bound state poles of the S-matrix correspond to the eigenstates of the hamiltonian of the system. The

Hilbert space cannot contain the states that might correspond to the complex poles of the S-matrix and this is the principal difficulty of the conventional description of unstable particles.

2. Hilbert Space Description of the $K^0 - \overline{K^0}$

The description of the $K^0 - \overline{K^0}$ system in quantum mechanics has been given by Lee, Oeheme and Yang [6]. They describe the time evolution of the neutral K meson system using the Wigner-Weisskopf approximation which leads to the truncation of the infinite set of equations to a closed finite system. For the $K^0 - \overline{K^0}$ system this leads to the Shrödinger equation with non hermitian hamiltonian

$$(2) \qquad i\frac{d\psi}{dt} = M - i\Gamma\psi,, \quad \psi = \begin{pmatrix} K^0 \\ \overline{K^0} \end{pmatrix}$$

where ψ is the wave function of the $K^0 - \overline{K^0}$ system. Here M and Γ are hermitian matrices with the properties

$$(3) \qquad M_{11} = M_{22}, \qquad \Gamma_{11} = \Gamma_{22}.$$

Eq. (2) can be solved by the diagonalization of the $M - i\Gamma$ matrix and the time evolution of any coherent combination of the K^0 and $\overline{K^0}$ is given by the following formula

$$(4) \qquad \phi = \alpha_S e^{-\lambda_{st}}\psi_S + \alpha_L e^{-\lambda_L t}\psi_L$$

where $\lambda_{S,L}$ and $\psi_{S,L}$ are the eigenvalues and eigenvectors of the $(\Gamma + iM)$ matrix

$$(5) \qquad (\Gamma + iM)\psi_{S,L} = \lambda_{S,L}\psi_{S,L} = (\frac{1}{2}\Gamma_{S,L} + iM_{S,L})$$

and the states $\psi_{S,L}$ are equal

$$(6) \qquad \psi_{S,L} = \frac{N_{S,L}}{\sqrt{2}}((1 + \epsilon_{S,L})|K^0\rangle \pm (1 - \epsilon_{S,L})|\overline{K^0}\rangle.$$

The states $\psi_{S,L}$ have definite lifetimes and are considered "true particles". Using Eq. (4) on can find the decay rate of the $K^0 - \overline{K^0}$ mixture into a given channel. For the $\pi\pi$ final state we obtain

$$(7) \quad R(t) = \frac{|\langle\pi\pi|H_{int}|\phi\rangle|^2}{|\langle\pi\pi|H_{int}|K_S\rangle|^2} = |\alpha_S|^2 e^{-\Gamma_{st}} + |\alpha_L\eta|^2 e^{-\Gamma_L t}$$

$$+ 2|\alpha_S||\alpha_L||\eta|e^{-\frac{1}{2}(\Gamma_s + \Gamma_L)t}\cos(\Delta m + \phi_0).$$

From Eq. (7) we see that for values of $t \geq 20\tau_S$ the second term prevails $(\Gamma_L/\Gamma_S \approx 1.72 \cdot 10^{-3})$ and one has a pure sample of K_L mesons. The experimental fact that $R(t \approx 20\tau_S) \neq 0$ indicates the existence of CP violation in the neutral K mesons decays.

3. Rigged Hilbert Space Quantum Mechanics

3.1 Basic Definitions

The Rigged Hilbert Space formulation of quantum mechanics [7] has a much richer mathematical structure and in particular admits a much wider set of physical states. The RHS allows the exact formulation of the Dirac formalism and also the exact derivation of the Golden Rule [8]

The construction of RHS is the following. One introduces a linear space of states Ψ with stronger topology than that of the Hilbert space. In this topology all the relevant physical operators are continuous and bounded. The space Ψ is completed in this topology and the resulting space is called Φ. The RHS is then the set of three spaces (Gel'fand triplet)

$$(8) \qquad \Phi \subset \mathcal{H} \subset \Phi^{\times}.$$

Here $\mathcal{H}$ is the Hilbert space of the system and $\Phi^{\times}$ is the space dual to Φ. The set of the admissible states is now $\Phi^{\times}$.

The hierarchy of spaces leads also to the hierarchy of the operators. If A is an observable (a bounded operator in Φ) then there exists a corresponding triplet of operators

$$(9) \qquad A^{+}\big|_{\Phi} \subset A^{+} \subset A^{\times},$$

where $A^{\times}$ is the extension of A^{+} to $\Phi^{\times}$ and is defined by

$$(10) \qquad \langle \phi | A^{\times} F \rangle = \langle A\phi | F \rangle$$

for all $\langle \phi | \in \Phi$ and all $|F\rangle \in \Phi^{\times}$.

Now one can define the generalized eigenvalue and the generalized eigenvector of an operator $A^{\times}$:

Definition
$|F\rangle$ *is the generalized eigenvector with the generalized eigenvalue ω of an operator $A^{\times}$ if*

$$(11) \qquad \langle \phi | A^{\times} F \rangle = \langle A\phi | F \rangle = \omega \langle \phi | F \rangle$$

for all $\langle \phi | \in \Phi$.

The remarkable fact is that the generalized eigenvalues of the self adjoint operator do not have to be real. In fact one can construct explicitly the exact eigenstates of the hamiltonian with complex eigenvalues (Gamow states) [9]

$$(12a) \qquad |z_R^-\rangle = -\frac{1}{2\pi i} \int_{-\infty_{II}}^{\infty} dE\, |E^-\rangle \frac{1}{E - z_R}$$

$$(12b) \qquad |z_R^{*+}\rangle = \frac{1}{2\pi i} \int_{-\infty_{II}}^{\infty} dE\, |E^+\rangle \frac{1}{E - z_R^*}$$

The Gamow states correspond to the states of the idealized Breit-Wigner resonances and are associated with a pole z_R on the second sheet of complex energy of the S matrix. Their time evolution is the following

(13a) $$e^{-iH^\times t}|z_R^-\rangle = e^{-i(E_R - i\frac{\Gamma}{2})t}|z_R^-\rangle, \qquad t \geq 0$$

(13b) $$e^{-iH^\times t}|z_R^{*+}\rangle = e^{-i(E_R + i\frac{\Gamma}{2})t}|z_R^{*+}\rangle, \qquad t \leq 0$$

Eqs. (13) show that in the RHS there exist the states with the time evolution required by the Lee-Oehme-Yang theory.

3.2 Scattering

The description of scattering in the RHS quantum mechanics makes the strong distinction between the process of preparation and the process of measurement [10].

The different initial and boundary conditions require the introduction of two RHS's, one for the initial states of the scattering experiment

(14a) $$\Phi_- \subset \mathcal{H} \subset \Phi_-^\times$$

and the other for the outgoing states (observables)

(14b) $$\Phi_+ \subset \mathcal{H} \subset \Phi_+^\times$$

The space Φ_- describes all the states that can be formed in the preparation process and Φ_+ describes all the possible observables. We have

(15) $$\Phi = \Phi_+ + \Phi_- \quad \text{and} \quad \Phi_+ \cap \Phi_- \neq \emptyset$$

For the resonance scattering the spaces Φ_+ and Φ_- are different and are equal

(16) $$\Phi_\pm = \mathcal{S} \cap \mathcal{H}_\mp^2$$

where $\mathcal{H}_\pm^2$ denote the Hardy class functions from above (below).

4. K Meson Decays in RHS Quantum Mechanics

Let us now consider the neutral K meson decay in the RHS quantum mechanics. The experiment for K meson decay consists of the two following phases

1. preparation, e.g. in the reaction

$$\pi^- p \to \Lambda K^0$$

2. measurement, e.g. the observation of the 2π final state.

For the quantum mechanical description one needs a complete set of commuting observables (CSCO). The CSCO depends on the system considered and its dynamics and for the neutral K meson system one can take the CSCO inspired either by the formation process or the decay process. The corresponding CSCO's (omitting all irrelevant variables) are

(17)   formation process (H_0, Y)
          decay process (with CP conservation) (H, CP),

where Y is the hypercharge. The states can be classified according to both CSCO's and at the end of the formation process they are the eigenstates of (H_0, Y). For the decay process it is the most convenient to expand the decaying state in terms of the complete set of the eigenstates of the corresponding CSCO, i.e. of (H, CP) and the expansion reads

$$(18) \qquad \phi^+ = \int_0^\infty dE\,(b_+|E, CP = +^+\rangle + b_-|E, CP = -^+\rangle)$$

where

$$(19) \qquad b_\pm = \langle {}^+E, CP = \pm|\phi^+\rangle.$$

In the Hilbert space approach one makes the assumption that $b_\pm \in L^2$ but then $b_\pm$ cannot be the Breit-Wigner amplitudes. To derive the Lee-Oeheme-Yang theory one might be tempted to define the states

$$(20) \qquad K_{1,2} = N_\pm \int_0^\infty dE\,b_\pm|E, CP = \pm^+\rangle$$

to get the identical expansion as in the Lee-Oeheme-Yang theory

$$(21) \qquad \phi^+ = \alpha_1|K_1\rangle + \alpha_2|K_2\rangle.$$

The expansion (21) is formally identical to the Lee-Oeheme-Yang expansion but they are not mathematically equivalent. The time evolution of the states $K_{1,2}$ is not equivalent to the time evolution (4) and therefore the straightforward application of the Dirac expansion in the Hilbert space cannot serve as a justification of the Lee-Oeheme-Yang theory.

If we use the boundary condition of the RHS quantum mechanics then

$$(22) \qquad b_\pm \in \mathcal{S} \cap \mathcal{H}^2_\mp$$

and one can use the generalized eigenfunction expansion which reads

$$(23) \qquad \phi^+ = b_1|K_1\rangle + b_2|K_2\rangle + F_+ + F_-.$$

The expansion (23) is exact and has the following properties

1. The states $K_{1,2}$ are the eigenstates of the CP operator with the eigen-values $\pm$.

2. the time evolution of the states is the following

(24) $$e^{-iHt}|K_{1,2}\rangle = e^{-i(m_{1,2}-i\frac{\Gamma_{1,2}}{2})t}|K_{1,2}\rangle.$$

3. $F_{\pm}$ are the background integrals over the negative real axis on the second sheet of the complex energy plane

(25) $$F_{\pm} = \int_0^{-\infty_{II}} dE\, b_{\pm}|E, CP = \pm^-\rangle$$

From these properties it follows that the expansion (23) apart from the background integral is identical to the Lee-Oehme-Yang expansion. In fact the time evolution of the state ϕ^+ given by Eq. (23) is the following

(26) $$e^{-iHt}\phi^+ = b_1 e^{-i(m_1-i\frac{\Gamma_1}{2})t}|K_1\rangle + b_2 e^{-i(m_2-i\frac{\Gamma_2}{2})t}|K_2\rangle$$
$$+ \int_0^{-\infty_{ii}} dE\, e^{-iEt}(b_+|E, CP = +^-\rangle + b_+|E, CP = -^-\rangle).$$

We thus see that the results of the Lee-Oeheme-Yang theory are reproduced if the the background integral is small and can be neglected. The confirmation of the Lee-Oehme-Yang theory can serve as the indication that the background term is small.

5. Interpretation of the Background Integral

The value of the background integral is unknown and we only know that it has to be present and non-vanishing. The time evolution of the background integral is different than the time evolution of the $K_{1,2}$ states. It means that the damping of F_+ can be much smaller than of K_1. This means that the 2π states observed at times $t \approx 20\tau_2$ may come from the background integral. In such a case the direct CP violation in the hamiltonian is not necessary. The behavior of the background integral should be such as to reproduce the interference experiments of the K mesons decays in vacuum [11] and with the regenerator [12] which measure the time dependence of the interference term. The Lee-Oeheme-Yang theory predicts the $\cos(\Delta mt - \phi)$ time dependence for this term. In the RHS approach there is the interference between the K_1 Gamow state and the background integral and its time dependence is governed by the the evolution of the Gamow state and of the background integral. The time evolution of the Gamow state is known and is given in Eq. (24) and the time evolution of the background integral is unknown and cannot be calculated without more specific assumptions. The hypothesis that there is no direct CP violation therefore needs further study of the size and time evolution of the background integral.

If in the RHS approach to the neutral K meson decays there is a direct CP violation in the hamiltonian then the earlier discussed experimental results can be explained if the background integral is small and can be neglected. In such a case the RHS approach and the Lee-Oeheme-Yang theory give the same result and are equivalent.

It is worth noting that the Lee-Oehme-Yang theory is not exact because it is based on the Wigner-Weisskopf approximation. Thus a question about the predictions of the *exact theory* is important.

The RHS approach is exact but its predictions are not conclusive because of the nature of the background integral but it offers another, more symmetric possibility of the interpretation of the effects observed in the K meson decays.

The other possibility is the exact Hilbert space approach strongly advocated by L. Khalfin [13]. In this approach all the classical results of the Lee-Oehme-Yang theory are reproduced but there are additional predictions that may be experimentally tested. The relation between this approach and the RHS is such that some of the results coincide while the others do not, e.g. the conclusion A of the **Main Theorem** in Ref. [13] holds in the RHS quantum mechanics while the conclusion B is not fulfilled. These two approaches are thus *not* equivalent.

Another approach is based on the quantum field theory analysis of the $K^0 - \overline{K^0}$ system [14]. This analysis is fully covariant and does not use the Wigner-Weisskopf approximation but still reproduces all the results of the Lee-Oehme-Yang theory.

From this discussion it is rather clear that the effects *not* predicted by the Lee-Oehme-Yang theory will play the most important role in the further development of the theory of CP-violation.

6. Conclusions

We have shown that the RHS quantum mechanics applied to the description of the $K^0 - \overline{K^0}$ system has the following properties

1. The description of the $K^0 - \overline{K^0}$ system is exact.
2. The states K_1 and K_2 (K_S and K_L) are the decaying Gamow states with the exact exponential decay.
3. The Lee-Oehme-Yang theory is reproduced if the background integrals can be neglected.
4. There is a possibility of reproducing the results of the CP violating experiments (Princeton effect) without the CP violating hamiltonians.

References

1. E.P. Wigner, Z. Phys. **43** (1927) 624.
2. T.D. Lee and C.N. Yang, Phys. Rev. **104** (1956) 254.
3. C.S. Wu, E. Ambler, R.W. Hayward, D.D. Hoppes and R.P. Hudson, Phys. Rev. **105** (1957) 1413.
4. J.H. Christenson, J.W. Cronin, V.L. Fitch and R. Turlay, Phys. Rev. Lett. **13** (1964) 138.
5. M. Gell-Mann and A. Pais, Phys. Rev. **97** (1955) 1387.
6. T.D. Lee, R. Oeheme and C.N. Yang, Phys. Rev. **106** (1957) 340.
7. E. Roberts, J. Math. Phys. **7** (1966) 1097; A. Bohm, *Boulder Lectures in Theoretical Physics 1966*, Volume **9A** (Gordon and Breach, New York, 1967).
8. A. Bohm, *Springer Lecture Notes in Physics*, vol. **94** (1978).
9. A. Bohm, J. Math. Physics **22** (1981) 2813.
10. A. Bohm, *Gamow vectors and the Arrow of Time* in *Proceedings on the Symposium on the Foundations of Modern Physics*, Cologne (1993), ed. P. Busch and P. Mittelstaedt.
11. see e.g. J.H. Christensen et al., Phys. Rev. Lett. **43** (1979) 1212.
12. see e.g. W.C. Carithers et al., Phys. Rev. Lett. **34** (1975) 1240; *ibid.* **34** (1975) 1244.
13. L. Khalfin, *CP-violation problem beyond the Standard Lee-Oeheme-Yang Theory* in this volume and also L. Khalfin, *On the CPT-invariant theory of CP-violation*, preprint PDMI-3/1995; L. Khalfin, *On the decay and regeneration amplitudes and the corresponding probabilities for the $K^0 - \overline{K^0}$ mesons from the correlation experiments in Φ-factories (**DAΦNE**)*, preprint PDMI-4/1995.
14. M. Beuthe, G. López Castro and J. Pestieu, *Field Theory Approach to $K^0 - \overline{K^0}$ and $B^0 - \overline{B^0}$ systems*, preprint UCL-IPT-97-09

Chapter VI

Lax-Phillips Semigroup

The Lax-Phillips Semigroup
of the Unstable Quantum System

E. Eisenberg[1], L.P. Horwitz[1,2], and Y. Strauss[2]

[1] Department of Physics, Bar-Ilan University, Ramat-Gan 52900, Israel
[2] School of Physics and Astronomy, Raymond and Beverly Sackler Faculty of Exact
Sciences Tel-Aviv University, Ramat-Aviv 69978, Israel

Summary. The scattering theory of Lax and Phillips, designed primarily for hyperbolic systems, such as electromagnetic or acoustic waves, is described. The embedding of the quantum theory into this structure, carried out by Flesia and Piron, is reviewed. We show how the density matrix for an effectively pure state can evolve to an effectively mixed state (decoherence) in this framework. Necessary conditions are given for the realization of the relation between the spectrum of the generator of the semigroup and the singularities of the S-matrix (in energy representation). It is shown that these conditions may be met in the Liouville space formulation of quantum evolution, and in the Hilbert space of relativistic quantum theory.

1. Introduction

Irreversible evolution in the quantum theory has been described by the addition of non-Hermitian terms to the Hamiltonian, such that it has complex eigenvalues, and the induced evolution is non-unitary [1][2]. In this method, the non-Hermitian terms in the Hamiltonian are introduced phenomenologically, and may only indirectly be associated with some known interaction in a more fundamental Hamiltonian.

Weisskopf and Wigner [3], in a well known paper in 1930, introduced an alternative approach to the decay problem. According to their approach, the evolution takes place in a Hilbert space which is a direct sum of two subspaces: the subspace of the decaying states and that of decay products. These two subspaces are stable under the "free" evolution induced by H_0, but are combined linearly under the full evolution induced by $H = H_0 + V$. In this Hilbert space, the evolution is unitary, and hence its generator, i.e., the Hamiltonian, is self-adjoint. The decay is described as the probability flow from the subspace of the decaying states to its complement, the subspace of the decay products. They studied perturbatively, for the single-channel case, what has become known as the *survival amplitude*

$$(1) \qquad\qquad A(t) = (\psi, e^{-iHt}\psi),$$

which is the probability amplitude for the system to remain in the discrete state until time t. In the following we will describe this approach, and pose critical problems, motivating the development of a more general theory.

By differentiating the absolute square of (1) and setting $t = 0$, one sees that the initial decay rate is necessarily zero (providing that the Hamiltonian is defined on the initial state); in fact, it is easy to show that the change in $p(t)$ is $O(t^2)$. The intermediate and long time behavior follow most simply by an examination of the Laplace transform, i.e., the resolvent kernel, of (1). Deforming the contour of integration, in the inverse transform, which runs below the real positive spectrum of H to the negative imaginary axis, where its contribution for large times is small, the remaining contribution of the contour running above the spectrum of H can be estimated by bringing this contour continuously through the cut. When resonances exist (as the only singularities), this contour will pass through simple poles on the way to the negative imaginary axis in the second sheet (as can be explicitly demonstrated [4] in the Lee-Friedrichs model [5]). The residues of the poles may dominate the time dependence for intermediate times, and give the approximate (due to the presence of residual contributions from the integrals along the negative imaginary axis) exponential decay behavior. For very long times, the pole contributions disappear, and the remaining integration around the branch cut results in an inverse power law asymptotic behavior [6].

It is not difficult to see that an irreversible motion must be defined by a semigroup [7], which is defined as follows:

Let $\{Z(t)\}$ be a family (over t) of operators on a Hilbert space; then $Z(t)$ is an element of a semigroup if

$$(2) \qquad Z(t_1)Z(t_2) = Z(t_1 + t_2), \qquad t_1, t_2 \geq 0$$

The semigroup is said to be *strongly contractive* if $\|Z(t)\| \to 0$, for $t \to \infty$, where $\|A\|$ is the operator norm of A. On the other hand, it can be shown that the reduced motion, as described above, cannot generate a semigroup [8].

There is, furthermore, another, perhaps more fundamental, problem associated with the general method of Wigner and Weisskopf; this is that the expression (1.1) for the survival amplitude implicitly assumes the existence of a linear superposition (we restrict our discussion here to the one-channel case)

$$(3) \qquad e^{-iHt}\psi = A(t)\psi + \chi(t),$$

where $\chi(t)$ represents the decayed system and $(\psi, \chi(t)) = 0$. In general this linear superposition does not correspond to any physical situation in our experience; a short-lived particle, for example, is seen as either the particle before the decay, or the decay products at a certain time, which can not be predicted. This linear superposition does not correspond to the object that we see experimentally in such a process.

In the framework of the theory of Weisskopf and Wigner, techniques have been developed which are capable of displaying the exact semigroup behavior of an unstable system [9]. These methods, involving analytic continuation, result in an exact eigenfunction of the (extended) Hamiltonian which is an element in the large (Banach) space of a Gel'fand triple (rigged Hilbert space).

The definition of this vector depends on the domain of analyticity chosen (in the construction of the bilinear form for analytic continuation); its physical interpretation is not clear, except for the fact that its evolution is that of exact exponential decay.

The problems discussed above are essentially related to the attempt to describe an unstable system in a framework more suitable to the description of reversible phenomena. In what follows we will discuss another approach to irreversible phenomena which attempts to solve these difficulties.

2. Lax-Phillips Theory and the Exact Semigroup

Lax-Phillips theory [10] assumes the existence of a one-parameter unitary group of evolution on a Hilbert space $\bar{\mathcal{H}}$, and incoming and outgoing subspaces $\mathcal{D}_-$ and $\mathcal{D}_+$ such that

$$U(\tau)\mathcal{D}_+ \subset \mathcal{D}_+ \text{, for all } \tau > 0$$

$$U(\tau)\mathcal{D}_- \subset \mathcal{D}_- \text{, for all } \tau < 0$$

$$\bigcap_\tau U(\tau)\mathcal{D}_\pm\}, = \{0\}$$

(4)
$$\overline{\bigcup_\tau U(\tau)\mathcal{D}_\pm} = \bar{\mathcal{H}}$$

where τ is the evolution parameter identified with the laboratory time. It follows from a theorem of Sinai [11] that $\bar{\mathcal{H}}$ can be foliated in such a way that it can be represented as a family of (auxiliary) Hilbert spaces in the form $L^2(-\infty, +\infty; \mathcal{H}_t)$, over Lebesgue measure in t, and all the $\mathcal{H}_t$ are isomorphic (we therefore sometimes refer to these spaces simply as $\mathcal{H}$) and determined up to unitary equivalence. The scalar product in $\bar{\mathcal{H}}$ is given by

(5)
$$(f, g) = \int_{-\infty}^\infty (f_t, g_t)_{\mathcal{H}_t} dt.$$

Lax and Phillips show that there are unitary operators W_+^{-1}, W_-^{-1} which map the elements of $\bar{\mathcal{H}}$ into representations, called the outgoing and incoming translation representations, for which the evolution is translation in t. The subspaces $\mathcal{D}_+, \mathcal{D}_-$ correspond to the sets of functions with, in these representations, support in semi-infinite segments of the positive and negative t-axis respectively. They define the S matrix abstractly as the map from the incoming translation representation to the outgoing one, i.e., $S = W_+^{-1}W_-$. This map is defined up to unitary transformations on the auxiliary spaces $\{\mathcal{H}_t\}$, and refers to the equivalence classes for which the incoming and outgoing representations have the property that the evolution is represented by translation.

Lax and Phillips furthermore define the operator

(6) $$\mathcal{Z}(\tau) = P_+ U(\tau) P_-$$

on $\bar{\mathcal{H}}$, where $P_\pm$ is the projection on the orthogonal complement of $\mathcal{D}_\pm$. This operator vanishes on $\mathcal{D}_\pm$ and maps the subspace

(7) $$\mathcal{K} = \bar{\mathcal{H}} \ominus (\mathcal{D}_+ \oplus \mathcal{D}_-),$$

into itself. These mappings form a semigroup [10], i.e., for $\tau_1, \tau_2 \geq 0$,

(8) $$\mathcal{Z}(\tau_1)\mathcal{Z}(\tau_2) = \mathcal{Z}(\tau_1 + \tau_2),$$

and this semigroup is strongly contractive, i.e., for each $\phi \in \mathcal{K}$ and any ϵ, there exists a τ_ϕ such that

(9) $$\|\mathcal{Z}(\tau)\phi\|_{\bar{\mathcal{H}}} < \epsilon$$

for $\tau > \tau_\phi$. It can be shown that $\mathcal{Z}(\tau)$ is just the unitary evolution $U(\tau)$ projected into the subspace $\mathcal{K}$. Since the states which lie in the subspaces $\mathcal{D}_\pm$, in the case of scattering, describe the incoming and outgoing waves which are not influenced by the interaction, the states which lie in $\mathcal{K}$ describe the unstable states, i.e., resonances of the scattering. From this point of view, the Lax-Phillips semigroup is analogous to the reduced motion discussed in the previous section.

Flesia and Piron [12] have shown that the quantum theory may be embedded in a Lax-Phillips theory by considering the family of Hilbert spaces of the usual quantum theory on the parameter t as the auxiliary spaces of Lax and Phillips; the large Hilbert space $\bar{\mathcal{H}}$ is then the direct integral of these quantum mechanical spaces over all values of the time t with Lebesgue measure. This appears to be the general solution of the problem of the unstable quantum system. The form of the theory adopted by Flesia and Piron [12] distinguishes the elements of these equivalence classes, and constructs an S-matrix which maps the auxiliary space in the incoming translation representation to the auxiliary space of the outgoing one.

In this construction, Flesia and Piron assume the form

(10) $$\psi^\tau_{t+\tau} = W_t(\tau)\psi^0_t,$$

where, since $W_t(\tau)$ represents an evolution, it follows that

(11) $$W_{t+\tau_1}(\tau_2)W_t(\tau_1) = W_t(\tau_1 + \tau_2).$$

Lax and Phillips prove that the S-matrix (in their construction) is a multiplicative operator in the spectral representation of the generator of the unitary evolution K (which is the Fourier transform of the translation representation), i.e.,

$$(S\psi)_\sigma = S(\sigma)\psi_\sigma,$$

and that the eigenvalues of the generator of the semigroup $\mathcal{Z}(\tau)$ correspond to the singularities of the analytic continuation of $S(\sigma)$. The eigenstates corresponding to these eigenvalues are analogous to the generalized eigenstates found in the framework of Weisskopf and Wigner, as discussed in Section 1. Thus, the S-matrix contains all the information about the unstable states . It can be seen [13], however, that the S matrix obtained from a model in which the evolution is given in the form (10) has no t-dependence, and hence its spectral representation is trivial. In this form, one therefore has no relation between the singularities of the S-matrix and the spectrum of the generator of the semigroup.

The most general linear evolution law has the form [13]

$$(12) \qquad (U(\tau)\psi)_{t+\tau} = \int_{-\infty}^{+\infty} W_{t,t'}(\tau)\psi_{t'}dt' \,.$$

This type of evolution, which goes beyond the formulation of Flesia and Piron [12], corresponds to unitary evolution (with some conditions on $W_{t,t'}$) in $\bar{\mathcal{H}}$ with a nontrivial S-matrix for which the singularities of its Fourier transform are associated with the spectrum of the generator of the Lax-Phillips semigroup. As we shall show below, the form of the evolution law (12) has a natural realization in Liouville space as well as in the framework of relativistic quantum theory.

Lax and Phillips [10] prove that the semigroup defined by (6) is strongly contractive. It was shown by Horwitz and Piron [14] that under the free motion, the generator of the semigroup is *dissipative* due to the defect subspaces associated with the non-self-adjointness of the operator dual to the time. The non-self-adjointness of $P_+K_0P_-$ corresponds to the restriction of $-i\partial_t$ to a finite interval, so that, in fact, the operator has imaginary eigenvalues [14].

There are three distinct types of time operator. The incoming time operator T^{in}, provides a spectral family in terms of which the incoming representation can be constructed, and in which functions in $\mathcal{D}_-$ have definite support and functions in $\bar{\mathcal{H}}$ evolve by translation. In this representation, the norm of the evolving states in $(-\infty, 0)$ must decrease. After sufficient laboratory time τ passes, the states evolve to $\mathcal{D}_+$, and in the outgoing representation, provided by the spectral family of the outgoing time operator T^{out}, they have definite support in (ρ, ∞), $\rho > 0$. The mapping of functions in the incoming representation to the outgoing representation is provided by the Lax-Phillips S-matrix, and the time operators are related by

$$(13) \qquad T^{out} = ST^{in}S^\dagger \,.$$

The third type of time operator corresponds to the "free" representation and is related to T^{in}, T^{out} by the Lax-Phillips wave operators. The spectral family for this operator provides the "standard" representation (analogous to Dirac's choice of "standard" spectral families), which we have used above.

The expectation value of the operator T^{in} in the state ψ^τ projected into $\mathcal{K} \oplus \mathcal{D}_+$ (corresponding to the projection P_-) can be interpreted as the *age* of the unstable system after creation at $t = 0$. The expectation value of T^{in} in the state $P_-\psi^\tau$ is

$$(14) \qquad < T^{in} >_\tau = \int t|\,_{in}\langle t|P_-\psi^\tau\rangle|^2 dt\,;$$

here, $|\,_{in}\langle t|P_-\psi^\tau\rangle|^2$ is the probability density for the age t at time τ, an intrinsic dynamical property of the system. The positive value that the expectation value develops corresponds to the average age. One can similarly compute the expected time after decay, the expected lifetime, and the expected value of any other observable of interest as a property of the unstable system [13].

The structure of the theory is somewhat similar to the Wigner-Weisskopf idea, in that a subspace is associated with the decaying system. In this framework, let us choose a vector ψ in the subspace $\mathcal{K}$ to represent the state of an unstable system. Then, under the full evolution,

$$
\begin{aligned}
(\psi, U(\tau)\psi) &= (\psi, P_\mathcal{K} U(\tau) P_\mathcal{K}\psi) \\
&= (\psi, \mathcal{Z}(\tau)\psi),
\end{aligned}
$$
(15)

so that *the reduced evolution is an exact semigroup.*

Moreover, in the Lax-Phillips theory, the expectation value of an observable which is decomposable in the free or outgoing representations, where $\mathcal{D}_+$ has definite support properties, necessarily reduces to the sum of the expectation values in the subspaces $\mathcal{K} \oplus \mathcal{D}_-$ and in the subspace $\mathcal{D}_+$ (the decay products), i.e.,

$$(16) \qquad \langle A \rangle = \int dt(\psi_t, A\psi_t) = \sum_{M=D_\pm,\mathcal{K}} \int dt\,(\psi_t^M, A\psi_t^M).$$

Note that there are no cross terms. *There is, therefore, an exact superselection rule for measurements of the system by means of such decomposable operators.*

3. Applications

3.1 Measurement According to Namiki and Machida

Recently, Machida and Namiki [15] have proposed a measurement theory based on a direct integral space of continuously many Hilbert spaces and a continuous superselection rule. As pointed out by Tasaki *et al* [16], although they had some success, their theory has a conceptual difficulty. Indeed, in their theory, while the apparatus is described by many Hilbert spaces, the system corresponds to a single Hilbert space as in the conventional theory. Thus, one needs to specify the boundary between the system and the apparatus. As discussed by von Neumann, this is impossible.

Most measurement processes are concerned with measurements of observables which are time-independent in the Schrödinger picture. Therefore, if two different Lax-Phillips states give the same expectation value for all time-independent observables, these two states are essentially indistinguishable. In this sense, we define the following:

1. A Lax-Phillips vector $\psi \in \bar{\mathcal{H}}$ is called " effectively pure" if there exists a pure state

$$\rho_0 = \phi_0 \phi_0^*, \qquad \phi_0 \in \mathcal{H},$$

   such that

   (17) $$\langle \hat{A} \rangle_\psi = \text{Tr} \rho_0 A = (\phi_0, A\phi_0),$$

   where $\hat{A}$ is the "lift" of A on $\mathcal{H}$ to $\bar{\mathcal{H}}$, for every element of the algebra of bounded linear operators associated with the spectral families of the time-independent observables[1] on the original space $\mathcal{H}$.
2. A Lax-Phillips vector is called "effectively mixed" if no such (pure) ρ_0 exists.

It can be shown [13][16] that $\psi = \{\psi_t\} \in \mathcal{H}$ is effectively pure if an only if it has the form

$$\psi_t = f(t)\phi_0.$$

We now discuss the possibility of decoherence, or the evolution from effectively pure to effectively mixed states. Consider the Schrödinger evolution for a time-dependent Hamiltonian. The solution of the time-dependent Schrödinger equation can always be written formally as $\psi_t = U(t, t')\psi_{t'}$, where $U(t, t')$ satisfies the chain property $U(t, t')U(t', t'') = U(t, t'')$, and can be expressed in terms of the integral of a time-ordered product. We define $W_t(\tau) = U(t + \tau, t)$, and lift the evolution to $\bar{\mathcal{H}}$ according to

(18) $$\psi_{t+\tau}^\tau = W_t(\tau)\psi_t ,$$

where $W_t(\tau)$ is given by (T implies the time-ordered product)

(19) $$W_t(\tau) = T\left(e^{-i \int_t^{t+\tau} H(t')dt'} \right).$$

For this kind of time evolution we obtain

(20) $$\langle \hat{A} \rangle_\psi = \int dt \, (W_t(\tau)\psi_t, A W_t(\tau)\psi_t)_{\mathcal{H}},$$

---

[1] We wish to emphasize that what is meant is *explicit* time-dependence in the Schrödinger picture; we do not refer here to the dynamical time-dependence that may arise in the Heisenberg picture if A is not a constant of the motion.

where we have taken the normalization as unity. For the effectively pure states
we have

$$(21) \qquad \langle \hat{A} \rangle_\psi = \int dt \; |f(t)|^2 (W_t(\tau)\phi_0, A W_t(\tau)\phi_0)_\mathcal{H}.$$

It follows from our previous argument that the effective state corresponding
to (21) is effectively mixed if $W_t(\tau)\phi_0 \neq W_{t'}(\tau)\phi_0$ (i.e., the state ρ_ψ induced
from $\psi^\tau_{t+\tau} = W_t(\tau)\psi_t = f(t)W_t(\tau)\phi_0$ is not pure in $\mathcal{H}$). This result is true for
the generalized evolution $W_{tt'}$ of (12) as well (in this case, also for a *closed*
system [13]).

3.2 Intrinsic Decoherence
in Classical and Quantum Liouville Evolution

It has long been emphasized by Prigogine and his co-workers [17] that the
natural description for the evolution of a system with many degrees of freedom
is that of the evolution of the density matrix ρ, through the Liouville equation,

$$(22) \qquad i\frac{d\rho}{dt} = [H, \rho].$$

The density matrix ρ ($\rho \geq 0, Tr\rho = 1$) has the property that $Tr\rho^2 \leq 1$, where
the equality is attained only for a pure state. In general, one considers the
space of Hilbert-Schmidt operators A for which

$$(23) \qquad Tr\, A^* A < \infty;$$

the positive (normalized) elements of such a space correspond to the physical states, the density matrices. On this space, the commutator with the
Hamiltonian H defines a linear operator $\mathcal{L}$, called the Liouvillian, for which

$$(24) \qquad i\frac{d\rho}{d\tau} = \mathcal{L}\rho,$$

where one assumes that $\mathcal{L}$ is self-adjoint in the Liouville space. In particular,
for a Hamiltonian of the form of the sum of an unperturbed operator H_0 and
a perturbation V, i.e., $H = H_0 + V$, the corresponding Liouvillian is

$$(25) \qquad \mathcal{L} = \mathcal{L}_0 + \mathcal{L}_I.$$

Now suppose we consider the "time operator" T_0, conjugate to $\mathcal{L}_0$ (with
spectrum $(-\infty, \infty)$; it satisfies

$$[T_0, \mathcal{L}_0] = i.$$

Then, in the spectral representation of T_0,

$$(26) \qquad {}_0\langle t|\mathcal{L}|t'\rangle_0 = -i\partial_t \delta(t - t') + {}_0\langle t|\mathcal{L}_I|t'\rangle_0,$$

where the last term is, in general, not diagonal, and leads to an evolution law
of type (12) [18].

3.3 Relativistic Quantum Mechanics

The form of relativistic quantum mechanics introduced by Stueckelberg [19], extended to the many-body case by Horwitz and Piron [20], covariantly describes the evolution of a system according to the Stueckelberg-Schrödinger equation

$$(27) \qquad i\frac{\partial \psi_\tau}{\partial \tau} = \frac{p_\mu p^\mu}{2M}\psi_\tau \equiv K_0 \psi_\tau,$$

where M is an intrinsic property of the particle ("on-shell" mass).

Since the d'Alembertian, corresponding to the operator K_0, has spectrum $(-\infty, \infty)$, there exists an operator ξ which satisfies [21]

$$(28) \qquad [K_0, \xi] = i.$$

Note that the operator t of the relativistic theory will not serve this purpose, since its commutator with K_0 is iE/M, which only approaches i in the non-relativistic limit.

If ξ is a function of $\mathbf{x}, t$, we may construct the transformation function $\langle \xi', \beta | x \rangle$ using the defining commutation relation; a short calculation results in

$$(29) \qquad i\frac{\partial}{\partial \xi'}\langle \xi', \beta | x \rangle = -\frac{\partial^\mu \partial_\mu}{2M}\langle \xi', \beta | x \rangle.$$

We thus see that the relativistic quantum theory provides a natural framework for the Lax-Phillips formulation of the description of an unstable system. It is interesting that the continuous spectrum of K_0 is essential to the construction; this implies that we must have both positive and negative mass-squared states in the spectrum, i.e., that the so-called tachyons, at least in the form of intermediate states, play a fundamental role in the relativistic description of unstable systems.

Acknowledgments: One of us (LPH) wishes to thank C. Piron for many discussions, and he is grateful for his hospitality in Geneva during several visits. We also wish to thank our colleagues in Brussels, I. Antoniou, B. Misra, I. Prigogine, and S. Tasaki, for helpful discussions and comments which contributed significantly to many of the results that we have presented.

References

1. G. Gamow, Zeits. f. Phys. **51**, 204 (1928).
2. T.D. Lee, R. Oehme and C.N. Yang, Phys. Rev. **106**, 340 (1957); T.T. Wu and C.N. Yang, Phys. Rev. Lett. **13**, 380 (1964).
3. V.F. Weisskopf and E.P.Wigner, Zeits. f. Phys. **63**, 54 (1930); **65**, 18 (1930).

4. L.P. Horwitz and J.P. Marchand, Helv. Phys. Acta, **42**, 1039 (1969); L.P. Horwitz and J.P. Marchand, Rocky Mtn. Jour. Math. **1**, 225 (1973).
5. K.O. Friedrichs, Comm. Pure Appl. Math. **1**, 361 (1948);T.D. Lee, Phys. Rev. **95**, 1329 (1956).
6. N. Bleistein, H. Neumann, R, Handelsman and L.P. Horwitz, Nuovo Cim. **41A**, 389 (1977).
7. C. Piron, *Foundations of Quantum Physics*, Benjamin/Cummings, Reading, Mass. (1976).
8. L.P. Horwitz, J.P. Marchand and J. LaVita, Jour. Math. Phys. **12**, 2537 (1971); D. Williams, Comm. Math. Phys. **21**, 314 (1971).
9. L.P. Horwitz and I.M. Sigal, Helv. Phys. Acta **51**, 685 (1980); W. Baumgartel, Math. Nachr. **75**, 133 (1978). See also, G. Parravicini, V. Gorini and E.C.G. Sudarshan, Jour. Math. Phys. **21**, 2208 (1980); A. Bohm, *The Rigged Hilbert Space and Quantum Mechanics*, Springer Lecture Notes on Physics **78**, Berlin (1978); A. Bohm, *Quantum Mechanics: Foundations and Applications*, Springer, Berlin (1986); A. Bohm, M. Gadella and G.B. Mainland, Am. Jour. Phys. **57**, 1103 (1989); T. Bailey and W.C. Schieve, Nuovo Cimento **47A**, 231 (1978).
10. P.D. Lax and R.S. Phillips, *Scattering Theory*, Academic Press, New York (1967).
11. I.P. Cornfeld, S.V. Fomin and Ya.G Sinai, *Ergodic Theory*, Springer, Berlin (1982).
12. C. Flesia and C. Piron, Helv. Phys. Acta **57**, 697 (1984).
13. E. Eisenberg and L.P. Horwitz, *Advances in Chemical Physics*, to be published (1996). New York (1990).
14. L.P. Horwitz and C. Piron, Helv. Phys. Acta **66**, 694 (1993).
15. Machida and Namiki, Prog. Theor. Phys., **63**, 1457 (1980); **63**, 1833 (1980); Namiki and Pascazio, Phys. Rev. **A44**, 39 (1991).
16. S. Tasaki, E. Eisenberg and L.P. Horwitz, Found. of Phys. **24**, 1179 (1994).
17. C. George, Physica **65**, 277 (1973); I. Prigogine, C. George, F. Henin and L. Rosenfeld, Chemica Scripta 4, 5 (1973); T. Petrosky, I. Prigogine and S. Tasaki, Physica **A173**, 175 (1991); T. Petrosky and I. Prigogine, Physica **A175**, 146 (1991), and references therein.
18. E. Eisenberg and L.P. Horwitz, Phys. Rev A **52**, 70 (1995).
19. E.C.G. Stueckelberg, Helv. Phys. Acta **14**, 372, 588 (1941); **15**, 23 (1942).
20. L.P. Horwitz and C. Piron, Helv. Phys. Acta **46**, 316 (1973).
21. Y. Strauss and L.P. Horwitz, in preparation.

A Geometrical Approach to Calculating Determinants of Wiener-Hopf Operators

J.P. MacCormick and B.S. Pavlov

Department of Mathematics, University of Auckland, Private Bag 92019, Auckland, New Zealand

Summary. Using a geometric idea from semigroup theory, we derive a formula for the determinants of certain Wiener-Hopf operators whose symbols are bounded and analytic in the upper half plane. For rational symbols, we propose a method for calculating the asymptotics of the eigenvalues of these operators.

1. Introduction

In [6] the general approach for asymptotic calculation of classical Szegö-Kac determinants [14, 9, 1] was proposed. This approach permits investigation of the asymptotics of the Fredholm determinants of the Wiener-Hopf operator T, defined on the finite interval $(0, a)$ by

$$T \equiv T_a(g) : \quad \begin{aligned} L_2(0,a) &\longrightarrow & L_2(0,a) \\ u(x) &\longmapsto & u(x) + \int_0^a g(x-s)u(s)\, ds. \end{aligned}$$

More specifically (see [4]), the asymptotics as $a \to \infty$ can be calculated provided the symbol $\sigma = 1 + \int e^{iks} g(s) ds$ possesses real roots. In [3], an elegant description is given for the oscillating terms in the asymptotics, in the special case that the symbol has only two real roots. Recovering similar asymptotics in the more general case of matrix integral operators (see for instance [15, 16]), requires a more direct and general approach to the problem, which is supplied by the Lax-Phillips version of analytic semigroup theory suggested in [10] for scattering problems. This approach is equivalent [2], to the construction of functional models of dissipative operators [7].

In this paper we consider Wiener-Hopf operators on the finite interval $(0, a)$ whose symbol σ is analytic in the upper half plane. We also consider the case where σ is a real rational function.

Let T be the above Wiener-Hopf operator, and set $\rho = \sigma - 1 = \mathcal{F}^* g$, where $\mathcal{F}^*$ is the inverse Fourier transform. (The operator of multiplication by σ will also be denoted by σ — the meaning is always clear from the context). Our approach to the calculation of the determinant of T is based on the fact that T can be approximated by some other operators W_β whose eigenvalues are known exactly.

Recall that the inverse Fourier transform $\mathcal{F}^*$ maps $L_2[0, \infty]$ unitarily to the Hardy space H_2^+ of the upper half plane. Further, $\mathcal{F}^*(L_2[0, a]) = H_2^+ \ominus e^{ika} H_2^+$, and of course the convolution operator becomes the multiplication operator. So denoting $H_2^+ \ominus e^{ika} H_2^+$ by K_a, and writing P_a for orthogonal projection $H_2^+ \to K_a$, we see that T is unitarily equivalent to $W \equiv W_a(\sigma) = P_a \sigma P_a$. In other words, we have reduced our original problem to the calculation of

(*) $\det (P_a \sigma P_a)$.

The exponential e^{ika} is a singular function, so we can find a sequence of Blaschke products tending uniformly to it on the upper half plane. A good choice turns out to be

$$\Pi_\beta(k) = \frac{e^{ika} - e^{-\beta}}{1 - e^{ika} e^{-\beta}},$$

which does indeed tend uniformly (on the upper half plane) to e^{ika} as $\beta \to \infty$. It's convenient to note here that the zeroes of $\Pi_\beta(k)$ occur at the points $k_l = 2\pi l/a + i\beta/a$, $l \in \mathbb{Z}$.

Let $K_\beta = H_2^+ \ominus \Pi_\beta H_2^+$, and P_β = orthogonal projection $H_2^+ \to K_\beta$. We consider the operator

$$W_\beta = P_\beta \sigma P_\beta$$

as an approximation for W, because in a sense to be made precise later, $K_\beta \to K_a$ and $P_\beta \to P_a$ as $\beta \to \infty$. The idea is that instead of (*), we can use

(†) $\lim_{\beta \to \infty} \det (P_\beta \sigma P_\beta)$,

provided that the W_β approximate W well enough. But the whole point of this approach is that the operator $P_\beta \sigma P_\beta$ turns out to have a remarkably simple form, provided the function σ is *analytic in the upper half plane*: its eigenvectors form a complete set (and even a Riesz basis) in K_β, and the eigenvalues are just $\sigma(k_l)$. This fact immediately gives an explicit expression for (†).

A similar approach is applicable to the case of a rational symbol, including the case of zeroes on the real axis. We show that the spectral analysis of such Wiener-Hopf operators can be reduced to the investigation of a finite matrix, and a procedure for deriving the asymptotics of the eigenvalues for fixed a is suggested.

The straightforward plan outlined here meets some minor obstacles, such as the fact that the operators W and W_β are close in operator norm but not in trace norm. Therefore we need an intermediate operator, which is similar[1] to W_β (and therefore has the same determinant), but close to W in trace norm. This intermediate operator will be constructed as the image of W_β under the multiplication operator of an entire function f_β, which is bounded and invertible as an operator in $L_2(\mathbb{R})$. In summary, the plan is realised as the following chain of statements, which sketch the way of using Semigroup Theory (or the functional calculus for shift operators) for calculating Szegö-Kac determinants.

---

[1] In the technical sense, i.e. equal to $A W_\beta A^{-1}$ for some A.

2. Proof of the Main Theorem

In all the following results, $a > 0$ and $\beta > 1$. We first state a well-known result from the theory of semigroups.

Proposition 1 *Let Π_β be the family of Blaschke products*

$$\Pi_\beta(k) = \frac{e^{ika} - e^{-\beta}}{1 - e^{ika}e^{-\beta}},$$

approaching the singular function $e^{ika} \equiv \theta_a$ uniformly in upper half plane as $\beta \to \infty$. Consider the generators B_β of the contracting semigroup

$$Z_\beta(t) = P_\beta e^{ikt} P_\beta \equiv e^{iB_\beta t}, \quad t > 0,$$

which arises as a compression of the shift group onto the coinvariant subspaces

$$K_\beta = H_+^2 \ominus \Pi_\beta H_+^2.$$

Then the B_β are simple dissipative operators, with eigenfunctions given by

$$\psi_l(k) = \frac{\Pi_\beta(k)}{k - k_l}, \quad l \in \mathbb{Z}$$

and corresponding eigenvalues $k_l = 2\pi l/a + i\beta/a$.

For a proof, see for example [11].

A similar statement is valid for the systems of eigenvectors of the adjoint operators B_β^*. Actually in this case the eigenvectors conveniently coincide with the H_2^+ reproducing kernels: $\varphi_l(k) = \frac{1}{k - \bar{k}_l}$. In other words, we have,

Proposition 2 *The eigenvectors of B_β^* are*

$$\varphi_l(k) = \frac{1}{k - \bar{k}_l}, \quad l \in \mathbb{Z}$$

with eigenvalues $\bar{k}_l = 2\pi l - i\beta$.

Proof: We need only check that $\{\varphi_l\}$ and $\{\psi_l\}$ are biorthogonal sets. The details are in [11], for example. ∎

We also have the following fact which will be crucial for our proofs later on.

Proposition 3 *The sets $\{\varphi_l\}$ and $\{\psi_l\}$ both form Riesz bases[2] for the subspace K_β.*

---

[2] By a Riesz basis, we mean a basis obtained from an orthonormal basis by an invertible, bounded, linear transformation.

Proof: This follows from [13], and full details are given in [11]. The interesting point of the proof is that this problem was solved for us many years ago by Carleson [5], in the context of interpolation by analytic functions. The well-known Carleson condition states that the family $\{\varphi_l\}$ is a Riesz basis[3] if

$$\inf_m \prod_{l \neq m} \left| \frac{k_m - k_l}{k_m - \bar{k}_l} \right| > 0,$$

and a quick calculation shows that this condition holds for our set $\{\varphi_l\}$. ∎

In the next theorem we describe an important automorphism of $L_2(\mathbb{R})$ which maps K_β to K_a.

Theorem 4 *Write $\theta(k) = e^{ika}$, and let $f_\beta(k)$ be the entire function of exponential type defined by*

$$f_\beta(k) = 1 - e^{-\beta} e^{ika}.$$

Then the multiplication operator $u \mapsto f_\beta u$ is a bounded and invertible operator on $L_2(\mathbb{R})$, transforming the orthogonal sum

$$L_2(\mathbb{R}) = H_-^2 \oplus K_\beta \oplus \Pi_\beta H_+^2$$

into the direct sum

$$\overline{\Pi}_\beta \theta H_-^2 + K_a + \Pi_\beta H_+^2,$$

where $K_a = H_+^2 \ominus \theta H_+^2$ is a coinvariant subspace of the shift group corresponding to the singular function θ. The entire functions

$$\Phi_l = f_\beta \varphi_l$$

form a Riesz basis in K_a for each $\beta > 1$.

Proof: The full proof is in [11]. Multiplication by f_β is clearly bounded and invertible since $1 - e^{-\beta} \leq |f_\beta| \leq 1 + e^{-\beta}$ on the real axis, and the proof proceeds by checking the transformation of each orthogonal subspace of $L_2(\mathbb{R})$ separately. Of course we exploit our knowledge from Propositions 1 and 2 to examine the map $f_\beta : K_\beta \longrightarrow K_a$. The final claim that the $\{\Phi_l\}$ form a Riesz basis for K_a follows immediately from Proposition 3, as they were obtained from the Riesz basis $\{\varphi_l\}$ of K_β by a bounded invertible linear transformation. ∎

It is interesting to note here that in fact the functions Φ_l are Fourier images of the projections of exponentials $e^{-ik_l x}$ in $L_2(\mathbb{R})$ onto $L_2(0, a)$.

As explained in the introduction, it will turn out that the operators $W_\beta \equiv P_\beta \sigma P_\beta$ do not approximate the operator $W_a \equiv P_a \sigma P_a$ well enough for our purposes. Therefore in the next theorem we introduce the operators $\mathcal{W}_\beta$ which have the same determinants as the W_β. Then in Theorem 6 we show that $\mathcal{W}_\beta$ is a good enough approximation to W_a.

---

[3] Actually, the Carleson condition guarantees only that we have a so-called *unconditional basis*. A Riesz basis must also satisfy $\inf \|\varphi_l\| > 0$ and $\sup \|\varphi_l\| < \infty$, but these conditions are clearly fulfilled here.

Theorem 5 *The operator*

$$\mathcal{P}_a^\beta = f_\beta P_\beta f_\beta^{-1}$$

is a skew (i.e. nonorthogonal) projection onto K_a parallel to the sum of subspaces

$$\overline{\Pi}_\beta \theta H_-^2 + \Pi_\beta H_+^2.$$

For each essentially bounded function σ defined on the real axis, the operator $W_\beta \equiv P_\beta \sigma P_\beta$ is bounded and is similar to the operator $\mathcal{W}_\beta \equiv \mathcal{P}_a^\beta \sigma \mathcal{P}_a^\beta$ acting in the subspace K_a; in particular, W_β and $\mathcal{W}_\beta$ have the same determinants.

Proof: The effect of f_β described in Theorem 4 means precisely that $\mathcal{P}_a^\beta$ is zero on $\Pi_\beta \theta H_2^- + \Pi_\beta H_2^+$, and the identity on K_a; this is the definition of a skew projection so the first statement is proved.

Now it is not generally true for infinite-dimensional determinants that

$$\det(PQP^{-1}) = \det Q,$$

but this formula does hold if P maps some Riesz basis of dom Q to a Riesz basis of dom PQP^{-1}. But we proved in Theorem 4 that f_β maps the Riesz basis $\{\varphi_l\}$ of K_β to the Riesz basis $\{\Phi_l\}$ of K_a. By the definition of $\mathcal{W}_\beta$ we have

$$\mathcal{W}_\beta|_{K_a} = f_\beta P_\beta \sigma P_\beta f_\beta^{-1}|_{K_a} = f_\beta W_a(\sigma) f_\beta^{-1}|_{K_a},$$

so the above remarks tell us that $\det \mathcal{W}_\beta = \det W_\beta$. ∎

Our next task is to estimate the norms of the various operators defined so far. Define

$$\epsilon_\beta = \sup_{k \in \mathbb{R}} \left(\Pi_\beta(k) - \theta(k) \right) = \sup_{k \in \mathbb{R}} \left(1 - \overline{\Pi}_\beta(k)\theta(k) \right).$$

(It is easy to see these suprema are equal, and in fact that $\epsilon_\beta \leq \frac{2e^{-\beta}}{1 - e^{-\beta}}$. In particular, we see that $\epsilon_\beta \to 0$ as $\beta \to \infty$.)

The next theorem states that the intermediate operator $\mathcal{W}_\beta$ is actually close to W in trace norm.

Theorem 6 *Let σ be a bounded analytic function in the upper half plane, set $\rho = \sigma - 1$ and suppose ρ can be expressed as the product of 3 functions, each in the intersection of the Hardy classes $H_\infty^+ \cap H_2^+$:*

(1) $$\rho = \rho_1 \rho_2 \rho_3, \quad \rho_j \in H_\infty^+ \cap H_2^+$$

Then

$$\left\| P_a \rho P_a - \mathcal{P}_a^\beta \rho \mathcal{P}_a^\beta \right\|_{Trace} \leq \frac{\sqrt{2}\epsilon_\beta}{\left(1 - \sqrt{2\epsilon_\beta}\right)^3} \, const,$$

where the constant depends only on the L_2 and L_∞ norms of the factors ρ_j. In particular, $\mathcal{P}_a^\beta \rho \mathcal{P}_a^\beta \to P_a \rho P_a$ in trace norm as $\beta \to \infty$.

Proof: Full details are in [11]. The proof involves carefully exploiting the relationship between operator, Hilbert-Schmidt, and trace norms, together with some straightforward estimation.   ∎

At last we are in a position to prove our main theorem, obtaining a formula for det W.

Theorem 7 *Suppose σ is an analytic function in the upper half plane, and $\sigma - 1$ satisfies the 3-factor condition (1) of the previous theorem. Then*

$$(2) \qquad \det W_a(\sigma) = \lim_{\beta \to \infty} \prod_{l \in \mathbb{Z}} \sigma(k_l),$$

where $k_l = 2\pi l/a + i\beta/a$.

Proof: We apply our previous results to carry out the plan outlined in the introduction. Note that in the first line we will use the fact that det is continuous with respect to the trace norm. (This is proved in [4], for example). We have

$$
\begin{aligned}
\det W_a(\sigma) &= \lim_{\beta \to \infty} \det \mathcal{W}_\beta, & \text{by Theorem 6} \\
&= \lim_{\beta \to \infty} \det W_\beta, & \text{by Theorem 5} \\
&= \lim_{\beta \to \infty} \prod (\text{evals of } W_\beta), & \text{by definition of det} \\
&= \lim_{\beta \to \infty} \prod_{l \in \mathbb{Z}} \sigma(k_l), & \text{by Proposition 1}
\end{aligned}
$$

∎

3. An Example

As an example, suppose $\sigma(z) = 1 + (z + i)^{-n}$. (This is a natural example since symbols of this form arise from kernels of W proportional to $x^{n-1}e^{-x}$). If $n \geq 3$, the 3-factor condition (1) is satisfied, so we can apply (2) to see

$$
\det W = \lim_{\beta \to \infty} \prod_{l=-\infty}^{\infty} \left\{ 1 + \left[\frac{2\pi l}{a} + i \left(1 + \frac{2\pi \beta}{a} \right) \right]^{-n} \right\}
$$

$$(3) \qquad\qquad = 1.$$

(To calculate the limit, use the fact that $1 + a_1 + \ldots + a_n \leq (1+a_1) \ldots (1+a_n) \leq e^{a_1 + \cdots + a_n}$, provided each $a_n > 0$.)

The Achiezer-Kac formula for the asymptotics of det W states that as $a \to \infty$,

$$(4) \qquad \det W \sim exp \left(\frac{a}{2\pi} \int_{-\infty}^{\infty} \log \sigma \, dz \right) \times$$

$$
exp \left(\frac{1}{4\pi^2} \int_0^{\infty} z \, \mathcal{F}(\log \sigma)(z) \, \sigma(\log \sigma)(-z) \, dz \right),
$$

and of course it should agree with (3). It is not hard to check this agreement directly; using contour integration on each factor of (4) shows that both are equal to 1.

Note that when $n = 2$, these calculations still work even though σ no longer satisfies (1). This provides some evidence that the condition (1) can be weakened without affecting formula (2).

If we are prepared to abandon rigour for a moment, we can make a direct connection between (2) and the Achiezer-Kac formula, in the case where σ is an outer function of the upper half plane. In this case we know that up to a constant of modulus 1, σ can be represented as

$$\sigma(z) = exp\left(\frac{1}{i\pi} \int_{-\infty}^{\infty} \frac{\log|\sigma(t)|}{t - z} dt\right).$$

After substituting this expression into (2), we obtain a $\sum_{l \in \mathbb{Z}}$ which is not absolutely convergent; nevertheless if we make the most obvious choice for ordering this sum — in order of increasing $|l|$ so that terms for $+l$ and $-l$ have a cancelling effect on each other — the result is absolutely convergent, and we can simplify the result using the Poisson Summation Formula, eventually obtaining

$$\det W \overset{?}{=} exp\left(\frac{a}{2\pi} \int_{-\infty}^{\infty} \log|\sigma(t)|\, dt\right)$$

(5)
$$= exp\left(\frac{a}{2\pi} \int_{-\infty}^{\infty} \log\sigma(t)\, dt\right).$$

The last line follows since it can be shown $\arg\sigma$ is an odd function[4], and hence $\int \log|\sigma| = \int \log\sigma$. When σ is an outer function, the second factor in (4) turns out to be 1, so (5) agrees with the asymptotic behaviour predicted by the Achiezer-Kac formula.

4. The Rational Symbol Case

In [6] the case of a rational symbol is carefully investigated. Now we demonstrate how to generalise our approach to real rational symbols. Calculating the eigenvalues of the corresponding Wiener-Hopf operators will be reduced to the investigation of a finite dimensional matrix, whose determinant vanishes at the eigenvalues of the original Wiener-Hopf operator.

So, suppose the symbol σ is real on the real axis and thus can be represented the in form of a finite sum of Cauchy kernels with poles at prescribed complex points. In what follows we deal only with the case that all the poles

---

[4] That $\arg\sigma$ is odd follows from the fact that $1 - \sigma$ is the Fourier Transform of a real-valued function.

are simple; generalising to multiple poles is fairly straightforward but makes the notation unnecessarily complicated. Thus σ has the form

$$
(6) \qquad \sigma(k) = 1 + \sum_{l=1}^{L} \frac{\alpha_l}{k - k_l} + \frac{\bar{\alpha}_l}{k - \bar{k}_l}
$$

$$
\equiv 1 + \varphi(k), \quad \Im k_l > 0.
$$

It is easy to see that the corresponding Wiener-Hopf operator $P_a \varphi|_{K_a}$ is compact. Let us denote by $z_n(\lambda)$, $n = -L, \dots -2, -1, 1, 2, \dots L$ the roots of the auxiliary equation $\varphi(k) = \lambda$ (counting multiplicity).

Theorem 8 *The eigenvalues of the Wiener-Hopf operator in the coinvariant subspace $K_a = H_2^+ \ominus \theta_a H_2^+$, $\theta_a(k) = e^{ika}$ coincide with the zeroes of the determinant of a finite square matrix:*

$$
\det \begin{pmatrix} \dots & \dots & \dots & \dots \\ \dots & \frac{\alpha_l}{z_n(\lambda) - k_l} & \frac{\bar{\alpha}_l}{z_n(\lambda) - \bar{k}_l} e^{iz_n(\lambda)a} & \dots \\ \dots & \dots & \dots & \dots \end{pmatrix}
$$

Here $n = -L, \dots -2, -1, 1, 2, \dots L$, and $l = 1, \dots L$.

Sketch of the proof. The proof is based on the following well-known statement (see [7]).

Proposition 9 *The following representations are true for the resolvents of the generators of compressions shifts and adjoint shifts semigroups onto the coinvariant subspace $K_a = H_2^+ \ominus \theta_a H_2^+$:*

$$
P_a \frac{1}{k - p} u(k) = \frac{u - u(p)}{k - p}, \quad \Im p > 0, \; u \in K_a
$$

$$
P_a \frac{1}{k - \bar{p}} u(k) = \frac{u - \theta(k)\bar{\theta}(\bar{p})u(\bar{p})}{k - \bar{p}}, \quad \Im \bar{p} < 0, \; u \in K_a.
$$

So for φ of the form (6) and $u \in K_a$ we have

$$
P_a(\phi u) = \phi(k)\, u - \sum_{l=1}^{L} \left(\alpha_l \frac{u(k_l)}{k - k_l} + \theta(k)\bar{\alpha}_l \frac{\bar{\theta}(\bar{k}_l)u(\bar{k}_l)}{k - \bar{k}_l} \right).
$$

In particular, if u is an eigenfunction of $P_a \phi|_{K_a}$ with eigenvalue λ, we must have

$$
(7) \qquad \sum_{l=1}^{L} \left(\alpha_l \frac{u(k_l)}{k - k_l} + \theta(k)\bar{\alpha}_l \frac{\bar{\theta}(\bar{k}_l)u(\bar{k}_l)}{k - \bar{k}_l} \right) = 0
$$

whenever $k = z_n(\lambda)$ for some n.

Fix λ, and consider $u(k_l) \equiv u_l$ and $\bar{\theta}(\bar{k}_l)u(\bar{k}_l) \equiv v_l$ as the $2L$ unknowns of the linear system (7). (Note that (7) really does consist of $2L$ equations in these unknowns as there are $2L$ possible values of k.) Since the system has a solution, the determinant of its coefficients must be zero. It is clear from our definitions that these coefficients have the form $\frac{\alpha_l}{z_n(\lambda)-k_l}$ or $\frac{\bar{\alpha}_l\theta(z_n(\lambda))}{z_n(\lambda)-k_l}$, exactly as stated in the theorem.

Conversely, suppose the determinant in the theorem is zero and choose $u_1, \ldots u_L, v_1, \ldots v_L$ to be a solution of (7). Define $u(k)$ by

$$ u(k) = \frac{1}{\varphi(k)-\lambda} \sum_{l=1}^{L} \left(\frac{\alpha_l}{k-k_l}u_l^r + \frac{\theta(k)\bar{\alpha}_l}{k-k_l}v_l^r \right). $$

Then $u(k)$ is in K_a, since the zeroes of the denominator are compensated by the roots of the sum counting multiplicity, and it is easy to see u is an eigenfunction of the original Wiener-Hopf operator, and λ is the corresponding eigenvalue.   ∎

The determinant of Theorem 8 can be simplified for small λ by using a Taylor series. If $z_n(0)$ is a simple zero of φ, then

$$ (8) \qquad z_n(\lambda) = \frac{1}{2\pi i} \oint_\gamma \frac{s\dot{\varphi}(s)}{\varphi(s)-\lambda}ds $$

$$ = z_n(0) + \sum_{j=1}^{\infty} \lambda^j \frac{1}{2\pi i} \oint_\gamma \frac{s\dot{\varphi}(s)}{\varphi^{j+1}(s)}ds, $$

where the integral is taken over a small contour γ, encircling the simple root z_n once in the positive direction.

As remarked earlier, all of the above can be generalised to the case where φ has multiple poles; the details are in [11]. Instead of (8) we can use the generalised Taylor series, with a properly chosen factor λ^δ, δ defined by the multiplicity of the root of the symbol. Then the asymptotic behaviour of the determinant for small λ^δ is described by a quasipolynomial in $\lambda^\delta, \exp(d/\lambda^\delta)$. The asymptotics of the zeroes of these quasipolynomials for small λ can be derived using a construction involving Newton polygons ([8, 12]).

It is obvious that the determinant of the finite matrix under consideration coincides with the determinant or the regularized determinant of the operator $1 - \frac{1}{\lambda}P_a\varphi|_{K_a}$ up to some analytic factor which has no zeroes on the complex plane. It is a challenging problem to describe this factor in explicit form.

Acknowledgments: We would like to thank Rowan Killip, who made very helpful comments on an early draft of this paper. We are also grateful to the Marsden Fund, which supported B. Pavlov during some of this work.

References

1. N. I. Achieser, *The continuous analogue of some theorems on Toeplitz Matrices.* Ukrain. Math. Zh. **16**:4 (1964) (Russian); Eng. transl. Amer. Math. Soc. Transl. Ser. 2, **50** (1966), 295-316.
2. V. Adamjan, D. Arov. *On Unitary couplings of semi-unitary operators.* Mat. Issled. **1**:2 (1966), 3-64; Eng. transl. Amer. Math. Soc. Transl. Ser. 2, **95** (1970), 75-129.
3. S. Albeverio, S. Lakaev, K. Makarov, *A remark on a Szegö-Kac limit theorem for convolution operators with rational symbols.* Manuscript.
4. A. Böttcher, *Wiener-Hopf Determinants with Rational Symbols.* Math. Nachrichten. **144** (1989), 39-64.
5. L. Carleson, *An interpolation problem for bounded analytic functions.* Amer. J. Math. **80**:4 (1958), 921-930.
6. K. M. Day, *Toeplitz matrices generated by the Laurent series expansion of an arbitrary rational function.* Trans. Amer. Math. Soc. **206** (1975), 294-312.
7. C. Foias, B. Sz. Nagy, *Analyse Harmonique des Operateurs de l'Espace de Hilbert*, Academie Kiado, Budapest (1967); Eng. transl. *Harmonic analysis of operators on Hilbert space*, North-Holland, Amsterdam: American Elsevier, New York: Kiado, Budapest (1970).
8. I. Horn. *Asymptotic distribution of roots of quasipolynomials.* Mathematische Annalen, Bd. 49 (1897).
9. M. Kac, *Toeplitz matrices, translation kernels and a related problem in probability theory.* Duke Math. J. **21** (1954), 501-509.
10. P. D. Lax, R. S. Phillips, *Scattering Theory*, Academic Press, New York-London (1967).
11. J. P. MacCormick, B. S. Pavlov, *A geometrical approach to calculating determinants of Wiener-Hopf operators.* Departmental Report, Department of Mathematics, University of Auckland, March 1997
12. B. S. Pavlov. *Spectral Analysis of the differential operator with a spread boundary condition.* in Problems of Mathematical Physics Vol. 6 ed. M. Birman , Leningrad University 1972, 101-119 (Russian).
13. B. S. Pavlov, *The Riesz-basis property of a system of exponentials and Muckenhoupt condition.* Dokl. Acad. Nauk USSR, **247** (1979) 37-40. Eng. transl. Sov. Math. Dokl. **20** (1979), 655-659.
14. Szegö, *On certain Hermitian forms, associated with the Fourier series of a positive function.* Communications du Seminaire mathematique de l'universite de Lund, tome supplementaire (1952) dedie a Marcel Riesz, 228-238.
15. H. Widom, *Toeplitz determinants with singular generating function.* Amer. J. Math. **95** (1973), 333-383.
16. H. Widom, *Szegö limit theorem: higher-dimensional matrix case.* Journ. Funk. Anal. **39** (1980), 182-198.

Chapter VII

Irreversibility and Measurement
in Quantum Mechanics

Time Scale, Objectivity and Irreversibility in Quantum Mechanics

L. Lanz[1] and O. Melsheimer[2]

[1] Dipartimento di Fisica dell'Università di Milano and Istituto Nazionale di Fisica Nucleare, Sezione di Milano, Via Celoria 16, I-20133, Milan, Italy
[2] Fachbereich Physik Philipps-Universität, Renthof 7,D-35032 Marburg, Germany

Summary. It is argued that setting isolated systems as primary scope of field theory and looking at particles as derived entities, the problem of an objective anchorage of quantum mechanics can be solved and irreversibility acquires a fundamental role. These general ideas are checked in the case of the Boltzmann description of a dilute gas.

1. Introduction

For the physical interpretation of the formalism of quantum mechanics (QM) a classical world of sources and detectors of microsystems must be invoked in order to get rid of the so-called E.R.P. paradoxon and the subjectivity within the measuring process. Substituting invocation with high mathematical effort, Ludwig obtained QM as the description of a microphysical interaction channel between macrosystems and showed that QM must be enriched by the mathematical tools that are now called POV measures, operations and instruments (for a more recent overview see [8]. It is very remarkable that these concepts also arose in more accurate descriptions of the measuring process and more profound thinking about the statistical structure of QM [2, 3], and [4].

In Ludwig's approach a microsystem is tied with the measuring processes giving evidence for it. The measuring process consists in the *selection of outcomes* during a *registration procedure* associated with a *preparation procedure*, so that a statistical regularity arises. The kind of complexity allowed for the registration procedure (e.g. sequences of correlated outcomes can be considered) enters in the definition of the microsystem; its quantum mechanical state is representative of the whole class of preparation procedures that are equivalent with respect to the envisaged registration procedures: E.P.R. paradox arises, due to a too naive linking of the concrete preparation of the microsystem with its representation by a quantum state, neglecting relativity of this representation with respect to the actually performed registrations [9]. By the careful analysis of Ludwig the concept of *physical microsystem* can be reconciled with the typical quantum mechanical feature of many-particle entanglement.

However Ludwig's foundation of quantum mechanics can appear unsatisfactory if one looks at QM as at the definitive theoretical scheme for reality;

to reach microsystems the detour through measurement is necessary and the phenomenological description of the systems building the measuring device must be used, in order to concretely perform what Ludwig calls a *selection procedure*: in this way the phenomenological specification of systems becomes the very starting point for QM. A satisfactory completion of this point of view requires that at least in principle, the phenomenological way of describing systems naturally fits inside quantum mechanics for their microphysical components; Ludwig himself has suggested that a more encompassing theory could perhaps be necessary to cope with the last problem.

Our aim is to reconsider this problem taking in an explicit way isolated systems as main elements of reality and looking at QM, already in its field formulation, as the basic theory of finite isolated systems; only afterwards one arrives at particles when the pecularities of finite systems, like boundary conditions, are negligible: in this philosophy the thermodynamic limit has just a reversed role, it is not used in order to reveal the classical behaviour of macrosystems, but is necessary to attain local covariance and universality of the theory. All the difficulties related to quantum mechanical inseparability, enter now in the very concept of isolated system: our point of view implies a weakening of the idea of an absolutely isolated system. Isolation is relative to a suitable set of "slow variables", whose dynamics restricted to expectations, as phenomenology indicates, has only a restricted memory of the previous values of these expectations.

So to achieve isolation one has to establish a suitable time scale, choose variables with expectations having a typical variation time of this scale, prepare the system inside some confined space region during a suitable time interval, controlling and measuring the relevant variables inside a suitable preparation time interval; we have to explicitly introduce in the formalism the fact that restricting to suitable variables and using some effective description of quasilocal field interaction, the too remote history of the system has to be neglected. On the contrary if one pretends to describe the local behaviour with completely sharp time specification one can expect that the full history of the whole universe would be involved.

By taking finite isolated systems, instead of particles as the main subjects, the typical ultraviolet and infrared problem of quantum field theory is absorbed inside the non universal features of the description. The opening to irreversibility entailed by this point of view, will be discussed in Section 3 to settle problems arising in the classical description of a macrosystem; such a description is given in Section 2, essentially by a suitable reconsideration of Zubarev's approach to non equilibrium thermodynamics. As an example Boltzmann description of a dilute gas is discussed in Section 4.

The implementation of the Ludwig setting of a quantum mechanical experiment within our formalism may then be obtained by taking two macrosystems whose "isolation" is broken in the most elementary way giving rise to a change of the dynamical behaviour of one of these systems, the so- called registration device. This change of the dynamics can then be interpreted as the action of a microsystem. For more details we refer to [7].

2. Description of a Finite Isolated System

The microphysical structure is described by a set of interacting quantum Schrödinger fields. Here we consider the simplest model: one interacting quantum Schrödinger field (QSF) $\hat{\psi}_{nc}(\mathbf{x})$ (not yet confined) , to which the following local Hamiltonian density is associated:

$$(1) \quad \hat{e}_{nc}(\mathbf{x}) = \frac{\hbar^2}{2m} \nabla \hat{\psi}^\dagger_{nc}(\mathbf{x}) \cdot \nabla \hat{\psi}_{nc}(\mathbf{x}) +$$

$$+ \tfrac{1}{2} \int d^3\mathbf{r}\, \hat{\psi}^\dagger_{nc}(\mathbf{x}\text{-}\mathbf{r}/2)\hat{\psi}^\dagger_{nc}(\mathbf{x}\text{+}\mathbf{r}/2) V(|\mathbf{r}|)\hat{\psi}_{nc}(\mathbf{x}\text{+}\mathbf{r}/2)\hat{\psi}_{nc}(\mathbf{x}\text{-}\mathbf{r}/2)$$

$$(2) \qquad\qquad \left[\hat{\psi}(\mathbf{x}), \hat{\psi}^\dagger(\mathbf{x}')\right]_\pm = \delta(\mathbf{x} - \mathbf{x}').$$

The basic local dynamical variable of this model

$$\hat{\psi}_{nc}(\mathbf{x}, t) = e^{\frac{i}{\hbar} H_{nc} t} \hat{\psi}_{nc}(\mathbf{x}) e^{-\frac{i}{\hbar} H_{nc} t} \quad , \quad \hat{H} = \int_\omega d^3\mathbf{x}\, \hat{e}_{nc}(\mathbf{x})$$

satisfies the Schrödinger field equation

$$(3) \qquad i\hbar \frac{\partial \hat{\psi}_{nc}(\mathbf{x}, t)}{\partial t} = -\frac{\hbar^2}{2m} \Delta \hat{\psi}_{nc}(\mathbf{x}, t) +$$

$$+ \int d^3\mathbf{y}\, \hat{\psi}^\dagger_{nc}(\mathbf{y}, t) V(|\mathbf{x} - \mathbf{y}|)\hat{\psi}_{nc}(\mathbf{y}, t)\hat{\psi}_{nc}(\mathbf{x}, t)$$

As it is well known and will appear in the results, choice (1) means in usual language, N-body system of structureless molecules interacting via the two body potential $V|\mathbf{x}_1 - \mathbf{x}_2|)$.

The field equation (3) that accounts for covariance under Galilei transformations, has for the massive continuum we are describing a similar role that Maxwell equations have for electromagnetism, however, due to the irrelevance of selfinteraction in the electromagnetic case, the classical field theory plays in the latter case a much more extended role.

Confinement of the system inside a region $\omega \subset \mathbf{R}^3$ ist obtained expanding $\hat{\psi}(\mathbf{x})$ on the normal modes $u_f(\mathbf{x})$ of the system, where

$$-\frac{\hbar^2}{2m}\Delta u_f(\mathbf{x}) = E_f u_f(\mathbf{x}) \qquad u_f(\mathbf{x}) = 0 \qquad \mathbf{x} \in \partial\omega$$

setting

(4)
$$\hat{\psi}(\mathbf{x}) = \sum_f u_f(\mathbf{x}) \hat{a}_f \qquad [\hat{a}_f, \hat{a}_g^\dagger]_\pm = \delta_{fg}$$

We replace $\hat{\psi}_{nc}(\mathbf{x})$ by $\hat{\psi}(\mathbf{x})$ for $\mathbf{x} \in \omega$ and by 0 for $\mathbf{x} \notin \omega$, in the energy density $\hat{e}(\mathbf{x})$ and in all relevant expressions built with the field operators. In this way peculiar confinement is superposed to the quasilocal universal (within the limits related to the effective potential $V(|\mathbf{r}|)$) behaviour.

Our aim is not at all a full description of the finite isolated system, but to give a description of it having negligible correlations with the environment; this description is related to suitable *slow variables*, linked to the fundamental constants of motion of the system: mass and energy.

The densities generating these observables are the *relevant variables*; as phenomenology indicates there are two meaningful descriptions:

A: the hydrodynamical one based on energy density and mass density,

B: the kinetic one based on energy density and phase-space density;

the energy density is given by equation (1) with $\hat{\psi}_{nc}(\mathbf{x})$ replaced by $\hat{\psi}(\mathbf{x})$, the mass density is given by

(5)
$$\hat{m}(\mathbf{x}) = m\hat{\psi}^\dagger(\mathbf{x})\hat{\psi}(\mathbf{x})$$

the phase-space density is given by:

(6)
$$\hat{f}(\mathbf{x},\mathbf{p}) = \sum_{hk} \hat{a}_h^\dagger < u_h|\hat{F}^{(1)}(\mathbf{x},\mathbf{p})|u_k > \hat{a}_k$$

$\hat{F}^{(1)}(\mathbf{x},\mathbf{p})$ being the density of a POV measure for the joint position-momentum measurement in one-particle quantum mechanics [5].

The first step towards a classical description is the axiomatic introduction of a velocity field of the continuum, so that the former observables can be referred to a local rest frame: denoting by an index $^{(0)}$ these observables, one has:

$$\hat{e}^{(0)}(\mathbf{x}) = \frac{1}{2m}(i\hbar\nabla - m\mathbf{v}(\mathbf{x},t))\,\hat{\psi}^\dagger(\mathbf{x}) \cdot (-i\hbar\nabla - m\mathbf{v}(\mathbf{x},t))\,\hat{\psi}(\mathbf{x}) +$$
$$+ \frac{1}{2}\int d^3\mathbf{r}\,\hat{\psi}^\dagger(\mathbf{x} - \frac{\mathbf{r}}{2})\hat{\psi}^\dagger(\mathbf{x} + \frac{\mathbf{r}}{2})V(|\mathbf{r}|)\hat{\psi}(\mathbf{x} + \frac{\mathbf{r}}{2})\hat{\psi}(\mathbf{x} - \frac{\mathbf{r}}{2})$$

(7)
$$\hat{m}^{(0)}(\mathbf{x}) = \hat{m}(\mathbf{x})$$

$$\hat{f}^{(0)}(\mathbf{x},\mathbf{p}) = \hat{f}(\mathbf{x}, \mathbf{p} - m\mathbf{v}(\mathbf{x},t))$$

The introduction of this external classical field allows to compensate a gauge transformation of the field $\hat{\psi}(\mathbf{x}) \rightarrow \hat{\psi}(\mathbf{x})e^{\frac{i}{\hbar}\Lambda(\mathbf{x})}$ with a transformation $\mathbf{v}(\mathbf{x},t) \rightarrow \mathbf{v}(\mathbf{x},t) - 1/m\nabla\Lambda(\mathbf{x},t)$ of the external parameter. $\mathbf{v}(\mathbf{x},t)$ is linked to the expectation at time t of the momentum density of the system:

(8) $\hat{\mathbf{p}}(\mathbf{x}) = \frac{1}{2}\Big\{\Big[(i\hbar\nabla - m\mathbf{v}(\mathbf{x},t))\,\hat{\psi}^\dagger(\mathbf{x})\Big]\hat{\psi}(\mathbf{x})-$

$\hat{\psi}^\dagger(\mathbf{x})\,(i\hbar\nabla + m\mathbf{v}(\mathbf{x},t))\,\hat{\psi}(\mathbf{x})\Big\}$

by the relation:

(9) $\langle\hat{\mathbf{p}}(\mathbf{x})\rangle_t = \langle\hat{m}(\mathbf{x})\rangle_t \cdot \mathbf{v}(\mathbf{x},t)$

or equivalently

(10) $\langle\hat{\mathbf{p}}^{(0)}(\mathbf{x})\rangle_t = 0$

The other classical state parameters are linked to the expectations $\langle\hat{e}^{(0)}(\mathbf{x})\rangle_t$ and $\langle\hat{m}^{(0)}(\mathbf{x})\rangle_t$ (or $\langle\hat{f}^{(0)}(\mathbf{x},\mathbf{p})\rangle_t$ in the kinetic case), as they enter into the structure of the most unbiased statistical operator, giving these assigned expectations. Such operator is characterized by the conditions:

(11) $\langle\hat{e}^{(0)}(\mathbf{x})_t = \mathrm{Tr}(\hat{e}^{(0)}(\mathbf{x})\hat{\rho}), \qquad \langle\hat{m}^{(0)}(\mathbf{x})\rangle_t = \mathrm{Tr}(\hat{m}^{(0)}(\mathbf{x})\hat{\rho})$

$0 = \mathrm{Tr}(\hat{\mathbf{p}}^{(0)}(\mathbf{x})\hat{\rho})$

$S(\hat{\rho}) = -\mathrm{Tr}(\hat{\rho}\log\hat{\rho})$ being maximal.. The solution of this problem of conditioned maximality is a generalized Gibbs state

(12) $\hat{w}[\beta(t),\mu(t),\mathbf{v}(t)] = \dfrac{e^{-\int_\omega d^3\mathbf{x}\,\beta(\mathbf{x},t)\left[\hat{e}^{(0)}(\mathbf{x})-\mu(\mathbf{x},t)\hat{m}^{(0)}(\mathbf{x})\right]}}{\mathrm{Tr}\,e^{-\int_\omega d^3\mathbf{x}\,\beta(\mathbf{x},t)\left[\hat{e}^{(0)}(\mathbf{x})-\mu(\mathbf{x},t)\hat{m}^{(0)}(\mathbf{x})\right]}}\,,$

where the fields $\beta(\mathbf{x},t)$, $\mu(\mathbf{x},t)$ are determined by the equations:

$\langle\hat{e}^{(0)}(\mathbf{x})\rangle_t = \mathrm{Tr}(\hat{e}^{(0)}(\mathbf{x})\hat{w}[\beta(t),\mu(t),\mathbf{v}(t)])$

'

$\langle\hat{m}^{(0)}(\mathbf{x})\rangle_t = \mathrm{Tr}(\hat{m}^{(0)}(\mathbf{x})\hat{w}[\beta(t),\mu(t),\mathbf{v}(t)])$

in terms of the assigned expectations $\langle\hat{e}^{(0)}(\mathbf{x})\rangle_t$, $\hat{m}^{(0)}(\mathbf{x})\rangle_t$ and the given field $\mathbf{v}(\mathbf{x},t)$ (for simplicity the dependence of $\hat{e}^{(0)}(\mathbf{x})$ on this field has not been made explicit).

In the kinetic case $\hat{m}^{(0)} \to \hat{f}^{(0)}$ and $\mu(\mathbf{x},t) \to \mu(\mathbf{x},\mathbf{p},t)$. The state function

$S(\beta(t),\mu(t),\mathbf{v}(t)) = \mathrm{Tr}(\hat{w}[\beta(t),\mu(t),\mathbf{v}(t)]\log\hat{w}[\beta(t),\mu(t),\mathbf{v}(t)])$

is the thermodynamic entropy of the system. Let us take at the moment for granted that the dynamics of the system is given by the unitary evolution, generated by the Hamiltonian:

(13) $\hat{H} = \displaystyle\int_\omega d^3\mathbf{x}\,\hat{e}(\mathbf{x})$

then the main point is to give at some time T the initial statistical operator $\hat{\rho}_T$. Let us investigate what happens if one takes:

(14) $\hat{\rho}_T = \hat{w}[\beta(T), \mu(T), \mathbf{v}(T)]$

then by a straightforward calculation [10], the statistical operator of the system is:

$$\hat{\rho}_t = e^{-\frac{i}{\hbar}\hat{H}(t-T)}\hat{\rho}_T e^{\frac{i}{\hbar}\hat{H}(t-T)},$$

which can be written in the following form:

(15) $\hat{\rho}_t = \dfrac{e^{-\langle\beta(t)\cdot\hat{e}^{(0)}\rangle+\langle[\beta(t)\mu(t)]\cdot\hat{m}^{(0)}\rangle+\int_T^t d\tau \hat{S}_t[\beta(\tau),\mu(\tau),\mathbf{v}(\tau)]}}{\text{Tr}\,e^{-\langle\beta(t)\cdot\hat{e}^{(0)}\rangle+\langle[\beta(t)\mu(t)]\cdot\hat{m}^{(0)}\rangle+\int_T^t d\tau \hat{S}_t[\beta(\tau),\mu(\tau),\mathbf{v}(\tau)]}}$

The first two terms in the exponent are a more compact notation to represent the typical exponent of a Gibbs state with parameters $\beta(t)$, $\mu(t)$, $\mathbf{v}(t)$ (the latter is subintended inside $\hat{e}^{(0)}$) referring to time t, e.g.:

$$\langle\beta(t)\cdot\hat{e}^{(0)}\rangle = \int_\omega d^3\mathbf{x}\beta(\mathbf{x},t)\hat{e}^{(0)}(\mathbf{x},\mathbf{v}(t))$$

The last term contains the history of the classical state parameters for $\tau \in [T, t]$:

$$\hat{S}_t[\beta(\tau),\mu(\tau),\mathbf{v}(\tau)] =$$
$$= \int_\omega d^3\mathbf{x}\left(\frac{\partial\beta(\mathbf{x},\tau)}{\partial\tau}\hat{e}(\mathbf{x},\tau-t) - \nabla\beta(\mathbf{x},\tau)\cdot\hat{\mathbf{J}}_e(\mathbf{x},\tau-t)\right) +$$

(16) $+ \int_{\partial\omega} d\sigma\mathbf{n}\beta(\mathbf{x},\tau)\hat{\mathbf{J}}_e(\mathbf{x},\tau-t) + \cdots$

where $\hat{\mathbf{J}}_e$ is the energy current:

$$\frac{i}{\hbar}[\hat{H},\hat{e}(\mathbf{x})] = -\text{div}\hat{\mathbf{J}}_e(\mathbf{x})$$

the time dependence of the operators, e.g.: $\hat{e}(\mathbf{x},t)$ means time dependence in Heisenberg picture: $\hat{e}(\mathbf{x},t) = e^{\frac{i}{\hbar}\hat{H}t}\hat{e}(\mathbf{x})e^{-\frac{i}{\hbar}\hat{H}t}$; at the r.h.s. of equation (16) similar terms related to the other corresponding densities $\hat{m}(\mathbf{x})$, $\mathbf{p}(\mathbf{x})$ have been omitted for brevity. In the framework of information thermodynamics expression (15) would already be taken as a reliable description of the system: no wonder at all about the different structure of $\hat{\rho}_T$ and $\hat{\rho}_t$ since at time t information on the history interval $[T, t]$ is available, while it was not for $t < T$. In our philosophy instead, $\hat{\rho}_t$ is an objective representation of the preparation of the system until time t; expression (15) indicates that the history can be relevant, so the choice (14) becomes highly critical: if the history of the system for $t < T$ is relevant, as one can expect looking at expression (15), the choice (14) (which is the most unbiased by the previous history) is wrong and also $\hat{\rho}_t$ given by (15) is meaningless. A way out could be to shift $T \to -\infty$, thus eliminating the previous history: however the infinite system limit must be taken before, just the contrary of our attitude; furthermore the classical parameters for remote times are not a practically available input; in this way, associating to $\hat{S}_t[\cdots]$ in (15) a factor $e^{(\tau-t)\epsilon}$ and taking $T = -\infty$ one obtains Zubarev's non-equilibrium statistical operator [11].

We propose another solution to the question whether (15) makes sense: by the very definition of the classical parameters, the history term in (15) is by construction irrelevant to calculate the expectations of the basic quantities $\hat{e}$, $\hat{m}$, $\hat{f}$ and $\hat{p}$; one can expect that its contributions to the expectations of the corresponding *time derivatives*, e.g.:

$$\dot{\hat{e}} = \frac{i}{\hbar}[\hat{H}, \hat{e}] = -\mathrm{div}\hat{\mathbf{J}}_e$$

is small enough to allow a pertubative expansion of the exponential in (15) with respect to the history term; then the classical parameters at time τ contribute to correlation functions of the type flow-flow or flow-density, e.g.: $\langle \hat{\mathbf{J}}_e(\mathbf{x}), \hat{e}(\mathbf{y}, \tau - t) \rangle$.

The short-time behaviour of such correlation functions is rapidly decaying, when the time separation of the two functions becomes of the order of a suitable decay time τ_c. Therefore if one considers $t > \tau_c$ only the part of the history referring to times $\tau > t - \tau_c$ does appreciably contribute to the dynamics of the relevant variables. Then to compute their dynamical behaviour for times t, $t - T > \tau_c$, the initial condition (14) is indeed the appropriate choice and consequently $\hat{\rho}_t$ given by (15) can be used to calculate the expectations of the time derivatives, thus yielding closed integrodifferential evolution equations for the classical variables

$$z(t) \equiv (\beta(t), \mu(t), \mathbf{v}(t)), \qquad t \geq T + \tau_c$$

$$\mathrm{Tr}\left(\frac{i}{\hbar}[\hat{H}, \hat{e}(\mathbf{x})]\hat{\rho}_t\right) = \frac{d}{dt}\mathrm{Tr}\left(\hat{e}(\mathbf{x})\hat{w}(\beta(t), \mu(t), \mathbf{v}(t))\right)$$

(17) $\qquad \mathrm{Tr}\left(\frac{i}{\hbar}[\hat{H}, \hat{m}(\mathbf{x})]\hat{\rho}_t\right) = \frac{d}{dt}\mathrm{Tr}\left(\hat{m}(\mathbf{x})\hat{w}(\beta(t), \mu(t), \mathbf{v}(t))\right)$

$$\mathrm{Tr}\left(\frac{i}{\hbar}[\hat{H}, \hat{p}(\mathbf{x})]\hat{\rho}_t\right) = \frac{i}{\hbar}\mathrm{Tr}\left(\hat{p}(\mathbf{x})\hat{w}(\beta(t), \mu(t), \mathbf{v}(t))\right) =$$

$$= \frac{d}{dt}\left(\mathbf{v}(\mathbf{x}, t)\mathrm{Tr}(\hat{m}(\mathbf{x})\hat{w}(\beta(t), \mu(t), \mathbf{v}(t)))\right)$$

With respect to these equations, the values of the state variables $z(t)$ within the time interval $[T, T+\tau_c]$ are prescribed parameters, related to the expectations of $\hat{e}(\mathbf{x})$, $\hat{m}(\mathbf{x})$ and $p(\mathbf{x})$: these expectations have the role of an input for the dynamics at times $t > T + \tau_c$. The time interval $[T, T+\tau_c]$ will be called *preparation time interval* of the system; during such time interval one might also assume that, due to the transition from $\hat{\psi}_{nc}$ to $\hat{\psi}$, the flows indicated in equation (16) do not vanish on the boundary $\partial\omega$, so that a surface contribution can arise. Neglecting mathematical problems about the existence of the solution of these evolution equations, it seems at first that one has solved in principle the question of a classical characterization of the finite isolated system: the parameters $z(t)$ establish the mathematical structure of the statistical operator that provides for times $t > T + \tau_c$ the expectations of the relevant observables. However a serious flaw is evident at long times: due to the pure point spectrum of $\hat{H}$ for a confined system, correlation functions have a quasiperiodical behaviour, so they cannot decay indefinitly and as soon as memory is recovered, choice (14) is no longer tenable and then also $\hat{\rho}_t$ looses its meaning. Practically the difficulty can be avoided if one approximates expression (15), replacing the integral $\int_T^t d\tau \cdots$ by $\int_{t-\tau_c}^t d\tau \cdots$ and reproposing at time $t - \tau_c$ the initial condition

$$\hat{\rho}_{t-\tau_c} \equiv \hat{w}[\beta(t-\tau_c), \mu(t-\tau_c), \mathbf{v}(t-\tau_c)]$$

but this is to resort to an expedient. However it is the basic assumption of unitary dynamics for the isolated system that leads to this difficulty and just this assumption becomes questionable inside the framework which sets isolated systems as basic elements of reality.

3. An Opening to Irreversibility

As it is clear from Section 2 the preparation of a finite system is described by a statistical operator, in our aim bearing the classical state parameters of the system; instead the quantum state vector $\psi \in \mathcal{H}$, bearing indexes related to measurements of observables is a very strong idealization which applies to highly controlled preparations typical of particle physics. If preparations are represented by statistical operators, i.e. elements of the set $\mathcal{K}$, the base of the positive cone in the space $T(\mathcal{H})$ of trace-class operators in $\mathcal{H}$, it becomes most natural to describe transformations of preparations by positive trace-preserving maps $\mathcal{A}$ on $T(\mathcal{H})$, these maps taking now the role that unitary transformations play in the theory based on microsystems. If such a map $\mathcal{A}$ has an inverse one can show that $\mathcal{A} \cdot = \hat{U} \cdot \hat{U}^\dagger$ with $\hat{U}$ unitary or antiunitary on $\mathcal{H}$ and in this way the two formulations become equivalent; but $\mathcal{A}$ need not have an inverse. Denoting by $\hat{\rho}_t$ a preparation performed until time t, let us consider for a system isolated during the time interval $[t_0, t_1]$, the family of spontaneous repreparations ρ_t, $t \in [t_0, t_1]$, which arise due to the time evolution. One assumes that two preparations $\hat{\rho}_{t'}$, $\hat{\rho}_{t''}$, $t_0 \leq t' \leq t'' \leq t_1$ are connected to each other by a map $\mathcal{A}(t'' - t')$:

$$(18) \qquad\qquad \hat{\rho}_{t''} = \mathcal{A}(t'' - t')\hat{\rho}_{t'}$$

the family $\mathcal{A}(\tau), \tau \geq 0$ being a semigroup of positive, trace-preserving maps. Actually taking the construction of section 2 into account, since one restricts to the relevant densities $\hat{e}(\mathbf{x})$, $\hat{\mathbf{p}}(\mathbf{x})$, $\hat{m}(\mathbf{x})$ ($\hat{f}(\mathbf{x}, \mathbf{p})$), looking at the time evolution in the Heisenberg picture, the properties of $\mathcal{A}'(\tau)$, which are mappings on $\mathcal{B}(\mathcal{H})$ into $\mathcal{B}(\mathcal{H})$, are important and one could assume only that $\mathcal{A}'(\tau)$ maps positive densities into positive operators and conserves basic constants of motion like mass and energy: $\mathcal{A}'(\tau)\hat{M} = \hat{M}$, $\mathcal{A}'(\tau)\hat{H} = \hat{H}$. Thus the basic feature of the description characterized by *systems first and particles afterwards* is irreversibility. In a sense we are now exploiting the arrow of time that is implicitly contained in the operative approach to QM, based on preparations and measurements of prepared systems [1]. On the other hand this almost trivial insertion of irreversibility into the formalism of QM can easily be pushed back, as it is shown in Ludwig's approach to QM of microsystems: in his approach a statistical operator represents an equivalence class of preparation procedures of a microsystem and an *effect operator* $\hat{F}$, $(0 \leq \hat{F} \leq \hat{1})$ represents an equivalence class of registration precedures; any time shift of these procedures is allowed and still one assumes that for any preparation procedure another one, shifted back in time, can be found, equivalent to it. Then one arrives at a unitary representation of time shifts (Comi et al., 1975) and as a consequence, at unitary time evolution generated by the Hamiltonian; then also the fundamental principle of conservation of energy is most directly settled. The strategy we are proposing is to start always with local universal microphysics related to unitary representations of the fundamental symmetries, i.e. one has reversibility, energy conservation and some model leading to an energy density; e.g. in this preliminary discussion: expression (1).

Then, as we did in Section 2, one turns to the description of a system, characterized by a suitable choice of relevant variables. Let us specialize our macroscopic system to the case of a dilute gas with short range interaction $V(|\mathbf{r}|)$; it is well known that for such a system a Boltzmann type of description is satisfactory: this description is characterized by a typical macroscopic variation time τ_1 much larger than the duration of a collision: the mean free path is much larger than the range of $V(|\mathbf{r}|)$. We shall see in Section 3 that the introduction of the time scale τ_1, i.e.: the detailed dynamics of the two-body interaction is replaced by collision, leads from the energy density (1) to a semigroup $\mathcal{A}'(\tau)$ of maps, that display a stronger form of positivity, called *complete positivity* which is reminiscent of the unitarity of the dynamics we started with; however this positivity will be *relative* to the relevant variables. We expect that the more general description based on equations (17) can be settled starting with a more fundamental expression than (1). Here $V(|\mathbf{x}|)$ is a phenomenological input that could be derived from a Hamiltonian describing the structure of the molecule as a bound state of charged particles: i.e. the very presence of $V(|\mathbf{r}|)$ indicates that more fundamental fields should be considered. When dealing with the Hamiltonian dynamics of the charged fields one introduces a time scale typical of center of mass motion of the bound states, one expects that a semigroup $\mathcal{A}'(\tau)$ can be derived, whose generator displays an irreversible contribution together with the Hamiltonian contributions like (1). In this way the difficulty we met in Section 2 with the long time behaviour of the correlation function, should eventually disappear.

4. Introduction of Time Scale and Scattering Map

The relevant variables of the hydrodynamic or kinetic description have the following general structure, cf. Section 2:

$$\sum_{hk} \hat{a}_h^\dagger A_{hk}(\xi)\hat{a}_k, \quad \sum_{k_1 k_2 h_1 h_2} \hat{a}_{h_1}^\dagger \hat{a}_{h_2}^\dagger A_{h_1 h_2 k_2 k_1}(\mathbf{x})\hat{a}_{k_2}\hat{a}_{k_1} \ ;$$

thus, in Heisenberg picture, we have to study expressions of the form:

(19)
$$\sum_{hk} e^{\frac{i}{\hbar}\hat{H}t}\hat{a}_h^\dagger \hat{a}_k e^{-\frac{i}{\hbar}\hat{H}t} A_{hk}(\xi) \quad \sum_{k_1 k_2 h_1 h_2} e^{\frac{i}{\hbar}\hat{H}t}\hat{a}_{h_1}^\dagger \hat{a}_{h_2}^\dagger \hat{a}_{k_2}\hat{a}_{k_1} e^{-\frac{i}{\hbar}\hat{H}t} A_{h_1 h_2 k_2 k_1}(\mathbf{x}) \ ;$$

where restriction to slow variables means that in the sums (19) only terms are considered such that:

(20) $$\frac{1}{\hbar}|E_h - E_k| < \frac{1}{\tau_1} \quad \frac{1}{\hbar}|E_{h_1} + E_{h_2} - E_{k_1} - E_{k_2}| < \frac{1}{\tau_1}$$

where τ_1 is the typical variation time of the Boltzmann description: $\tau_1 \sim 10^{-13}$sec, the time interval between two collisions. The Hamiltonian $\hat{H}$, given by (13) generates an isomorphism $\mathcal{U}'_H(t)$ of $\mathcal{B}(\mathcal{H})$:

(21) $$\mathcal{U}'_H(t)\cdot = e^{\frac{i}{\hbar}\hat{H}t}\cdot e^{-\frac{i}{\hbar}\hat{H}t} = \int_{-i\infty+\epsilon}^{i\infty+\epsilon} dz \frac{e^{zt}}{2\pi i}\frac{1}{z - \mathcal{H}'},$$

where $\mathcal{H}'\cdot = \frac{i}{\hbar}[\hat{H},\cdot]$. We shall introduce a formalism reminiscent of usual scattering theory shifting the space $\mathcal{H}$ to $\mathcal{B}(\mathcal{H})$ and operators in $\mathcal{H}$ to maps in $\mathcal{B}(\mathcal{H})$; for brevity only the main steps of the treatment are indicated:

(22) $$\frac{1}{z - \mathcal{H}'} = \frac{1}{z - \mathcal{H}'_0} + \frac{1}{z - \mathcal{H}'_0}T(z)\frac{1}{z - \mathcal{H}'_0}$$

(23) $$T(z) \equiv V' + V'\frac{1}{z - \mathcal{H}'}V',$$

where $\mathcal{H}'_0 = \frac{i}{\hbar}[\hat{H}_0,\cdot]$, $V' = \frac{i}{\hbar}[\hat{H} - \hat{H}_0,\cdot]$, $\hat{H}_0 = \sum_f E_f \hat{a}_f^\dagger \hat{a}_f$. The operators

(24) $$\hat{a}_{h_1}^\dagger \hat{a}_{h_2}^\dagger \cdots \hat{a}_{h_r}^\dagger \hat{a}_{k_s} \cdots \hat{a}_{k_2}\hat{a}_{k_1}$$

are *eigenstates* of $\mathcal{H}'_0$ with eigenvalues $\frac{i}{\hbar}(E_{h_1} + E_{h_2} \cdots + E_{h_r} - E_{k_1} - E_{k_2} - \cdots - E_{k_s})$. By the basic algebraic property:

(25) $$\mathcal{U}'(t)\hat{a}_h^\dagger \hat{a}_k = \left(\mathcal{U}'(t)\hat{a}_h\right)^\dagger \left(\mathcal{U}'(t)\hat{a}_k\right),$$

it is clear that the main formal tool to treat expressions 19 is the representation of the operator $\mathcal{T}(z)\hat{a}_k$, in terms of the basis 24. By conservation of total mass one has the general structure:

$$\mathcal{T}(z)\hat{a}_k = \sum_f A_{kf}(z,\hat{n})\hat{a}_f + \sum_{lf_2f_1} \hat{a}_l^\dagger A_{lkf_2f_1}(z,\hat{n})\hat{a}_{f_2}\hat{a}_{f_1} +$$

(26)
$$+ \sum_{l_1l_2f_3f_2f_1} \hat{a}_{l_1}^\dagger \hat{a}_{l_2}^\dagger A_{l_1l_2kf_3f_2f_1}(z,\hat{n})\hat{a}_{f_3}\hat{a}_{f_2}\hat{a}_{f_1} + \cdots$$

where the coefficients $A_{lkf_2f_1}(z,\hat{n})$, $A_{l_1l_2f_3f_2f_1}(z,\hat{n})$ are operator functions of the set of number operators $\hat{n}_h = \hat{a}_h^\dagger \hat{a}_h$, i.e.: they are diagonal with respect to the basis in Fock space, generated by the creation operators. A very natural approximation in usual kinetic theory is the evolution by two-particle collisions. In our field description the corresponding approximation seems to be the following: evolution involving *only one* other field mode; i.e. one would break up expansion (26) after the first two terms; however all the *spectator modes* are also relevant through the $\hat{n}$ dependence of the coefficients and provide the Pauli principle corrections. Then also the third term in (26) contributes to these Pauli principle corrections inside an expression of the form $\hat{a}_h^\dagger \mathcal{T}(z)\hat{a}_k$ when some index f_1, f_2, f_3 is equal to h. Let us indicate briefly the structure of the coefficient $A_{lkf_1f_2}(z,\hat{n})$ in (26); it is given essentially by the matrix elements of a two-particle scattering operator, bearing Pauli-principle corrections, defined as follows:

(27) $$\hat{T}^{(2)}(z) = \hat{V}^{(2)} + \hat{V}^{(2)}\frac{1}{z - \hat{H}_L^{(2)}}\hat{V}_L^{(2)} \quad , \quad \hat{H}_L^{(2)} = \hat{H}_0^{(2)} + \hat{V}_L^{(2)}$$

the operators labelled by the index (2) are defined in the Hilbert space $\mathcal{H}^{(2)}$ of two identical particles by matrix elements in the two particle (symmetric or antisymmetric) basis $|l_1 l_2 >$:

$$< l_2 l_1 | \hat{H}_0^{(2)} | f_1 f_2 > = (E_{f_1} + E_{f_2})\frac{1}{2!}(\delta_{l_2 f_2}\delta_{l_1 f_1} \pm \delta_{l_2 f_1}\delta_{l_1 f_2})$$

$$< l_2 l_1 | \hat{V}^{(2)} | f_1 f_2 > = V_{l_1 l_2 f_2 f_1} \quad ,$$

(28) $$< l_2 l_1 | \hat{V}_L^{(2)} | f_1 f_2 > = (1 \pm \hat{n}_{l_1} \pm \hat{n}_{l_2}) V_{l_1 l_2 f_2 f_1}$$

These *two-particle* quantum mechanical elements, are produced in a natural way by the quantum field structure and are constructed with the coefficients E_f, $V_{l_1 l_2 f_2 f_1}$, arising in the Hamiltonian (13), written in terms of $\hat{a}_l^\dagger, \hat{a}_f$:

(29) $$\hat{H} = \sum_f E_f \hat{a}_f^\dagger \hat{a}_f + \frac{1}{2}\sum_{l_1 l_2 f_1 f_2} \hat{a}_{l_1}^\dagger \hat{a}_{l_2}^\dagger V_{l_1 l_2 f_2 f_1} \hat{a}_{f_2}\hat{a}_{f_1}$$

The factor $(1 \pm \hat{n}_{l_1} \pm \hat{n}_{l_2})$ in the last of equations (28) represents the Pauli-principle correction, it is an operator valued expression, but this makes no problem for the definition (27) of $\hat{T}^{(2)}(z)$ since for all l_1, l_2 they commute. We assume for simplicity that no bound states between the molecules can be formed, this means that in the thermodynamic limit the coefficients in (26) have no singularities on the imaginary axis. Then the time scale is introduced by the following modifications. First the expression $\mathcal{T}(z)\hat{a}_k$ is replaced by $\mathcal{T}(z+\eta)\hat{a}_k$, with $\eta \approx \frac{\hbar}{\tau_0}$, τ_0 being of the order of the collision time. Final results for expectations of the relevant variables, having a typical variation time $\tau_1 \gg \tau_0 = \frac{\hbar}{\eta}$ are practically independent on η: actually only these η independent results are significant in our essentially incomplete description of the finite system; η dependence would mean dependence on the distribution of the huge set of poles that $\mathcal{T}(z)$ has on the imaginary axis, which in turn is related to the confinement of the sytem, only roughly represented by the boundary conditions we assumed in Section 2; so η dependence is more an artefact of the idealized confinement than a physical feature. A second change concerns the external variables E_h, E_k ; we set for them $E_h = E_{hk} + \frac{1}{2}\xi_{hk}$, $E_k = E_{hk} - \frac{1}{2}\xi_{hk}$ and the replace $\xi_{hk} \rightarrow \xi_{hk} - 2i\epsilon$, with $\eta > \epsilon \gg \frac{\hbar}{\tau_1}$, thus implying some smoothness property of the dependence on the variable ξ_{hk}, as the existence of an analytic continuation into the lower half-plane. Now expressions (25) are calculated taking into account equations (21), (22) and the representation (26) where $z \rightarrow z + \eta$. Then one can separate in the calculation of (21) the contribution of the singularities of $\frac{1}{z-\mathcal{H}_0'}$ from the singularities at points z, with $\mathrm{Re}\, z \leq -\eta$ and neglect the last ones: their contribution is negligible for $t \gg \tau_0$ if we consider only relevant variables. Finally one arrives with some calculations to the following very perspicuous representation of expression (25):

$$\mathcal{U}'(t)\hat{a}_h^\dagger \hat{a}_k = \hat{a}_h^\dagger \hat{a}_k + t\mathcal{L}'(\hat{a}_h^\dagger \hat{a}_k) \quad t \gg \tau_o, \quad t \sim \tau_1$$

where

(30)

$$\mathcal{L}'\hat{a}_h^\dagger \hat{a}_k = \frac{i}{\hbar}\left[\hat{H}_{\mathrm{eff}}, \hat{a}_h^\dagger \hat{a}_k\right] - \frac{1}{\hbar}\left(\left[\hat{\Gamma}, \hat{a}_h^\dagger\right]\hat{a}_k - \frac{1}{\hbar}\hat{a}_h^\dagger\left[\hat{\Gamma}, \hat{a}_k\right]\right) + \frac{1}{\hbar}\sum_\lambda \hat{R}_{h\lambda}^\dagger \hat{R}_{k\lambda};$$

In the first term at the r.h.s. of equation (30) an *effective* Hamiltonian appears, given by equation (29) with $V_{l_1 l_2 f_2 f_1}$ replaced by $V_{l_1 l_2 f_2 f_1}^{\mathrm{eff}}$,

$$V_{l_1 l_2 f_2 f_1}^{\mathrm{eff}} = \langle l_2 l_1 | \frac{1}{2}\left(\hat{T}^{(2)}(E_{f_1} + E_{f_2} + i\eta\hbar) + \hat{T}^{(2)}(E_{l_1} + E_{l_2} + i\eta\hbar)^\dagger\right)|f_2 f_1\rangle$$

Thus the interaction potential is replaced by the selfadjoint part of the scattering operator; the remaining part of the scattering operator is not zero if one goes beyond the Born approximation and yields the second term at r.h.s. of equation (30), which is no longer of the form $[\cdot, \hat{a}_h^\dagger \hat{a}_k]$,

$$\hat{\Gamma} = \frac{1}{2} \sum_{f_1 f_2 l_1 l_2} \hat{a}^\dagger_{l_1} \hat{a}^\dagger_{l_2}$$

$$\times <l_2 l_1| \frac{i}{2} \left(\hat{T}^{(2)}(E_{f_1} + E_{f_2} + i\eta\hbar) - \hat{T}^{(2)}(E_{l_1} + E_{l_2} + i\eta\hbar)^\dagger \right) |f_2 f_1 > \hat{a}_{f_2} \hat{a}_{f_1}$$

The operators $\hat{R}_{k\lambda}$ are given by:

(31)

$$\hat{R}_{k\lambda} = -i\sqrt{2\epsilon(1 \pm \hat{n}_\lambda \pm \hat{n}_k)} \sum_{f_1 f_2} \frac{< k\lambda | \hat{T}^{(2)}(E_{f_1} + E_{f_2} + i\hbar(\eta - \epsilon))}{E_k + E_\lambda - E_{f_1} - E_{f_2}) - i\hbar\epsilon} \hat{a}_{f_2} \hat{a}_{f_1},$$

the factor $\sqrt{2\epsilon(1 \pm \hat{n}_\lambda \pm \hat{n}_k)}$ arising by the approximation:

$$2\epsilon(1 \pm \hat{n}_\lambda \pm (\hat{n}_k \pm \hat{n}_k)) \approx \sqrt{2\epsilon(1 \pm \hat{n}_\lambda \pm \hat{n}_h)}\sqrt{2\epsilon(1 \pm \hat{n}_\lambda \pm \hat{n}_k)}$$

which holds in the case of not too large Pauli principle corrections. Because of mass conservation one has within our approximations:

(32)
$$\hat{\Gamma}^{(2)} \approx \frac{1}{4} \sum_{h\lambda} \hat{R}^\dagger_{h\lambda} \hat{R}_{k\lambda}$$

or in order to have exactly $\mathcal{L}'\hat{M} = 0$, one can set in equation (30):

$$\hat{\Gamma}^{(2)} = \frac{1}{4} \sum_{h\lambda} \hat{R}^\dagger_{h\lambda} \hat{R}_{h\lambda} .$$

Let us notice that the general structure of $\mathcal{L}'$, arising from the factorized form shown by (25), indicates a form of complete positivity *relative* to the operators $\hat{a}^\dagger_h \hat{a}_k$; one can easily see that:

$$0 \leq \sum_{hk} < \psi_h | ([1 + t \cdot \mathcal{L}'] \hat{a}^\dagger_h \hat{a}_k | \psi_k > \qquad \forall \quad \{\psi_h\} \subset \mathcal{H}$$

to first order in τ. Assuming (1) as *fundamental* energy density, introducing confinement and a time scale much larger than τ_0, we arrived by a systematic procedure to the generator $\mathcal{L}'$: the effective Hamiltonian is now associated with a non Hamiltonian contribution. By a similar treatment for the system consisting of a particle interacting with a medium, one can obtain $\mathcal{L}'$ for the particle variables [6], then considering the time evolution of the statistical operator for the sole particle, the typical quantum master equation describing Brownian motion is obtained.

In our present case we can expect that, on the time scale ruled by $\mathcal{L}'$, no memory of the classical state variables introduced in Section 2 is relevant, so that one has a closed set of evolution equations:

(33) $$\mathrm{Tr}\left((\mathcal{L}'\hat{f}(\mathbf{x})) \hat{w}(\beta(t), \mu(t), \mathbf{v}(t)) \right) = \frac{\mathrm{d}}{\mathrm{d}t} \mathrm{Tr}(\hat{f}(\mathbf{x}) \hat{w}(\beta(t), \mu(t) \mathbf{v}(t)))$$

where $\hat{f}(\mathbf{x})$ are the relevant fields $\hat{e}(\mathbf{x})$, $\mathbf{p}(\mathbf{x})$, $\hat{m}(\mathbf{x})$. Looking at $\mathcal{L}'(\hat{a}_h^\dagger \hat{a}_h)$ one can see that $\sum \hat{R}_{h\lambda}^\dagger \hat{R}_{h\lambda}$ has the typical structure of the gain contribution by a collision ending up in the two-particle state $h\lambda$,which is present in the Boltzmann collision term, while $-\frac{1}{\hbar}([\Gamma, \hat{a}_h^\dagger]\hat{a}_h - \hat{a}_h^\dagger[\hat{\Gamma}, \hat{a}_h])$ yields the loss term by a collision involving a particle in the state h. Therefore one can expect that the description based on equations (33) is an improvement of the usual Boltzmann equation, because no factorization hypothesis of two-particle distribution function must be used and $\hat{H}^{\text{eff}}$ is not purely kinetic. We hope that the procedure used to obtain $\mathcal{L}'$, which is based on smoothness properties of the scattering map $\mathcal{T}(z)$ can be extended to the case of singularities of $\mathcal{T}(z)$ related to bound states: so one could also make that the energy density (1) from which we started, fits in a suitable $\mathcal{L}'$, derived by a more fundamental model.

References

1. Bohm, A. (1993). Gamow vectors and the arrow of time, in *Symposium On the Foundations of Modern Physics*, (1993) Busch, P., Lahti, P. J., and Mittelstaedt, P., eds., World Scientific, Singapore, p.77-97
2. Davies, E. B. (1976). *Quantum Theory of Open Systems*, Academic Press, London.
3. Holevo, A. S. (1982). *Probabilistic and Statistical Aspects of Quantum Theory*, North Holland, Amsterdam.
4. Kraus, K. (1983). States, Effects and Operations, in *Lecture Notes in Physics*, Volume 190, Springer, Berlin.
5. Lanz, L., Melsheimer, O. and Wacker, E. (1985) *Physica*, **131A**, 520
6. Lanz, L., and Vacchini, B. (1997). *International Journal of Theoretical Physics*, **36**, 67.
7. Lanz, L. ,and Melsheimer, O. (1993) *Il Nuovo Cim.* , **108 B**, 511.
8. Ludwig, G. (1983). *Foundations of Quantum Mechanics I*, Springer, Berlin.
9. Ludwig, G. (1985) *Foundations of Quantum Mechanics II*, Springer , Berlin.
10. Robin, W. A. (1990). *Journal of Physics A*, **23**, 2065.
11. Zubarev, D. N. (1974). *Non-equilibrium statistical thermodynamics*, Consultant Bureau, New York.

Indirect Measurements and the Mirror Theorem: A Liouville Formulation of Quantum Mechanics

E. Brändas and B. Hessmo

Department of Quantum Chemistry. Uppsala University, Box 518, S-751 20 Uppsala, Sweden

Summary. It is argued that the indirect measurement in Quantum Mechanics can be objectively formulated via the standard mirror property of noncommuting mappings. In this Liouville like formulation the primary, zero order step, regulates that the quantum probe and the measuring device entangle through objectively defined correlations uniquely prescribed from the mirror theorem. The secondary step of the indirect measurement process is then pictured as a resonance forming and decaying process which couples the phenomenon with the relevant time scales. It is suggested that these correlations lead to a description that guides our intuition towards a more satisfactory comprehension of quantum phenomena. The Liouville like formulation also assumes a stochastic component reminiscent of stochastic resonances in driven nonlinear dynamical systems. The possible relation to quantum aspects of chaos via existence of nonlocal hidden variables further promotes an ontological setting. We illustrate this in some concrete examples.

1. Introduction

The tools of modern physics extend into many new areas which soon will find industrial applications. For instance, quantum physics is becoming essential for miniaturized objects and for advanced computer and data communications. New developments are further promoted in non-linear dynamic systems operating in noisy environments bringing into focus cooperative phenomena with characteristic deterministic and stochastic time scales. The new complexity concept (sometimes expressed as a new paradigm) endure the supposed limits as regards the appropriate phenomenological level of description and, most interesting, the boundary between classical and quantum physics. This is sometimes also phrased as the paradox of irreversibility versus dynamics. Therefore it is quite ironic that the interpretation and understanding of quantum mechanics even today defeats a general feeling of agreement or consensus.

At present the usual picture of an applied quantum scientist seems to be essentially that of an orthodox consumer of quantum mechanics. In other words, and this concerns particularly the older generation, this perspective points back to a strict and devoted education in the giant shadow of Bohr. Nevertheless, the partisans of the system could, when necessary, trade more or less consciously the epistemological view for an ontological one. The growing awareness of an inconsistent orthodox thinking, had, however, started to emerge rather early, mainly due to Bohm [1]. The infringing consequences of classical predictions was further developed by Bell [2] and this excited fundamental and concrete examinations on how we comprehend physical phenomena. As we know today the Bell's inequalities were not supported by experimental tests, see e.g. [3]. It was indeed to the contrary and the counter intuitive nature of quantum mechanics is not only reconfirmed but it is also the starting point for new developments in quantum optics [4], e.g. optical spectroscopy [5], photon polarization entanglement [6], free-electron lasing without inversion [7] and quantum cryptography [8]. For various theoretical developments which serve to cope with the new situation we refer to [9]–[13]. For a brief account of the indirect measurement process, see the excellent monograph by Braginski and Khalil [13].

Obviously the new development does not bring to rest existing confusion regarding the correct frame for a consistent belief and quantum mechanical interpretation. Many of these arguments concern either so-called direct measurements or the various steps of the indirect process, see Ref. [13] mentioned above. Thus, on one hand, one may prefer an ontological description at least for philosophical reasons, but with notable difficulties continually inducing epistemological reverberations. On the other hand, the adherence to the strict Bohr doctrine inevitably leads, sooner or later, to controversial questions around quantum theory. The paradox may be phrased in several different ways. Oversimplifying a bit one might rebound on the key element of many standard schemes namely at what point should a measurement be recorded or at least be regarded as completed. In other words, *when does it start and when does it end.*

2. Liouville Formulation

In order to suggest a realistic alternative to current views, we will develop a Liouville like formulation which, in indirect measurement terminology, considers the relevant steps in the same algebraic and analytic mode, see below, albeit with some important distinctions. This will be shown to lead to some very simple but powerful consequences. There is at the outset also a rather trivial motive for the present formulation. For instance, the Schrödinger equation portraying an apparatus or device entangled with a microscopic quantum system should in principle be represented by a general basis of all possible product functions. This design provokes the inducement of "counter intuitive" vectors of type "the device read spin up" (tensor) multiplied with "the system is spin down". The only way, as far as we know, to cope with such "anomalies" is to consider mixed tensor products between the ket's and the bra's to be defined below. The basic criterion then becomes that of establishing objective correlation patterns associated with the primary phase of the indirect measurement. The Liouville like equation as well as the mirror theorem, see below, will provide us with these characteristics.

For simplicity we start with a mapping between a microscopic system defined by the system and device state vectors $|\psi_{\text{syst}}\rangle$, $|\psi_{\text{dev}}\rangle$ entangled through the matrix R

$$(1) \qquad R_{\text{M}} = |\psi_{\text{syst}}\rangle R \langle \psi_{\text{dev}}|$$

In general R_{M} is represented as ket-bra vectors with a matrix R "in between". In principle this representation may account for very general situations. For instance in an X-ray Photoelectron Spectroscopy (XPS) experiment [14], $|\psi_{\text{syst}}\rangle$ could refer to a substrate consisting of particular atoms or molecules together with monochromatised X-rays originating from e.g. an ordinary electron gun-rotating anode construction. The experiment then consists of illumination of the substrate, after which the sample absorbs radiation and - if the main result is electron ionization - expelled particles are gathered in an electron lens and then released through a spherical sector analyser from which space distributions and energy spectra can be obtained. Thus $|\psi_{\text{dev}}\rangle$ represents the electron gun and the spherical sector analyser.

The key to this setting lies in the way the device is correlated with the microscopic system. Before proceeding further we make some brief comments on the algebraic formulation. Note that we would need a dynamical framework, according to Eq. (1), which naturally leads to the other steps. First we demonstrate that in the weakly interactive or zero order picture the Liouville like formulation [15] becomes

$$(2) \qquad i\hbar \frac{\partial \rho}{\partial t} = \hat{L} \rho$$

where $\rho = R_{\text{M}} = |\psi_{\text{syst}}\rangle R \langle \psi_{\text{dev}}|$ and further

$$(3) \qquad \hat{P} = A \times + \times B$$

with $\hat{P} = -\frac{i}{\hbar}\hat{L}t$, $A = -\frac{i}{\hbar}H_{\text{syst}}t$, $B^{\dagger} = -\frac{i}{\hbar}H_{\text{dev}}t$ consisting of the relevant Hamiltonian operators for system and device respectively. Note that the $\times$ construction is short for the algebraic relation

$$(4) \qquad e^{\hat{P}} = e^A \times e^B; \quad e^{\hat{P}}\rho = e^A \rho\, e^B$$

and that this construction is also valid in the classical limit. For completeness we further refer to statistical probabilities obtained from the scalar product

$$(5) \qquad \text{Tr}\{\rho^{\dagger}\rho\}$$

Obviously the representation above *must be embedded in an extended description*. First of all the interaction between system and device $H_{\text{syst-dev}}$ must be appropriately included. In principle one could solve the associated

Schrödinger equation and then reduce the corresponding density matrix or alternatively find the desired projected component via a suitable master equation. ¿From the viewpoint of Eqs. (1–4) it amounts to the introduction of mixing terms in the Liouvillian, compare for instance the so-called Lindblad term [16]. *As far as the evolution can be generated by a Hamiltonian we will speak of intrinsic developments*, in contrast to more general stochastic processes in open dissipative systems.

Next, an attempted measurement must have a cognate time scale compatible with the dynamics of the system. We emphasize that the details here refers to the second part of the indirect measuring process, and we therefore need some generalizations. The emergence of a fundamental time scales is, to say the least, a complex process. Nevertheless the situation is not hopeless. Since we are dealing with an open system in a noisy environment, compare e.g. the modern notion of stochastic resonances[17], there should exist *a priori* characteristic stochastic time scales. This can be effectively modeled within the algebraic formulation Eqs. (1-4) appropriately extended with a mixing contribution in the general case. This leads to broken symmetries and the incorporation of objective time scales. For instance by inserting a complex conjugate on the bra-side above we may for instance include resonance Gamow waves via dilation analytic parameters. The formulation imposes, already on the intrinsic level, a new order of description originating from general principles and possibly exposing new phenomenological structures [15]. Then, in the most simplified case, Eq. (4) gives essentially

$$(6) \qquad e^{\hat{P}}\rho = e^{-\Gamma\frac{t}{\hbar}}\rho$$

where Γ (at zero order) by definition is the sum of the actual imaginary parts $\epsilon_{\text{syst}} + \epsilon_{\text{dev}}$ of the system och device resonance energies. In the general case one should add mixing terms to $\hat{P}$ with correspondent perturbations in Γ. This would in addition include entangled representations containing linear, quadratic etc., time dependent factors [15] as well as the correct mathematical framework for this setting, but we leave these comments for the moment. We will first return to the primary time independent step below.

3. Mirror Theorem

The relevant correlations for a realistic measurement analysis comes from the so-called mirror theorem [18]–[19] and Ref. 6 in [20] to be stated below.

Assume to begin with that R_{M} in Eq. (1) and its companion mapping

$$(7) \qquad S_{\text{M}} = |\psi_{\text{dev}}\rangle S \langle \psi_{\text{syst}}|$$

correspond to rectangular matrices which means that $R_{\text{M}}S_{\text{M}}$ and $S_{\text{M}}R_{\text{M}}$ are square dimensional matrices of orders m and n respectively (m and n is also the dimension of the finite subspace of the system and device Hilbert spaces). The mirror theorem now states

Theorem: The mappings defined above, $R_M S_M$ and $S_M R_M$, have identical nonvanishing eigenvalues. Furthermore the two matrices corresponding to the mappings and associated with the mentioned nonzero eigenvalues can be brought to the same classical canonical forms.

Proof: Trivially one finds from the Cayleigh-Hamilton equation for $R_M S_M$ and $S_M R_M$ that the corresponding characteristic polynomial have the same eigenvalues $a_k \neq 0$, where $k = 1, 2, \ldots, \min(m, n)$. The complete proof can be traced in Ref.[20]. For the diagonal case it means that we can introduce the conjugate pairs $|u_k\rangle$ and $|v_k\rangle$, from (choose $m < n$ for simplicity)

$$(8) \qquad R_M S_M = \sum_{k=1}^{m} |u_k\rangle a_k \langle u_k| = |\mathbf{u}\rangle \mathbf{a} \langle \mathbf{u}|$$

with $|u_k\rangle$, the relevant correlated system eigenfunctions with eigenvalue a_k, where $k = 1, 2, \ldots, m$ and

$$(9) \qquad S_M R_M = \sum_{k=1}^{m} |v_k\rangle a_k \langle v_k| = |\mathbf{v}\rangle \mathbf{a} \langle \mathbf{v}|$$

with v_k corresponding to the conjugate pair device eigenvector. Note that $n \geq k > m$ never occurs in Eq. (9), according to the mirror theorem.

For completeness we now discuss the nondiagonal case. It is convenient to introduce the matrices $\mathbf{J_R}$ and $\mathbf{J_S}$, which in the standard representation have zero elements everywhere except that, independent of R and S, $\mathbf{J}_{k,k+1} = 1$ for $k = 1, 2 \ldots, m - 1$. We know from [18]–[20] that they are Jordan blocks of the same order $\min(m, n)$, and in this case we can also, without loss of generality, choose the so-called Segrè characteristic to be m. Eqs. (8) and (9) would then be modified as

$$(10) \qquad R_M S_M = |\mathbf{u}\rangle \mathbf{a} \exp(\alpha \mathbf{J_R}) \exp(\beta \mathbf{J_S}) \langle \mathbf{u}|$$

and

$$(11) \qquad S_M R_M = |\mathbf{v}\rangle \mathbf{a} \exp(\beta \mathbf{J_S}) \exp(\alpha \mathbf{J_R}) \langle \mathbf{v}|$$

where α and β are parameters to be defined by the physical problem.

We can now supervene the mapping between the system and device as follows.

Corollary: From the mirror theorem we can establish a direct representation for the mappings R_M and S_M.

Proof: It follows directly from the above mentioned conjugate pairing property that the diagonal case becomes

$$(12) \qquad R_{\mathrm{M}} = \sum_{k=1}^{m} |u_k\rangle a_k^{\frac{1}{2}} \langle v_k| = |\mathbf{u}\rangle \mathbf{a}^{\frac{1}{2}} \langle \mathbf{v}|$$

where the diagonal matrix $\mathbf{a}^{\frac{1}{2}}$ contains the elements $a_k^{\frac{1}{2}}$. Similarly we find

$$(13) \qquad S_{\mathrm{M}} = \sum_{k=1}^{m} |v_k\rangle a_k^{\frac{1}{2}} \langle u_k| = |\mathbf{v}\rangle \mathbf{a}^{\frac{1}{2}} \langle \mathbf{u}|$$

However, the conjugate pairing property is lost in the non-diagonal case. We also note that our eigenvalue here must be degenerate so that $a_k = a$ for all k. This gives therefore

$$(14) \qquad R_{\mathrm{M}} = |\mathbf{u}\rangle \mathbf{a}^{\frac{1}{2}} \exp(\alpha \mathbf{J}_{\mathrm{R}}) \langle \mathbf{v}|$$

and

$$(15) \qquad S_{\mathrm{M}} = |\mathbf{v}\rangle \mathbf{a}^{\frac{1}{2}} \exp(\beta \mathbf{J}_{\mathrm{S}}) \langle \mathbf{u}|$$

Note that the nilpotent character of the matrices $\mathbf{J}$ permits a finite expansion of the corresponding exponential. Hence one obtains for instance

$$\exp(\alpha \mathbf{J}_{\mathrm{R}}) = 1 + \alpha \mathbf{J}_{\mathrm{R}} + \cdots + \frac{\alpha^{m-1}}{(m-1)!} \mathbf{J}_{\mathrm{R}}^{m-1}$$

The second term in Eq. (14) thus becomes

$$|\mathbf{u}\rangle a^{\frac{1}{2}} \alpha \mathbf{J}_{\mathrm{R}}) \langle \mathbf{v}| = a^{\frac{1}{2}} \alpha \sum_{k=1}^{m-1} |u_k\rangle \langle v_{k+1}|$$

displaying the build up of the mixing of the pairs $|u_k\rangle$ and $|v_k\rangle$, $k = 1, 2, \ldots m$. As previously stated we have not yet attempted to define the relevant mathematical framework, see below, for the emergence of nilpotent mappings, but we will see the convenience of this representation in the following section. Note also that there might, in general, exist Jordan blocks associated with the eigenvalue zero. Since they fall outside the present formulation they have to be treated separately.

4. Application to the Bose Condensate

A realistic interpretation corresponding to the primary step in the indirect measurement process can now be projected from either a given master equation or an objectively devised reduced density matrix. It is important to emphasize that *there is no arbitrariness or subjectivity here once the physical process has been defined*. In fact we may consider fundamentally different phenomena, e.g. well defined spectroscopic measurements like in [14], or the less contoured condensed matter applications (the high-temperature super-conductivity, the integer and the fractional quantum Hall effect, etc.) or more complex situations in biochemical environments where one recently has discussed, from a fundamental point of view, how macroscopically discernible states may be traced back to microscopic changes in a DNA molecule [21].

As a first application of our suggested strategy we will consider macroscopic quantum mechanical coherences such as those of a Bose condensate. There is an enormous literature dedicated to this field so it is obviously impossible to support all traits except some important characteristics, see e.g. [15] for some examples. To be specific we will consider a density matrix displaying a "large" coherent component and a "small" background of competing correlations, i.e.

$$(16) \qquad \Gamma^{(2)} = \lambda_L |g\rangle\langle g| + \lambda_S \sum_{k=1}^{m} \sum_{l=1}^{m} |h_k\rangle (\delta_{kl} - \frac{1}{m})\langle h_l|$$

where $|g\rangle = \frac{1}{\sqrt{m}} \sum_{k=1}^{m} |h_k\rangle$ is the coherent component expressed in the localized geminal (pair) basis h_k, $k = 1, 2, \ldots, m$, complete for $m \to \infty$. If we deal with Fermi statistics we must consider pairs of fermions appropriately coupled and this motivates the superscript (2) in Eq. (16). In this case the trace of $\Gamma^{(2)}$, over the pair space and leaving the unpaired contribution for the moment, is equal to $\frac{N}{2}$ or the number of fermion pairs in the system. The coherent state yields the eigenvalues as $\lambda_L = \frac{N}{2} - \frac{N(N-2)}{4m}$ and $\lambda_S = \frac{N(N-2)}{4m(m-1)}$, see e.g. Yang [22], Coleman [23] and Sasaki [24].

Obviously a measurement of a fermionic property, like spin, or, let us say, some general local property of the individual pairs, would completely destroy the condensate. All the troublesome features of quantum mechanics as an ontological theory containing in its description *elements of reality* are immediately provoked. To cope with this perplexity we will appeal to the mirror theorem and see how the coherent description emerges from the viewpoint of a local representation. Let us introduce, for each h_k, a site or "nucleus" with the coordinate $\vec{r}_k$, $k = 1, 2, \ldots, m$. The local point could be e.g. the center of mass for a composite atom or Cooper pair. One may also construct a local pairfunction with coupled spins. Although the Dirac bra $|\vec{r}_k\rangle$ should be pictured as a distribution, it is technically simple to represent it here as suitably localized smooth function which, when necessary, can be taken to its δ-function limits. Alternatively we can make this rigorous by the machinery of Gelfand triplets, see e.g. [25] for a review.

Note that $|\vec{r}_k\rangle$, $k = 1, 2, \ldots, m$ are vectors in a Hilbert space (or appropriate generalization) and should not be confused with the geometric coordinate $\vec{r}_k$. Hence associating each "nucleus" with the same effective mass so that the center of mass of all sites above becomes

$$\vec{r}_{CM} = \frac{1}{m} \sum_{k=1}^{m} \vec{r}_k$$

this relation refers to geometric space and not Hilbert space. To avoid confusion we will introduce the transformation

$$|\vec{r}_\mathrm{M}\rangle = \frac{1}{\sqrt{m}} \sum_{k=1}^{m} |\vec{r}_k\rangle$$

demonstrating that the state vector $|\vec{r}_\mathrm{M}\rangle$ in the limit approaches a sum of delta functions at the sites $\vec{r}_k$. The mirror theorem then yields

$$(17) \qquad R_\mathrm{M} = |g\rangle \lambda_\mathrm{L}^{\frac{1}{2}} \langle \vec{r}_\mathrm{M}| + \sum_{k=1}^{m} |h_k\rangle \lambda_\mathrm{S}^{\frac{1}{2}} \langle \vec{r}_k| - |g\rangle \lambda_\mathrm{S}^{\frac{1}{2}} \langle \vec{r}_\mathrm{M}|$$

One can also write it, remembering the warning above

$$(18) \qquad R_\mathrm{M} = \sum_{k=1}^{m} |h_k\rangle \lambda_\mathrm{L}^{\frac{1}{2}} \langle \vec{r}_\mathrm{CM}| + \sum_{k=1}^{m} |h_k\rangle \lambda_\mathrm{S}^{\frac{1}{2}} \langle \vec{r}_k - \vec{r}_\mathrm{CM}|$$

This construction is established from the identification $\Gamma^{(2)} = R_\mathrm{M} S_\mathrm{M}$. Similarly we also find, in the local representation

$$(19) \qquad S_\mathrm{M} R_\mathrm{M} = |\vec{r}_\mathrm{M}\rangle \lambda_\mathrm{L} \langle \vec{r}_\mathrm{M}| + \lambda_\mathrm{S} \sum_{k=1}^{m} \sum_{l=1}^{m} |\vec{r}_k\rangle (\delta_{kl} - \frac{1}{m}) \langle \vec{r}_l|$$

It is important to emphasize that the knowledge of the so-called off-diagonal long-range order (ODLRO) [22] permits an identification of the objective quantum mechanical correlation patterns, even if any measurement, if carried out, would completely destroy the condensate. Eqs. (17–19) display the origin and nature of the correlations. For instance taking the general trace of Eq. (17) gives the contribution

$$(20) \quad \lambda_\mathrm{L}^{\frac{1}{2}} \sum_{k=1}^{m} \frac{1}{m} [h_k(\vec{r}_k) + \sum_{k \neq l} h_k(\vec{r}_l)] + \lambda_\mathrm{S}^{\frac{1}{2}} [\sum_{k=1}^{m} h_k(\vec{r}_k) - \sum_{k=1}^{m} \sum_{l=1}^{m} \frac{1}{m} h_k(\vec{r}_l)]$$

Since the measurement device for local interactions refers to the basis $|\vec{r}_k\rangle$, we can identify $\langle \vec{r}_l | h_k \rangle = h_k(\vec{r}_l)$ for all k and l, where $h_k(\vec{r}) = f(\vec{r} - \vec{r}_k)$ and f is a function localized around the origin. If there is a different function for every site we will replace $\frac{1}{m} \sum h_k(\vec{r}_k)$ with the average $f_\mathrm{Av}(0)$. In general coherent situations, cf. ordinary superconductivity, off-diagonal terms $h_k(\vec{r}_l)$ for different k and l contribute to Yang's coherent phase known as ODLRO [22]. It is also interesting to speculate whether atomically localized properties, contributing to atomic Bose condensates, would get dominantly large contributions from $f_\mathrm{Av}(0)$ in Eq. (20). In such cases one would prefer s-pairing - a controversial question in the mechanistic approach to HTSC - and, in the atomic case, this leads to hyperfine couplings between electrons and nuclei (both fermions!), the latter mechanism thus serving a two fold purpose both as a coherence driver and a Bose converter. Noting the indirect importance of spin, it is quite surprising to see how the delocalised nature of the coherent state influences the ontological character of an individual fermion.

Note that our description respect pairs as units. In this case we might divide λ_L and λ_S by $N/2$ so that our trace relation, Eq. (16), sums up to one. However, since we know that $\Gamma^{(2)}$ by definition fulfills

$$\mathrm{Tr}\{\Gamma^{(2)}\} = \binom{N}{2}$$

the normalization should respect the number of pairings instead of pairs. The reason for this discrepancy is easy to understand and also easy to correct. In Eq. (16) we have neglected a usually uninteresting "noise" component, namely the unpaired degenerate contribution of dimension $2m(m-1)$ and with trace $\frac{N(N-2)}{2}$. If one wants to honor unit normalized density matrices one thus divides accordingly. The mirror theorem naturally adjusts.

Returning to the characterization of a fermion in the extreme state, see below, we sense the next conundrum. Obviously the full many particle description, here denoted as an extreme configuration by Coleman [23], pictures all electrons as *equally* paired with every other electron in a most democratic and orderly fashion. It is therefore obvious that, in addition to complete delocalization, we also have a Pandemonium spin property for the individual members of the condensate! The individual fermions are completely delocalised over a macroscopic domain with objective local correlations predicted by the mirror theorem. However, even if the condensate as a whole begins with the Fermi statistics of the the individual subunits of the many particle system, there is absolutely no way to assign individual spin directions for more than two fermions. The new (bosonic) phase appears, in the notation of Yang [22], as ODLRO.

5. Time Scales

As the final step in the indirect measurement process we must obtain relevant time scales for the process under study. To indicate a possible route to this general and difficult problem we will particularly refer to the formation and the destruction of the coherent state. For additional discussions on this point see e.g. Ref.[15]. Remember also that we have repeatedly pointed out that complex correlation patterns may sometimes lead to a new phase or structure. For example, in the previous section we did refer extensively to the occurrence of off-diagonal long-range order and Bose condensation.

Although a new phase may involve some kind of thermodynamic limit it is important to realize that this is not mandatory or even wanted. Rather, the new structure should include some other limits or analyticity requirements and, most importantly, should not be maintained at equilibrium. Furthermore, since time scales refers to irreversible time directed processes, which by necessity lies outside the unitary formulation of Schrödinger and Liouville type descriptions, we must also be prepared to embed the associated dynamics in an appropriate framework. A popular candidate for such an embedding can be found in the theory of dilation analyticity [26]–[28]. It is true that quantum mechanics does not impose invariance under complex scaling - a feature provoked by the possibility of quantum mechanical operators displaying singularly continuous spectra [26] - but there are, nevertheless, obvious qualities to such a dilatation analytic formulation. Not only are standard many particle Coulombic Hamiltonians dilation analytic - a feature of great utility in quantum chemistry - but there are also many situations where building new models and concepts, including also the possible breakdown of analyticity, have led to notable improvements. To demonstrate what these statements measure up to, we will proceed to discuss the coherent model introduced in the previous sections.

Since we will prepare the formulation via the Corollary to the mirror theorem, it will be convenient to display the Jordan block as a complex symmetric form. The general theorem is due to Gantmacher [29]. Here we will briefly refer to the form found by Reid and Brändas [30], i.e.

$$(21) \qquad Q_{kl} = (\delta_{kl} - \frac{1}{m}) \exp(i\pi \frac{k+l-2}{m}),$$

where m as before is the dimensionality of the Jordan block and $1 \leq k, l \leq m$. $\mathbf{Q}$, defined in Eq. (21) thus has the property that it is <u>similar</u> to

$$(22) \qquad \mathbf{J} = \begin{pmatrix} 0 & 1 & 0 & \ldots & 0 \\ \vdots & \ddots & \ddots & \ldots & \vdots \\ & & \ddots & \ddots & 0 \\ \vdots & & & \ddots & 1 \\ 0 & \ldots & & \ldots & 0 \end{pmatrix}$$

It is straightforward to prove this statement by considering the matrix $\mathbf{B}$ given by

$$(23) \qquad \mathbf{B} = \frac{1}{\sqrt{m}} \begin{pmatrix} 1 & \omega & \omega^2 & \ldots & \omega^{m-1} \\ 1 & \omega^3 & \omega^6 & \ldots & \omega^{3(m-1)} \\ \ldots & \ldots & \ldots & \ldots & \ldots \\ 1 & \omega^{2m-1} & \omega^{2(2m-1)} & \ldots & \omega^{(m-1)(2m-1)} \end{pmatrix}$$

where $\omega = \exp(\frac{i\pi}{m})$, and then show that $\mathbf{Q} = \mathbf{B}^{-1}\mathbf{JB}$. The present construction hence explicitly demonstrates that every matrix can be brought to a complex symmetric form via a similarity (here unitary) transformation.

With this transformation in mind and the full machinery of the Liouville like formulation, including appropriate dilation analytic extensions to resonance phenomena, it will be easy to find the appropriate structure for using the mirror theorem for general classical canonical forms.

The first step in this embedding process is to couple the system to a thermal environment, but since we are not at equilibrium we need to keep track of the off-diagonal terms in the density matrix in Eq. (16), see also the discussion in section 4. This "thermalization" is easily incorporated by noting that the reduction of the full density matrix description from N fermions to interacting pairs also comes with a well-defined reduced Hamiltonian H_2, or H_2^c in its scaled version, see e.g. [15] for further references and discussion. Leaving out the technicalities one obtains a complex symmetric construction with the complex conjugate scaled component in the bra position. The second term in Eq. (16) becomes after thermalization, i.e. using the construction completely analogous to Eqs. (3,4) (for details see [15]), $e^{-\frac{\beta}{2}H_2^c}\Gamma^{(2)c}e^{-\frac{\beta}{2}H_2^c}$, where the inverse of β equals Boltzmann's constant times the absolute temperature T. After the transformation the term may be written as

$$(24) \qquad \lambda_S \sum_{k=1}^{m}\sum_{l=1}^{m}|h_k^c\rangle(\delta_{kl}-\frac{1}{m})e^{i\frac{\beta}{2}(\epsilon_k+\epsilon_l)}\langle h_l^{c^*}|$$

where the, as yet undefined, complex energies and corresponding life times τ_k appear, i.e.

$$2\epsilon_k = \hbar\tau_k^{-1}$$

Even if the Liouville construction should include a mixing contribution, see section 2, it is surprising to realise that the dilated formulation allows for complex resonance solutions even at zero order. The scaled formulation can therefore be viewed in two complementary ways, either as a restriction of the dynamics via hidden analytic dilation parameter or as generalization of the emerging dynamics to a higher order phenomenological level.

It is an interesting coincidence that, assuming a harmonic oscillator type of structure for the epsilons or associated frequencies and time scales, i.e.

$$(25) \qquad 2\epsilon_{k+1} = \hbar k\tau_{rel}^{-1}, \quad k=1,2,\ldots,m$$

is tantamount to the crucial observation that the matrix elements defined in Eq. (24) constitutes a matrix which is proportional to $\mathbf{Q}$, see Eq. (21), or in other words the matrix in Eq. (24) corresponds to a Jordan block of order m. One readily sees that the first term in Eq. (20) after thermalization also corresponds to a Jordan block, namely of order 2.

The simple imposition of a harmonic spectrum leading to the relation Eq. (25), makes it a simple matter to apply the mirror theorem.

$$\tilde{R}_M = \lambda^{\frac{1}{2}}\sum_{k=1}^{m}\sum_{l=1}^{m}|h_k\rangle[e^{\alpha\mathbf{Q}}]_{kl}\langle\vec{r}_l|$$

from which one may easily obtain the general trace, cf. Eq. (16). The eigen-value λ equals the total trace divided by m, and the tilde, $\tilde{\ }$, denotes that R_M has been subject to a nonunitary transformation. Note that we can return to the unscaled representation by a similarity, i.e. we drop the c in h_k^c. Furthermore we can choose our localized reference basis real without loss of generality except that specific phases associated with distinct sites would correspond to a rearrangement of the vectors in Eq. (23). From Eq. (25) one may extract the longest survival time $\tau_{\mathrm{rel}} = \tau_2$, i.e. the smallest width or energy separation

$$(26) \qquad 2\epsilon_2 = \hbar\tau_{\mathrm{rel}}^{-1}; \quad \tau_{\mathrm{rel}} = \frac{\hbar}{4\pi k_B T} \cdot m$$

where k_B is Boltzmann's constant and T is the absolute temperature.

The proof of Eq. (26) follows from a simple scattering argument. Assume that a site or scattering center in the system has by definition a phenomeno-logical relaxation time τ_{rel} corresponding to m degrees of freedom and total cross section area of σ_{tot} matched with a detection sensitivity, in the solid angle Ω, of $\frac{1}{\tau_{\mathrm{lim}}}$ where from the uncertainty relation $\tau_{\mathrm{lim}} = \frac{\hbar}{k_B T}$. A simple calculation of the total cross section yields precisely the relation found in Eq. (26).

We have thus showed that complex correlation patterns, including thermal perturbations on off-diagonal long-range order, between system and device can be obtained in a precise way, save the definition of α. The key to this mapping is offered by the possibility to organize these patterns in terms of general Jordan block structures. For instance, utilizing the energy quantiza-tion in Eq. (26), it is possible to find precise energy-phase relations so that one can follow the break-down (or formation) of a specific phenomenological phase. In somewhat sumptuous terms one might say that Nature, in aptly mixing time scales, parents Jordan blocks in reward. Note also the generality and objectivity of the mapping despite our subjective choice of sites - a sign of a new emerging phase.

6. Time Evolution

Before we end this discussion we must also try to control the dynamics and accordingly understand the time evolution and the emergence of associated time scales as expressed in the present Liouville formulation. Note that the description here crosses into a higher order phenomenological level thereby introducing correlations not generally representable in a finite dimensional state space compatible with standard quantum theory.

Since the nilpotent terms above seem to be more than imaginable their effect on the evolution must be specifically considered. Using for instance the representation

$$(27) \qquad \tilde{S}_{\mathrm{M}} \tilde{R}_{\mathrm{M}} = \lambda \sum_{kljn}^{m} |\vec{r}_k \omega^{-(2l-1)(k-1)} \rangle e^{(\alpha+\beta)\mathbf{J}} \langle \vec{r}_n \omega^{-(2j-1)(n-1)}|$$

one observes that the particular correlation pattern, projected above, display sites which have progressed out of phase with the coherent motion depicted in Eq. (16), cf. the phasons appearing in the analysis of the exceptional properties of quasi crystals. Of course, to destroy the extreme configuration, we need to satisfy Eq. (25) for $k = m$, which means that we have to "destroy" all particle clusters including also the pairs - a possible characteristic of the new cuprite superconductors [15].

We now need to specify what governs the actual time evolution of the correlations just derived. Obviously the general evolution is already indicated in section 2. However, since there is no specific information given, except the indefinite reference to H_{syst} and H_{dev} and possible mixing terms, we must lay down a more concrete proposition referring specifically to the degenerate subspace associated with the Jordan block of, let us say, order m. It is convenient to work with the basis set $|\mathbf{f}\rangle$, see below and Eqs. (21-23), since it corresponds to emerging phase motion in the system, cf. Eq. (27)

$$|\mathbf{h}\rangle = |\mathbf{f}\rangle \mathbf{B}$$

The mirror theorem then prescribes for $\tilde{R}_{\mathrm{M}} \tilde{S}_{\mathrm{M}}$ a form equivalent to Eq. (27). As we will see the largest Jordan block appearing in the mapping will determine the specific characteristics of the dynamical evolution. Thus, we will only be interested in the matrix part $\mathbf{1} + (\alpha + \beta)\mathbf{J}$. The general (slow) time scale has already been obtained as $\tau = \frac{\hbar}{2\epsilon}$, where $2\epsilon = \Gamma$ (not to be confused with $\Gamma^{(2)}$!) is the resonance width. The complex resonance energy eigenvalue thus becomes $\hbar\omega_0 - i\Gamma$ where ω_0^{-1} is the fast time scale, for instance an enhanced signal in stochastic resonance theory (e.g. Array Enhanced Stochastic Resonance), or perhaps an evaporative cycle in the case of a Bose condensate. We can thus write for the t-evolution

$$(28) \qquad \hat{\mathcal{P}} = (\omega_0 \tau - i)\hat{\mathcal{I}} + \hat{\mathcal{J}}$$

and its propagator

$$(29) \qquad \exp(-i\hat{\mathcal{P}}t/\tau)$$

where $\hat{\mathcal{I}} = \sum_{k=1}^{m} |h_k\rangle\langle h_k| = |\mathbf{f}\rangle\langle\mathbf{f}|$ and $\hat{\mathcal{J}} = |\mathbf{f}\rangle\mathbf{J}\langle\mathbf{f}|$. It follows directly that

$$(30) \qquad \exp(-i\hat{\mathcal{P}}t/\tau) = \exp(-i\omega t)\exp(-t/\tau) \sum_{k=0}^{m-1} (-it/\tau)^k \frac{1}{k!} \hat{\mathcal{J}}^k$$

and

$$(31) \qquad (y\hat{\mathcal{I}} - \hat{\mathcal{P}})^{-1} = \sum_{k=1}^{m} (y+i)^{-k} \hat{\mathcal{J}}^{k-1}; \quad y = (\omega - \omega_0)\tau$$

Above, we have also assumed that the factor in front of $\hat{\mathcal{J}}$ is unity which may be justified in some cases, see Ref.[15] but this may be different in other situations. Thus the operator $\hat{\mathcal{J}}$ "condenses" m singular points into one higher order singularity. In principle this set of points could have an infinite number of elements condensed to an essential singularity and as such it has been speculated whether one could interpret this as the sought-after connection between chaos in the classical and quantum domains.

Let us briefly look at the transitions induced by the Jordan block, namely $|f_1\rangle$ correlated with the transitions $|f_k\rangle$ characterized by the operator $\hat{J}^{k-1}$ with $k = 2, 3, \ldots m$. Remembering that the characteristic lifetime τ is given by $\frac{\hbar}{\Gamma}$, one gets directly from Eq. (30) with $\hat{\mathcal{P}} = \frac{\tau}{\hbar}\hat{L}$

$$N(t) = |\langle f_1| \exp(-\frac{i\hat{L}t}{\hbar})|f_k\rangle| \propto t^{k-1} \exp(-\frac{\Gamma t}{\hbar})$$

which yields

$$dN \propto t^{k-2}(k - 1 - \frac{\Gamma t}{\hbar}) \exp(-\frac{\Gamma t}{\hbar}) \cdot dt.$$

This shows that $dN > 0$ for $t < (k-1) \cdot \frac{\hbar}{\Gamma}$, $k = 2, 3, \ldots m$. Thus the occurrence of Jordan blocks in the Liouville picture leads to increase of $N(t)$ for a new macroscopic time scale about $m\tau$. There is hence an increase of the number of particles in the correlated state given by $|f_1\rangle$ from correlated transitions from all other states $|f_k\rangle$, $k = 2, \ldots m$ during the macroscopic (if m is large) time $(m-1)\tau$. Since there is, at the same time, an overall decay out of $\hat{I}$ given by

$$N(t) = |\langle f_1| \exp(-\frac{i\hat{L}t}{\hbar})|f_1\rangle| \propto \exp(-\frac{\Gamma t}{\hbar})$$

yielding the standard exponential decay rule

$$dN \propto (-\frac{\Gamma}{\hbar}) \exp(-\frac{\Gamma t}{\hbar}) \cdot dt$$

the summing up of all contributions to (or from) the correlated state $|f_1\rangle$ may lead to dissipative structures which during certain timespans of order $m\tau$ have increasing order etc., and in this sense we may speak of self-organization on a microscopic level "created" by the Jordan block term in $\hat{L}$.

7. Conclusions

Finally some words regarding the technical details needed for a rigorous mathematical formulation related to the present development. Even if we, for simplicity, here often have referred to intrinsic evolutions there is in principle no restriction to generalize the discussion to open systems reduced to an appropriate level of relevant degrees of freedom [15], [32]. Thus we can ask -

in addition to those questions connected with the existence of solutions to dissipative stochastic differential equations - e.g. how properties of specific generators of dynamical semigroups and the subsequent domain characteristics will influence the evolution. We may also speculate how Nature responds to semigroups and/or dilatation analyticity?

There have been many studies concerned with the development of infinitesimal generators of any dynamic semigroups corresponding to general quantum stochastic processes, see e.g. [16] and [32], to quote a few results. In particular there are situations where minimal extensions of Markov semigroups do not capture the physics of the problem. Further we also know that atomic and molecular problems respect Coulomb interactions and that the dilated Coulomb Hamiltonian does not generate a dynamical contractive semigroup. It is also well know that the Coulomb Hamiltonian perturbed by an electric field is not dilatation analytic. Irrespective of this "negative" countenance the above mentioned examples sustain well defined resonance structures - operationally resolved by complex scaling through asymptotic adjustment and fine tuning. Analytic continuation is a strong and comprehensive tool and yet quantum physics does not guarantee such analytic structures in general, i.e. cases involving singularly continuous spectra. The point is not here to exclude such cases, but rather to embed the relevant dynamics in a vaster framework so that new analytic structures, if actualized, can be separated out. The latter point is particularly important, since the question of semigroups in dilation analytic theories, as concluded above, does not yield affirmative answers. In special cases one can restrict the unitary time evolution to an isometric group which via the scaling transformation is converted to a contractive evolution, see e.g. Ref.[31]. The present Jordan block formation, should therefore be seen in this mathematical perspective. In particular one finds that the emergence of a specific Jordan block structure is controlled by a super selection rule in that the conjugate block is associated with the opposite time arrow, or equivalently the mirroring complex energy plane. The associated resolvent expansions yields further higher order spectral or gainfunctions of great interest in many concrete applications, see below. In a sense one may say that Nature represents a complex dynamical system commensurate with many fundamental time scales, which may appear from general analytic structures not necessarily compatible with a quantum stochastic process where the central object is a dynamical semigroup.

The present approach, see e.g. Ref.[15] and [33] for more details, has been applied to many different situations in condensed matter with some surprising agreements between theory and experiments, $viz.$ proton transfer processes in water and aqueous solutions, ionic conductance of molten alkali chlorides, quantum correlations in high-T_c Cu-O-superconductors, see also [34], quantum correlations as shown by the spin-waves of Gd far above T_c, the fractional quantum Hall effect, conductance background effects in high-quality tunnel junctions, proton dynamics in DNA and spontaneous and stimulated emission of radiation in masers [35].

As we have pointed out repeatedly, the viewpoint advocated here centers around the indirect measurement hypothesis. The primary step consists in analyzing the correlation pattern between the system and the device. There is no subjectivity involved here as this preparation is governed by the mirror theorem. The other step involves the appearance of time scales and associated dynamical evolution. Here we find various types of stochastic time scales and, which not only leads to decay and a one sided approach to equilibrium, but also to the formation of new structures and microscopically driven self-organization, c.f. other developments in nonlinear dynamics [36], [37]. The new phenomenological structures are here modeled via so-called dilatation analytic imbeddings which in some sense can be interpreted as a hidden non-local parameter. In principle we could, despite the conundrums experienced by the example of a Bose condensate, accept that ontological features in a general way allows the elevation of the formulation to a higher order phenomenological level. It has also been speculated whether higher order singularities and cuts associated with different layers of interconnected Riemann sheets would be a quantum analogue of classical chaos.

References

1. D. Bohm, Phys. Rev. **85**, 166 (1952).
2. J. Bell, Physics 1, 195 (1964).
3. L. Mandel and E. Wolf, *Optical coherence and quantum optics* (Cambridge University Press, Cambridge, 1995).
4. A. Aspect, P. Grangier and C. Roger, Phys. Rev. Lett. **47**, 460 (1981); **49**, 91 (1982); A.Aspect, J. Dalibard and C. Roger, Phys. Rev. Lett. **49**, 1804 (1982).
5. J. Lavall, S.Kulin, B. Saubamea, N. Bigelow, M. Leduc, and C. Cohen-Tannoudji, Phys. Rev. Lett. **75**, 4194 (1995).
6. P G. Kwiat, K. Mattle, H. Weinfurter, and A. Zeilinger; A. V. Sergienko and Y, Shih, Phys. Rev. Lett. **75**, 4337 (1995).
7. B. Sherman and G. Kurizki, D. E. Nikonov and M. O. Scully, Phys. Rev. Lett. **75**, 4602 (1995).
8. C. Monroe, D. M. Meekhof, B. E. King, W. M. Itano, and D. Wineland, Phys. Rev. Lett. **75**, 4714 (1995).
9. B. D'Espagnat, *Veiled Reality - An Analysis of Present Day Quantum Mechanical Concepts* (Addison-Wesley, Reading, 1994).
10. R. Omnés, *The Interpretation of Quantum Mechanics* (Princeton University Press, Princeton, 1994).
11. A. J. Legget, in *Quantum Implications*, edited by J. de Boer *et al.*, (North Holland, Amsterdam, 1986).
12. W. H. Zurek, Prog. Theor. Phys. **89**, 281 (1993).
13. V. B.Braginski and F. Ya. Khalili, *Quantum Measurement*, (Cambridge University Press, Cambridge, 1992); see also M. B. Mensky, *Continuous Quantum Measurements and Path Integrals*, Institute of Physics, Publishing Ltd, Techno House, Redcliffe Way, (Bristol and Philadelphia, 1993).

14. N. Mårtensson, Treat. Mat. Sci. Tech., **27**,, 65 (1988).
15. E. J. Brändas, in *Dynamics During Spectroscopic Transitions*, edited by E. Lippert and J. D. Macomber, p. 148 (Springer Verlag, 1995).
16. G. Lindblad, Comm. Math. Phys. **48**, 119 (1976).
17. A. R. Bulsara and L. Gammaitoni, Phys. Today **49**, (1996).
18. E. Schmidt, Math Ann. **63**, 433 (1907).
19. B. C. Carlson and J. M. Keller, Phys. Rev. **121**, 659 (1961).
20. P.-O. Löwdin, Int. J. Quant. Chem. **21**, 269 (1982); Isr. J. Chem. **31**, 297 (1991).
21. D. Home and R. Chattopadhyaya, Phys. Rev. Lett. **76**, 2836 (1996).
22. C. N. Yang, Rev. Mod. Phys. **34**, 694 (1962).
23. A. J. Coleman, Rev. Mod. Phys. **35**, 668 (1963).
24. F. Sasaki, Phys. Rev. **138**, B 1338 (1965).
25. A. Bohm and M. Gadella, *Dirac Kets, Gamow Vectors and Gel'fand Triplets*, (Springer Verlag, Berlin, 1989).
26. E. Balslev and J. M. Combes, Commun. Math. Phys. **22**, 280 (1971).
27. C. van Winter, J. Math. Anal. **47**, 633 (1974).
28. B. Simon, Ann. Math. **97**, 247 (1973).
29. F. R. Gantmacher, *The theory of matrices.* Vol. II, Chelsea Publishing Company, New York (1959); H. Baumgärtel, *Endlichdimensionale analytische Störungstheorie,* (Akademie-Verlag Berlin, 1972); H. Baumgärtel, *Analytic Perturbation Theory for Matrices and Operators,* (Akademie-Verlag Berlin, 1984); Extended English Version, Series: *Advances in Operator Theory,* (Birkhauser Verlag, Basel, 1985).
30. C. E. Reid and E. J. Brändas, Lecture Notes in Physics **325**, p. 475 (1989); see also E. Brändas, in *Quantum Science Methods and Structure A Tribute to Per-Olov Löwdin,* J.-L. Calais. O. Goscinski, J. Linderberg and Y. Öhrn, Eds. (Plenum, New York), P.381 (1976).
31. G. G. Emch J. Functional Analysis **19**, 1 (1975); Commun. Math. Phys. **49**, 191 (1976).
32. J. Kumičák and E. Brändas, Int. J. Quant. Chem. **46**, 391 (1993); see also C. H. Obcemea and E. J. Brändas, Ann. Phys. **151**, 383 (1983).
33. C. A. Chatzidimitriou-Dreismann, Adv. Chem. Phys. **80**, 201 (1991).
34. L. J. Dunne, J. Mol. Struct. (Theochem), **341**, 101 (1995).
35. E. J. Brändas, Adv. Chem. Phys. 99, 211 (1997).
36. I. Prigogine, *From Being To Becoming.* W. H. Freeman and Company, San Francisco (1980).
37. G. Nicolis and I. Prigogine, *Exploring Complexity.* W. H. Freeman and Company, New York (1989).
38. E. J. Brändas and C. A. Chatzidimitriou-Dreismann, Lecture Notes in Physics **325**, p. 485 (1989).

Chapter VIII

Semigroups Operator Theory

Semigroups and Antieigenvalues

K. Gustafson

Department of Mathematics, CB 395, University of Colorado, Boulder, Colorado 80309–0395, USA

Summary. It is of historical interest that questions in semigroup perturbation theory led some years ago to a theory of antieigenvalues. That theory and those origins will be explained. Simultaneously we propose and pursue by similar considerations a new mechanism of irreversibility for quantum theory, based upon multiplicative disturbances.

1. Introduction

Given a time-dependent Schrödinger equation

$$\frac{du}{dt} = iHu \tag{1}$$

where $u(t)$ is to be an evolving entity in a Hilbert space $\mathcal{H}$ and where H is a selfadjoint operator in $\mathcal{H}$, by exponentiation we know the solution to be the one parameter unitary group

$$u(t) = e^{iHt}u_0, \qquad -\infty < t < \infty \tag{2}$$

applied to the initial state u_0. Due to the importance of the Schrödinger equation in quantum mechanics, considerable attention has been given to the case of *additive* perturbation of a given infinitesimal generator H_0, viz., one wants to treat the hydrogen atom through the expression $H = H_0 + V$ where H_0 is the Laplacian $-\Delta$ and where V is a Coulomb potential Z/r, similarly the higher elements. Thus a rather substantial theory of selfadjoint operators additively perturbed by selfadjoint or at least symmetric operators has been built up, now comprising an essential part of quantum scattering theory.[1] Within such a theory, selfadjointness of the infinitesimal generators is preserved. As a consequence, the order of time in (2) is physically reversible.

If we now take as ansatz that the physical operators with which we are concerned are always subject to small disturbances, e.g., slight changes of Hamiltonian due to measurement, distant interactions with other atoms or with radiation, or just some inherent stochastic fluctuations, then at least for symmetric additive disturbances the perturbation theory is comforting: selfadjointness is preserved, and reversibility is sustained. However, if we suppose that multiplicative disturbances come into play, then there is less reason

---

[1] See, for instance, T. Kato, *Perturbation Theory for Linear Operators* (Springer, Berlin, 1980).

for comfort: as we will show, even for bounded selfadjoint disturbances, generally self-adjointness is not preserved, irreversibility is induced, and whether the perturbed operator remains dissipative depends intrinsically upon its antieigenvalues and those of the disturbance.

2. Semigroups and Antieigenvalues

By the Lumer–Phillips version of the Hille–Yosida Theorem,[2] a densely defined operator A in a real or complex Banach space X generates a contraction semigroup T_t on X if and only if the range $\mathcal{R}(\lambda I - A) = X$, for some λ with $\mathrm{Re}\,\lambda > 0$, and A is *dissipative*: $\mathrm{Re}\langle Ax, x \rangle \leqq 0$ for all x in the domain of A. Here $\langle Ax, x \rangle$ denotes some semi-inner product on X. The property of contraction semigroup, i.e., $\|T_t\| \leqq 1$ for all $t \geq 0$, generalizes unitary evolutions $U_t = e^{iHt}$ for conservative systems to contracting evolutions for dissipative systems. The property of generating a contraction semigroup is preserved under dissipative relatively bounded additive perturbations, i.e., $A + B$ is still a generator if $\|Bx\| \leqq a\|x\| + b\|Ax\|$, $b < 1$. This latter fact generalizes the preservation of selfadjointness under the same relative boundedness additive perturbation condition, which we alluded to above.

In the later 1960's, the question of preservation of the property of contraction semigroup generator under multiplicative perturbation came under treatment, motivated principally by questions in stochastic evolution equations. In particular, the following theorem was shown:[3] if A is a generator and B is a bounded strongly accretive operator, then BA is a contraction semigroup generator if and only if BA is dissipative. Accretive, i.e., $\mathrm{Re}\langle Bx, x \rangle \geqq 0$, is the opposite of dissipative, and strong accretivity requires a strictly positive lower bound: $\mathrm{Re}\langle Bx, x \rangle \geqq m > 0$ for all $\|x\| = 1$. The multiplicative perturbation result was obtained from the additive perturbation result as follows. Write, for any $\epsilon > 0$,

$$(3) \qquad \epsilon BA = A + (\epsilon B - I)A.$$

Then strong accretivity of B ensures that $\|\epsilon B - I\| < 1$ for all small enough positive ϵ, and ϵBA is a generator by the additive perturbation relative bound criteria, provided that ϵBA is dissipative. The ϵ is easily removed, leaving the more interesting question: when is BA dissipative when A is dissipative and B is strongly accretive? This question led to the formulation of the notion of antieigenvalue in 1967. Elaborating (3) to

$$(4) \qquad \begin{aligned} \mathrm{Re}\langle \epsilon BAx, x \rangle &= \mathrm{Re}\langle (\epsilon B - I)Ax, x \rangle + \mathrm{Re}\langle Ax, x \rangle \\ &\leqq \|\epsilon B - I\|\|Ax\|\|x\| + \mathrm{Re}\langle Ax, x \rangle \end{aligned}$$

[2] See, for instance, K. Yosida, *Functional Analysis* (Springer, Berlin, 1966).
[3] K. Gustafson, *Pacific J. Math.* 24 (1968), 463.

reveals that ϵBA, and hence BA, will be dissipative provided that

$$(5) \qquad \inf_{\epsilon>0} \|\epsilon B - I\| \overset{\leq}{=} \inf_{x \in \mathcal{D}(A)} \frac{\mathrm{Re}\langle(-A)x, x\rangle}{\|Ax\|\|x\|}.$$

It was found that (in Hilbert space) for A unbounded, the right side of (5) is zero, so hereafter we restrict attention to both $-A$ and B bounded (strongly) accretive operators on a Hilbert space. Then (5) leads to an operator trigonometry[4] in which (5) becomes

$$(6) \qquad \sin B \overset{\leq}{=} \cos(-A).$$

Examples show that this is a sharp criteria to ensure the dissipativeness of BA. For any bounded strongly accretive operator T, the first antieigenvalue μ is defined by

$$(7) \qquad \mu = \inf_{x} \frac{\mathrm{Re}\langle Tx, x\rangle}{\|Tx\|\|x\|}.$$

The geometric interpretation is clear: $\mu = \cos T$ determines the operator angle $\phi(T)$, namely, the maximum angle through which T may turn a vector. Corresponding first antieigenvectors attain this turning angle. Higher antieigenvalues and their corresponding antieigenvectors may be defined as well.

For simplicity in the following we specialize attention to positive definite selfadjoint operators T on a Hilbert Space $\mathcal{H}$. Let m and M denote the positive lower and upper bounds of the spectrum $\sigma(T)$. Then it is known that

$$(8) \qquad \mu_1 = \cos T = \frac{2\sqrt{mM}}{M + m}, \quad \sin T = \frac{M - m}{M + m}.$$

When m and M are eigenvalues of T, the first antieigenvectors come in pairs

$$(9) \qquad x_{\pm}^1 = \pm \left(\frac{M}{M + m}\right)^{1/2} x_1 + \left(\frac{m}{M + m}\right)^{1/2} x_n$$

expressible in terms of the normalized eigenvectors x_1 and x_n corresponding to the eigenvalues m and M, respectively. It should perhaps be emphasized that $\cos T$ and $\sin T$ within the operator trigonometry[5] of the antieigenvalue theory are not the same as the $\cos(T)$ and $\sin(T)$ one has in the functional calculus of an operator T.

---

[4] A good accounting of the antieigenvalue theory may be found in the recent books: K. Gustafson, *Lectures on Computational Fluid Dynamics, Mathematical Physics, and Linear Algebra* (Kaigai, Tokyo, 1996), K. Gustafson and D. Rao, *Numerical Range: The Field of Values of Linear Operators and Matrices* (Springer, Berlin, 1997).

[5] The operator trigonometry and antieigenvalue theory lay mostly dormant from 1970 to 1990. Then the author found that the famous Kantorovich convergence rate for descent algorithms in numerical optimization schemes was trigonomet-

3. Multiplicative Perturbation and Irreversibility

For simplicity let us consider only the case of Hilbert Space and bounded positive definite operators H and B. Let $A = iH$. Then e^{At} is a unitary evolution with spectrum $\sigma(e^{itH}) = e^{it\sigma(H)} = \cos(t\sigma(H)) + i\sin(t\sigma(H))$ consisting only of phase and no decay. Consider now a multiplicative disturbance to H represented by a selfadjoint positive definite operator B. The perturbed operator BH will remain selfadjoint iff B commutes with H, a highly unlikely event if B represents something extraneous to H's self-dynamics, or some random effect. However e^{BAt} is still a semigroup evolution, and again by the spectral mapping theorem $\sigma(e^{BAt}) = e^{it\sigma(BH)} = \cos(t\sigma(BH)) + i\sin(t\sigma(BH))$ because $\sigma(BH)$ remains real.[6] Furthermore the phase time-direction is not changed because $\sigma(BH)$ remains positive. However, the rate of time change in the evolution is now different, loosely speaking, by a factor of B. Let us summarize and elaborate the situation as follows.

Theorem 3.1 *With noncommuting positive definite H and B as above, the multiplicatively disturbed operator BH is no longer selfadjoint, the evolution e^{itBH} is no longer a group, and although the direction of time has not changed, the speed of evolving events has been altered. The disturbed generator $T = BH$ may be written in the form $T = B_1T_1B_1^{-1}$ where all three factors are positive definite. Conversely, any generator of the latter form may be written as a multiplicatively disturbed $T = BH$ in which both B and H are positive definite.*

Proof First, in augmentation to what was stated above, BH being bounded still exponentiates, and $\sigma(BH)$ being real and positive leaves time in $\mathbb{R}^1$ in the same direction but at altered speed. Second, it follows from results of Wigner[7] on weakly positive operators that the three characterizations $T = BH$, $T = B_1T_1B_1^{-1}$, and $T = B_2T_2B_2^{-1}$ where B_2 need not be selfadjoint, are equivalent.   ∎

From Theorem 3.1 we see that a noncommuting disturbance B causes $A = iH$ to lose the group generation property but BA still remains a semigroup

---

ric. See K. Gustafson, *Proc. Fourth International Workshop in Analysis and Applications* (eds.: C. Stanojevic and O. Hadzic), Novi Sad, Yugoslavia (1991), 57. This discovery has led to a fundamental new conceptualization of a wide class of iterative algorithms of wide use for solving large sparse systems $Ax = b$. See K. Gustafson, *Num. Lin. Alg. with Applic.* (1997), to appear.

[6] By a general result of J.P. Williams, *J. Math. Anal. Appl.* 17 (1967), 214, $\sigma(T^{-1}S) \subset \overline{W(S)}/\overline{W(T)}$ whenever $0 \notin W(T)$. Here $W(T)$ denotes the numerical range of an operator T. For selfadjoint T, $\overline{W(T)}$ is the convex hull of the spectrum $\sigma(T)$. Thus in our situation, $\sigma(BH)$ is real and positive.

[7] E.P. Wigner, *Canadian J. Math.* 15 (1963), 313. The essential meaning of weakly positive operators T is best seen in the finite dimensional matrix case: $\sigma(T)$ is real and positive and the eigenvectors of T form a complete set.

generator for time $t \geq 0$. Now we may come to the question of irreversibility:[8] when is BA a *dissipative* semigroup generator? To determine this, we may now use the operator trigonometry of the antieigenvalue theory to produce a sufficient condition.

Theorem 3.2 *Let bounded $A = iH$ be the infinitesimal generator of the unitary group $U_t = e^{iHt}$. Let $B = C + iD$ be a bounded multiplicative perturbation of A, where C and D are B's real and imaginary parts, respectively. Assume that C commutes with H, that H and D are positive definite, and that either D commutes with H or that*

$$\sin D \stackrel{\leq}{=} \cos H. \tag{10}$$

Then $W_t = e^{BAt}$ is a dissipative contraction semigroup.

Proof First we note from

$$BA = (C + iD)(iH) = iCH - DH \tag{11}$$

that when C commutes with H (a condition we took for simplicity), BA will be a dissipative contraction semigroup generator iff DH is accretive, i.e.,

$$\text{Re}\langle DHx, x \rangle \stackrel{\geq}{=} 0. \tag{12}$$

By the accretive product condition (6), that will be the case when $\sin D \stackrel{\leq}{=} \cos H$. In the case that D commutes with H, DH is always positive. ∎

For application to quantum mechanical situations that following corollary may be of interest.

Corollary 3.1 *Let $B = iH_2$ be a bounded positive cut-off portion of a self-adjoint disturbance-Hamiltonian H_2 projected onto its spectral subspace over a positive interval $[m_2, M_2]$ of its scattering (i.e., continuous) spectrum. Let $A = iH_1$ be the given Hamiltonian, similarly projected to its spectral subspace over a positive interval $[m_1, M_1]$. Then $W_t = e^{BAt}$ is a dissipative contraction semigroup provided that*

$$\frac{M_2 - m_2}{M_2 + m_2} \stackrel{\leq}{=} \frac{2\sqrt{m_1 M_1}}{M_1 + m_1}. \tag{13}$$

---

[8] To be clear what we mean by this, we take the point of view of Nicolis and Prigogine, *Exploring Complexity* (Freeman, New York, 1989), 164, from which we quote "··· the signature of irreversibility lies in the emergence of a dissipative semigroup description of an appropriately defined markovian process." Here we only deal with the semigroup issue. For the corresponding markov processes, see I. Antoniou and K. Gustafson, *Physica A* 197, (1993), 153; *Physica A* (1997), to appear.

Proof We have taken $C = 0$ in Theorem 2, and then used (8).   ∎

We remark that the accretive product criteria (6), (13) is sharp in two ways. First, the trigonometric condition $\sin B \overset{\leq}{=} \cos A$ can be seen to guarantee the accretivity of a great many operator products. For example, if in Corollary 3.1 we take $[m_1, M_1]$ to be $[0.5, 1]$, then $H_2 H_1$ will be accretive for any H_2 over the interval $[0.0295, 1]$. Secondly, it should be noted that given H_1 positive and $H_2 H_1$ accretive, it can be shown that it is necessary that H_2 be positive.

We close with a connection to Davies' Theorem[9] on completely nonunitary semigroups. Recall that every one parameter contraction semigroup W_t has a unique orthogonal decomposition into a unitary part and a completely nonunitary part, and that the completely nonunitary part has no nontrivial subspaces which remain invariant under the semigroup for all $t \geq 0$,

Corollary 3.2 *Under the conditions of Theorem 2, there always exists a small disturbance such that $W_t = e^{BAt}$ is a completely nonunitary contraction semigroup.*

Proof If we take $C = I$ in Theorem 3.2, then $BA = (I + iD)iH = iH - DH$ and DH is positive definite, so the result follows from Davies' Theorem. Thus a multiplicative disturbance has converted a reversible unitary group into a completely nonunitary semigroup. Notice that for every strictly positive interval in H's spectrum (as in Corollary 3.1), there exists a small positive disturbance D which changes H's evolution from unitary to completely nonunitary.   ∎

4. Conclusion

Antieigenvalues originated from multiplicative perturbation questions in semigroup operator theory. Here we have traced those origins, related them to Wigner's theory of weakly positive operators, and proposed their application to irreversibility propositions in quantum theory.

---

[9] See E.B. Davies, *One Parameter Semigroups* (Academic Press, London, 1980), p. 155. By Davies' Theorem, any Hamiltonian of the form $iH - V$ with V bounded positive definite will generate a completely nonunitary contraction semigroup.

Author Index

Lecture Notes in Physics

For information about Vols. 1–469
please contact your bookseller or Springer-Verlag

New Series m: Monographs